Marine Anthropogenic Litter

Melanie Bergmann · Lars Gutow
Michael Klages
Editors

Marine Anthropogenic Litter

UNIVERSITY OF
GOTHENBURG

Editors
Melanie Bergmann
HGF-MPG Group for Deep-Sea Ecology
  and Technology
Alfred-Wegener-Institut
  Helmholtz-Zentrum für Polar- und
  Meeresforschung
Bremerhaven
Germany

Michael Klages
Sven Lovén Centre for Marine Sciences
University of Gothenburg
Fiskebäckskil
Sweden

Lars Gutow
Biosciences I Functional Ecology
Alfred-Wegener-Institut
  Helmholtz-Zentrum für Polar- und
  Meeresforschung
Bremerhaven
Germany

This publication is Eprint ID 37207 of the Alfred-Wegener-Institut Helmholtz-Zentrum für Polar- und Meeresforschung.

Permission for photo on cover: Crab *Paromola cuvieri* walking over plastic litter at a deep-water coral reef off Santa Maria di Leuca (582 m depth), Italy. Also shown: the coral *Madrepora oculata* and a sponge carried by the fifth pereiopods of the crab as a defence. The image was recorded during dive 728 of the remotely operated vehicle QUEST (MARUM, Bremen University). Reprinted with permission from A. Freiwald, L. Beuck, A. Rüggeberg, M. Taviani, D. Hebbeln, and R/V Meteor Cruise M70-1 Participants. 2009. The white coral community in the central Mediterranean Sea revealed by ROV surveys. Oceanography 22(1):58–74, http://dx.doi.org/10.5670/oceanog.2009.06.

ISBN 978-3-319-16509-7   ISBN 978-3-319-16510-3   (eBook)
DOI 10.1007/978-3-319-16510-3

Library of Congress Control Number: 2015935215

Springer Cham Heidelberg New York Dordrecht London

Printed on acid-free paper

Springer International Publishing AG Switzerland is part of Springer Science+Business Media
(www.springer.com)

For Rosa, Frida and Piet.

Foreword

Synthetic polymers, commonly known as plastics, have made themselves a permanent part of the marine environment for the first time in the long history of planetary seas. No sediment or ice core will reveal ancient deposits of these materials or the biological consequences associated with high concentrations of synthetic polymers in the planet's prehistoric ocean. However, current ice and sediment cores do reveal an abundance of this material. Only a broad combination of traditional fields of scientific inquiry is adequate to uncover the effects of this new pollutant, and it seems a pity that a field of study, rather than springing from insights into natural phenomena, arises from new ways that natural phenomena are compromised.

Reports of plastics in the marine environment began to appear in the early 1970s. At the time, Edward Carpenter of the Woods Hole Oceanographic Institution speculated that the problem was likely to get worse and that toxic, non-polymeric compounds in plastics known as plasticizers could be delivered to marine organisms as a potential effect. Carpenter's speculations were correct and probably more so than he imagined. The quantity of plastics in ocean waters has increased enormously, and toxic plastic additives, as well as toxicants concentrated by plastics from the surrounding sea water, have been documented in many marine species.

The rapid expansion of the use of synthetic polymers over the last half century has been such that the characterization of the current era as the "Age of Plastics", seems appropriate. There is no real mystery as to why plastics have become the predominant material of the current epoch. The use value of the material is truly surprising. It can substitute for nearly every traditional material from millinery to metal and offers qualities unknown in naturally occurring substances, so that it now feeds a worldwide industry. The plastic industry creates new applications and products with growth trending sharply upward and showing no signs of slowing in the foreseeable future. Laser printing using plastic "ink" will guarantee expanded use of polymeric feedstocks.

Although the majority of plastics produced today use petroleum resources which are finite, the carbon backbone of synthetic polymers can be fashioned from switchgrass, soya beans, corn, sugar cane or other renewable resources—price alone determines industry's preference. The fact that synthetic polymers can be made from row crops (so-called biopolymers) need have nothing to do with their biodegradability. Olefins are still olefins and acrylates are still acrylates, and behave like their petroleum-fabricated counterparts. Furthermore, biodegradability standards are not applicable in the marine environment and marine degradability requires a separate standard. Marine degradable plastics have a negligible market share and are not poised to make headway into the consumer plastics market at the present time. The difficulty of recycling plastics has made their profitable recovery a problem, which in turn results in failure to provide take-back infrastructure and results in accelerated pollution.

Given the proliferation of plastics into all spheres of human activity, and their increasing use value in the developing world, the phenomena associated with plastic pollution of the marine environment will continue to merit scientific investigation. These studies, however, are hampered by the lack of basic geospatial and quantitative data. Estimates abound based on limited sampling and modeling, but the ocean is the biggest habitat on the planet by far and knowledge of its plastic pollution will require new methods of data acquisition. The role of citizens in the monitoring of plastic pollution will increase in the coming years, and the truly "big" data they document must become part of the science of plastic pollution. For the present, it is fortunate that a few pioneering scientists around the world are engaged in attempting to understand the consequences of the plague of plastic that contaminates our precious ocean.

Long Beach   Captain Charles James Moore
                                                            http://www.algalita.org

Preface

The ocean is of eminent importance to mankind. Twenty-three per cent of the world's population (~1.2 billion people) live within 100 km of the coast (Small and Nicholls 2003), a figure, which is likely to rise up to 50 % by 2030 (Adger et al. 2005). Furthermore, the ocean sustains nearly half of the global primary production (Field et al. 1998), a great share of which fuels global fisheries (Pauly and Christensen 1995). The marine environment hosts a substantial biodiversity, and tourism is an important and constantly growing economic sector for many coastal countries. Although human welfare is intricately linked with the sea and its natural resources, people have substantially altered the face of the ocean within only a few centuries. Fisheries, pollution, eutrophication, deep-sea hydrocarbon exploration, ocean acidification and global ocean warming accompanied by sea-level rise as a consequence of rapid glacier melting and thermal expansion of sea water (IPCC 2014) are prominent examples of man-made pressures exerted on the oceans with severe ecological and socio-economic repercussions. As a result, marine environmental protection and management have become integral political and societal issues in many countries worldwide. However, effective environmental management requires a proper understanding of the ecological implications of human activities and should, therefore, be accompanied by sound multidisciplinary research, scientific advice, education and public outreach.

In recent decades, the pollution of the oceans by anthropogenic litter has been recognized as a serious global environmental concern. Marine litter is defined as "any persistent, manufactured or processed solid material discarded, disposed of or abandoned in the marine and coastal environment" (UNEP 2009). Since its first mention in the scientific literature in the 1960s, research efforts addressing marine litter have constantly grown as has the amount of litter in the oceans. Many studies have shown that it consists primarily of plastics with a continuously increasing global annual production of 299 million t (PlasticsEurope 2015). It has been estimated that 10 % of all plastic debris ends up in the oceans (Thompson 2006), and Barnes (2005) suggested that the 1982 figure of 8 million litter items entering the oceans every day probably needs to be multiplied several fold. Eriksen et al. (2014) estimate a minimum of 5.25 trillion plastic particles weighing 268,940 tons afloat in the sea, but this figure does not include debris on the seafloor or beaches. The increasing use of single-use products, uncontrolled disposal of litter along with poor waste management and recycling practices is the main reason for the accumulation of litter in the sea. Increasing quantities of litter are lost from municipal waste streams and enter the oceans (Barnes et al. 2009). The ubiquity of litter in the open ocean is prominently illustrated by numerous images of floating debris from the ocean garbage patches and by the fact that the search for the missing Malaysia Airlines flight MH370 in March 2014 produced quite a few misidentifications caused by litter floating at the water surface.

Since plastic accounts for the majority of litter items in the sea, the chapters of this book primarily focus on plastic litter and its implications for the marine environment. Numerous quantitative reports on marine anthropogenic litter from various parts of the world's oceans indicate that anthropogenic litter is ubiquitous at the shores as well as in the pelagic and benthic realms. Global surveys revealed that plastics have already reached the shores of the remotest islands (Barnes 2005) and even polar waters far off urban centres (Barnes et al. 2010; Bergmann and Klages 2012). The use of advanced technology, such as remotely and autonomously operated vehicles, revealed that anthropogenic litter has conquered the deep sea before mankind set eye upon it suggesting that the deep seafloor may constitute the ultimate sink for marine litter (Pham et al. 2014). However, we are just beginning to understand how litter actually "behaves" at sea and to identify the drivers of the temporal and the spatial distribution of litter in the oceans. Still, we have already started to generate a mankind memory made out of plastic in the world ocean.

Marine anthropogenic litter causes harm to a wide range of marine biota. Seabirds, fish, turtles and marine mammals suffer from entanglement with and ingestion of marine litter items as illustrated by countless pictures of animals injured and strangled by discarded fishing gear in the public media. However, we have only limited knowledge about the implications of marine litter for the many less charismatic invertebrate species that easily escape public perception but play important roles in marine ecosystems. Although already mentioned in the late 1980s (Ryan 1988), it took Thompson's time series (Thompson et al. 2004) to raise public awareness of the widespread presence of microplastics, which are used in industrial production processes, cosmetics and toothpaste or generated through degradation of larger items. Indeed, substantial concentrations of microplastics were recently reported from remote and presumably unspoiled environments such as the deep seafloor (Woodall et al. 2014) and Arctic sea ice, which is considered a historic global sink at least until its plastic load is released into the ocean during the projected increase of ice melts (Obbard et al. 2014). Microplastics are available for ingestion by a wide range of organisms, and there are indications that microplastics are propagated over trophic levels of the marine food web (Farrell and Nelson 2013; Setälä et al. 2014). However, scientists have only recently started to investigate whether the contamination of marine organisms with plastics and associated chemicals is causing harm to ecosystems and human health (Browne et al. 2013; Bakir et al. 2014; De Witte et al. 2014; Van Cauwenberghe and Janssen 2014).

The accumulation of litter at sea and along coastlines worldwide and the many open questions concerning the amount, distribution and fate of marine litter and potential implications for marine wildlife and humans have raised public awareness, stimulated scientific research and initiated political action to tackle this environmental problem (UNEP 2014). Identification, quantification and sampling of marine litter do not necessarily require professional scientific skills so that NGOs as well as committed citizens and other stakeholders have contributed substantially to the collection of data on marine litter pollution and to the global perception of the problem (Rosevelt et al. 2013; Anderson and Alford 2014; Smith and Edgar 2014). Scientists, politicians, authorities, NGOs and industries have started to share knowledge at international conferences aimed at developing managerial solutions. These joint activities, public awareness and, finally, the scientific curiosity of numerous committed researchers have stimulated a rapidly increasing number of publications from various scientific disciplines in dedicated volumes (Coe and Rogers 1997; Thompson et al. 2009). This latest volume on *Marine Anthropogenic Litter* was inspired by the remarkable recent progress in marine litter research. A large proportion of the references reviewed in this book was published in the last three years demonstrating the topicality of this book and the issue as a whole. Because of the high dynamics in this field of research, this volume may already be outdated when published.

This book consists of five major sections. In the first section, Peter Ryan gives a historical synopsis of marine litter research starting from the first mention of floating debris in the famous novel *20,000 Leagues Under the Sea* by Jules Verne in 1870 but with a focus on the past 50 years, which have seen a strong increase in the production of plastics. The reader will learn about the rapid development of this research field, and a series of international key conferences such as the "Honolulu Conferences", which brought together scientists, environmentalists, industry, NGOs and policy makers and fuelled numerous publications and new research and management schemes.

The second section of the book addresses abiotic aspects of marine litter pollution. François Galgani, Georg Hanke and Thomas Maes portray the abundance, global distribution and composition of marine litter, which illustrates the ubiquity of litter in the oceans from the urban centres of human activity to the Earth's remotest sites. Anthony L. Andrady describes the physical and chemical processes involved in the degradation of plastics in the marine environment.

The third section of the book covers the biological and ecological implications of marine litter. Susanne Kühn, Elisa L. Bravo Rebolledo and Jan A. van Franeker summarize the deleterious effects of litter on marine wildlife. The authors compiled an extensive list of 580 species, ranging from invertebrates to fish, turtles, birds and mammals that have been shown to suffer from the effects of marine litter. Toxicity of contaminants associated with marine plastic debris as well as health implications is described by Chelsea Rochman who demonstrates that plastics are more than a mechanic threat to marine biota. Tim Kiessling, Lars Gutow and Martin Thiel show how marine litter facilitates the dispersal of marine organisms, which are capable of colonizing litter items floating at the sea surface. The authors compiled a list of 387 taxa that have been found rafting on floating litter, and they evaluate how marine litter might facilitate the spread of invasive species. This may alter the face of biodiversity with yet unknown consequences for ecosystem functioning.

The fourth section of this book is dedicated entirely to the young but rapidly expanding field of microplastic research. Since the recent rise in public awareness of microplastics in the marine environment, intensive research on this topic has yielded a considerable amount of important scientific results. Accordingly, this topic deserves an entire section, which is introduced by a synopsis of microplastic research by Richard C. Thompson. The various primary and secondary sources of microplastics and the pathways through the environment to the biota are outlined by Mark A. Browne who also highlights the need for hypothesis-driven approaches in microplastic research. Because of the small size and the diversity of plastic polymers, the detection, proper identification and quantification of microplastics are challenging, which hampers the comparability of results from different studies. Therefore, Martin G.J. Löder and Gunnar Gerdts composed a critical appraisal of methods and procedures applied in this field including a case study that demonstrates how improper methodology easily leads to a misevaluation of the contamination of habitats and organisms. The global distribution and the environmental effects of microplastics are summarized by Amy Lusher. She compiled a list of 172 taxa, which have been found to ingest microplastics either in the field (131) or in laboratory experiments (46) with variable effects on the behavior and health status of the organisms.

Although deleterious effects of microplastics have been demonstrated for a considerable number of marine organisms, the role of these particles as vectors for chemicals from the environment to the organisms is subject to intense debate. Albert A. Koelmans used a modeling approach to critically evaluate the transfer of environmental contaminants to marine organisms. Nanoparticles are of even smaller particle size (<1 μm). They are of particular concern as they are more likely to pass biological membranes and affect the functioning of cells including blood cells and photosynthesis. Albert A. Koelmans, Ellen Besseling and Won J. Shim summarize what little is known about this litter fraction, whose significance in the marine environment is just coming to light.

The final section of this book moves away from natural science towards the socio-economic implications of marine anthropogenic litter. Tamara S. Galloway reviews the current knowledge on how chemicals associated with plastics may affect human health. As top consumers of ocean-based food webs, humans likely accumulate contaminants, which may compromise fecundity, reproduction and other somatic processes. The accumulation of litter in the oceans can be considered a result of market failure on land. The root of the problem is probably—as so often—that producers/manufacturers of goods (plastics) are not economically held responsible for the products they sell. Stephanie Newman, Emma Watkins, Andrew Farmer, Patrick ten Brink and Jean-Pierre Schweitzer describe economic instruments that were used in different parts of the world to reduce litter inputs to the sea. Although a number of international policies have been in place for quite some time to manage the input of litter to the sea, their shortcomings make them unlikely to result in significant reductions of marine litter (Gold et al. 2013). Chung-Ling Chen describes and assesses key multilateral and national regulative measures with respect to their sufficiency to tackle marine litter pollution. Another way to reduce the input of litter to the ocean is suasion of citizens and stakeholders, which requires public awareness of the problem through education and outreach activities (Hartley et al. 2015). Ideally, such initiatives also generate data that can be used for assessments of marine litter pollution and distribution. In the last chapter, Valeria Hidalgo-Ruz and Martin Thiel review the potential of "citizen science" initiatives for supporting research on this global environmental issue.

The solution of the marine litter problem requires expertise from various sectors, including industries, science, policy, authorities, NGOs and citizens. We hope that this book will facilitate the exchange of knowledge amongst the various actors and contribute to finding solutions to this challenge.

Bremerhaven, Fiskebäckskil, 2015   Melanie Bergmann
                                                                        Lars Gutow
                                                                        Michael Klages

References

Adger, W. N., Hughes, T. P., Folke, C., Carpenter, S. R., & Rockström, J. (2005). Social-ecological resilience to coastal disasters. *Science, 309*, 1036–1039.

Anderson, J. A., & Alford, A. B. (2014). Ghost fishing activity in derelict blue crab traps in Louisiana. *Marine Pollution Bulletin, 79*, 261–267.

Bakir, A., Rowland, S. J., & Thompson, R. C. (2014). Enhanced desorption of persistent organic pollutants from microplastics under simulated physiological conditions. *Environmental Pollution, 185*, 16–23.

Barnes, D. K. A. (2005). Remote Islands reveal rapid rise of Southern Hemisphere, sea debris. *The Scientific World Journal, 5*, 915–921.

Barnes, D. K. A., Galgani, F., Thompson, R. C., & Barlaz, M. (2009). Accumulation and fragmentation of plastic debris in global environments. *Philosophical Transactions of the Royal Society B, 364*, 1985–1998.

Barnes, D. K. A., Walters, A., & Gonçalves, L. (2010). Macroplastics at sea around Antarctica. *Marine Environmental Research, 70*, 250–252.

Bergmann, M., & Klages, M. (2012). Increase of litter at the Arctic deep-sea observatory HAUSGARTEN. *Marine Pollution Bulletin, 64*, 2734–2741.

Browne, M. A., Niven, S. J., Galloway, T. S., Rowland, S. J., & Thompson, R. C. (2013). Microplastic moves pollutants and additives to worms, reducing functions linked to health and biodiversity. *Current Biology, 23*, 2388–2392.

Coe, J. M., & Rogers, D. (1997). *Marine debris. Sources, impacts, and solutions*. New York: Springer.

De Witte, B., Devriese, L., Bekaert, K., Hoffman, S., Vandermeersch, G., Cooreman, K., et al. (2014). Quality assessment of the blue mussel (*Mytilus edulis*): Comparison between commercial and wild types. *Marine Pollution Bulletin, 85*, 146–155.

Eriksen, M., Lebreton, L. C. M., Carson, H. S., Thiel, M., Moore, C. J., Borerro, J. C., et al. (2014). Plastic pollution in the world's oceans: More than 5 trillion plastic pieces weighing over 250,000 tons afloat at sea.

Farrell, P., & Nelson, K. (2013). Trophic level transfer of microplastic: *Mytilus edulis* (L.) to *Carcinus maenas* (L.). *Environmental Pollution, 177*, 1–3.

Field, C. B., Behrenfeld, M. J., Randerson, J. T. & Falkowski P. (1998). Primary production of the biosphere: Integrating terrestrial and oceanic components. *Science, 281*, 237–240.

Gold, M., Mika, K., Horowitz, C., Herzog, M., & Leitner, L. (2013). Stemming the tide of plastic marine litter: A global action agenda. *Pritzker Policy Brief, 5*.

Hartley, B. L., Thompson, R. C., & Pahl, S. (2015). Marine litter education boosts children's understanding and self-reported actions. *Marine Pollution Bulletin, 90*, 209–217.

IPCC (2014). Climate change: Synthesis report. Contribution of working groups I, II and III to the fifth assessment report of the intergovernmental panel on climate change (pp. 151). Geneva, Switzerland.

Obbard, R. W., Sadri, S., Wong, Y. Q., Khitun, A. A., Baker, I., & Thompson, R. C. (2014). Global warming releases microplastic legacy frozen in Arctic Sea ice. *Earth's Future, 2*, 315–320.

Pauly, D. & Christensen, V. (1995). Primary production required to sustain global fisheries. *Nature, 374*, 255–257.

Pham, C. K., Ramirez-Llodra, E., Alt, C. H. S., Amaro, T., Bergmann, M., Canals, M., et al. (2014). Marine litter distribution and density in European seas, from the shelves to deep basins. *PLoS ONE, 9*, e95839.

PlasticsEurope. (2015). Plastics—the Facts 2014/2015. http://www.plasticseurope.fr/Document/plastics—the-facts-2013.aspx?Page=DOCUMENT&FolID=2, http://issuu.com/plasticseuro peebook/docs/final_plastics_the_facts_2014_19122.

Rosevelt, C., Los Huertos, M., Garza, C., & Nevins, H. M. (2013). Marine debris in central California: Quantifying type and abundance of beach litter in Monterey Bay, CA. *Marine Pollution Bulletin, 71*, 299–306.

Ryan, P. G. (1988). The characteristics and distribution of plastic particles at the sea-surface off the southwestern Cape Province, South Africa. *Marine Environmental Research, 25*, 249–273.

Setälä, O., Fleming-Lehtinen, V., & Lehtiniemi, M. (2014). Ingestion and transfer of microplastics in the planktonic food web. *Environmental Pollution, 185*, 77–83.

Small, C. & Nicholls, R. J. (2003). A global analysis of human settlement in coastal zones. *Journal of Coastal Research, 19*, 584–599.

Smith, S. D. A., & Edgar, R. J. (2014). Documenting the density of subtidal marine debris across multiple marine and coastal habitats. *PLoS ONE, 9*, e94593.

Thompson, R. C. (2006). Plastic debris in the marine environment: Consequences and solutions. In J. C. Krause, H. Nordheim, & S. Bräger (Eds.), *Marine nature conservation in Europe* (pp. 107–115). Stralsund, Germany: Bundesamt für Naturschutz.

Thompson, R. C., Moore, C. J., vom Saal, F. S., & Swan, S. H. (2009). Theme issue 'Plastics, the environment and human health: Current consensus and future trends'. *Philosophical Transactions of the Royal Society B: Biological Sciences, 364*, 2153–2166.

Thompson, R. C., Olsen, Y., Mitchell, R. P., Davis, A., Rowland, S. J., John, A. W. G., et al. (2004). Lost at sea: where is all the plastic? *Science, 304*, 838.

UNEP. (2009). *Marine litter: A global challenge*. Nairobi.

Van Cauwenberghe, L., & Janssen, C. R. (2014). Microplastics in bivalves cultured for human consumption. *Environmental Pollution, 193*, 65–70.

Woodall, L. C., Sanchez-Vidal, A., Canals, M., Paterson, G. L. J., Coppock, R., Sleight, V., et al. (2014). The deep sea is a major sink for microplastic debris. *Royal Society Open Science, 1*, 140137.

Acknowledgments

We are grateful for financial support from the Alfred-Wegener-Institut Helmholtz-Zentrum für Polar- und Meeresforschung, University of Exeter, University of Gothenburg, Wageningen University, IMARES, Galway Mayo Institute of Technology and IFREMER, which enabled open access to this book. This is publication number 37207 of the Alfred-Wegener-Institut Helmholtz-Zentrum für Polar- und Meeresforschung. We would like to acknowledge the invaluable support of the reviewers whose comments helped to improve the quality of the book:

Emmett Clarkin (Queen's University, UK)
Monica F. da Costa (Federal University of Pernambuco, Brazil)
Satoshi Endo (Osaka City University, Japan)
David Fleet (The Schleswig-Holstein Agency for Coastal Defence, National Park and Marine Conservation, Germany)
François Galgani (IFREMER, France)
Miriam Goldstein (formerly of Scripps Institution of Oceanography, USA)
Rolf Halden (Arizona State University, USA)
Claudia Halsband (Akvaplan-niva, Norway)
Jesse Harrison (UK Centre for Astrobiology, UK)
Emily Hastings (James Hutton Institute, UK)
Iván A. Hinojosa (University of Tasmania, Australia)
Patricia Holm (Universität Basel, Suisse)
Jörg Klasmeier (University of Osnabrück, Germany)
Albert A. Koelmans (IMARES, The Netherlands)
Angela Köhler (Alfred Wegener Institute, Helmholtz Centre for Polar and Marine Research, Germany)
Scott Lambert (Scientific Consulting Company GmbH, UK)
Michael Matthies (University of Osnabrück, Germany)
Alistair McIlgorm (University of Wollongong, Australia)
Luca Monticelli (CNRS Institute of Protein Biology and Chemistry, France)
Captain Charles James Moore (Algalita, USA)
Ramani Narayan (Michigan State University, USA)

Jennifer Provencher (Carleton University, Canada)
Carolyn Rosevelt (NASA-Ames/CSU Monterey Bay Cooperative Agreement, USA)
Reinhard Saborowski (Alfred Wegener Institute, Helmholtz Centre for Polar and
    Marine Research, Germany)
Rob Tinch (Economics for the Environment Consultancy, UK)
Laura N. Vandenberg (University of Massachusetts, USA)
Stefanie Werner (German Federal Environment Agency, Germany)
Christiane Zarfl (Leibniz-Institute of Freshwater Ecology and Inland Fisheries,
    Germany)

Contents

Part III   Microplastics

Part IV   Socio-economic Implications of Marine Anthropogenic Litter

Chapter 1
A Brief History of Marine Litter Research

Peter G. Ryan

Abstract This chapter traces the history of marine litter research from anecdotal reports of entanglement and plastic ingestion in the 1960s to the current focus on microplastics and their role in the transfer of persistent organic pollutants to marine food webs. The reports in *Science* of large numbers of plastic pellets in the North Atlantic in the early 1970s stimulated research interest in plastic litter at sea, with papers reporting plastics on the seafloor and impacting a variety of marine animals. The focus then shifted to high concentrations of plastic litter in the North Pacific, where novel studies reported the dynamics of stranded beach litter, the factors influencing plastic ingestion by seabirds, and trends in fur seal entanglement. By the early 1980s, growing concern about the potential impacts of marine litter resulted in a series of meetings on marine debris. The first two international conferences held in Honolulu by the US National Marine Fisheries Service played a key role in setting the research agenda for the next decade. By the end of the 1980s, most impacts of marine litter were reasonably well understood, and attention shifted to seeking effective solutions to tackle the marine litter problem. Research was largely restricted to monitoring trends in litter to assess the effectiveness of mitigation measures, until the last decade, when concern about microplastics coupled with the discovery of alarming densities of small plastic particles in the North Pacific 'garbage patch' (and other mid-ocean gyres) stimulated the current wave of research.

Keywords Plastic · History · Environmental impact · Entanglement · Ingestion · Microplastics

P.G. Ryan (✉)
Percy FitzPatrick Institute, DST/NRF Centre of Excellence,
University of Cape Town, Rondebosch 7701, South Africa
e-mail: pryan31@gmail.com

© The Author(s) 2015
M. Bergmann et al. (eds.), *Marine Anthropogenic Litter*,
DOI 10.1007/978-3-319-16510-3_1

1

1.1 Introduction

From messages in bottles to exotic tropical seeds washing up on temperate shores (Guppy 1917; Muir 1937), the dispersal of floating debris at sea has long fascinated people. As early as 1870 Jules Verne provided a graphic description of how floating debris accumulates in ocean gyres in the chapter on the Sargasso Sea in his famous novel *Twenty Thousand Leagues under the Sea*. However, this review focuses on the last 50 years because from the perspective of environmental impacts the history of marine litter research is closely linked to the development of plastics. Plastics are a diverse group of synthetic polymers that have their origins in the late 19th century, but which really came to the fore in the mid-twentieth century. Their low density, durability, excellent barrier properties and relatively low cost make plastics ideal materials for a wide range of manufacturing and packaging applications. Their versatility has seen the amount of plastic produced annually increase rapidly over the last few decades to an estimated 288 million tonnes in 2012 (Fig. 1.1), and this total continues to grow at about 4 % per year (PlasticsEurope 2013). However, the properties that make plastics so useful also make inappropriately handled waste plastics a significant environmental threat. Their durability means that they persist in the environment for many years, and their low density means that they are readily dispersed by water and wind, sometimes travelling thousands of kilometres from source areas (Ryan et al. 2009). As a result, plastic wastes are now ubiquitous pollutants in even the most remote areas of the world (Barnes et al. 2009).

Over the last 60 years we have seen a major shift in perception surrounding the use of plastics, especially in one-off applications. Once seen as the savior of the American housewife (Life Magazine 1955), there are now calls to treat waste plastics as hazardous materials (Rochman et al. 2013a), reiterating a point first made by Bean (1987) that persistent plastic wastes qualify as hazardous wastes under the US Resource Conservation and Recovery Act. Most of the threats posed by plastics occur at sea (Gregory 2009; Thompson et al. 2009), where waste plastics tend to accumulate (Barnes et al. 2009; Ryan et al. 2009). This chapter briefly summarises the history of marine litter research. Trends in the numbers of

Fig. 1.1 Growth in global plastic production from 1950 to 2012 (millions of tonnes, adapted from PlasticsEurope 2013)

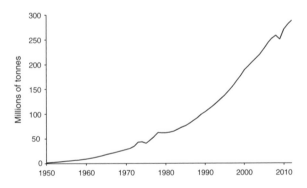

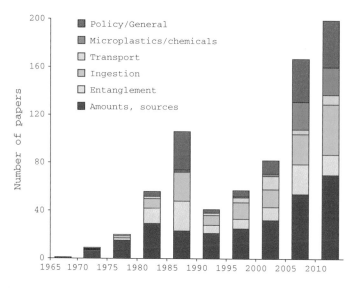

Fig. 1.2 Numbers of papers on different aspects of the marine litter issue published in five-year intervals over the last 50 years (based on a Web of Science search and unpublished bibliography; note that the final column only covers three years, 2011–2013)

papers on the marine litter problem (Fig. 1.2) show the growth in research from its infancy in the late 1960s, when it was still treated largely as a curiosity, through the 1970s and 1980s, when most of the threats to marine systems were identified, baseline data were collected on the distribution, abundance and impacts of marine litter, and policies were formulated to tackle the problem. Research tapered off in the 1990s, despite ongoing increases in the amounts of marine litter (Ryan and Moloney 1990, 1993), and it is only in the last decade or so that there has been a resurgence in research interest, following alarming reports of mid-ocean 'garbage patches' (Moore et al. 2001) and increasing appreciation of the pervasive nature of very small 'microplastic' particles (<0.5 mm) and their potential impacts on the health of marine ecosystems (Oehlmann et al. 2009; Thompson et al. 2009).

1.2 Seabirds and Seals—The First Signs of Trouble

Interactions between marine organisms and persistent litter were first recorded in the scientific literature in the late 1960s, when Kenyon and Kridler (1969) reported the ingestion of plastic items by Laysan Albatrosses (*Phoebastria immutabilis*) on the northwest Hawaiian Islands. They found plastic in the stomachs of 74 of 100 albatross chicks that died prior to fledging in 1966, with up to 8 items and an average of 2 g plastic per bird. However, this was an order of magnitude less than the average mass of pumice, seeds, charcoal and wood that the chicks also were fed by their parents. Kenyon and Kridler (1969) inferred that these indigestible items were swallowed inadvertently at sea, because virtually all items floated in seawater. They also speculated that the large size of many of the items might have contributed to the chicks' deaths by blocking their digestive tracts.

In fact, there were earlier records of seabirds ingesting plastics, with plastic found in stranded prions (*Pachyptila* spp.) in New Zealand as early as 1960 (Harper and Fowler 1987), and in Leach's storm petrels (*Oceanodroma leucorhoa*) from Newfoundland, Canada, in 1962 (Rothstein 1973). Non-breeding Atlantic puffins (*Fratercula arctica*) collected from 1969 to 1971 were reported to contain elastic threads in their stomachs (Berland 1971; Parslow and Jefferies 1972). In some birds, these threads had formed tight balls up to 10 mm across, filling the gizzard and possibly partially blocking the pyloric valve leading into the intestine (Parslow and Jefferies 1972). Parslow and Jefferies (1972) noted that ingesting rubber and elastic was common among scavenging birds such as gulls, but that they regularly regurgitated such items along with other indigestible prey remains, implying that this was not a problem for such birds. And it was not just seabirds at risk. By the late 1950s there were records of marine turtles ingesting plastic bags, sometimes resulting in their deaths (Cornelius 1975; Balazs 1985). A mass of fishing line and other fishing gear blocked the intestine of a manatee (*Trichechus manatus*) in 1974 (Forrester et al. 1975), and stranded cetaceans were found to have eaten plastic by the mid-1970s (Cawthorn 1985).

Records of entanglement of marine organisms in plastic litter also started to increase in the 1960s. There were reports of birds and seals entangled in man-made items before this (e.g. Jacobson 1947), but they tended to remain in the gray literature (Fowler 1985; Wallace 1985). By 1964 northern fur seals (*Callorhinus ursinus*) were often reported entangled in netting and other artefacts in the Bering Sea, and the incidence of entangled seals harvested in the Pribilof Islands showed a steady increase from less than 0.2 % of the population in 1967 to a peak of over 0.7 % in 1975 (Fowler 1987). The entanglement rate then stabilized at around 0.4 % through the late 1970s and early 1980s (Fig. 1.3), but this was still sufficient to help to drive a population decrease in this species (Fowler 1987; Fowler et al. 1990). By comparison, entanglement rates of three seal species at the Farallon Islands off central California showed a marked increase in the early 1980s (Hanni and Pyle 2000).

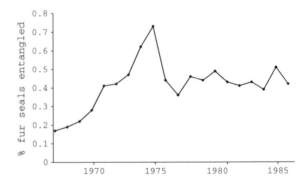

Fig. 1.3 Trends in the percentage of northern fur seals entangled on St. Paul Island, Alaska (adapted from Fowler et al. 1990)

Entanglement of fish and dogfish in rubber bands was reported in 1971 (Anon 1971; Berland 1971), and Gochfeld (1973) highlighted the entanglement threat posed by marine litter to coastal birds. Based on observations on Long Island in 1970 and 1971, Gochfeld (1973) reported how adult and chick black skimmers (*Rhynchops niger*) and two species of terns died after being entangled in nylon fishing line, kite strings, six-pack holders, bags and bottles. Although the numbers of birds affected were not great, Gochfeld (1973) argued that they might be sufficient to cause at least some populations to decrease, especially when combined with other human impacts in the region. Subsequently, Bourne (1976, 1977) summarised what was known about the threat posed by plastic ingestion and entanglement to seabirds, and reported how the incorporation of rope and netting in seabird nests can entangle and kill seabird chicks. He also highlighted the threat posed by the switch to manufacturing nets and other fishing gear from persistent polymers, including ghost fishing by lost or discarded gear (Bourne 1977). Entanglement was a significant cause of mortality for northern gannets (*Morus bassanus*), affecting roughly a quarter of birds found dead in the North Sea in the 1980s (Schrey and Vauk 1987), and remains a problem for this species today (Rodríguez et al. 2013).

1.3 The Early 1970s—Pellets and Other Problems in the North Atlantic

Many of these early records of ingestion and entanglement only came to light after two seminal papers on the occurrence of plastic particles at sea in the northwest Atlantic Ocean appeared in the leading journal *Science* in 1972. In the first paper, Carpenter and Smith (1972) reported the presence of plastic pellets and fragments in all 11 surface net samples collected in the western Sargasso Sea in late 1971, at an average density of around 3500 particles km^{-2} (290 g km^{-2}). Interestingly, the density of plastic was lowest towards the edge of the Sargasso Sea, where it bordered the Gulf Stream, suggesting that these particles had been accumulating in the North Atlantic gyre for some time (cf. Law et al. 2010; Lebreton et al. 2012; Maximenko et al. 2012). Carpenter and Smith (1972) noted that the plastic particles provided attachment sites for epibionts, including hydroids and diatoms, and speculated that such particles could become a significant problem if plastic production continued to increase. They also suggested that plastic particles could be a source of toxic compounds such as plasticisers and polychlorinated biphenyls (PCBs) into marine food webs.

In the second paper, Carpenter et al. (1972) reported high densities of polystyrene pellets in coastal waters off southern New England, east of Long Island (average 0.0–2.6 pellets m^{-3}, exceptionally reaching 14 pellets m^{-3}). Polystyrene is denser than seawater, so the pellets were not expected to disperse far from source areas, but some contained air-filled vacuoles, allowing them to float. The pellets supported communities of bacteria, and were found to have absorbed polychlorinated biphenyls (PCBs) from seawater. Pellets were recorded in the stomachs of eight of 14 fish species and one chaetognath (*Sagitta elegans*) sampled in the area. The fish ignored translucent pellets, only eating opaque white pellets, which suggested selective feeding on the more visible pellets. With up to 33 % of individuals of some fish species affected, Carpenter et al. (1972) raised concerns about the possible impacts due to intestinal blockage of smaller individuals as well as pellets being a source of PCBs.

In fact, Carpenter's two *Science* papers were not the first papers to describe small pieces of plastic litter at sea. Buchanan (1971) reported densities of up to 10^5 synthetic fibres m^{-3} in water samples from the North Sea, and larger fragments were reported to occur in "embarrassing proportions" in plankton samples. And although Heyerdahl (1971) mainly concentrated on oil and tar pollution, he reported sightings of plastic containers throughout the second *Ra* expedition across the North Atlantic. However, Carpenter's papers focused scientific attention on the ubiquitous nature of small plastic particle pollution at sea, and identified three possible impacts: intestinal blockage and a source of toxic compounds from ingested plastic, and the transport of epibionts.

Following Carpenter et al. (1972), large numbers of polystyrene pellets were reported from coastal waters in the United Kingdom (Kartar et al. 1973, 1976; Morris and Hamilton 1974) where they were ingested by three species of fish and a marine snailfish (*Liparis liparis*). More than 20 % of juvenile flounder (*Platichthys flesus*) contained ingested plastics, with up to 30 pellets in some individuals. Hays and Cormons (1974) found polystyrene pellets in gull and tern regurgitations collected on Long Island, New York, in 1971. Although the gulls may have consumed the plastic pellets directly while scavenging, their presence in the diet of terns almost certainly indicated that they were consumed in contaminated fish prey, providing the first evidence of trophic transfers of small plastic items. Sampling close to wastewater outfalls confirmed that the pellets came from plastic manufacturing plants (Hays and Cormons 1974). Fortunately, these point sources were fairly easy to identify and address. By 1975 the incidence of plastic ingestion by fish and snails in the UK's Severn Estuary had fallen to zero, indicating that the release of polystyrene pellets had virtually ceased from the manufacturing plants (Kartar et al. 1976). However, spillage of pellets by converters and during transport proved more difficult to contain.

Carpenter's two *Science* papers in 1972 stimulated a broader interest in marine litter and its impacts. Colton et al. (1974) reported a much more extensive survey of floating plastics in the North Atlantic and Caribbean. They showed that both industrial pellets and fragments of manufactured items occurred throughout the region, but were concentrated close to major land-based sources along the US eastern seaboard. Unlike Carpenter et al. (1972), they failed to find any plastics in fish sampled. Feeding trials with polystyrene pellets showed that juvenile fish seldom ingested plastics, and those pellets that were ingested seemingly passed through the fish with little impact.

Beach litter also came under increased scrutiny. Scott (1972) debunked the notion that beach users were responsible for most litter. He examined the litter found on inaccessible Scottish beaches that have few if any visitors, and inferred that most litter came from shipping and fisheries operating in the area. Initial studies of beach litter simply assessed standing stocks (Ryan et al. 2009); Cundell (1973) was the first researcher to report the rate of plastic accumulation. Working on a beach in Narragansett Bay, USA, he assessed the amount of litter washing ashore over one month. The first study of beach litter dynamics was conducted in Kent, United Kingdom, from 1973 to 1976. Dixon and Cooke (1977) showed that the weekly retention rate of marked bottles and other containers varied depending on the type of beach, and that plastic bottles remained on beaches longer than glass bottles. Strong tidal currents resulted in low retention rates (11–29 % per week) and transported litter throughout the southern North Sea. Some marked bottles travelled >100 km in one week, and others reached Germany and Denmark within 3–6 weeks. Dixon and Cooke (1977) also used manufacturer's codes to assess the longevity of containers and found that few (<20 %) were manufactured more than two years prior to stranding.

In addition to the growing awareness of plastic litter at the sea surface and stranded on beaches, the mid-1970s also saw the first records of plastics on the seabed. Holmström (1975) reported how Swedish fishermen "almost invariably" caught plastic sheets in their trawl nets when fishing in the Skagerrak. Subsequent analysis showed this to be low-density polyethylene, similar to that used for packaging. The samples, obtained from the seabed 180–400 m deep, were encrusted with a calcareous bryozoan and a brown alga (*Lithoderma* sp.). Holmström (1975) surmised that these encrusting biota had increased the density of the plastic sheets sufficiently to cause them to sink to the seabed. The bryozoan and brown alga typically occur in water <25 m deep, and the size of bryozoan colonies suggested that the plastic sheets had spent 3–4 months drifting in the euphotic zone close to the sea surface before sinking to the seabed. Subsequent trials confirmed that most plastics sink due to fouling (Ye and Andrady 1991), and trawl surveys and direct observations have confirmed that plastics and other persistent artefacts now occur on the seabed throughout the world's oceans (Barnes et al. 2009). Indeed, Goldberg (1994, 1997) suggested that the seabed is the ultimate sink for plastics in the environment, and plastic items typically comprise >70 % of seabed artefacts (Galgani et al. 2000). The Mediterranean Sea supports particularly high densities of litter on the seafloor, locally exceeding 100,000 items km^{-2}, and has been the subject of numerous studies to ascertain the factors determining the distribution and abundance of this litter (e.g. Bingel et al. 1987; Galil et al. 1995; Galgani et al. 1995, 1996). Interestingly, although benthic litter tends to concentrate around coastal cities and river mouths, the density of litter is often greater in deep waters along the continental shelf edge than in shallow, inshore waters due to the decrease in bottom currents offshore (Galgani et al. 1995, 2000; Barnes et al. 2009; Keller et al. 2010).

Winston (1982) elaborated on Carpenter and Smith's (1972) suggestion that plastic debris greatly increased settlement opportunities for organisms that live on objects floating at the sea surface. In particular, the bryozoan *Electra tenella* appeared to have extended its range and greatly increased in abundance in the western Atlantic Ocean. Subsequent research has highlighted the potential threat posed by drifting litter transporting organisms outside their native ranges (Barnes 2002; Barnes and Milner 2005; Gregory 2009). This is a serious problem, especially in remote regions, and can result in the transfer of potentially harmful organisms (Masó et al. 2003). However, it probably pales into insignificance in most regions compared to the transport by shipping and other human-mediated vectors (Bax et al. 2003), which in extreme cases can transfer entire communities across ocean basins (Wanless et al. 2010).

1.4 Shifting Focus to the North Pacific Ocean

Indications that the North Pacific was a hot spot for plastic litter date back to Kenyon and Kridler's (1969) paper on plastic ingestion by Laysan albatross. Subsequently, Bond (1971) found plastic pellets in all 20 red phalaropes (*Phalaropus fulicarius*) examined when many individuals of this species came ashore along the coasts of southern California and Mexico in 1969. The birds apparently starved due to a shortage of surface plankton, and some were observed feeding along the strand line where plastic pellets were abundant (Bond 1971). It was unclear whether this had contributed to the high incidence of plastic in these birds, but Connors and Smith (1982) found plastic in six of seven red phalaropes killed by colliding with powerlines on their northward migration in central California. Birds with large volumes of ingested plastic had smaller fat reserves, raising concerns that ingested plastic reduced digestive efficiency or meal size.

Baltz and Morejohn (1976) reported plastic in nine species of seabirds stranded in Monterey Bay, central California, during 1974–1975. All individuals of two species contained plastic: northern fulmar (*Fulmarus glacialis*) and short-tailed shearwater (*Puffinus tenuirostris*). Industrial pellets predominated in these birds, but they were also found to contain pieces of food wrap, foamed polystyrene, synthetic sponge and pieces of rigid plastic. Baltz and Morejohn (1976) speculated that having large volumes of plastic in their stomachs could interfere with the birds' digestion, although they considered that toxic chemicals adsorbed to the plastics posed the greatest threat to bird health. Ohlendorf et al. (1978) showed that plastic ingestion also occurred among Alaskan seabirds.

In the same year that Colton et al. (1974) showed the ubiquitous nature of plastic particles floating in the northwest Atlantic, Wong et al. (1974) reported that plastic pellets were widespread in the North Pacific Ocean. Sampling in 1972, they found that pellets occurred at lower densities (average 300 g km^{-2}) than tar balls, but they outnumbered tar balls northeast of Hawaii, with up to 34,000 pellets km^{-2} (3500 g km^{-2}). Even before this, however, Venrick et al. (1973) had shown that large litter items, at least half of which were made of plastic, were commonly encountered in the North Pacific gyre northeast of Hawaii (roughly 4.2 items km^{-2}) in the area of the now notorious 'North Pacific Garbage Patch'. This is where Moore et al. (2001) recorded densities of more than

300,000 particles km^{-2} in 1999, and where the weight of the plastic was six times that of the associated zooplankton.

Merrell (1980) conducted one of the first detailed studies of beach litter. Working on remote Alaskan beaches, he reported how the amount of plastic litter more than doubled in abundance between 1972 and 1974, increasing from an average density of 122 to 345 kg km^{-1}. Most of this litter came from fisheries operating in the area, but some apparently had drifted more than 1500 km from Asia. At the same time, Jewett (1976) and Feder et al. (1978) found that litter was common on the seabed off Alaska, with plastic items predominating. Merrell (1980) considered that the most obvious impact of beach litter was its aesthetic impact. In terms of biological threats, he speculated that plastic litter might account for the elevated levels of PCBs recorded in rats and intertidal organisms on Amchitka Island, and also suggested that plastics might be a source of phthalates and other toxic compounds into marine systems. Litter also entangled animals, especially seals and seabirds (Merrell 1980), and even terrestrial species were not immune from this problem (Beach et al. 1976).

Merrell (1980) reported the first long-term study of litter accumulation from a 1-km beach on Amchitka Island, Aleutians. He showed that the accumulation rate of litter (average 0.9 kg km^{-1} d^{-1}) varied considerably between sample periods (0.6–2.3 kg km^{-1} d^{-1}), and at a fine temporal scale the amount of litter stranded was a function of recent weather conditions. He also estimated the annual turnover rate of plastic items on the beach by marking gillnet floats, the most abundant litter item on the island, in two successive years. During the intervening year, 41 % of marked floats disappeared (25 % at one beach and 70 % at another beach), but this loss was more than compensated for by new arrivals, with a net increase of 130 %. Merrell (1980) discussed the various factors causing the loss of plastic items from beaches (burial, export inland or out to sea, etc.), and noted the bias introduced by selective beachcombing. Even on remote Amchitka Island, the small Atomic Energy Commission workforce removed certain types of fishing floats within a few days of the floats washing ashore.

The large amounts of litter found in Alaska, coupled with ingestion by seabirds (Ohlendorf et al. 1978) and entanglement of seals (Fowler 1985, 1987), stimulated the first post-graduate thesis on the marine litter problem. Bob Day (1980) studied the amounts of plastic ingested by Alaskan seabirds, in the first community-level study of plastic ingestion. Of the almost 2000 birds from 37 species collected off Alaska from 1969 to 1977, plastic was found in 40 % of species and 23 % of individuals. His main findings were presented in a review paper at the first marine debris conference in 1984 that summarized what was known about plastic ingestion by birds (Day et al. 1985). By that stage, it was clear that the incidence of plastic ingestion varied greatly among taxa, with high rates typically recorded among petrels and shearwaters (Procellariidae), phalaropes (*Phalaropus*) and some auks (Alcidae). Unsurprisingly, generalist foragers that fed near the water surface tended to have the highest plastic loads, although some pursuit-diving shearwaters and auks also contained large amounts of plastic. Plastic items were only found in the stomachs of birds; no visible items passed into the intestines. There was some evidence that at least some species retained plastic particles in their stomachs for considerable periods (up to 15 months), where they slowly eroded. Almost all particles floated in seawater, and comparison of the colors of ingested plastics with observations of the colors of litter items at sea demonstrated that all species favoured more conspicuous items, suggesting they were consumed deliberately. Industrial pellets comprised the majority of plastic items in most species sampled, possibly due to their similarity to fish eggs.

Day et al. (1985) also showed that the incidence of plastic ingestion generally increased over the study period, but patterns were affected by seasonal and age-related differences in plastic loads. Sex had no effect on plastic loads, but immature birds contained more plastic than adults in two of three species where this could be tested. There were also regional differences in plastic loads, with birds from the Aleutian Islands containing more plastic than birds from the Gulf of Alaska, and even lower loads in birds from the Bering and Chukchi Seas. Surveys in the North Atlantic confirmed regional differences in plastic loads in northern fulmars (Bourne 1976; Furness 1985a; van Franeker 1985), paving the way for the use of this species to monitor the abundance and distribution of plastic litter at sea (Ryan et al. 2009; van Franeker et al. 2011; Kühn and van Franeker 2012).

Like Connors and Smith (1982), Day (1980) found weak negative correlations between the amount of ingested plastic and body mass or fat reserves in some species, suggesting a sub-lethal effect on birds. And among parakeet auklets (*Cyclorrhynchus psittacula*), non-breeding adults contained twice as much plastic as breeding adults. However, Day (1980) was quick to point out that the differences in plastic loads could be a consequence of poor body condition or breeding status rather than vice versa. Harper and Fowler (1987) assumed that the negative correlation between the amount of ingested plastic and body mass of juvenile Salvin's prions (*Pachyptila salvini*) stranded in New Zealand in 1966 resulted from starving birds resorting to eat inedible objects such as pumice and plastic pellets. Spear et al. (1995) reported that among a large series of birds collected in the tropical Pacific, heavier birds were more likely to contain plastic, and attributed this to the fact that they fed in productive frontal areas where plastic tends to accumulate (cf. Bourne and Clarke 1984). Among birds that contained plastic, there was a negative correlation between the amount of plastic and body weight, which they interpreted as providing the first solid evidence of a negative relationship between plastic ingestion and body condition (Spear et al. 1995). However, caution must be exercised in such comparisons, given the effects of age and breeding status on the amounts of plastic in seabirds such as petrels that regurgitate accumulated plastic to their chicks (Ryan 1988a).

1.5 Into the Southern Hemisphere

Despite the fact that the first record of plastic ingestion came from the Southern Hemisphere in 1960 (Harper and Fowler 1987), reports of the occurrence of plastics at sea in the Southern Hemisphere generally lagged somewhat behind that in the north. Notable exceptions were the reports of plastic ingestion by turtles in South Africa, where plastic pellets were found in juvenile loggerhead turtles (*Caretta caretta*) in 1968 (Hughes 1970) and a large sheet of plastic was found blocking the intestine of a leatherback turtle (*Dermochelys coriacea*) that died in 1970 (Hughes 1974). The paucity of records of plastic litter from the Southern Hemisphere did not mean that the problem was not as severe in the less industrialized south. Gregory (1977, 1978) reported plastic pellets from virtually all New Zealand beaches, with densities at some beaches estimated at >100,000 pellets m^{-1}, which probably are the highest estimates of industrial pellet densities from any beach. Quite why such high densities were found in a country with a relatively small manufacturing base is unclear. Plastic pellets were also recorded in oceanic waters of the South Atlantic off the Cape in 1979, an area far removed from major shipping lanes and with little industrial activity in adjacent coastal regions (Morris 1980). There was a suggestion that pellets were more abundant west of 12°E (1500–3600 km^{-2}) than closer to the Cape coast (0–2000 km^{-2}), possibly linked to their aggregation in the South Atlantic gyre (cf. Lebreton et al. 2012; Maximenko et al. 2012; Ryan 2014). However, the average density of pellets and other plastic fragments close to the Cape coast was more than 3600 particles km^{-2} (Ryan 1988b), similar to densities reported in oceanic waters of the North Atlantic (Carpenter and Smith 1972; Colton et al. 1974) and North Pacific (Wong et al. 1974). By comparison, the density of pellets and other plastic litter in sub-Antarctic waters south of New Zealand was very low (<100 items km^{-2}, Gregory et al. 1984).

In addition to plastic pellet ingestion by New Zealand prions since the 1960s (Harper and Fowler 1987), rubber bands were found in Antarctic fulmars (*Fulmarus glacialoides*) stranded on New Zealand beaches in 1975 (Crockett and Reed 1976), and during an irruption of Southern Ocean petrels to New Zealand in 1981 all blue petrels (*Halobaena caerulea*) but very few Kerguelen petrels (*Lugensa brevirostris*) contained plastic (Reed 1981). Subsequent studies confirmed the high levels of plastic in blue petrels, despite the species rarely foraging north of the Subtropical Convergence (Ryan 1987a). Sampling in 1981 also showed that at least three petrel species collected in the South Atlantic Ocean contained plastics (Bourne and Imber 1982; Furness 1983; Randall et al. 1983). The incidence was greatest in great shearwaters (*Puffinus gravis*), with 90 % of individuals of this trans-equatorial migrant containing plastic particles, sometimes in large volumes (up to 78 pellets and fragments; Furness 1983). Further surveys even found plastics in Antarctic seabirds, but they were scarce in species that remained south of the Antarctic Polar Front year round compared to migrants that ventured farther north in the non-breeding season (Ryan 1987a; van Franeker and Bell 1988). Beach litter surveys confirmed the presence of plastic wastes in the far south, although the amounts of litter decreased from south temperate to sub-Antarctic and Antarctic locations (Gregory et al. 1984; Gregory 1987; Ryan 1987b).

Bob Furness (1985b) reported the first systematic survey of plastic ingestion by Southern Hemisphere birds for the seabirds of Gough Island, central South Atlantic Ocean. Of the 15 species sampled, 10 contained plastic, and two species had plastic in more than 80 % of individuals sampled. Petrels were again the most affected species, and Furness (1985b) was able to show that this was linked to the structure of their stomachs. The angled constriction between the fore-stomach and gizzard apparently prevents petrels regurgitating indigestible prey remains (except when feeding their chicks). Once again body mass was inversely correlated with the amount of ingested plastic in some species, but Furness (1985b) highlighted the need for controlled experiments to demonstrate an adverse impact of plastic ingestion. Building on this study, Ryan (1987a) showed that 40 of 60 Southern Hemisphere seabird species ingested plastic. Controlling for age and breeding status there was no correlation between plastic load and body condition (Ryan 1987c), but there was a correlation with PCB concentrations (Ryan et al. 1988), and chicks experimentally fed plastic grew more slowly than control birds, because they ate smaller meals (Ryan 1988c). A subsequent experiment showed that marine turtle hatchlings did not increase their food intake sufficiently to offset dietary dilution by an inert substance used to mimic the presence of plastic in their diet (McCauley and Bjorndal 1999).

Although most plastic apparently was ingested directly by the marine vertebrates studied, there was some evidence of secondary ingestion. Eriksson and Burton (2003) collected plastic particles from fur seal scat on Macquarie Island and speculated that they were ingested by lantern fish (*Electrona subaspera*), which were then eaten by the seals. And ingestion was not the only issue reported from the Southern Hemisphere. During the 1970s the rates of entanglement of Cape fur seals (*Arctocephalus pusillus*) in southern Africa (Shaughnessy 1980) were similar to those of northern fur seals in Alaska. The first entangled New Zealand fur seal (*Arctocephalus forsteri*) was observed in 1975 (Cawthorn 1985), and by the late 1970s entanglements of fur seals were recorded as far south as South Georgia (Bonner and McCann 1982). The first entanglements of cetaceans and sharks also were recorded from New Zealand in the 1970s (Cawthorn 1985).

1.6 Aloha—The Marine Debris Conferences

The growing awareness of the accumulation of plastic wastes in marine systems, and their impacts on marine biota, resulted in the Marine Mammal Commission approaching the US National Marine Fisheries Service in 1982 to arrange a workshop on the issue. Given the severity of the problem in the North Pacific Ocean, the task devolved to the Southwest Fisheries Center's Honolulu Laboratory. The Workshop on the Fate and Impact of Marine Debris took place in late November 1984 and was attended by 125 people from eight countries (91 % from the USA, 4 % from Asia, 3 % from Europe and 1 % each from Canada and New Zealand). Given the geographic bias of delegates, most of the 31 papers dealt with the North Pacific, but there were more general papers on the distribution and dynamics of floating litter as well as reviews of entanglement (Wallace 1985), and ingestion by seabirds (Day et al. 1985). The 580-page proceedings, edited by Richard Shomura and Howard Yoshida, appeared laudably fast as a NOAA Technical Memorandum in July 1985. Papers presented at the workshop were divided into three themes: the origins and amounts of marine debris (12 papers), impacts on marine resources (13 papers), and its fate (4 papers). The proceedings concluded with summary documents from working groups addressing each of the three main themes. The workshop emphasized the need to raise awareness of the threat posed by marine litter, and recommended three mitigation initiatives: to regulate the disposal of high-risk plastic items, to promote recycling of fishing nets, and to investigate the use of biodegradable material in fishing gear.

The success of the first marine debris workshop led to plans for a Second International Conference on Marine Debris. However, before this could occur the Sixth International Ocean Disposal Symposium took place in Pacific Grove, California, in April 1986. This was the first symposium in this series to address the dumping of persistent plastic wastes (Wolfe 1987). It was attended by 160 delegates from 10 countries and resulted in a special issue of *Marine Pollution Bulletin* (1987, volume 18, issue 6B). The focus was largely on ship-based sources of marine debris and their impacts, but also addressed incidental bycatch in fishing gear as well as land-based sources of debris. A few papers were repeated from the 1984 Honolulu workshop, and apart from Pruter's (1987) review of litter sources and amounts and Laist's (1987) review of the biological impacts of marine plastics, two of the most important papers dealt with legal approaches and strategies to reduce the amount of plastic entering the sea (Bean 1987; Lentz 1987).

The Second International Conference on Marine Debris was again held in Honolulu in April 1989, attracting over 170 delegates from 10 countries (USA 83 %, Japan 6 %, Canada and New Zealand 3 % each, UK 2 %; all other countries <1 %). It had a more ambitious scope than the first conference, with seven themed sessions following a series of regional overview papers. Whereas the focus of the first meeting was largely on the amounts and impacts of debris, the second conference concentrated more on tackling the problem, with sessions on solutions through technology, law and policy, and education, as well as the first estimates of the economic costs of marine litter. The two-volume, 1274-page proceedings, edited by Richard Shomura and Mary Lynne Godfrey, was again published as a NOAA Technical Memorandum in December 1990 and contained 76 papers plus eight working group reports. The proceedings made numerous recommendations, including nine priority recommendations. Both the first and second conference proceedings are available as internet downloads.

The first two Marine Debris Conferences played a major role in collating information on the marine debris issue. The large numbers of papers in the two proceedings resulted in a spike in publications on the subject (Fig. 1.2). Three further conferences have taken place. The Third International Conference on Marine Debris was held in Miami in May 1994 and had a more Caribbean flavor. It also differed from the two earlier conferences in having only selected papers published from the meeting in a book that aimed to provide a definitive treatment of the marine debris problem (Coe and Rogers 1997). The theme of the conference was 'Seeking Global Solutions', and two-thirds of the papers were devoted to mitigation, with four chapters on the socioeconomics of marine litter, eight chapters addressing at-sea sources, and ten chapters on land-based sources. This reflected the increasing appreciation that not only were diffuse, land-based inputs the major source of marine litter, but that in many ways they were harder to tackle than ship-based sources.

The two most recent Marine Debris Conferences were again held in Honolulu. The fourth conference (August 2000), which focused on the problems posed by derelict fishing gear, attracted 235 people from more than 20 countries, all but one in the Pacific region. The fifth meeting (March 2011) was the largest yet, with more than 450 delegates from across the world, reflecting the mounting concern among civil society regarding the threats posed by marine litter. Entitled 'Waves of Change: Global Lessons to Inspire Local Action', the conference concluded that despite the challenges inherent in tackling marine debris, the problem is preventable. The summary proceedings, released on the internet after the meeting, included reports from the three working groups established to address the prevention, reduction and management of land-based sources, of at-sea sources, and the removal and processing of accumulated marine debris. The reports highlighted progress made in each of these areas over the last decade, identified remaining challenges, and made recommendations for future action. The conference concluded with the Honolulu Commitment, which called on governmental and non-governmental organisations, industry and other stakeholders to commit to 12 action points, including formulating the Honolulu Strategy to prevent, reduce and manage marine debris. This framework document, sponsored by United Nations Environment Programme and the US National Oceanic and Atmospheric Administration, was released in 2012.

1.7 Mitigation Measures and Long-Term Changes in Marine Litter

One of the major challenges in addressing the marine plastics problem is the diverse nature of plastic products, and the many routes they can follow to enter marine systems (Pruter 1987; Ryan et al. 2009). As a result, a diversity of mitigation measures is needed to tackle the problem. Initial efforts focused on two specific user groups, shipping/fisheries and the plastics industry, at least in part because they are relatively discrete user groups, and thus are more easily addressed (at least in theory). Shipping was a major source of marine litter (Scott 1972; Horsman 1982). Dumping persistent plastic wastes from land-based sources at sea was banned under the Convention on the Prevention of Pollution by Dumping of Wastes and Other Matter (London Dumping Convention, promulgated in 1972; Lentz 1987), but operational wastes generated by vessels were exempt until Annex V of the International Convention for the Prevention of Pollution from Ships (MARPOL, promulagated in 1973) came into force at the end of 1988 (www.imo.org). Since then considerable effort has been expended to ensure there are adequate port facilities to receive wastes from ships (Coe and Rogers 1997). Current signatories to MARPOL Annex V are responsible for more than 97 % of the world's shipping tonnage, but compliance and enforcement remain significant problems (Carpenter and MacGill 2005).

Industrial pellets were another target for early mitigation measures because they were abundant in the environment, often ingested by marine birds and turtles, and only handled by a relatively small group of manufacturers and converters. As early as the 1970s it was clear that improving controls in manufacturing plants could significantly reduce the numbers of pellets entering coastal waters (Kartar et al. 1976). The loss of pellets in wastewater should fall under national water quality control measures, but in most countries the issue has been ignored in favour of chemical pollutants (Bean 1987). As a result, it was left to the plastics industry to initiate efforts to reduce losses of industrial pellets such as Operation Clean Sweep, established in the USA in 1992, and adopted in various guises by many other plastics industry organisations around the world (Redford et al. 1997).

How effective were these measures in reducing litter entering the sea? Although there were some exceptions (e.g. Merrell 1984), amounts of plastic litter at sea increased up to the 1990s, and then appeared to stabilize, whereas quantities on beaches and on the seabed have continued to increase (Barnes et al. 2009; Law et al. 2010). This could result from a decrease in the amounts of litter entering the sea (Barnes et al. 2009), but interpretation is complicated by the difficulty of monitoring marine litter loads, and our rather poor understanding of the rates of degradation and transport between habitats and regions (Ryan et al. 2009). Part of the problem is that mitigation measures may be effective in reducing the proportion of the waste stream reaching the sea, but this decrease may be insufficient to decrease the absolute amount of litter entering the sea, given the ongoing increase in plastic production (Fig. 1.1).

Interaction rates with marine biota provide one way to track the impacts of marine litter, and several studies have focused on the effects of specific mitigation initiatives. For example, the rate of entanglement in Antarctic fur seals (*Arctocephalus gazella*) at South Georgia decreased over the last two decades following active steps to prevent dumping of persistent wastes by vessels operating in the waters around the island. However, some of the decrease can be attributed to changes in seal numbers (Arnould and Croxall 1995; Waluda and Staniland 2013). A similar conclusion was reached by Boren et al. (2006) for New Zealand fur seals, where the decrease in the entanglement rate after 1997 was more likely a result of increasing seal numbers than a decrease in the amounts of litter at sea. Henderson (2001) showed no change in entanglement rates of Hawaiian monk seals (*Monachus schauinslandi*) before and after the implementation of MARPOL Annex V, nor was there a decrease in the rate at which netting washed ashore at the northwest Hawaiian Islands. Page et al. (2004) also showed no change in seal entanglement rates in southeast Australia despite efforts by government and fishing organisations to reduce the amount of litter discarded at sea. However, beach surveys in this region suggested that the implementation of MARPOL Annex V reduced the amounts of litter washed ashore (Edyvane et al. 2004). Ribic et al. (2010)

showed how carefully designed beach litter surveys can detect regional differences in long-term trends in the amounts of stranded litter, with consistent trends in land- and ship-based sources of litter.

Long-term studies of plastic ingestion by seabirds also indicate limited success in tackling the marine litter problem. The rapid increase in the amount of ingested plastic through the 1960s and 1970s (Harper and Fowler 1987; Moser and Lee 1992) stabilized during the 1980s and 1990s (Vlietstra and Parga 2002; Ryan 2008; Bond et al. 2013), but only studies of North Atlantic fulmars show a recent decrease in the amount of ingested plastic (van Franeker et al. 2011). Although the total amount of ingested plastic has tended to remain fairly constant over the last few decades, there has been a marked change in the composition of ingested plastic from pellets to plastic fragments (Vlietstra and Parga 2002; Ryan 2008; van Franeker et al. 2011), suggesting that efforts to reduce the numbers of pellets entering the sea have been at least partly successful. These results mirror the findings of net-samples of plastic litter at sea, which have seen a major increase in the proportion of user fragments and a corresponding decrease in industrial pellets relative to surveys conducted in the 1970s and 1980s (Moore et al. 2001; Law et al. 2010).

1.8 Plastic Degradation and the Microplastic Boom

Although many plastics are remarkably persistent, they are not immune to degradation. Indeed the plastics industry goes to considerable effort to slow the rate of degradation in many applications (Andrady et al. 2003). Ultraviolet (UV) radiation plays a key role in plastic degradation, and because UV light is absorbed rapidly by water, plastics generally take much longer to degrade at sea than on land (Andrady 2003). However, the rate of degradation depends on the ambient temperature as well as polymer type, additives and fillers (Andrady et al. 2003). Carpenter and Smith (1972) observed some degradation in polyethylene pellets collected at sea, but Gregory (1987) inferred degradation occurred more rapidly in stranded plastics, where they were exposed to high levels of UV radiation. The proportion of degraded pellets increased higher up the beach, away from the most recent strandline (Gregory 1987). Little is known about the fate of plastic that sinks to the seafloor; it is widely assumed that plastic is largely impervious to degradation once shielded from UV radiation (Goldberg 1997). However, there is some evidence that plastic fragments may be susceptible to bacterial decay at sea (Harshvardhan and Jha 2013; Zettler et al. 2013).

At the same time that plastics were being recognized as a significant marine pollutant, it was recognized that plastic litter was broken down by photodegradation and oxidation (Scott 1972; Cundell 1974). Scott (1972) reported how some beach litter items became embrittled and were reduced to small particles by very slight pressure. The apparent lack of disintegrated plastic around such items led him to conclude that the particles "had clearly been absorbed rapidly by the environment"

(Scott 1972, p. 36). Gregory (1983) also assumed that this process led to "complete degradation of the plastic pellets and dispersal as dust" (p. 82). However, it was a case of out of sight, out of mind. Thompson et al. (2004) showed that microscopic plastic fragments and fibres are ubiquitous marine pollutants. Together with the high media profile given the Pacific 'garbage patch' (Moore et al. 2001) and similar litter aggregations in other mid-ocean gyres (e.g. Law et al. 2010; Eriksen et al. 2013a), the research by Thompson et al. (2004) was largely responsible for the recent resurgence in interest in the marine litter problem (Fig. 1.2). Like larger plastic items, 'microplastics' (Ryan and Moloney 1990) are now found throughout the world's oceans, including in deep-sea sediments (van Cauwenberghe et al. 2013).

There is ongoing debate as to the size limit for 'microplastics' (Thompson 2015). Some authors take a broad view, including items <5 mm diameter (Arthur et al. 2009), whereas others restrict the term to items <2 mm, <1 mm or even <500 μm (Cole et al. 2011). Andrady (2011) argued the need for three terms: mesoplastics (500 μm–5 mm), microplastics (50–500 μm) and nanoplastics (<50 μm), each with their own set of physical characteristics and biological impacts. Depending on the upper size limit, industrial pellets may or may not be included in the term. But even if we adopt a narrow view, not all microplastics derive from degradation of larger plastic items. Some cosmetics, hand cleaners and air blast cleaning media contain small (<500 μm) plastic beads manufactured specifically for this purpose (Zitko and Hanlon 1991; Gregory 1996), the so-called primary microplastics (Cole et al. 2011). The proportion of primary microplastics in the environment probably is small compared to secondary microplastics, except for some areas of the Great Lakes in the United States (Eriksen et al. 2013b), but it is a largely avoidable source of pollution. Public pressure has already forced one major chemical company to commit to phasing out the use of plastic scrubbers in their products by 2015.

Much of the concern around microplastics concerns their role in introducing persistent organic pollutants (POPs) into marine foodwebs (Cole et al. 2011; Ivar do Sul and Costa 2014). Some of the additives used to modify the properties of plastics are biologically active, potentially affecting development and reproduction (Oehlmann et al. 2009; Meeker et al. 2009). Also, hydrophobic POPs in seawater are adsorbed onto plastic items (Carpenter et al. 1972; Mato et al. 2001; Teuten et al. 2009), and the smaller the particle, the more efficiently they accumulate toxins (Andrady 2011). Thompson et al. (2004) showed that invertebrates from three feeding guilds (detritivores, deposit feeders and filter feeders) all consumed microscopic plastic particles, reinforcing the results of early selectivity experiments demonstrating that filter feeders can consume small plastic particles (De Mott 1988; Bern 1990). Small particles also are eaten by myctophid fish (Boerger et al. 2010), which are an important trophic link in many oceanic ecosystems (Davison and Asch 2011). The subject of POP transfer is explored in more detail by Rochman (2015), but it is worth noting that strict controls on the use of several POPs (e.g. PCBs, HCHs, DDT and its derivatives) have decreased their concentrations on plastic pellets over the last few decades (Ryan et al. 2012). There remain concerns about the health impacts of other compounds whose use is not as strictly regulated (e.g. PBDE, BPA, phthalates, nonylphenol, etc.; Meeker et al. 2009; Oehlmann et al. 2009; Gassel et al. 2013), and even the ingestion of uncontaminated microplastic particles can induce stress responses in fish (Rochman et al. 2013b).

1.9 Summary and Conclusions

Awareness of the threats posed by waste plastics to marine ecosystems developed gradually through the 1960s and 1970s. Most of the environmental impacts of plastic litter were identified in the 1970s and 1980s, resulting in numerous policy discussions and recommendations to decrease the amount of waste plastic entering the environment (Chen 2015). Tightened controls by plastic manufacturers and converters reduced losses of industrial pellets and legislation such as MARPOL Annex V reduced disposal of plastic wastes at sea (although compliance remains problematic in at least some sectors). However, it also became apparent that most litter entering the sea did so from diffuse, land-based sources that are more difficult to control. The rapid increase in global plastic production has resulted in an increase in the amount of plastic items and fragments in marine systems, which in many cases has offset the gains made by reducing losses of industrial pellets and dumping of ship-generated wastes. Plastic is becoming so abundant in some marine systems that it is actually altering the physical properties of the environment (e.g. Carson et al. 2011).

There was a lull in research activity in the 1990s, but the confirmation that microplastics were a ubiquitous marine pollutant in the early 2000s, coupled with publicity around the formation of mid-ocean garbage patches, has stimulated renewed research interest and increased public awareness of the marine litter problem. One of the most urgent current challenges is the need to develop techniques to trace the smallest plastic particles through marine ecosystems, including uptake and release from marine organisms. We also need an improved understanding of the dynamics of waste plastics if we are to monitor the efficacy of mitigation measures (Ryan et al. 2009). Just as we can't interpret the significance of plastic loads in organisms without assessing their turnover rates (Ryan 1988a), we need estimates of transport rates between environments and their biota, and of plastic degradation rates under different environmental conditions. However, we already know enough to say with certainty that the release of waste plastics into the environment is already impacting adversely on marine systems, and affecting human quality of life. Given that plastic litter is, at least theoretically, a wholly avoidable problem, increased effort is needed to stop the inappropriate disposal of waste plastics through a combination of education, product design, incentives, legislation and enforcement.

Acknowledgments I thank my many colleagues, and especially Coleen Moloney, for sharing my plastic-related adventures over the last 30 years. Bill Naude from the Plastics Federation of South Africa supported our research. The South African National Antarctic Programme and Tristan's Conservation Department provided logistical support as well as permission to visit some of the world's most spectacular islands.

References

Andrady, A. L. (Ed.). (2003). *Plastics and the environment*. New York: Wiley.
Andrady, A. L., Hamid, H. S., & Torikai, A. (2003). Effects of climate change and UV-B on materials. *Photochemical and Photobiological Sciences, 2*, 68–72.
Andrady, A. L. (2011). Microplastics in the marine environment. *Marine Pollution Bulletin, 62*, 1596–1605.
Anon. (1971). Elastic band pollution. *Marine Pollution Bulletin, 2*, 165.
Arnould, J. P. Y., & Croxall, J. P. (1995). Trends in entanglement of Antarctic fur seals (*Arctocephalus gazella*) in man-made debris at South Georgia. *Marine Pollution Bulletin, 30*, 707–712.
Arthur, C., Baker, J. & Bamford, H. (2009). *Proceedings of the International Research Workshop on the Occurrence, Effects and Fate of Micro-plastic Marine Debris*, September 9–11, 2008. NOAA Technical Memorandum NOS-OR&R-30.
Balazs, G. H. (1985). Impact of ocean debris on marine turtles: entanglement and ingestion. In *Proceedings of the Workshop on the Fate and Impact of Marine Debris* (pp. 387–429). NOAA Technical Memorandum, NMFS, SWFC 54.
Baltz, D. M., & Morejohn, G. V. (1976). Evidence from seabirds of plastic particle pollution off central California. *Western Birds, 7*, 111–112.
Barnes, D. K. A. (2002). Invasions by marine life on plastic debris. *Nature, 416*, 808–809.
Barnes, D. K. A., & Milner, P. (2005). Drifting plastic and its consequences for sessile organism dispersal in the Atlantic Ocean. *Marine Biology, 146*, 815–825.
Barnes, D. K. A., Galgani, F., Thompson, R. C., & Barlaz, M. (2009). Accumulation and fragmentation of plastic debris in global environments. *Philosophical Transactions of the Royal Society B, 364*, 1985–1998.
Bax, N., Williamson, A., Aguero, M., Gonzalez, E., & Geeves, W. (2003). Marine invasive alien species: A threat to global biodiversity. *Marine Policy, 27*, 313–323.
Beach, R. J., Newby, T. C., Larson, R. O., Pedersen, M., & Juris, J. (1976). Entanglement of an Aleutian reindeer in a Japanese fish net. *Murrelet, 57*, 66.
Bean, M. J. (1987). Legal strategies for reducing persistent plastics in marine environments. *Marine Pollution Bulletin, 18*, 357–360.
Berland, B. (1971). Piggha og lundefugl med gummistrik. *Fauna, 24*, 35–37.
Bern, L. (1990). Size-related discrimination of nutritive and inert particles by freshwater zooplankton. *Journal of Plankton Research, 12*, 1059–1067.
Bingel, E., Avsar, D., & Unsal, M. (1987). A note on plastic materials in trawl catches in the north-eastern Mediterranean. *Meeresforschung, 31*, 227–233.
Boerger, C. M., Lattin, G. L., Moore, S. L., & Moore, C. J. (2010). Plastic ingestion by planktivorous fishes in the North Pacific Central Gyre. *Marine Pollution Bulletin, 60*, 2275–2278.
Bond, S. I. (1971). Red Phalarope mortality in Southern California. *California Birds, 2*, 97.
Bond, A. L., Provencher, J. F., Elliot, R. D., Ryan, P. C., Rowe, S., Jones, I. L., et al. (2013). Ingestion of plastic marine debris by Common and Thick-billed Murres in the northwestern Atlantic from 1985 to 2012. *Marine Pollution Bulletin, 77*, 192–195.
Bonner, W. N., & McCann, T. S. (1982). Neck collars on fur seals, *Arctocephalus gazella*, at South Georgia. *British Antarctic Survey Bulletin, 57*, 73–77.
Boren, L. J., Morrissey, M., Muller, C. G., & Gemmell, N. J. (2006). Entanglement of New Zealand Fur Seals in man-made debris at Kaikoura, New Zealand. *Marine Pollution Bulletin, 52*, 442–446.
Bourne, W. R. P. (1976). Seabirds and pollution. In R. Johnson (Ed.), *Marine pollution* (pp. 403–502). London: Academic Press.

Bourne, W. R. P. (1977). Nylon netting as a hazard to birds. *Marine Pollution Bulletin, 8*, 75–76.

Bourne, W. R. P. & Imber, M. J. (1982). Plastic pellets collected by a prion on Gough Island, central South Atlantic Ocean. *Marine Pollution Bulletin, 13*, 20–21.

Bourne, W. R. P., & Clarke, G. C. (1984). The occurrence of birds and garbage at the Humboldt front off Valparaiso, Chile. *Marine Pollution Bulletin, 15*, 343–345.

Buchanan, J. B. (1971). Pollution by synthetic fibres. *Marine Pollution Bulletin, 2*, 23.

Carpenter, A., & Macgill, S. M. (2005). The EU Directive on port reception facilities for ship-generated waste and cargo residues: The results of a second survey on the provision and uptake of facilities in North Sea ports. *Marine Pollution Bulletin, 50*, 1541–1547.

Carpenter, E. J., & Smith, K. L, Jr. (1972). Plastic on the Sargasso Sea surface. *Science, 175*, 1240–1241.

Carpenter, E. J., Anderson, S. J., Harvey, G. R., Miklas, H. P., & Peck, B. B. (1972). Polystyrene spherules in coastal waters. *Science, 178*, 749–750.

Carson, H. S., Colbert, S. L., Kaylor, M. J., & McDermid, K. J. (2011). Small plastic debris changes water movement and heat transfer through beach sediments. *Marine Pollution Bulletin, 62*, 1708–1713.

Cawthorn, M. W. (1985). Entanglement in, and ingestion of, plastic litter by marine mammals, sharks, and turtles in New Zealand waters. In *Proceedings of the Workshop on the Fate and Impact of Marine Debris* (pp. 336–343). NOAA Technical Memorandum, NMFS, SWFC 54.

Chen, C.-L. (2015). Regulation and management of marine litter. In M. Bergmann, L. Gutow & M. Klages (Eds.), *Marine anthropogenic litter* (pp. 398–432). Berlin: Springer.

Coe, J. M., & Rogers, D. B. (1997). *Marine debris: Sources, impacts and solutions*. New York: Springer.

Cole, M., Lindeque, P., Halsband, C., & Galloway, T. S. (2011). Microplastics as contaminants in the marine environment: A review. *Marine Pollution Bulletin, 62*, 2588–2597.

Colton, J. B, Jr, Knapp, F. D., & Burns, B. R. (1974). Plastic particles in surface waters of the northwestern Atlantic. *Science, 185*, 491–497.

Connors, P. G., & Smith, K. G. (1982). Oceanic plastic particle pollution: Suspected effect on fat deposition in Red Phalaropes. *Marine Pollution Bulletin, 13*, 18–20.

Cornelius, S. H. (1975). Marine turtle mortalities along the Pacific coast of Costa Rica. *Copeia, 1975*, 186–187.

Crockett, D. E., & Reed, S. M. (1976). Phenomenal Antarctic Fulmar wreck. *Notornis, 23*, 250–252.

Cundell, A. M. (1973). Plastic materials accumulating in Narragansett Bay. *Marine Pollution Bulletin, 4*, 187–188.

Cundell, A. M. (1974). Plastic in the marine environment. *Environmental Conservation, 1*, 63–68.

Davison, P., & Asch, R. G. (2011). Plastic ingestion by mesopelagic fishes in the North Pacific Subtropical Gyre. *Marine Ecology Progress Series, 432*, 173–180.

Day, R. H. (1980). The occurrence and characteristics of plastic pollution in Alaska's marine birds. M.Sc. Thesis, University of Alaska, Fairbanks (111 pp).

Day, R. H., Wehle, D. H. S. & Coleman, F. C. (1985). Ingestion of plastic pollutants by marine birds. In *Proceedings of the Workshop on the Fate and Impact of Marine Debris* (pp. 344–386). NOAA Technical Memorandum, NMFS, SWFC 54.

De Mott, W. (1988). Discrimination between algae and artificial particles by freshwater and marine copepods. *Limnology and Oceanography, 33*, 397–408.

Dixon, T. R., & Cooke, A. J. (1977). Discarded containers on a Kent Beach. *Marine Pollution Bulletin, 8*, 105–109.

Edyvane, K. S., Dalgetty, A., Hone, P. W., Higham, J. S., & Wace, N. M. (2004). Long-term marine litter monitoring in the remote Great Australian Bight, South Australia. *Marine Pollution Bulletin, 48*, 1060–1075.

Eriksen, M., Maximenko, N., Thiel, M., Cummins, A., Lattin, G., Wilson, S., et al. (2013a). Plastic marine pollution in the South Pacific subtropical gyre. *Marine Pollution Bulletin, 68*, 71–76.

Eriksen, M., Mason, S., Wilson, S., Box, C., Zellers, A., Edwards, W., et al. (2013b). Microplastic pollution in the surface waters of the Laurentian Great Lakes. *Marine Pollution Bulletin, 77*, 177–182.

Eriksson, C., & Burton, H. (2003). Origins and biological accumulation of small plastic particles in fur seals from Macquarie Island. *Ambio, 32*, 380–384.

Feder, H. M., Jewett, S. C., & Hilsinger, J. R. (1978). Man-made debris on the Bering Sea floor. *Marine Pollution Bulletin, 9*, 52–53.

Forrester, D. J., White, F. H., Woodard, J. C., & Thompson, N. P. (1975). Intussusception in a Florida Manatee. *Journal of Wildlife Diseases, 11*, 566–568.

Fowler, C. W. (1985). An evaluation of the role of entanglement in the population dynamics of Northern Fur Seals in the Pribilof Islands. In *Proceedings of the Workshop on the Fate and Impact of Marine Debris* (pp. 291–307). NOAA Technical Memorandum, NMFS, SWFC 54.

Fowler, C. W. (1987). Marine debris and northern fur seals: A case study. *Marine Pollution Bulletin, 18*, 326–335.

Fowler, C. W., Merrick, R. & Baker, J. D. (1990). Studies of the population level effects of entanglement on Northern Fur Seals. In *Proceedings of the Second International Conference on Marine Debris* (pp. 453–474). NOAA Technical Memorandum, NMFS, SWFC 154.

Furness, B. L. (1983). Plastic particles in three procellariiform seabirds from the Benguela Current, South Africa. *Marine Pollution Bulletin, 14*, 307–308.

Furness, R. W. (1985a). Plastic particle pollution: Accumulation by procellariiform seabirds at Scottish colonies. *Marine Pollution Bulletin, 16*, 103–106.

Furness, R. W. (1985b). Ingestion of plastic particles by seabirds at Gough Island, South Atlantic Ocean. *Environmental Pollution, 38*, 261–272.

Galgani, F., Jaunet, S., Campillo, A., Guenegen, X., & His, E. (1995). Distribution and abundance of debris on the continental shelf of the north-western Mediterranean Sea. *Marine Pollution Bulletin, 11*, 713–717.

Galgani, F., Souplet, A., & Cadiou, Y. (1996). Accumulation of debris on the deep sea floor off the French Mediterranean coast. *Marine Ecology Progress Series, 142*, 225–234.

Galgani, F., Leaute, J. P., Moguedet, P., Souplet, A., Verin, Y., Carpentier, A., et al. (2000). Litter on the sea floor along European coasts. *Marine Pollution Bulletin, 40*, 516–527.

Galil, B. S., Golik, A., & Türkay, M. (1995). Litter at the bottom of the sea: A sea bed survey in the Eastern Mediterranean. *Marine Pollution Bulletin, 30*, 22–24.

Gassel, M., Harwani, S., Park, J.-S., & Jahn, A. (2013). Detection of nonylphenol and persistent organic pollutants in fish from the North Pacific Central Gyre. *Marine Pollution Bulletin, 73*, 231–242.

Gochfeld, M. (1973). Effect of artefact pollution on the viability of seabird colonies on Long Island, New York. *Environmental Pollution, 4*, 1–6.

Goldberg, E. D. (1994). Diamonds and plastics are forever? *Marine Pollution Bulletin, 28*, 466.

Goldberg, E. D. (1997). Plasticizing the seafloor: An overview. *Environmental Technology, 18*, 195–201.

Gregory, M. R. (1977). Plastic pellets on New Zealand beaches. *Marine Pollution Bulletin, 9*, 82–84.

Gregory, M. R. (1978). Accumulation and distribution of virgin plastic granules on New Zealand beaches. *New Zealand Journal of Marine and Freshwater Research, 12*, 399–414.

Gregory, M. R. (1983). Virgin plastic granules on some beaches of eastern Canada and Bermuda. *Marine Environmental Research, 10*, 73–92.

Gregory, M. R. (1987). Plastics and other seaborne litter on the shores of New Zealand's sub-Antarctic islands. *New Zealand Antarctic Record, 7*, 32–47.

Gregory, M. R. (1996). Plastic 'scrubbers' in hand cleansers: A further (and minor) source for marine pollution identified. *Marine Pollution Bulletin, 32*, 867–871.

Gregory, M. R. (2009). Environmental implications of plastic debris in marine settings—entanglement, ingestion, smothering, hangers-on, hitch-hiking and alien invasions. *Philosophical Transactions of the Royal Society B, 364*, 2013–2025.

Gregory, M. R., Kirk, R. M., & Mabin, M. C. G. (1984). Pelagic tar, oil, plastics and other litter in surface waters of the New Zealand sector of the Southern Ocean, and on Ross Dependency shores. *New Zealand Antarctic Record, 6*, 12–28.

Guppy, H. B. (1917). *Plants, seeds, and currents in the West Indies and Azores*. London: Williams & Norgate.

Hanni, K. D., & Pyle, P. (2000). Entanglement of pinnipeds in synthetic materials at South-east Farallon Island, California, 1976–1998. *Marine Pollution Bulletin, 40*, 1076–1081.

Harper, P. C., & Fowler, J. A. (1987). Plastic pellets in New Zealand storm-killed prions (*Pachyptila* spp.) 1958–1977. *Notornis, 34*, 65–70.

Harshvardhan, K., & Jha, B. (2013). Biodegradation of low-density polyethylene by marine bacteria from pelagic waters, Arabian Sea, India. *Marine Pollution Bulletin, 77*, 100–106.

Hays, H., & Cormons, G. (1974). Plastic particles found in tern pellets, on coastal beaches and at factory sites. *Marine Pollution Bulletin, 5*, 44–46.

Henderson, J. R. (2001). A pre- and post-MARPOL Annex V summary of Hawaiian Monk Seal entanglements and marine debris accumulations in the northwestern Hawaiian Islands, 1982–1998. *Marine Pollution Bulletin, 42*, 584–589.

Heyerdahl, H. (1971). Atlantic Ocean pollution and biota observed by the Ra Expeditions. *Biological Conservation, 3*, 164–167.

Holmström, A. (1975). Plastic films on the bottom of the Skagerrak. *Nature, 255*, 622–623.

Horsman, P. V. (1982). The amount of garbage pollution from merchant ships. *Marine Pollution Bulletin, 13*, 167–169.

Hughes, G. R. (1970). Further studies on marine turtles in Tongaland, III. *Lammergeyer, 12*, 7–25.

Hughes, G. R. (1974). The sea turtles of south-east Africa. I. Status, morphology and distribution. *Oceanographic Research Institute Investigational Report, 35*, 1–114.

Ivar do Sul, J. A., & Costa, M. F. (2014). The present and future of microplastic pollution in the marine environment. *Environmental Pollution, 185*, 352–364.

Jacobson, M. A. (1947). An impeded Herring Gull. *Auk, 64*, 619.

Jewett, S. C. (1976). Pollutants of the northeast Gulf of Alaska. *Marine Pollution Bulletin, 7*, 169.

Kartar, S., Milne, R. A., & Sainsbury, M. (1973). Polystyrene waste in the Severn Estuary. *Marine Pollution Bulletin, 4*, 144.

Kartar, S., Abou-Seedou, F., & Sainsbury, M. (1976). Polystyrene spherules in the Severn Estuary—a progress report. *Marine Pollution Bulletin, 7*, 52.

Keller, A. A., Fruh, E. L., Johnson, M. M., Simon, V., & McGourty, C. (2010). Distribution and abundance of anthropogenic marine debris along the shelf and slope of the US west coast. *Marine Pollution Bulletin, 60*, 692–700.

Kenyon, K. W., & Kridler, E. (1969). Laysan Albatrosses swallow indigestible matter. *Auk, 86*, 339–343.

Kühn, S., & van Franeker, J. A. (2012). Plastic ingestion by the Northern Fulmar (*Fulmarus glacialis*) in Iceland. *Marine Pollution Bulletin, 64*, 1252–1254.

Laist, D. W. (1987). Overview of the biological effects of lost and discarded plastic debris in the marine environment. *Marine Pollution Bulletin, 18*, 319–326.

Law, K. L., Morét-Ferguson, S., Maximenko, N. A., Proskurowski, G., Peacock, E. E., Hafner, J., et al. (2010). Plastic accumulation in the North Atlantic subtropical gyre. *Science, 329*, 1185–1188.

Lebreton, L. C.-M., Greer, S. D., & Borrero, J. C. (2012). Numerical modelling of floating debris in the world's oceans. *Marine Pollution Bulletin, 64*, 653–661.

Lentz, S. A. (1987). Plastics in the marine environment: Legal approaches for international action. *Marine Pollution Bulletin, 18*, 361–365.

Magazine, Life. (1955). Throwaway living: Disposable items cut down household chores. *Life, 39*, 43–44.

Masó, M., Garcés, E., Pagès, F., & Camp, J. (2003). Drifting plastic debris as a potential vector for dispersing Harmful Algal Bloom (HAB) species. *Scientia Marina, 67*, 107–111.

Mato, Y., Isobe, T., Takada, H., Kanehiro, H., Ohtake, C., & Kaminuma, T. (2001). Plastic resin pellets as a transport medium for toxic chemicals in the marine environment. *Environmental Science and Technology, 35*, 318–324.

Maximenko, N., Hafner, J., & Niiler, P. (2012). Pathways of marine debris derived from trajectories of Lagrangian drifters. *Marine Pollution Bulletin, 65*, 51–62.

McCauley, S. J., & Bjorndal, K. A. (1999). Conservation implications of dietary dilution from debris ingestion: Sublethal effects in post-hatchling loggerhead sea turtles. *Conservation Biology, 13*, 925–929.

Meeker, J. D., Sathyanarayana, S., & Swan, S. H. (2009). Phthalates and other additives in plastics: Human exposure and associated health outcomes. *Philosophical Transactions of the Royal Society B, 364*, 2097–2113.

Merrell, T. R, Jr. (1980). Accumulation of plastic litter on beaches of Amchitka Island, Alaska. *Marine Environmental Research, 3*, 171–184.

Merrell, T. R, Jr. (1984). A decade of change in nets and plastic litter from fisheries off Alaska. *Marine Pollution Bulletin, 15*, 378–384.

Moore, C. J., Moore, S. L., Leecaster, M. K., & Weisberg, S. B. (2001). A comparison of plastic and plankton in the North Pacific central gyre. *Marine Pollution Bulletin, 42*, 1297–1300.

Morris, R. J. (1980). Plastic debris in the surface waters of the South Atlantic. *Marine Pollution Bulletin, 11*, 164–166.

Morris, A. W., & Hamilton, E. I. (1974). Polystyrene spherules in the Bristol Channel. *Marine Pollution Bulletin, 5*, 26–27.

Moser, M. L., & Lee, D. S. (1992). A fourteen-year survey of plastic ingestion by western North Atlantic seabirds. *Colonial Waterbirds, 15*, 83–94.

Muir, J. (1937). The seed-drift of South Africa and some influences of ocean currents on the strand vegetation. *Union of South Africa Botanical Survey Memoir, 16*, 1–108.

Oehlmann, J., Schulte-Oehlmann, U., Kloas, W., Jagnytsch, O., Lutz, I., Kusk, K. O., et al. (2009). A critical analysis of the biological impacts of plasticizers on wildlife. *Philosophical Transactions of the Royal Society B, 364*, 2047–2062.

Ohlendorf, H. M., Risebrough, R. W. & Vermeer, K. (1978). Exposure of marine birds to environmental pollutants. U.S. Fish and Wildlife Service Research Report 9, Washington D.C.

Page, B., McKenzie, J., McIntosh, R., Baylis, A., Morrissey, A., Calvert, N., et al. (2004). Entanglement of Australian Sea Lions and New Zealand Fur Seals in lost fishing gear and other marine debris before and after government and industry attempts to reduce the problem. *Marine Pollution Bulletin, 49*, 33–42.

Parslow, J. L. F., & Jefferies, D. J. (1972). Elastic thread pollution of puffins. *Marine Pollution Bulletin, 3*, 43–45.

PlasticsEurope (2013). Plastics—the Facts 2013: An analysis of European latest plastics production, demand and waste data. PlasticsEurope (www.plasticseurope.org).

Pruter, A. T. (1987). Sources, quantities and distribution of persistent plastics in the marine environment. *Marine Pollution Bulletin, 18*, 305–310.

Randall, B. M., Randall, R. M., & Rossouw, G. J. (1983). Plastic particle pollution in Great Shearwaters (*Puffinus gravis*) from Gough Island. *South African Journal of Antarctic Research, 13*, 49–50.

Redford, D. P., Trulli, H. K., & Trulli, W. R. (1997). Sources of plastic pellets in the aquatic environment. In J. M. Coe & D. B. Rogers (Eds.), *Marine debris: Sources, impacts and solutions* (pp. 335–343). New York: Springer.

Reed, S. (1981). Wreck of Kerguelen and Blue Petrels. *Notornis, 28*, 239–241.

Ribic, C. A., Sheavley, S. B., Rugg, D. J., & Erdmann, E. S. (2010). Trends and drivers of marine debris on the Atlantic coast of the United States 1997–2007. *Marine Pollution Bulletin, 60*, 1231–1242.

Rochman, C. M. (2015). The complex mixture, fate and toxicity of chemicals associated with plastic debris in the marine environment. In M. Bergmann, L. Gutow & M. Klages (Eds.), *Marine anthropogenic litter* (pp. 117–140). Berlin: Springer.

Rochman, C. M., Browne, M. A., Halpern, B. S., Hentschel, B. T., Hoh, E., Karapanagioti, H. K., et al. (2013a). Policy: Classify plastic waste as hazardous. *Nature, 494*, 169–171.

Rochman, C. M., Hoh, E., Kurobe, T., & Teh, S. J. (2013b). Ingested plastic transfers hazardous chemicals to fish and induces hepatic stress. *Scientific Reports, 3*, 3263.

Rodríguez, B., Bécares, J., Rodríguez, A., & Arcos, J. M. (2013). Incidence of entanglements with marine debris by Northern Gannets (*Morus bassanus*) in the non-breeding grounds. *Marine Pollution Bulletin, 75*, 259–263.

Rothstein, S. I. (1973). Particle pollution of the surface of the Atlantic Ocean: Evidence from a seabird. *Condor, 73*, 344–345.

Ryan, P. G. (1987a). The incidence and characteristics of plastic particles ingested by seabirds. *Marine Environmental Research, 23*, 175–206.

Ryan, P. G. (1987b). The origin and fate of artefacts stranded on islands in the African sector of the Southern Ocean. *Environmental Conservation, 14*, 341–346.

Ryan, P. G. (1987c). The effects of ingested plastic on seabirds: Correlations between plastic load and body condition. *Environmental Pollution, 46*, 119–125.

Ryan, P. G. (1988a). Intraspecific variation in plastic ingestion by seabirds and the flux of plastic through seabird populations. *Condor, 90*, 446–452.

Ryan, P. G. (1988b). The characteristics and distribution of plastic particles at the sea-surface off the southwestern Cape Province, South Africa. *Marine Environmental Research, 25*, 249–273.

Ryan, P. G. (1988c). Effects of plastic ingestion on seabird feeding: Evidence from chickens. *Marine Pollution Bulletin, 19*, 125–128.

Ryan, P. G. (2008). Seabirds indicate decreases in plastic pellet litter in the Atlantic and south-western Indian Ocean. *Marine Pollution Bulletin, 56*, 1406–1409.

Ryan, P. G. (2014). Litter survey detects the South Atlantic 'garbage patch'. *Marine Pollution Bulletin, 79*, 220–224.

Ryan, P. G. & Moloney, C. L. (1990). Plastic and other artefacts on South African beaches: temporal trends in abundance and composition. *South African Journal of Science, 86*, 450–452.

Ryan, P. G., & Moloney, C. L. (1993). Marine litter keeps increasing. *Nature, 361*, 23.

Ryan, P. G., Connell, A. D., & Gardner, B. D. (1988). Plastic ingestion and PCBs in seabirds: Is there a relationship? *Marine Pollution Bulletin, 19*, 174–176.

Ryan, P. G., Moore, J. M., van Franeker, J. A., & Moloney, C. L. (2009). Monitoring the abundance of plastic debris in the marine environment. *Philosophical Transactions of the Royal Society B, 364*, 1999–2012.

Ryan, P. G., Bouwman, H., Moloney, C. L., Yuyama, M., & Takada, H. (2012). Long-term decreases in persistent organic pollutants in South African coastal waters detected from beached polyethylene pellets. *Marine Pollution Bulletin, 64*, 2756–2760.

Schrey, E., & Vauk, G. J. M. (1987). Records of entangled gannets (*Sula bassana*) at Helgoland, German Bight. *Marine Pollution Bulletin, 18*, 350–352.

Scott, G. (1972). Plastics packaging and coastal pollution. *International Journal of Environmental Studies, 3*, 35–36.

Shaughnessy, P. D. (1980). Entanglement of Cape fur seals with man-made objects. *Marine Pollution Bulletin, 11*, 332–336.

Spear, L. B., Ainley, D. G., & Ribic, C. A. (1995). Incidence of plastic in seabirds from the tropical Pacific 1984–91: Relation with distribution of species, sex, age, season, year and body weight. *Marine Environmental Research, 40*, 123–146.

Teuten, E. L., Saquing, J. M., Knappe, D. R. U., Barlaz, M. A., Jonsson, S., Björn, A., et al. (2009). Transport and release of chemicals from plastics to the environment and to wildlife. *Philosophical Transactions of the Royal Society B, 364*, 2027–2045.

Thompson, R. C. (2015). Microplastics in the marine environment: Sources, consequences and solutions. In: M. Bergmann, L. Gutow, M. Klages (Eds.) *Marine anthropogenic litter* (pp. 185–200). Berlin: Springer.

Thompson, R. C., Olsen, Y., Mitchell, R. P., Davis, A., Rowland, S. J., John, A. W. G., et al. (2004). Lost at sea: Where is all the plastic? *Science, 304*, 838.

Thompson, R. C., Swan, S. H., Moore, C. J., & vom Saal, F. S. (2009). Our plastic age. *Philosophical Transactions of the Royal Society B, 364*, 1973–1976.

van Cauwenberghe, L., Vanreusel, A., Mees, J., & Janssen, C. R. (2013). Microplastic pollution in deep-sea sediments. *Environmental Pollution, 182*, 495–499.

van Franeker, J. A. (1985). Plastic ingestion in the North Atlantic Fulmar. *Marine Pollution Bulletin, 16*, 367–369.

van Franeker, J. A., & Bell, P. J. (1988). Plastic ingestion by petrels breeding in Antarctica. *Marine Pollution Bulletin, 19*, 672–674.

van Franeker, J. A., Blaize, C., Danielsen, J., Fairclough, K., Gollan, J., Guse, N., et al. (2011). Monitoring plastic ingestion by the Northern Fulmar *Fulmarus glacialis* in the North Sea. *Environmental Pollution, 159*, 2609–2615.

Venrick, E. L., Backman, T. W., Bartram, W. C., Platt, C. J., Thornhill, M. S., & Yates, R. E. (1973). Man-made objects on the surface of the central North Pacific Ocean. *Nature, 241*, 271.

Vlietstra, L. S., & Parga, J. A. (2002). Long-term changes in the type, but not amount, of ingested plastic particles in Short-tailed Shearwaters in the southeastern Bering Sea. *Marine Pollution Bulletin, 44*, 945–955.

Wallace, N. (1985). Debris entanglement in the marine environment: A review. In *Proceedings of the Workshop on the Fate and Impact of Marine Debris* (pp. 259–277). NOAA Technical Memorandum, NMFS, SWFC 54.

Waluda, C. M. & Staniland, I. J. (2013). Entanglement of Antarctic fur seals at Bird Island, South Georgia. *Marine Pollution Bulletin, 74*, 244–252.

Wanless, R. M., Scott, S., Sauer, W. H. H., Andrew, T. G., Glass, J. P., Godfrey, B., et al. (2010). Semi-submersible rigs: A vector transporting entire marine communities around the world. *Biological Invasions, 12*, 2573–2583.

Winston, J. E. (1982). Drift plastic—an expanding niche for a marine invertebrate? *Marine Pollution Bulletin, 13*, 348–351.

Wolfe, D. A. (1987). Persistent plastics and debris in the ocean: An international problem of ocean disposal. *Marine Pollution Bulletin, 18*, 303–305.

Wong, C. S., Green, D. R., & Cretney, W. J. (1974). Quantitative tar and plastic waste distributions in the Pacific Ocean. *Nature, 247*, 31–32.

Ye, S., & Andrady, A. L. (1991). Fouling of floating plastic debris under Biscayne Bay exposure conditions. *Marine Pollution Bulletin, 22*, 608–613.

Zettler, E. R., Mincer, T. J., & Amaral-Zettler, L. A. (2013). Life in the "plastisphere": Microbial communities on plastic marine debris. *Environmental Science and Technology, 47*, 7137–7146.

Zitko, V., & Hanlon, M. (1991). Another source of pollution by plastics: Skin cleaners with plastic scrubbers. *Marine Pollution Bulletin, 22*, 41–42.

Part I
Abiotic Aspects of Marine Litter Pollution

Chapter 2
Global Distribution, Composition and Abundance of Marine Litter

François Galgani, Georg Hanke and Thomas Maes

Abstract Marine debris is commonly observed everywhere in the oceans. Litter enters the seas from both land-based sources, from ships and other installations at sea, from point and diffuse sources, and can travel long distances before being stranded. Plastics typically constitute the most important part of marine litter sometimes accounting for up to 100 % of floating litter. On beaches, most studies have demonstrated densities in the 1 item m^{-2} range except for very high concentrations because of local conditions, after typhoons or flooding events. Floating marine debris ranges from 0 to beyond 600 items km^{-2}. On the sea bed, the abundance of plastic debris is very dependent on location, with densities ranging from 0 to >7700 items km^{-2}, mainly in coastal areas. Recent studies have demonstrated that pollution of microplastics, particles <5 mm, has spread at the surface of oceans, in the water column and in sediments, even in the deep sea. Concentrations at the water surface ranged from thousands to hundred thousands of particles km^{-2}. Fluxes vary widely with factors such as proximity of urban activities, shore and coastal uses, wind and ocean currents. These enable the presence of accumulation areas in oceanic convergence zones and on the seafloor, notably in coastal canyons. Temporal trends are not clear with evidences for increases, decreases or without changes, depending on locations and environmental conditions. In terms of distribution and quantities, proper global estimations based on standardized approaches are still needed before considering efficient management and reduction measures.

Keywords Marine litter · Plastic · Distribution · Beaches · Seafloor · Microplastics · Floating litter

F. Galgani (✉)
IFREMER, LER/PAC, ZI furiani, 20600 Bastia, France
e-mail: Francois.galgani@ifremer.fr

G. Hanke
EC JRC, IES, European Commission Joint Research Centre,
Via Enrico Fermi 2749, 21027 Ispra, VA, Italy

T. Maes
CEFAS, Pakefield Road, Lowestoft, Suffolk NR330HT, UK

© The Author(s) 2015
M. Bergmann et al. (eds.), *Marine Anthropogenic Litter*,
DOI 10.1007/978-3-319-16510-3_2

2.1 Introduction

Anthropogenic litter on the sea surface, beaches and seafloor has significantly increased over recent decades. Initially described in the marine environment in the 1960s, marine litter is nowadays commonly observed across all oceans (Ryan 2015). Together with its breakdown products, meso-particles (5–2.5 cm) and micro-particles (<5 mm), they have become more numerous and floating litter items can be transported over long distances by prevailing winds and currents (Barnes et al. 2009).

Humans generate considerable amounts of waste and global quantities are continuously increasing, although waste production varies between countries. Plastic, the main component of litter, has become ubiquitous and forms sometimes up to 95 % of the waste that accumulates on shorelines, the sea surface and the seafloor. Plastic bags, fishing equipment, food and beverage containers are the most common items and constitute more than 80 % of litter stranded on beaches (Topçu et al. 2013; Thiel et al. 2013). A large part of these materials decomposes only slowly or not at all. This phenomenon can also be observed on the seafloor where 90 % of litter caught in benthic trawls is plastic (Galil et al. 1995; Galgani et al. 1995, 2000; Ramirez-Llodra et al. 2013).

Even with standardized monitoring approaches, the abundance and distribution of anthropogenic litter show considerable spatial variability. Strandline surveys and cleanings as well as regular surveys at sea are now starting to be organized in many countries in order to generate information about temporal and spatial distribution of marine litter (Hidalgo-Ruz and Thiel 2015). Accumulation rates vary widely and are influenced by many factors such as the presence of large cities, shore use, hydrodynamics and maritime activities. As a general pattern, accumulation rates appear to be lower in the southern than in the northern hemisphere. Enclosed seas such as the Mediterranean or Black Sea may harbor some of the highest densities of marine litter on the seafloor, reaching more than 100,000 items km^{-2} (Galgani et al. 2000). In surface waters, the problem of plastic fragments has increased in the last few decades. From the first reports in 1972 (Wong et al. 1974), the quantities of microparticles in European seas have grown in comparison to data from 2000 (Thompson et al. 2004). Recent data suggest that quantities of microparticles appear to have stabilized in the North Atlantic Ocean over the last decade (Law et al. 2010). Little is known about trends in accumulation of debris in the deep sea. Debris densities on the deep seafloor decreased in some areas, such as in the Bay of Tokyo from 1996 to 2003 and in the Gulf of Lion between 1994 and 2009 (Kuriyama et al. 2003; Galgani et al. 2011a, b). By contrast, in some areas around Greece, the abundance of debris in deep waters has substantially increased over a period of eight years (Stefatos et al. 1999; Koutsodendris et al. 2008) and on the deep Arctic seafloor of the HAUSGARTEN observatory over aperiod of ten years (Bergmann and Klages 2012). Interpretation of temporal trends is complicated by seasonal changes in the flow rate of rivers, currents, wave action, winds etc. Decreasing trends of macroplastics (>2.5 cm) on beaches of remote islands suggest that regulations to reduce dumping at sea have been successful to some extent (Eriksson et al. 2013). However, both the demand and the production of plastics reached 299 million tons in 2013 and are continuing to increase (PlasticsEurope 2015).

2.2 Composition

Analysis of the composition of marine litter is important as it provides vital information on individual litter items, which, in most cases, can be traced back to their sources. Sources of litter can be characterised in several ways (see also Browne 2015). One common method is to classify marine litter sources as either land-based or ocean-based, depending on where the litter entered the sea. Some items can be attributed with a high level of confidence to certain sources such as fishing gear, sewage-related debris and tourist litter. So-called use-categories provide valuable information for developing reduction measures (Galgani et al. 2011a).

Land-based sources include mainly recreational use of the coast, general public litter, industry, harbors and unprotected landfills and dumps located near the coast, but also sewage overflows, introduction by accidental loss and extreme events. Marine litter can be transported to the sea by rivers (Rech et al. 2014; Sadri and Thompson 2014) and other industrial discharges and run-offs or can even be blown into the marine environment by winds. Ocean-based sources of marine litter include commercial shipping, ferries and liners, both commercial and recreational fishing vessels, military and research fleets, pleasure boats and offshore installations such as platforms, rigs and aquaculture sites. Factors such as ocean current patterns, climate and tides, the proximity to urban, industrial and recreational areas, shipping lanes and fishing grounds also influence the types and amount of litter that are found in the open ocean or along beaches.

Assessments of the composition of litter in different marine regions show that "plastics", which include all petroleum-based synthetic materials, make up the largest proportion of overall litter pollution (e.g. Pham et al. 2014). Packaging, fishing nets and pieces thereof, as well as small pieces of unidentifiable plastic or polystyrene account for the majority of the litter items recorded in this category (Galgani et al. 2013). Some of this can take hundreds of years to break down or may never truly degrade (Barnes et al. 2009).

Whether or not visual observations from ships and airplanes, observations using underwater vehicles, manned or not, acoustics and finally trawling will provide the necessary detail to characterise litter and eventually define sources is not always clear. Previous notions that at a global scale most of the marine litter is from land-based sources rather than from ships, were confirmed (Galgani et al. 2011b). Marine litter found on beaches consists primarily of plastics (bottles, bags, caps/lids, etc.), aluminium (cans, pull tabs) and glass (bottles) and mainly originates from shoreline recreational activities but is also transported by the sea by currents. In some cases, specific activities account for local litter densities well above the global average (Pham et al. 2014). For example, marine litter densities on beaches can be increased by up to 40 % in summer because of high tourist numbers (Galgani et al. 2013). In some tourist areas, more than 75 % of the annual waste is generated in summer, when tourists produce on average 10–15 % more waste than the inhabitants; although not all of this waste enters the marine environment (Galgani et al. 2011b).

In some areas such as the North Sea or the Baltic Sea, the large diversity of items and the composition of the litter recorded indicate that shipping, fisheries and offshore installations are the main sources of litter found on beaches (Fleet et al. 2009). In some cases, litter can clearly be attributed to shipping, sometimes accounting for up to 95 % of all litter items in a given region, a large proportion of which originates from fishing activities often coming in the form of derelict nets (Van Franeker et al. 2011). In the North Sea, this percentage has been temporally stable (Galgani et al. 2011a) but litter may be supplemented by coastal recreational activities and riverine input (Lechner et al. 2014; Morritt et al. 2014). Studies along the US west coast, specifically off the coast of the southern California Bight (Moore and Allen 2000; Watters et al. 2010; Keller et al. 2010; Schlining et al. 2013) have shown that ocean-based sources are the major contributors to marine debris in the eastern North Pacific with, for example, fishing gear being the most abundant debris off Oregon (June 1990). Investigations in coastal waters and beaches around the northern South China Sea in 2009 and 2010 indicated that plastics (45 %) and Styrofoam (23 %) accounted for more than 90 % of floating debris and 95 % of beached debris. The sources were primarily land-based and mostly attributed to coastal recreational activities (Lee et al. 2013). In the Mediterranean, reports from Greece classify land-based (69 % of the litter) and vessel-based (26 %) waste as the two predominant sources of litter (Koutsodendris et al. 2008).

2.3 Distribution

2.3.1 Beaches

Marine debris is commonly found at the sea surface or washed up on shorelines, and much of the work on marine litter has focussed on coastal areas because of the presence of sources, ease of access/assessment and for aesthetic reasons (McGranahan et al. 2007). Marine litter stranded on beaches is found along all coasts and has become a permanent reason for concern. Beach-litter data are derived from various approaches based on measurements of quantities or fluxes, considering various litter categories, and sampling on transects of variable width and length parallel or perpendicular to the shore. This makes it difficult to draw a quantitative global picture of beach litter distribution. In general, methods that are used for estimating amounts of marine debris on beaches are considered cheap and fairly reliable, but it is not clear how it relates to litter at sea, floating or not. Moreover, in some coastal habitats, litter may be of terrestrial origin and may never actually enter the sea. Most surveys are done with a focus on cleaning, thereby missing proper classification of litter items. When studies are not dedicated to specific items, litter is categorized by the type of material, function or both. Studies record the numbers, some the mass of litter and some do both (Galgani et al. 2013). Evaluations of beach litter reflect the long-term balance between inputs, land-based sources or stranding, and outputs from export, burial, degradation and cleanups. Then, measures of stocks may reflect the presence and amounts of debris. Factors influencing densities such as cleanups, storm events, rain fall, tides, hydrological changes may alter counts, evaluations of fluxes and, even if surveys can track changes in the composition of beach litter, they may not be sensitive enough to monitor changes in the abundance (Ryan et al. 2009). This problem can be circumvented by recording the rate, at which litter accumulates on beaches through regular surveys that are performed weekly, monthly or annually after an initial cleanup (Ryan et al. 2009). This is actually the most common approach, revealing long-term patterns and cycles in accumulation, requiring nonetheless much effort to do surveys. However, past studies may have vastly underestimated the quantity of available debris because sampling was too infrequent (Smith and Markic 2013).

It is unfeasible to review the hundreds of papers on beach macro-debris, which often apply different approaches and lack sufficient detail (see also Hidalgo-Ruz and Thiel 2015). Most studies range from a local (Lee et al. 2013) to a regional scale (Bravo et al. 2009) and cover a broad temporal range. Information on sources, composition, amounts, usages, baseline data and environmental significance are often also gathered (Cordeiro and Costa 2010; Debrot et al. 2013; Rosevelt et al. 2013) as such data are easier collected. Most studies record all litter items encountered between the sea and the highest strandline on the upper shore. Sites are often chosen because of their ecological relevance, accessibility and particular anthropogenic activities and sources. Factors influencing the accumulation of debris in coastal areas include the shape of the beach, location and the nature of debris (Turra et al. 2014). In addition, most sediment-surface counts do not take buried litter into account and clearly underestimate abundance, which biases composition studies. However, raking of beach sediments for litter may disturb the resident fauna. Apparently, a good correlation exists between accumulated litter and the amount arriving, indicating regular inputs and processes. Recent experiments with drift models in Japan indicate good correlation of flux with litter abundances on beaches (Yoon et al. 2010; Kataoka et al. 2013).

It appears that glass and hard plastics are accumulating more easily on rocky shores (Moore et al. 2001a). Litter often strands on beaches that lack strong prevalent winds, which may blow them offshore (Galgani et al. 2000; Costa et al. 2011). Abundance or composition of litter often varies even among different parts of an individual beach (Claereboudt 2004) with higher amounts found frequently at high-tide or storm-level lines (Oigman-Pszczol and Creed 2007). Because of this and beach topography, patchiness is a common distribution pattern on beaches, especially for smaller and lighter items that are more easily dispersed or buried (Debrot et al. 1999).

It is very difficult to compare litter concentrations of various coastal areas (with different population densities, hydrographic and geological conditions) obtained from various studies with different methodologies, especially when the sizes of debris items that are taken into account are also different. Nevertheless, common patterns indicate the prevalence of plastics, greater loads close to urban areas and touristic regions (Barnes et al. 2009). Data expressed as items m^{-2} or larger areas are more convenient for comparisons. Most studies have reported densities in the m^{-2} range (Table 2.1). High concentrations of up to 37,000 items per 50-m beach line (78.3 items m^{-2}) were recorded in Bootless Bay, Papua New Guinea (Smith 2012) because of specific local conditions, following typhoons (3,227 items m^{-2}; Liu et al. 2013) or flooding events (5,058 items m^{-2}; Topçu et al. 2013). Data expressed as quantities per linear distance are more difficult to compare because the results depend on beach size/width. Plastic accounts for a large part of litter on beaches from many areas with up to 68 % in California (Rosevelt et al. 2013), 77 % in the south east of Taiwan (Liu et al. 2013),

Table 2.1 Comparison of mean litter densities from recent data worldwide (non-exhaustive list)

Region	Density (m^{-2})	Density (linear m^{-1})	Plastic (%)	References
SW Black Sea	0.88 (0.008–5.06)	24 (1.7–197)	91	Topçu et al. (2013)
Costa do Dende, Brazil	n.d.	9.1	75	Santos et al. (2009)
Cassina, Brazil	n.d.	5.3–10.7	48	Tourinho and Fillmann (2011)
Gulf of Aqaba	2 (1–6)	n.d.	n.d.	Al-Najjar and Al-Shiyabet (2011)
Monterey, USA	1 ± 2.1	n.d.	68	Rosevelt et al. (2013)
North Atlantic, USA	n.d.	0.10 (0.2)	n.d.	Ribic et al. (2010)
North Atlantic, USA	n.d.	0.42 (0.1)	n.d.	Ribic et al. (2010)
North Atlantic, USA	n.d.	0.08 (0.2)	n.d.	Ribic et al. (2010)
South Caribbean, Bonaire	1.4 (max. 115)	n.d.	n.d.	Debrot et al. (2013)
Bootless Bay, Papua New Guinea	15.3 (1.2–78.3)	n.d.	89	Smith (2012)
Nakdong, South Korea	0.97–1.03	n.d.	n.d.	Lee et al. (2013)
Kaosiung, Taiwan	0.9 (max. 3,227)	n.d.	77	Liu et al. (2013)
Tasmania	0.016–2.03	n.d.	n.d.	Slavin et al. (2012)
Midway, North Pacific	n.d.	0.60–3.52	91	Ribic et al. (2012a)
Chile	n.d.	0.01–0.25	n.d.	Thiel et al. (2013)
Heard Island, Antarctica	n.d.	0–0.132	n.d.	Eriksson et al. (2013)

Ranges of values are given in parentheses

86 % in Chile (Thiel et al. 2013), and 91 % in the southern Black Sea (Topçu et al. 2013). However, other types of litter or specific types of plastic may also be important in some areas, in terms of type (Styrofoam, crafted wood) or use (fishing gear).

For trends in the amount of litter washed ashore and/or deposited on coastlines, beach litter monitoring schemes provide the most comprehensive data on individual litter items. Large data sets have already been held by institutions (Ribic et al. 2010) or NGO's such as the Ocean Conservancy through their International Coastal Cleanup scheme for 25 years, or the EU OSPAR marine litter monitoring program, which started over 10 years ago and covers 78 beaches (Schultz et al. 2013). The lack of large-scale trends in the OSPAR-regions is probably due to small-scale heterogeneity of near-shore currents, which evoke small-scale heterogeneity in deposition patterns on beaches (Schulz et al. 2013).

Ribic et al. (2010, 2012b) derived several nonlinear models to describe the development of pollution of coastal areas with marine litter. There were long-term changes in indicator debris on the Pacific Coast of the U.S. and Hawaii over the nine-year period of the study. Ocean-based indicator debris loads declined substantially while at the same time land-based indicator items had also declined, except for the North Pacific coast region where no change was observed. Variation in debris loads was associated with land- and ocean-based processes with higher land-based debris loads being related to larger local populations. Overall and at the local scale, drivers included fishing activities and oceanic current systems for ocean-based debris and human population density and land use status for land-based debris.

At local scales, concentrations of specific items may be largely driven by specific activities or new sources. For example, 41 % of the total debris from beaches in California was of Styrofoam origin, with no other explanation than an increased use of packaging, which degrades very easily (Ribic et al. 2012b). Small-sized items may form an important fraction of debris on beaches. For example, up to 75 % of total debris from the southern Black Sea was smaller than 10 cm (Topçu et al. 2013). Small-sized particles include fragments smaller than 2.5 cm (Galgani et al. 2011b), the so-called meso-particles or mesodebris, which is, unlike macrodebris, often buried and not always targeted by cleanups. Stranding fluxes are then difficult to evaluate and a decrease in the amount of litter at sea will only slow the rate of stranding. Little attention has been paid to sampling design and statistical power even though optimal sampling strategies have been proposed (Ryan et al. 2009). Densities of small-sized debris were found to be very high in some areas where, in addition to floating debris, they can pose a direct threat to wildlife, especially to birds that are known to ingest plastic (Kühn et al. 2015; Lusher 2015).

2.3.2 Floating Marine Debris

Floating debris constitutes the fraction of debris in the marine environment, which is transported by wind and currents at the sea surface, and is thus directly related to the pathways of litter at sea. Floating litter items can be transported by the currents until they sink to the seafloor, be deposited on the shore or degrade over time (Andrady 2015). While the occurrence of anthropogenic litter items floating in the world oceans was reported already decades ago (Venrick et al. 1972; Morris 1980), the existence of accumulation zones of Floating Marine Debris (FMD) in oceanic gyres has only recently gained worldwide attention (Moore et al. 2001b).

Synthetic polymers constitute the major part of floating marine debris, the fate of which depends on their physico-chemical properties and the environmental conditions. As high-production volume polymers such as polyethylene and polypropylene have lower densities than seawater, they float until they are washed ashore or sink because their density changes due to biofouling and leaching of additives. While being subject to biological, photic or chemical degradation processes, they can be physically degraded gradually into smaller fragments until becoming microplastics, which is often defined as the size fraction <5 mm. This fraction requires different monitoring techniques, such as surface net trawls, and is therefore treated elsewhere (Löder and Gerdts 2015; Lusher 2015). Floating macrolitter is typically monitored by visual observation from ships, though results from net trawls are also being reported. The spatial coverage and thus the representativeness of the quantification depends on the methodology applied. Also, observation conditions, such as sea state, elevation of the observation position and ship speed affect results.

Existing datasets indicate substantial spatial variability and persistent gradients in floating marine litter concentrations (e.g. Erikssen et al. 2014). The variations can be attributed to differential release pathways or specific litter accumulation areas. Because of inconsistent reporting schemes used in scientific publications, data sets are often not comparable. Typically, item numbers are reported per surface area. Mass-based concentrations can then only be derived through estimates. Differences are found between studies in size ranges, concentration units and item categories used. As the number of pieces increases drastically with decreasing size of the observed litter items, the reporting of corresponding size classes is of high importance for comparing debris abundances among studies. Apart from the difficulty in reporting sizes correctly from shipboard observations, many publications use different size-range categories.

In addition to research activities, the quantification of floating litter is part of the assessment schemes of national and international monitoring frameworks. Monitoring of the quantity, composition and pathways of floating litter can contribute to an efficient management of waste streams and the protection of the marine environment. The European Marine Strategy Framework Directive, national programs, the Regional Sea Conventions and international agreements such as the United Nations Environmental Programme consider the monitoring of floating litter (Chen 2015). Visual assessment approaches include the use of research vessels, marine mammal surveys, commercial shipping carriers and dedicated litter observation surveys. Aerial surveys are often conducted for larger items (Pichel et al. 2012). However, available data for floating litter are currently difficult to compare because existing observation schemes (NOAA, UNEP, Hellenic Marine Environment Protection Association—HELMEPA, etc.) apply different approaches, observation schemes and category lists (Galgani et al. 2011a, b). Some approaches involve the reporting by volunteers (HELMEPA, Arthur et al. 2011). While the main principle of monitoring floating debris through visual observation is very simple there are not many data sets, which allow a comparison of debris abundance. Some data sets are accessible as peer-reviewed publications or through reports from international organizations. However, the regions covered are very limited and monitoring occurs only sporadically.

Globally, the reported densities of floating marine debris pieces >2 cm ranges from 0 to beyond 600 items km^{-2}. Ship-based visual surveys in the North Sea German Bight yielded 32 items km^{-2} on average (Thiel et al. 2011). The integration over different surveys and seasons resulted in litter densities of 25 items km^{-2} at the White Bank area, 28 items km^{-2} around the island of Helgoland and 39 items km^{-2} in the East Frisian part of the German Bight. More than 70 % of the observed items were identified as plastics. From 2002 to 2006, aerial marine mammals surveys were used for the quantification of floating litter. Results were reported as sightings km^{-1}, ranging from 0 to beyond 1 item km^{-1}. Concentrations in coastal waters appeared to be lower than in offshore regions (Herr 2009).

In the northern Mediterranean Sea, in an offshore area of ca. 100×200 km between Marseille and Nice and also in the Corsican Channel, floating debris was quantified during marine mammals surveys. A maximum of 55 pieces km^{-2} was recorded with strong spatial variability (Gerigny et al. 2011). In the Ligurian Sea, data were collected through ship-based visual observation in 1997 and 2000. Between 15 and 25 objects and between 1.5 and 3.0 objects km^{-2} were found in 1997 and 2000, respectively, without specification of the size ranges used (Aliani and Molcard 2003). Voluntary surveys through HELMEPA made from commercial shipping vessels in the Mediterranean Sea revealed a concentration of 2 items km^{-2} with higher concentrations in coastal areas but also longer transects without any litter encounters. While plastic material accounted for the highest proportion (83 %) of litter, textiles, paper, metal and wood comprised 17 % (UNEP 2009). No size ranges were given, but the described conditions during observation indicate that only larger items were considered. A large-scale survey in the Mediterranean Sea found 78 % of the observed objects larger than 2 cm to be of anthropogenic origin (Suaria and Aliani 2014). Plastic constituted 96 % of these. While highest densities (>52 items km^{-2}) were reported from the Adriatic Sea and Algerian basin, lowest densities (<6.3 items km^{-2}) were recorded in the central Thyrrenian and Sicilian Sea. Densities in other areas ranged between 11 and 31 items km^{-2} (Suaria and Aliani 2014).

Visual aerial surveys were conducted in the Black Sea, flying slow at low altitude above the Kerch Strait, the southern part of the Azov Sea and on the coastal Russian Black Sea. Concentrations in the Kerch Strait and the Azov Sea were comparable at 66 items km^{-2} and twice as high as those from the Black Sea (BSC 2007).

In a visual observation study in the north Pacific, ca. 56 km off Japan, Shiomoto and Kameda (2005) found densities of 0.1–0.8 items km^{-2} at a size >5 cm.

A study at the east coast of Japan utilized surface trawl nets for sampling on transects of 10 min at 2 knots with a net opening of 50 cm and a mesh size of 333 μm. The size of plastic pieces captured ranged from 1 to 280 mm. Pieces >11 mm accounted only for 8 % and particles of 1–3 mm accounted for 62 % at total average litter mass of 3600 g km^{-2} (Yamashita and Tanimura 2007).

Visual observation studies in southern Chilean fjords revealed 1–250 items km^{-2} >2 cm during seven oceanographic cruises from 2002 to 2005 (Hinojosa and Thiel 2009; Hinojosa et al. 2011; Thiel et al. 2013). Typically, densities in the northern areas ranged from 10 to 50 items km^{-2}. Matsumara and Nasu (1997) reported 0.5 items km^{-2} in the waters northwest of Hawaii, close to the so-called Pacific garbage patch, compared with 9 pieces km^{-2} in southeast Asia. Debris densities in the waters off British Columbia (Canada), comprised 0.9–2.3 pieces km^{-2} with a mean of 1.5 items km^{-2} (Williams et al. 2011), but no size range was given. In the Gulf of Mexico, Lecke-Mitchell and Mullin (1997) recorded 1.0–2.4 pieces km^{-2} during cetacean survey flights (Table 2.2).

FMD density in the northern South China Sea was quantified by net trawls at 4.9 (0.3–16.9) items km^{-2}, with Styrofoam (23 %) and other plastics (45 %) dominating (Zhou et al. 2011). More than 99 % of FMD was small- (<2.5 cm) or medium-sized (2.5–10 cm). Large items (10–100 cm) were detected by visual

Table 2.2 Comparison of mean litter densities on the sea surface from worldwide data (non-exhaustive list)

Region	Density (item km^{-2}) (max)	Size range (cm)	Plastic (%)	References
North Sea	25–38	>2	70	Thiel et al. (2011)
Belgian coast	0.7	n.d.	95	Van Cauwenberghe et al. (2013)
Ligurian coast	1.5–25	n.d.	n.d.	Aliani and Molcard (2003)
Mediterranean Sea	10.9 → 52 (194.6)	>2	95.6	Suaria and Aliani (2014)
North Sea	2 (1–6)	n.d.	n.d.	Herr (2009)
Kerch Strait/Black Sea	66	n.d.	n.d.	BSC (2007)
Chile	10–50 (250)	>2	>80	Hinojosa and Thiel (2009)
West of Hawaii	0.5	0.08 (0.2)	n.d.	Matsumura and Nasu (1997)
British Columbia	1.48 (2.3)	n.d.	92	Williams et al. (2011)
South China Sea	4.9 (0.3–16.9)	<2.5–10	68	Zhou et al. (2011)
North Pacific	459	2	95	Titmus and Hyrenbach (2011)
Strait of Malacca	579	>1–2	98.8	Ryan (2013)
Bay of Bengal	8.8	>1–2	95.5	Ryan (2013)
Southern Ocean	0.032–6	>1	96	Ryan et al. (2014)

observation resulting in mean concentrations of 0.025 items km^{-2} (Zhou et al. 2011). In the northeast Indian Ocean, Ryan (2013) reported a large difference in the concentration of marine debris between the Strait of Malacca (578 ± 219 items km^{-2}) and the Bengal Sea (8.8 ± 1.4 items km^{-2}). By contrast, Uneputty and Evans (1997) reported concentrations >375 items km^{-2} in Amon Bay, east Indonesia.

In 2009, a 4,400-km cruise from the American west coast to the North Pacific subtropical gyre and back to the coast provided data during 74 h of observation corresponding to a transect length of 1,343 km (Titmus and Hyrenbach 2011). A single observer at 10 m above the sea level recorded a total of 3,868 pieces, of which 90 % were fragments and 96 % of these were plastic. Eighty-one percent of the items had a size of 2–10 cm, 14 % of 10–30 cm and 5 % of >30 cm. The density of debris increased towards the centre of the gyre where smaller, probably older and weathered pieces were found. The authors note that visual observations are constrained by the inability to detect smaller fragments (<20 mm) and to retrieve the observed items for further analysis and concluded that visual observations can be easily conducted from ships of opportunity, which provide a useful and inexpensive tool for monitoring debris accumulation and distribution at sea.

A specific case of floating marine litter is abandoned or lost fishing gear, such as nets or longlines. These items cause significant harm when abandoned, as they continue to catch marine wildlife (Kühn et al. 2015). In 2003, a major effort, including the identification of possible accumulation areas by satellite imaging and ocean current modelling, was made to select appropriate areas for aerial surveys in search for abandoned fishing gear in the Gulf of Alaska (Pichel et al. 2012). Employing a wide range of methodologies including visual video, infrared video and Lidar imaging during 14 days of observation, 102 items of anthropogenic origin were sighted.

Modelling of oceanographic currents can help to identify pathways and accumulation areas, thus enabling source attribution (Martinez et al. 2009; Maximenko et al. 2012). A modelling approach in the North Sea identified seasonal signals in litter reaching the coasts (Neumann et al. 2014). The concentrations and distribution patterns of floating marine debris can be expected to change according to climatic changes (Howell et al. 2012). Lebreton et al. (2012) modelled the global oceanic currents in view of the cycling and distribution of introduced debris. Input scenarios were based on population density and major shipping lanes. A 30-year projection showed the accumulation of floating debris in ocean gyres and enclosed seas. These studies have the potential to investigate pathways and to guide monitoring to enable effective implementation of management measures and the assessment of their efficiency. Modelling is also used to predict the pathways and impacts of large quantities of debris introduced through natural events such as tsunamis and related run-offs (Lebreton and Borrero 2013). Single events may drastically increase local debris concentrations. A study combining available worldwide data with a modelling approach estimated the weight of the global plastic pollution to comprise 75 % macroplastic (>200 mm), 11 % mesoplastic (4.75–200 mm), and 11 and 3 % in two microplastic size classes, respectively (Erikssen et al. 2014).

The data suggest that a minimum of 233,400 tons of larger plastic items are adrift in the world's oceans compared to 35,540 tons of microplastics.

Floating marine litter can be considered as ubiquitous, occurring even in the most remote areas of the planet such as the Arctic (Bergmann and Klages 2012). Floating litter items are also present in the remote Antarctic Ocean, although densities are low and cannot be expressed as concentrations (Barnes et al. 2010). Some 42 % of the observed 120 objects south of 63°S consisted of plastic. Debris items were observed even as far south as 73°S. However, the small number of surveys and low total object counts do not allow for trend assessments. In the African part of the Southern Ocean, 52 items (>1 cm) were recorded during a 10,467 km transect survey, yielding densities ranging from 0.03 to 6 items km^{-2} (Ryan et al. 2014).

The diversity and non-comparability of monitoring approaches used currently hinders a comparison of absolute pollution indicators and spatial or temporal assessments. The development and widespread implementation of protocols for monitoring, such as the ongoing efforts for the implementation of the MSFD (Galgani et al. 2013), could improve the quality of data gathered. Established protocols should be accompanied by training schemes, quality assurance and control procedures. The implementation of standardized protocols in the monitoring of riverine litter may enable source allocation.

Unfortunately, data acquired by NGOs or authorities are often not published in peer-reviewed journals and are therefore not readily accessible. A joint international database would facilitate the collection of such data and improve standardization and comparability. The collection of data, e.g. on-site through tablet computer applications, the standardization of reporting formats and the streamlining of data flows would facilitate data treatment. More easily accessible data sets can then help to prioritize activities and to monitor the success of litter reduction measures.

While monitoring by human observers is a simple and straightforward approach, in particular for large-scale and frequent surveys, automatized approaches are promising. Developing technologies may lead to the use of digital imaging and image recognition techniques for the autonomous large-scale monitoring of litter (Hanke and Piha 2011).

The implementation of international frameworks such as the EU MSFD, Regional Action Plans against Marine Litter and the agreements of the Rio +20 Conference (United Nations 2012) require improvement of data availability and quality and can therefore be expected to provide the basis for coordinated assessments in the future.

2.3.3 Seafloor

Change in the nature, presence or abundance of anthropogenic debris on the seafloor is much less widely investigated than sea surface patterns. Studies typically focus on continental shelves, as sampling difficulties, inaccessibility and costs rarely allow for research in deeper waters, which accounts for almost half of the planet's surface. Deep-sea surveys are important because ca. 50 % of plastic litter items sink to the seafloor and even low-density polymers such as polyethylene and propylene may lose buoyancy under the weight of fouling (Engler 2012). While acoustic approaches do not enable discrimination of different types of debris on the seafloor except for metals and may not record smaller objects, trawling was considered the most adequate method when taking into account mesh sizes and net opening width (Galgani et al. 2011b) (Fig. 2.1). However, nets were primarily designed to collect specific biota leading to sample bias and underestimation of benthic litter quantities. Therefore, pole trawling has been suggested as the most consistent survey method for the assessment of benthic marine litter (Galgani and Andral 1998), although rather destructive to seafloor habitats because of the scraping of sediments and inhabiting biota. However, trawls cannot be used in rocky habitats or on hard substrates and they do not allow for a precise localization of individual items. Samples from trawls are likely to underestimate debris abundance and may miss some types of debris altogether such as monofilaments because of variability in the sampling efficiency for different debris items (Watters et al. 2010). Fibres from the trawl nets themselves (Murray and Cowie 2011) may contaminate samples. Finally, it does not enable the assessment of impacts of litter on habitats when it contributes its own impacts on the seafloor, which are more severe for the benthic fauna and habitats than the litter items caught by trawl.

Fig. 2.1 Litter collected by trawling in the Mediterranean Sea, France. 10 min experiment (credit Barbaroux and Galgani, IFREMER)

Strategies to investigate seabed debris are similar to those for evaluating the abundance and composition of benthic species. Mass is less often determined for marine debris, because very large items may increase variability in measures. Although floating debris, such as that found in the highly publicized "gyres" and/ or convergence zones, is currently the focus of attention, debris accumulating on the seafloor has a high potential to impact benthic habitats and organisms. Fourty-three studies were published between 2000 and 2013. Until recently, only few of them covered greater geographic areas or depths. The majority of these studies utilized a bottom trawl for sampling as part of fish stock assessments. More recently, remotely operated vehicles and towed camera systems were increasingly used for deep-sea surveys (e.g. Pham et al. 2014, see Fig. 2.2).

The geographic distribution of debris on the ocean floor is strongly influenced by hydrodynamics, geomorphology and human factors (Galgani et al. 1996; Pham et al. 2014). Moreover, there are notable temporal variations, particularly seasonal, with tendencies for accumulation and concentration of marine litter in particular

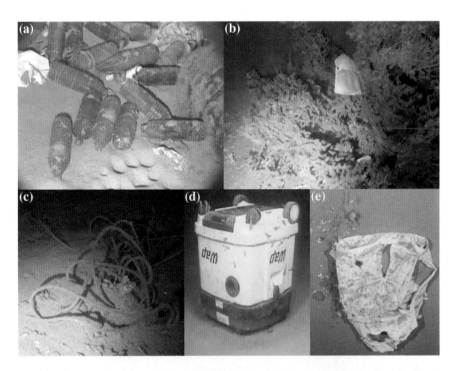

Fig. 2.2 Litter on the deep seafloor. **a** Plastic bags and bottles dumped 20 km off the French Mediterranean coast at 1,000 m in close vicinity to burrow holes (F. Galgani, IFREMER); **b** food package entrapped at 1,058 m in deep-water coral colony; **c** rope at 1,041 m depth, both from Darwin Mounds (courtesy of V. Huvenne, National Oceanography Centre Southampton (NOCS)); **d** waste disposal bin or a vaccum cleaner with prawns on the seafloor off Mauritania at 1,312 m depth (courtesy of D. Jones, SERPENT Project, NOCS); **e** plastic carrier bag found at ~2,500 m depth at the HAUSGARTEN observatory (Arctic) colonised by hormathiid anemones and surrounded by dead tests of irregular sea urchins (courtesy of M. Bergmann, AWI)

geographic areas (Galgani et al. 1995). Interpretation of trends is, however, difficult because the ageing of plastics at depth is unknown and the accumulation of debris on the seafloor certainly began before scientific investigations started in the 1990s.

In estuaries, large rivers are responsible for substantial input of debris to the seabed (Lechner et al. 2014; Rech et al. 2014). Rivers can also transport waste far offshore because of their high flow rate and strong currents (Galgani et al. 1995, 1996, 2000). Alternatively, small rivers and estuaries can also act as a sink for litter, when weak currents facilitate deposition on shores and banks (Galgani et al. 2000). In addition, litter may accumulate upstream of salinity fronts being transported to the sea later, when river flow velocity is increasing.

Plastics were found on the seabed of all seas and oceans and the presence of large amounts has been reported (Galil et al. 1995; Galgani et al. 2000; Barnes et al. 2009) but remains uncommon in remote areas such as Antarctica, particularly in deep waters (Barnes et al. 2009). So far, sampling has been limited to some dozens of trawls and van Cauwenberghe et al. (2013) and Fischer et al. (2015) found pieces of microplastics in deep-sea sediments from the southern Atlantic and Kuril-Kamchatka-trench area, respectively. Large-scale evaluations of seabed debris distribution and densities are more common in other regions (Galgani et al. 2000). However, these studies mostly involve extrapolations from small-scale investigations mainly in coastal areas such as bays, estuaries and sounds. The abundance of plastic debris shows strong spatial variations, with mean densities ranging from 0 to more than 7,700 items km^{-2} (Table 2.3). Mediterranean sites show the greatest densities owing to the combination of a densely populated coastline, shipping in coastal waters and negligible tidal flow. Moreover, the Mediterranean is a closed basin with limited water exchange through the Strait of Gibraltar. Generally, litter densities are higher in coastal seas (Lee et al. 2006) because of large-scale residual ocean circulation patterns but also because of extensive riverine input (Wei et al. 2012). However, debris that reaches the seabed may have been transported over considerable distances before sinking to the seafloor, e.g. as a consequence of heavy fouling. Indeed, some accumulation zones were identified far from coasts (Galgani and Lecornu 2004; Bergmann and Klages 2012; Woodall et al. 2014, 2015). Accordingly, even in the shallow subtidal abundance and distribution patterns can differ substantially from the adjacent strandlines with plastics being the most important fraction at sea. In general, bottom debris tends to become trapped in areas of low circulation where sediments are accumulating (Galgani et al. 1996; Schlining et al. 2013; Pham et al. 2014). The consequence is an accumulation of plastic debris in bays, including lagoons of coral reefs, rather than in the open sea. These are the locations where large amounts of derelict fishing gear accumulate and cause damage to shallow-water biota and habitats (Dameron et al. 2007; Kühn et al. 2015).

Continental shelves are considered as accumulation zones for marine debris (Lee et al. 2006), however, often with lower concentrations of debris than adjacent canyons because debris is not retained but washed offshore by currents associated with

Table 2.3 Comparison of litter densities on the seafloor from recent data worldwide (non-exhaustive list)

Location	Habitat	Date	Sampling	Depth (m)	Density (min-max)	Plastic (%)	References
Southern China	Benthic	2009–2010	4 trawl (mesh not available)/1 dive	0–10	693 (147–5,000) items km^{-2}	47	Zhou et al. (2011)
France-Mediterranean	Slope	2009	17 canyons, 101 ROV dives	80–700	3.01 km^{-1} survey (0–12)	12 (0–100)	Fabri et al. (2014)
Thyrenian Sea	Fishing ground	2009	6 × 1.5 ha samples, trawl, 10 mm mesh	40–80	5,960 ± 3,023 km^{-2}	76	Sanchez et al. (2013)
Spain-Mediterranean	Fishing ground	2009		40–80	4,424 ± 3,743 km^{-2}	37	Sanchez et al. (2013)
Mediterranean Sea	Bathyal/abyssal	2007–2010	292 tows, otter/ Agassiz trawl, 12 mm mesh	900–3,000	0.02– 3,264.6 kg km^{-2} (incl. clinker)	n.d.	Ramirez-Llodra et al. (2013)
Malta	Shelf	2005	Trawl (44 hauls, 20 mm mesh)	50–700	102	47	Misfud et al. (2013)
Turkey/Levantin Basin	Bottom/bathyal	2012	32 hauls (trawl, 24 mm mesh)	200–800	290 litter (3,264.6 kg km^{-2})	81.1	Güven et al. (2013)
Azores, Portugal	Condor seamount	2010–2011	45 dives	185–256	1,439 items km^{-2}	No plastic/89 % fishing gear	Pham et al. (2013)
Goringe Bank, NE Atlantic	Gettysburg and Ormonde seamounts	2011	4 ROV dives (124 h video, 4,832 photographs), total distance of 80.6 km	60–3,015	1–4 items·km^{-1}	9.9/56 fishing gear	Vieira et al. (2014)
US west coast	Shelf	2007–2008	1,347 sites (total, trawling, 38 mm mesh)	55–183	30 items km^{-2}	23	Keller et al. (2010)
	Slope	2007–2008		183–550	59 items km^{-2}	n.d.	Keller et al. (2010)
	Slope/bathyal	2007–2008		550–1.280	129 items km^{-2}	n.d.	Keller et al. (2010)
Mediterranean Sea, France	Shelf/canyon	1994–2009 (16 years study)	90 sites (trawls, 0.045 km^2/tow, 20 mm mesh)	0–800	76–146 km^{-2} (0–2,540)	29.5–74	Galgani et al. (2000) and unpublished data

(continued)

Table 2.3 (continued)

Location	Habitat	Date	Sampling	Depth (m)	Density (min-max)	Plastic (%)	References
Japan, offshore Iwate	Trench	Jamstek database	3 dives on 4,861 available,	299–400, 1,086–1,147, 1,682–1,753	15.9 items h^{-1}	42.8	Miyake et al. (2011)
Kuril-Kamchatka area (NW Pacific)	Trench/bathyal plain	2012	20 box cores (0.25 m^2) (Agassiz trawl, camera epi-benthic sledge)	4,869–5,766	60 → 2,000 micro-plastics m^{-2}	(Trawl samples: mostly fishing gear)	Fischer et al. (2015)
Fram Strait, Arctic	Slope	2002–2011 (5 surveys)	One OFOS camera tow $year^{-1}$, 5 transects (1,427–2,747 m^2)	2,500	3,635 (2002)–7,710 (2011) items km^{-2}	59	Bergman and Klages (2012)
Northern Antarctic Peninsula and Scotia Arc	Slopes/bathyal	2006	32 Agassiz trawls	200–1,500	2 pieces only	1 plastic	Barnes et al. (2009)
Monterey Canyon, California	From margin to abyssal	1989–2011	ROVs, 2,429 km^2 in total	25–3,971	632 items km^{-2}	33	Schlining et al. (2013)
ABC islands, Dutch Caribbean	Sandy bottoms to rocky slopes	2000	24 video transects, submersibles	80–900	2,700 items km^{-2} (0–4590)	29	Debrot et al. (2014)

offshore winds and river plumes. Only few studies have assessed debris below 500 m depth (June 1990; Galil et al. 1995; Galgani et al. 1996, 2000; Galgani and Lecornu 2004; Keller et al. 2010; Miyake et al. 2011; Mordecai et al. 2011; Bergmann and Klages 2012; Wei et al. 2012; Pham et al. 2013, 2014; Ramirez-Llodra et al. 2013, Schlining et al. 2013; Fischer et al. 2015; Vieira et al. 2014); Galgani et al. (2000) observed trends in deep-sea pollution over time (1992–98) off the European coast with an extremely variable distribution and debris accumulating in submarine canyons. Miyake et al. (2011) recorded debris down to 7,216 m depth in video surveys from the Ryukyu Trench. Litter was primarily composed of plastic and accumulated in deep-sea trenches and depressions. Accordingly, several authors (Galgani et al. 1996; Mordecai et al. 2011; Pham et al. 2014) concluded that submarine canyons may act as a conduit for the transport of marine debris into the deep sea. Recent studies conducted in coastal deep-sea areas along California and the Gulf of Mexico (Watters et al. 2010; Schlining et al. 2013; Wei et al. 2012) confirmed this pattern. Also, an analysis of the composition and abundance of man-made, benthic marine debris collected in bottom trawl surveys at 1,347 randomly-selected stations along the US west coast in 2007 and 2008 indicated that densities increased significantly with depth, ranging from 30 items km^{-2} in shallow (55–183 m) to 128 items km^{-2} in the deepest waters surveyed (550–1,280 m) (Keller et al. 2010). Higher densities at the bottom were also found in particular areas such as those around rocks, wrecks as well as in depressions or channels (Galgani et al. 1996). Deep submarine extensions of coastal rivers influence the distribution of seabed debris. In some areas, local water movements transport debris away from the coast to accumulate in zones of high sedimentation. In the case of the Mississippi river, for example, the front canyon was a focal point for litter, probably due to bottom topography and currents (Wei et al. 2012). Under these conditions, the distal deltas of rivers can fan out in deeper waters, creating areas of high accumulation. Many authors (Galgani et al. 1996; Moore and Allen 2000; Wei et al. 2012) show that circulation may be influenced by strong currents occurring in the upper part of canyons, which decrease rapidly in deeper areas resulting in an increased confinement with a litter distribution that seems to be temporally more stable as a consequence.

A great variety of human activities such as fishing, urban development and tourism contribute to the distribution pattern of debris on the seabed. Debris from the fishing industry is prevalent in fishing areas (Watters et al. 2010; Schlining et al. 2013; Vieira et al. 2014). This type of material may account for a high proportion of debris. In the eastern China Sea (Lee et al. 2006), for example, 72 % of debris is made of plastic, mainly pots, nets, *Octopus* jars, and fishing lines. Investigations using submersibles at depths beyond the continental shelf and canyons have revealed substantial quantities of debris in remote areas. Galgani and Lecornu (2004) counted 0.2–0.9 pieces of plastic per linear kilometre at the HAUSGARTEN observatory (2500 m) in the Fram Strait (Arctic). Fifteen items, of which 13 were plastic, were observed during one dive between 5,330 and 5,552 m ('Molloy Hole'), which reflects the local funnel-like topography and downwards directed eddies acting as particle trap. Bergmann and Klages (2012) reported doubled litter quantities between 2002 and 2011 in the HAUSGARTEN area. The accumulation trends reported in that study raise particular concern as degradation rates of most polymers in deep-sea environments are assumed to be even slower due to the absence of light, low temperature and oxygen concentrations.

2.3.4 Microplastics

Similar to large debris, there is growing concern about the implications of the diverse microparticles in the marine environment, which are particles ≤ 1 μm (Galgani et al. 2012; Thompson et al. 2004). Most microparticles are tiny plastic fragments known as microplastics, although other types of microparticles exist, such as fine fly ash particles emitted with flue gases from combustion, rubber from tyre wear and tear as well as glass and metal particles, all of which constantly enter the marine environment. The abundance and global distribution of microplastics in the oceans appeared to have steadily increased over past decades (Cole et al. 2011; Claessens et al. 2011; Thompson 2015), while a decrease in the average size of plastic litter has been observed over this time period (Barnes et al. 2009). In recent years, the existence of microplastics and their potential impact on wildlife and human health has received increased public and scientific attention (Betts 2008; Galloway 2015; Lusher 2015).

Microplastics comprise a very heterogeneous assemblage of particles that vary in size, shape, color, chemical composition, density, and other characteristics. They can be subdivided by usage and source as (i) 'primary' microplastics, produced either for indirect use as precursors (nurdles or virgin resin pellets) for the production of polymer consumer products, or for direct use, such as in cosmetics, scrubs and abrasives and (ii) 'secondary' microplastics, resulting from the breakdown of larger plastic material into smaller fragments. Fragmentation is caused by a combination of mechanical forces, e.g. waves and/or photochemical processes triggered by sunlight. Some 'degradable' plastics are even designed to fragment quickly into small particles, however, the resulting material does not necessarily biodegrade (Roy et al. 2011). The various sources of microplastics and the pathways into the oceans are summarized in detail by Browne (2015).

In order to understand the environmental impacts of microplastics, many studies have quantified their abundance in the marine environment. One of the major difficulties in making large-scale spatial and temporal comparisons between existing studies is the wide variety of methods that have been applied to isolate, identify and quantify marine microplastics (Hidalgo-Ruz et al. 2012). For meaningful comparisons to be made and robust monitoring studies to be conducted, it is therefore important to define common methodological criteria for estimating abundance, distribution and composition of microplastics (Löder and Gerdts 2015).

Microplastics normally float at the sea surface because they are less dense than seawater. However, the buoyancy and specific gravity of plastics may change during their time at sea due to weathering and biofouling, which results in their distribution across the sea surface, the deeper water column, the seabed, beaches and sea ice (Colton and

Knapp 1974; Barnes et al. 2009; Law et al. 2010; Browne et al. 2010; Claessens et al. 2011; Collignon et al. 2012; Obbard et al. 2014). Until now, only a limited number of global surveys have been conducted on the quantity and distribution of microplastics in the oceans (Lusher 2015). Most surveys focused on specific oceanic regions and habitats, such as coastal areas, regional seas, gyres or the poles (Thompson et al. 2004, Collignon et al. 2012; Rios and Moore 2007). Concentrations of microplastics at sea vary from thousands to hundreds of thousands of particles km^{-2} and latest reports suggest that microplastic pollution has spread throughout the world's oceans from the water column (Lattin et al. 2010; Cole et al. 2011) to sediments even of the deep sea (Moore et al. 2001b; Law et al. 2010; Claessens et al. 2011; Cole et al. 2011; Collignon et al. 2012; Erikssen et al. 2014; Reisser et al. 2013; van Cauwenberghe et al. 2013; Woodall et al. 2014; Fischer et al. 2015). Recently, microplastics were also recorded from Arctic sea ice in densities two orders of magnitude higher than those previously reported from highly contaminated surface waters, such as those of the Pacific gyre (Obbard et al. 2014). This has important implications considering the projected acceleration in sea ice melting due to global climate change and concomitant release of microplastics to the Arctic marine ecosystem.

Time-series data on the composition and abundance of microplastics are sparse. However, available evidence on long-term trends suggests various patterns in microplastic concentrations. A decade ago, Thompson et al. (2004) demonstrated the broad spatial extent and accumulation of this type of contamination. They found plastic particles in sediments from U.K. beaches and archived among the plankton in samples dating back to the 1960s with a significant increase in abundance over time. More recent evidence indicated that microplastic concentrations in the North Pacific subtropical gyre have increased by two orders of magnitude in the past four decades (Goldstein et al. 2013). However, no change in microplastic concentration was observed at the surface of the North Atlantic gyre for a period of 30 years (Law et al. 2010).

Less is known about the composition of microplastics in the oceans. Evidence suggests a temporal decrease in the average size of plastic litter (Barnes et al. 2009; Erikssen et al. 2014). Studies based on the stomach contents of shearwaters (*Puffinus tenuirostris*) in the Bering Sea also indicated a decrease in 'industrial' primary pellets and an increase in 'user' plastic between the 1970s and the late 1990s (Vlietstra and Parga 2002) but constant levels over the last decade (Van Franeker et al. 2011). Similarly, long-term data from The Netherlands since the 1980s show a decrease of industrial plastics and an increase in user plastics, with shipping and fisheries being the main sources (van Franeker 2012).

2.4 Summary and Conclusions

Marine debris is now commonly observed everywhere in the oceans and available information suggests that marine debris is highly dynamic in space and time. However, we need standardized methodologies for quantification and characterisation of marine litter to be able to achieve global estimates. Litter enters the sea from land-based sources, from ships and other installations at sea, from point and diffuse sources, and can travel long distances before being deposited. While plastic typically constitutes a lower proportion of the discarded waste, it represents the most important part of marine litter with sometimes up to 95 % of the waste, and has become ubiquitous even in remote polar regions. However, trends are not clear with quantities having slightly decreased over the last 20 years in some locations, notably in the western Mediterranean. At the same time no change in litter quantities are evident in the convergence zones from oceanic basins or beaches. In other locations, however, including the deep seafloor, densities have increased.

Accumulation rates vary widely with factors such as proximity of urban activities, shore and coastal uses, wind and ocean currents. These enable the accumulation of litter in specific areas at the sea surface, on beaches or on the seafloor. Before an accurate estimate of global debris quantities can be made, basic information is still needed on sources, inputs, degradation processes and fluxes. For this and because there is considerable variation in methodology between regions and investigators, more valuable and comparable data have to be obtained from standardized sampling programs. In terms of distribution and quantities, important questions concerning the balance between the increase of waste and plastic productions, reduction measures and the quantities found at the surface and on shorelines remain unanswered. Potentially, important accumulation areas with high densities of debris are still to be discovered. It is now clear that managers and policy makers will need to better understand the distribution of litter in order to assess and evaluate precisely the effectiveness of measures implemented to reduce marine litter pollution.

References

Aliani, S., & Molcard, A. (2003). Floating debris in the Ligurian Sea, north-western Mediterranean. *Marine Pollution Bulletin, 46*, 1142–1149.

Al-Najjar, T., & Al-Shiyab, A. (2011). Marine litter at (Al-Ghandoor area) the most northern part of the Jordanian coast of the Gulf of Aqaba, Red Sea. *Natural Science, 3*, 921–926.

Andrady, A. L. (2015). Persistence of plastic litter in the oceans. In M. Bergmann, L. Gutow & M. Klages (Eds.), *Marine anthropogenic litter* (pp. 57–72). Berlin: Springer.

Arthur, C., Murphy, P., Opfer, S., & Morishige, C. (2011). Bringing together the marine debris community using "ships of opportunity" and a Federal marine debris information clearinghouse. In *Technical Proceedings of the Fifth International Marine Debris Conference*. NOAA Technical Memorandum NOS-OR&R-38 (pp. 449–453), March 20–25, 2009.

Barnes, D. K. A., Galgani, F., Thompson, R. C., & Barlaz, M. (2009). Accumulation and fragmentation of plastic debris in global environments. *Philosophical Transactions of the Royal Society Series B, 364*, 1985–1998.

Barnes, D. K. A., Walters, A., & Gonçalves, L. (2010). Macroplastics at sea around Antarctica. *Marine Environmental Research, 70*, 250–252.

Bergmann, M., & Klages, M. (2012). Increase of litter at the Arctic deep-sea observatory HAUSGARTEN. *Marine Pollution Bulletin, 64*, 2734–2741.

Betts, K. (2008). Why small plastic particles may pose a big problem in the oceans. *Environmental Science and Technology, 42*, 8995.

Boerger, C. M., Lattin, G. L., Moore, S. L., & Moore, C. J. (2010). Plastic ingestion by planktivorous fishes in the North Pacific Central Gyre. *Marine Pollution Bulletin, 60*, 2275–2278.

Bravo, M., Gallardo, D. L. A., Luna-Jorquera, G., Núñez, P., Vásquez, N., & Thiel, M. (2009). Anthropogenic debris on beaches in the SE Pacific (Chile): Results from a national survey supported by volunteers. *Marine Pollution Bulletin, 58*, 1718–1726.

Browne, M. A. (2015). Sources and pathways of microplastic to habitats. In M. Bergmann, L. Gutow & M. Klages (Eds.), *Marine anthropogenic litter* (pp. 229–244). Berlin: Springer.

Browne, M. A., Galloway, T. S., & Thompson, R. C. (2010). Spatial patterns of plastic debris along estuarine shorelines. *Environmental Science and Technology, 44*, 3404–3409.

BSC [Black Sea Commission]. (2007). *Marine litter in the Black Sea region: A review of the problem* (p. 172). Black Sea Commission Publications 2007-1, Istanbul, Turkey.

Chen, C.-L. (2015). Regulation and management of marine litter. In M. Bergmann, L. Gutow & M. Klages (Eds.), *Marine anthropogenic litter* (pp. 399–432). Berlin: Springer.

Claereboudt, M. R. (2004). Shore litter along sandy beaches of the Gulf of Oman. *Marine Pollution Bulletin, 49*, 770–777.

Claessens, M., Meester, S. D., Landuyt, L. V., Clerck, K. D., & Janssen, C. R. (2011). Occurrence and distribution of microplastics in marine sediments along the Belgian coast. *Marine Pollution Bulletin, 62*, 2199–2204.

Cole, M., Lindeque, P., Halsband, C., & Galloway, T. S. (2011). Microplastics as contaminants in the marine environment: A review. *Marine Pollution Bulletin, 62*, 2588–2597.

Collignon, A., Hecq, J., Galgani, F., Voisin, P., & Goffard, A. (2012). Neustonic microlastics and zooplankton in the western Mediterranean sea. *Marine Pollution Bulletin, 64*, 861–864.

Colton, J. B., & Knapp, F. D. (1974). Plastic particles in surface waters of the northwestern Atlantic. *Science, 185*, 491–497.

Cordeiro, C. A., & Costa, T. M. (2010). Evaluation of solid residues removed from a mangrove swamp in the São Vicente Estuary, SP, Brazil. *Marine Pollution Bulletin, 60*, 1762–1767.

Costa, M. F., Silva-Cavalcanti, J. S., Barbosa, C. C., Portugal, J. L., & Barletta, M. (2011). Plastics buried in the inter-tidal plain of a tropical estuarine ecosystem. *Journal of Coastal Research, 64*, 339–343.

Dameron, O. J., Parke, M., Albins, M. A., & Brainard, R. (2007). Marine debris accumulation in the Northwestern Hawaiian Islands: An examination of rates and processes. *Marine Pollution Bulletin, 54*, 423–433.

Debrot, A. O., Meesters, H. W. G., Bron, P. S., & de León, R. (2013). Marine debris in mangroves and on the seabed: Largely-neglected litter problems. *Marine Pollution Bulletin, 72*, 1.

Debrot, A. O., Tiel, A. B., & Bradshaw, J. E. (1999). Beach debris in Curaçao. *Marine Pollution Bulletin, 38*, 795–801.

Debrot, A., Vinke, E., van der Wende, G., Hylkema, A., & Reed, J. (2014). Deepwater marine litter densities and composition from submersible video-transects around the ABC-islands, Dutch Caribbean. *Marine Pollution Bulletin, 88*(1–2), 361–365.

Engler, R. E. (2012). The complex Interaction between marine debris and toxic chemicals in the Ocean. *Environmental Science and Technology, 46*, 12302–12315.

Erikssen, M., Lebreton, L. C. M., Carson, H. S., Thiel, M., Moore, C. J., et al. (2014). Plastic pollution in the world's oceans: More than 5 trillion plastic pieces weighing over 250,000 tons afloat at sea. *PLoS ONE, 9*, e111913.

Eriksson, C., Burton, H., Fitch, S., Schulz, M., & van den Hoff, J. (2013). Daily accumulation rates of marine debris on sub-Antarctic island beaches. *Marine Pollution Bulletin, 66*, 199–208.

Fabri, M. C., Pedel, L., Beuck, L., Galgani, F., Hebbeln, D., & Freiwald, A. (2014). Megafauna of vulnerable marine ecosystems in French mediterranean submarine canyons: Spatial distribution and anthropogenic impacts. *Deep-Sea Research II, 104*, 184–207.

Fischer, V., Elsner, N. O., Brenke, N., Schwabe, E., & Brandt, A. (2015). Plastic pollution of the Kuril–Kamchatka Trench area (NW pacific). *Deep-Sea Research II, 111*, 399–405.

Fleet, D., van Franeker, J. A., Dagevos, J., & Hougee, M. (2009). Marine litter. Thematic Report No. 3.8. In H. Marencic & J. Vlas, de (Eds.), *Quality Status Report 2009. Wadden Sea Ecosystem No. 25. Common Wadden Sea Secretariat* (p. 12). Wilhelmshaven, Germany: Trilateral Monitoring and Assessment Group.

Galgani, F., & Andral, B. (1998). *Methods for evaluating debris on the deep sea floor. OCEANS'98/IEEE/OEC Conference* (Vol. 3(3), pp. 1512–1521). Nice 28/09-01/10/98.

Galgani, F., Burgeot, T., Bocquene, G., Vincent, F., Leaute, J., Labastie, J., et al. (1995). Distribution and abundance of debris on the continental shelf of the Bay of Biscay and in Seine Bay. *Marine Pollution Bulletin, 30*, 58–62.

Galgani, F., Fleet, D., van Franeker, J. A., Hanke, G., De Vrees, L., Katsanevakis, S., et al. (2011a). *Monitoring marine litter within the European marine strategy framework directive (MSFD): Scientific and technical basis. 5th International Marine Debris Conference*, Honolulu, Hawaii (pp. 164–168). 20–25 March, 2011, Oral Presentation Extended Abstracts 4.c.5.

Galgani, F., Hanke, G., Werner, S., Oosterbaan, L., Nilsson, P., Fleet, D., et al. (2013). *Monitoring guidance for marine litter in European Seas, JRC Scientific and Policy Reports, Report EUR 26113 EN, (p. 120).* (https://circabc.europa.eu/w/browse/85264644-ef32-401b-b9f1-f640a1c459c2).

Galgani, F., Leaute, J. P., Moguedet, P., Souplet, A., Verin, Y., Carpentier, A., et al. (2000). Litter on the sea floor along European coasts. *Marine Pollution Bulletin, 40*, 516–527.

Galgani, F., & Lecornu, F. (2004). Debris on the seafloor at "Hausgarten". In M. Klages, J. Thiede & J.-P. Foucher (Eds.), The expedition ARK XIX/3 of the Research Vessel POLARSTERN in 2003: Reports of legs 3a, 3b and 3c. *Reports on Polar and Marine Research, 488*, 260–262.

Galgani, F., Piha, H., Hanke, G., Werner, S., & GES MSFD group. (2011b). Marine litter: Technical recommendations for the implementation of MSFD requirements. EUR 25009 EN. Luxembourg (Luxembourg): Publications Office of the European Union; 2011. JRC67300. (http://publications.jrc.ec.europa.eu/repository/handle/111111111/22826).

Galgani, F., Souplet, A., & Cadiou, Y. (1996). Accumulation of debris on the deep sea floor off the French Mediterranean coast. *Marine Ecology Progress Series, 142*, 225–234.

Galil, B. S., Golik, A., & Türkay, M. (1995). Litter at the bottom of the sea: A sea bed survey in the eastern Mediterranean. *Marine Pollution Bulletin, 30*, 22–24.

Galloway, T. S. (2015). Micro- and nano-plastics and human health. In M. Bergmann, L. Gutow & M. Klages (Eds.), *Marine anthropogenic litter* (pp. 347–370). Berlin: Springer.

Gerigny, O., Henry, M., Tomasino, C., & Galgani, F. (2011). Déchets en mer et sur le fond. in rapport de l'évalution initiale, Plan d'action pour le milieu marin - Mediterranée Occidentale, rapport PI Déchets en mer V2 MO, (pp. 241–246). (http://www.affairesmaritimes.mediterrane e.equipement.gouv.fr/IMG/pdf/Evaluation_initiale_des_eaux_marines_web-2.pdf).

Goldstein, M., Titmus, A. J., & Ford, A. M. (2013). Scales of spatial heterogeneity of plastic marine debris in the Northeast Pacific ocean. *Plos One*, e184.

Güven, O., Gülyavuz, H., & Deval, M. C. (2013). Benthic debris accumulation in bathyal grounds in the Antalya Bay, Eastern Mediterranean. *Turkish Journal of Fisheries and Aquatic Sciences, 13*, 43–49.

Hanke, G., & Piha, H. (2011). Large-scale monitoring of surface floating marine litter by high-resolution imagery. In *Presentation and extended abstract, 5th International Marine DEBRIS Conference, Hawaii, Honolulu*. 20–25 March 2011.

Herr, H. (2009). Vorkommen von Schweinswalen (*Phocoena phocoena*) in Nord- und Ostsee – im Konflikt mit Schifffahrt und Fischerei? *Dissertation*. Universität Hamburg, 120.

Hidalgo-Ruz, V., Gutow, L., Thompson, R. C., & Thiel, M. (2012). Microplastics in the marine environment: A review of the methods used for identification and quantification. *Environmental Science and Technology, 46*, 3060–3075.

Hidalgo-Ruz, V., & Thiel, M. (2015). The contribution of citizen scientists to the monitoring of marine litter. In M. Bergmann, L. Gutow & M. Klages (Eds.), *Marine anthropogenic litter* (pp. 433–451). Berlin: Springer.

Hinojosa, I., Rivadeneira, M. M., & Thiel, M. (2011). Temporal and spatial distribution of floating objects in coastal waters of central–southern Chile and Patagonian fjords. *Continental Shelf Research, 31*, 172–186.

Hinojosa, I. A., & Thiel, M. (2009). Floating marine debris in fjords, gulfs and channels of southern Chile. *Marine Pollution Bulletin, 58*, 341–350.

Howell, E., Bogad, S., Morishige, C., Seki, M., & Polovina, J. (2012). On North Pacific circulation and associated marine debris concentration. *Marine Pollution Bulletin, 65*, 16–22.

June, J. A. (1990). Type, source, and abundance of trawl caught debris of Oregon, in the Eastern Bering Sea, and in Norton Sound in 1988. In R. S. Shomura & M. L. Godfrey (Eds.), *Proceedings of the Second International Conference on Marine Debris, NMFS-SWF-SC-154*, US Department of Commerce, NOAA Technical Memo, 279–301.

Kataoka, T., Hinata, H., & Kato, S. (2013). Analysis of a beach as a time-invariant linear input/output system of marine litter. *Marine Pollution Bulletin, 77*, 266–273.

Keller, A. A., Fruh, E. L., Johnson, M. M., Simon, V. & McGourty, C. (2010). Distribution and abundance of anthropogenic marine debris along the shelf and slope of the US west coast. *Marine Pollution Bulletin 60*, 692–700.

Koutsodendris, A., Papatheodorou, G., Kougiourouki, O., & Georgiadis, M. (2008). Benthic marine litter in four Gulfs in Greece, Eastern Mediterranean; abundance, composition and source identification. *Estuarine, Coastal and Shelf Science, 77*, 501–512.

Kühn, S., Bravo Rebolledo, E. L., & van Franeker, J. A. (2015). Deleterious effects of litter on marine life. In M. Bergmann, L. Gutow & M. Klages (Eds.), *Marine anthropogenic litter* (pp. 75–116). Berlin: Springer.

Kuriyama, Y., Tokai, T., Tabata, K., & Kanehiro, H. (2003). Distribution and composition of litter on seabed of Tokyo Gulf and its age analysis. *Nippon Suisan Gakkaishi, 69*, 770–781.

Law, K. L., Morét-Ferguson, S., Maximenko, N. A., Proskurowski, G., Peacock, E. E., Hafner, J., & Reddy, C. M. (2010). Plastic accumulation in the North Atlantic subtropical gyre. *Science, 329*, 1185–1188.

Lebreton, L. C. M., & Borrero, J. C. (2013). Modeling the transport and accumulation floating debris generated by the 11 March 2011 Tohoku tsunami. *Marine Pollution Bulletin, 66*, 53–58.

Lebreton, L., Greer, S., & Borrero, J. C. (2012). Numerical modelling of floating debris in the world's oceans. *Marine Pollution Bulletin, 64*, 653–661.

Lechner, A., Keckeis, H., Lumesberger-Loisl, F., Zens, B., Krusch, R., Tritthart, M., et al. (2014). The Danube so colourful: A potpourri of plastic litter outnumbers fish larvae in Europe's second largest river. *Environmental Pollution, 188*, 177–181.

Lecke-Mitchell, K. M., & Mullin, K. (1997). Floating marine debris in the US Gulf of Mexico. *Marine Pollution Bulletin, 34*, 702–705.

Lee, D. I., Cho, H. S., & Jeong, S. B. (2006). Distribution characteristics of marine litter on the sea bed of the East China Sea and the South Sea of Korea. *Estuarine, Coastal and Shelf Science, 70*, 187–194.

Lee, J., Hong, S., Song, Y., Hong, S., Janga, Y., Jang, M., et al. (2013). Relationships among the abundances of plastic debris in different size classes on beaches in South Korea. *Marine Pollution Bulletin, 77*, 349–354.

Liu, T., Wang, M. W., & Chen, P. (2013). Influence of waste management policy on the characteristics of beach litter in Kaohsiung, Taiwan. *Marine Pollution Bulletin, 72*, 99–106.

Löder, M. G. J., & Gerdts, G. (2015). Methodology used for the detection and identification of microplastics—A critical appraisal. In M. Bergmann, L. Gutow & M. Klages (Eds.), *Marine anthropogenic litter* (pp. 201–227). Berlin: Springer.

Lusher, A. (2015). Microplastics in the marine environment: Distribution, interactions and effects. In M. Bergmann, L. Gutow & M. Klages (Eds.), *Marine anthropogenic litter* (pp. 245–308). Berlin: Springer.

MacGranahan, G., Balk, D., & Anderson, B. (2007). The rising tide: Assessing the risks of climate change and human settlements in low elevation coastal zones. *Environment and Urbanization, 19*, 17–37.

Martinez, E., Maamaatuaiahutapu, K., & Taillandier, V. (2009). Floating marine debris surface drift: Convergence and accumulation toward the South Pacific subtropical gyre. *Marine Pollution Bulletin, 58*, 1347–1355.

Matsumura, S., & Nasu, K. (1997). Distribution of floating debris in the North Pacific Ocean: Sighting surveys 1986–1991. In J. M. Coe & D. B. Rogers (Eds.), *Marine debris: Sources, impact, and solution* (pp. 15–24). New York: Springer.

Maximenko, N. A., Hafner, J., & Niiler, P. (2012). Pathways of marine debris from trajectories of Lagrangian drifters. *Marine Pollution Bulletin, 65*, 51–62.

Misfud, R., Dimech, M., & Schembri, P. J. (2013). Marine litter from circalittoral and deeper bottoms off the Maltese islands (Central Mediterranean). *Mediterranean Marine Science, 14*, 298–308.

Miyake, H., Shibata, H., & Furushima, Y. (2011). Deep-sea litter study using deep-sea observation tools. In K. Omori, X. Guo, N. Yoshie, N. Fujii, I. C. Handoh, A. Isobe & S. Tanabe (Eds.), *Interdisciplinary studies on environmental Chemistry-Marine environmental modeling and analysis Terrapub* (pp. 261–269).

Moore, S. L., & Allen, M. J. (2000). Distribution of anthropogenic and natural debris on the mainland shelf of the Southern California bight. *Marine Pollution Bulletin, 40*, 83–88.

Moore, S. L., Gregorio, D., Carreon, M., Weisberg, S. B., & Leecaster, M. K. (2001a). Composition and distribution of beach debris in Orange County, California. *Marine Pollution Bulletin, 42*(3), 241–245.

Moore, C. J., Moore, S. L., Leecaster, M. K., & Weisberg, S. B. (2001b). A comparison of plastic and plankton in the North Pacific central gyre. *Marine Pollution Bulletin, 42*, 1297–1300.

Mordecai, G., Tyler, P. A., Masson, D. G., & Huvenne, V. A. I. (2011). Litter in submarine canyons off the west coast of Portugal. *Deep-Sea Research II, 58*, 2489–2496.

Morris, J. R. (1980). Floating plastic debris in the Mediterranean. *Marine Pollution Bulletin, 11*, 125.

Morritt, D., Stefanoudis, P. V., Pearce, D., Crimmen, O. A., & Clark, P. F. (2014). Plastic in the Thames: A river runs through it. *Marine Pollution Bulletin, 78*, 196–200.

Murray, F., & Cowie, P. (2011). Plastic contamination in the decapod crustacean *Nephrops norvegicus* (Linnaeus 1758). *Marine Pollution Bulletin, 62*, 1207–1217.

Neumann, D., Callies, U., & Matthies, M. (2014). Marine litter ensemble transport simulations in the southern North Sea. *Marine Pollution Bulletin, 86*, 219–228.

Obbard, R. W., Sadri, S., Wong, Y. Q., Khitun, A. A., Baker, I., & Thompson, R. C. (2014). Global warming releases microplastic legacy frozen in Arctic Sea ice. *Earth's Future, 2*, 2014EF000240.

Oigman-Pszczol, S. S., & Creed, J. C. (2007). Quantification and classification of marine litter on beaches along Armação dos Búzios, Rio de Janeiro, Brazil. *Journal of Coastal Research, 232*, 421–428.

Pham, C. K., Gomes-Pereira, J. N., Isidro, E. J., Santos, R. S., & Morato, T. (2013). Abundance of litter on condor seamount (Azores, Portugal, Northeast Atlantic). *Deep-Sea Research II, 98*, 204–208.

Pham, C. K., Ramirez-Llodra, E., Alt, C. H. S., Amaro, T., Bergmann, M., Canals, M., et al. (2014). Marine litter distribution and density in European Seas, from the shelves to deep basins. *PLoS ONE, 9*, e95839.

Pichel, W., Churnside, J., Veenstra, T., Foley, D., Friedman, K., Brainard, R., et al. (2012). GhostNet marine debris survey in the Gulf of Alaska—Satellite guidance and aircraft observations. *Marine Pollution Bulletin, 65*, 28–41.

PlasticsEurope. (2015). Plastics – the Facts 2014/2015. http://issuu.com/plasticseuropeebook/docs/final_plastics_the_facts_2014_19122

Ramirez-Llodra, E., De Mol, B., Company, J. B., Coll, M., & Sardà, F. (2013). Effects of natural and anthropogenic processes in the distribution of marine litter in the deep Mediterranean Sea. *Progress in Oceanography, 118*, 273–287.

Rech, S., Macaya-Caquilpán, V., Pantoja, J. F., Rivadeneira, M. M., Jofre Madariaga, D., & Thiel, M. (2014). Rivers as a source of marine litter—A study from the SE Pacific. *Marine Pollution Bulletin, 82*, 66–75.

Reisser, J., Shaw, J., Wilcox, C., Hardesty, B. D., Poreitti, M., Turms, M., et al. (2013). Marine plastic pollution in waters around Australia: Characteristics, concentrations, and pathways. *PLoS ONE, 8*(11), e80466.

Ribic, C., Sheavly, S., & Klavitter, J. (2012a). Baseline for beached marine debris on Sand Island, Midway Atoll. *Marine Pollution Bulletin, 64*, 726–1729.

Ribic, C., Sheavly, S., Rugg, D. J., & Erdmann, E. (2010). Trends and drivers of marine debris on the Atlantic coast of the United States 1997–2007. *Marine Pollution Bulletin, 60*, 1231–1242.

Ribic, C., Sheavly, S., Rugg, D., & Erdmann, E. (2012b). Trends in marine debris along the U.S. Pacific Coast and Hawai'i 1998–2007. *Marine Pollution Bulletin, 64*, 994–1004.

Rios, L. M., & Moore, C. (2007). Persistent organic pollutants carried by synthetic polymers in the ocean environment. *Marine Pollution Bulletin, 54*(8), 1230–1237.

Rosevelt, C., Los Huertos, M., Garza, C., & Nevins, H. M. (2013). Marine debris in central California: Quantifying type and abundance of beach litter in Monterey Bay, CA. *Marine Pollution Bulletin, 71*(1–2), 299–306.

Roy, P. K., Hakkarainen, M., Varma, I. K., & Albertsson, A.-C. (2011). Degradable polyethylene: Fantasy or reality. *Environmental Science and Technology, 45*, 4217–4227.

Ryan, P. G. (2013). A simple technique for counting marine debris at sea reveals steep litter gradients between the Straits of Malacca and the Bay of Bengal. *Marine Pollution Bulletin, 69*, 128–136.

Ryan, P. G. (2015). A brief history of marine litter research. In M. Bergmann, L. Gutow & M. Klages (Eds.), *Marine anthropogenic litter* (pp. 1–25). Berlin: Springer.

Ryan, P. G., Moore, C. J., vanFraneker, J. A., & Moloney, C. L. (2009). Monitoring the abundance of plastic debris in the marine environment, *Philosophical Transactions of the Royal Society B, 364*(1526), 1999–2012

Ryan, P. G., Musker, S., & Rink, A. (2014). Low densities of drifting litter in the African sector of the Southern Ocean. *Marine Pollution Bulletin, 89*, 16–19.

Sadri, S. S., & Thompson, R. C. (2014). On the quantity and composition of floating plastic debris entering and leaving the Tamar Estuary, Southwest England. *Marine Pollution Bulletin, 81*, 55–60.

Sanchez, P., Maso, M., Saez, R., De Juan, S., & Muntadas, A. (2013). Baseline study of the distribution of marine debris on soft-bottom habitats associated with trawling grounds in the northern Mediterranean. *Scientia Marina, 77*, 247–2255.

Santos, I. R., Friedrich, A. C., & Ivar do Sul, J. A. (2009). Marine debris contamination along undeveloped tropical beaches from northeast Brazil. *Environmental Monitoring and Assessment, 148*, 455–462.

Schlining, K., Von Thun, S., Kuhnz, L., Schlining, B., Lundsten, L., Jacobsen Stout, N., et al. (2013). Debris in the deep: Using a 22-year video annotation database to survey marine litter in Monterey Canyon, central California, USA. *Deep-Sea Research I, 79*, 96–105.

Schulz, M., Neumann, D., Fleet, D. M., & Matthies, M. (2013). A multi-criteria evaluation system for marine litter pollution based on statistical analyses of OSPAR beach litter monitoring time series. *Marine Environmental Research, 92*, 61–70.

Shiomoto, A., & Kameda, T. (2005). Distribution of manufactured floating marine debris in nearshore areas around Japan. *Marine Pollution Bulletin, 50*, 1430–1432.

Slavin, C., Grage, A., & Campbell, M. (2012). Linking social drivers of marine debris with actual marine debris on beaches. *Marine Pollution Bulletin, 64*, 1580–1588.

Smith, S. D. A. (2012). Marine debris: A proximate threat to marine sustainability in Bootless Bay, Papua New Guinea. *Marine Pollution Bulletin, 64*(9), 1880–1883.

Smith, S. D. A., & Markic, A. (2013). Estimates of marine debris accumulation on beaches are strongly affected by the temporal scale of sampling. *PLOS One, 8*(12), e83694.

Suaria, G., & Aliani, S. (2014). Floating debris in the Mediterranean Sea. *Marine Pollution Bulletin, 86*(1–2), 494–504.

Thiel, M., Hinojosa, I., Joschko, T., & Gutow, L. (2011). Spatio-temporal distribution of floating objects in the German Bight (North Sea). *Journal of Sea Research, 65*, 368–379.

Thiel, M., Hinojosa, I. A., Miranda, L., Pantoja, J. F., Rivadeneira, M. M., & Vásquez, N. (2013). Anthropogenic marine debris in the coastal environment: A multi-year comparison between coastal waters and local shores. *Marine Pollution Bulletin, 71*, 307–316.

Thompson, R. C. (2015). Microplastics in the marine environment: Sources, consequences and solutions. In M. Bergmann, L. Gutow & M. Klages (Eds.), *Marine anthropogenic litter* (pp. 185–200). Berlin: Springer.

Thompson, R. C., Olsen, Y., Mitchell, R. P., Davis, A., Rowland, S. J., John, A. W. G., et al. (2004). Lost at sea: Where is all the plastic? *Science, 304*, 838.

Titmus, A. J., & Hyrenbach, K. D. (2011). Habitat associations of floating debris and marine birds in the North East Pacific Ocean at coarse and meso spatial scales. *Marine Pollution Bulletin, 62*, 2496–2506.

Topçu, E. N., Tonay, A. M., Dede, A., Öztürk, A. A., & Öztürk, B. (2013). Origin and abundance of marine litter along sandy beaches of the Turkish Western Black Sea Coast. *Marine Environmental Research, 85*, 21–28.

Tourinho, P. S., & Fillmann, G. (2011). Temporal trend of litter contamination at Cassino Beach, Southern Brazil. *Journal of Integrated Coastal Zone Management, 11*, 97–102.

Turra, A., Manzano, A., Dias, R., Mahiques, M., Silva, D., & Moreira, F. (2014). Three-dimensional distribution of plastic pellets in sandy beaches: Shifting paradigms. *Nature Scientific Reports, 4*, 4435.

UNEP. (2009). *Marine litter a global challenge* (p. 232). Nairobi: UNEP.

United Nations. (2012). *Report of the United Nations Conference on Sustainable Development*, Rio de Janeiro, Brazil, 20–22 June 2012 A/CONF.216/16.

Uneputty, P., & Evans, S. M. (1997). The impact of plastic debris on the biota of tidal flats in Ambon Bay (eastern Indonesia). *Marine Environmental Research, 44*, 233–242.

Van Cauwenberghe, L., Vanreusel, A., Mees, J., & Janssen, C. R. (2013). Microplastic pollution in deep-sea sediments. *Environmental Pollution, 182*, 495–499.

Van Franeker, J. A., Blaize, C., Danielsen, J., Fairclough, K., Gollan, J., Guse, N., et al. (2011). Monitoring plastic ingestion by the northern fulmar *Fulmarus glacialis* in the North Sea. *Environmental Pollution, 159*, 2609–2615.

Venrick, E. L., Backman, T. W., Bartram, W. C., Platt, C. J., Thornhill, M. S., & Yates, R. E. (1972). Man-made objects on the surface of the Central North Pacific Ocean. *Nature, 241*, 271.

Vieira, R. P., Raposo, I. P., Sobral, P., Gonçalves, J. M. S., Bell, K. L. C., & Cunha, M. R. (2014). Lost fishing gear and litter at Gorringe Bank (NE Atlantic). *Journal of Sea Research, in press*, doi:http://dx.doi.org/10.1016/j.seares.2014.10.005.

Vlietstra, L. S., & Parga, J. A. (2002). Long-term changes in the type, but not amount, of ingested plastic particles in short-tailed shearwaters in the southeastern Bering Sea. *Marine Pollution Bulletin, 44*(9), 945–955.

Watters, D. L., Yoklavich, M. M., Love, M. S., & Schroeder, D. M. (2010). Assessing marine debris in deep seafloor habitats off California. *Marine Pollution Bulletin, 60*, 131–138.

Wei, C. L., Rowe, G. T., Nunnally, C. C., & Wicksten, M. K. (2012). Anthropogenic "litter" and macrophyte detritus in the deep Northern Gulf of Mexico. *Marine Pollution Bulletin, 64*, 966–973.

Williams, R., Ashe, E., & Ohara, P. D. (2011). Marine mammals and debris in coastal waters of British Columbia, Canada. *Marine Pollution Bulletin, 62*, 1303–1316.

Wong, C. S., Green, D. R., & Cretney, W. J. (1974). Quantitative tar and plastic waste distributions in the Pacific Ocean. *Nature, 247*, 30–32.

Woodall, L., Rogers, A., Packer, M., Robinson, L., & Paterson C. (2014). Extreme litter picking: Comparison of litter across seamounts. *International Marine Conservation Congress 2014 symposium SY70 (Science-based solutions to tackle marine debris impacts on wildlife)*, Glasgow (p. 1). August 14–18, 2014. Presentation and abstract.

Woodall L., Robinson, L., Rodgers, A., Narayanaswamy, B., & Paterson G. (2015). Deep sea litter: A comparison of seamounts, banks and a ridge in the Atlantic and Indian Oceans reveals both environmental and anthropogenic factors impact accumulation and composition. *Frontiers in Marine Science, 2*, doi:10.3389/fmars.2015.00003

Woodall, L. C., Sanchez-Vidal, A., Canals, M., Paterson, G. L. J., Coppock, R., Sleight, V., et al. (2014). The deep sea is a major sink for microplastic debris. *Royal Society Open Science, 1*, 140317.

Yamashita, R., & Tanimura, A. (2007). Floating plastic in the Kuroshio current area, western North Pacific Ocean. *Marine Pollution Bulletin, 54*, 485–488.

Yoon, J., Kawano, S., & Igawa, S. (2010). Modeling of marine litter drift and beaching in the Japan Sea. *Marine Pollution Bulletin, 60*, 448–463.

Zhou, P., Huang, C., Fang, H., Cai, W., Dongmei, L., Xiaomin, L., et al. (2011). The abundance, composition and sources of marine debris in coastal seawaters or beaches around the northern South China Sea (China). *Marine Pollution Bulletin, 62*(9), 1998–2007.

Chapter 3
Persistence of Plastic Litter in the Oceans

Anthony L. Andrady

Abstract The increasing global production and use of plastics has led to an accumulation of enormous amounts of plastic litter in the world's oceans. Characteristics such as low density, good mechanical properties and low cost allow for successful use of plastics in industries and everyday life but the high durability leads to persistence of the synthetic polymers in the marine environment where they cause harm to a great variety of organisms. In the diverse marine habitats, including beaches, the sea surface, the water column, and the seafloor, plastics are exposed to different environmental conditions that either accelerate or decelerate the physical, chemical and biological degradation of plastics. Degradation of plastics occurs primarily through solar UV-radiation induced photo oxidation reactions and is, thus, most intensive in photic environments such as the sea surface and on beaches. The rate of degradation is temperature-dependent resulting in considerable deceleration of the processes in seawater, which is a good heat sink. Below the photic zone in the water column, plastics degrade very slowly resulting in high persistence of plastic litter especially at the seafloor. Biological decomposition of plastics by microorganisms is negligible in the marine environment because the kinetics of biodegradation at sea is particularly slow and oxygen supply for these processes limited. Degradation of larger plastic items leads to the formation of abundant small microplastics. The transport of small particles to the seafloor and their deposition in the benthic environment is facilitated by the colonization of the material by fouling organisms, which increase the density of the particles and force them to sink.

Keywords Synthetic polymer · Mechanical properties · Weathering · Embrittlement · Photo oxidation · Microplastics

A.L. Andrady (✉)
Department of Chemical and Biomolecular Engineering,
North Carolina State University, Raleigh, NC 27695, USA
e-mail: andrady@andrady.com

© The Author(s) 2015
M. Bergmann et al. (eds.), *Marine Anthropogenic Litter*,
DOI 10.1007/978-3-319-16510-3_3

3.1 Introduction

Studies on the occurrence of marine litter on beaches and as flotsam generally find plastics to be the major component of the mix of debris (Galgani et al. 2015). Plastics have diverse uses and are gaining popularity in building and packaging applications because of their ease of processing, durability and relatively low cost (Andrady and Neal 2009). However, this predominance of plastics in litter is not the result of relatively more plastics being littered compared to paper, paperboard or wood products reaching the oceans, but because of the exceptional durability or persistence of plastics in the environment. Data on plastic debris on sediments are more limited (Spengler and Costa 2008) but suggest that plastics represent a significant fraction of the benthic debris as well (Watters et al. 2010). Quantitative information on the density of litter on beaches or in the ocean classified according to the class of plastic, are not available. Usual classification is by geometry (e.g. fiber) or by product type (e.g. cigarette butts). Also the surveys of water-borne plastic debris collected via neuston net sampling of surface waters (Hidalgo-Ruz et al. 2012) and even beach studies (Ng and Obbard 2006; Browne et al. 2011) close to the water line, seriously underestimate the magnitude of plastic litter. Not only do these exclude the negatively buoyant plastics but also fragments smaller than the mesh-size of the nets used.

3.2 Buoyancy and Sampling Errors

Of the five classes of the commonly used plastics (or commodity thermoplastics), polyethylenes (PE) and polypropylenes (PP) as well as the expanded form of poly-styrene or polystyrene foam (EPS) are less dense than sea water while others such as poly (vinyl chloride) (PVC) and poly(ethylene terephthalate) (PET) are nega-tively buoyant and sink into the mid water column or to the sediment (Andrady 2011). Significantly, one of the key fishing-gear related plastics, nylon or pol-yamide (PA), also belongs to this category and hence the negative buoyancy of these items likely explains their virtual absence in beach litter or flotsam surveys, despite their high volume use at sea. However, there are exceptions to this general expectation that is based on the properties of the pure resins such as with virgin resin pellets or prils found commonly in sampled debris. Some plastic products are compounded with fillers and other additives that alter the density of the virgin plastic material. These additives are needed to ensure ease of processing the plas-tic as well as to obtain the mechanical properties demanded of the final product. Where the density is increased because additives, such as fillers, are incorporated, the material may not float in surface water and, therefore, not be counted in net sampling. Accordingly, plastics such as PS, PET and PVC, which are denser than sea water, should be missing from floating samples as well. In fact, however, they might be included in flotsam samples because products such as bottles, bags and

Table 3.1 Marine debris items removed from the global coastline and waterways during the 2009 international coastal cleanup

Rank	Debris item	Count (millions)	Plastic used
1	Cigarette filter	2.19	CA
2	Plastic bags	1.13	PE
3	Food wrapper/container	0.94	PE, PP
4	Caps and lids	0.91	PP and HDPE
5	Beverage bottles	0.88	PET
6	Cups, plates and cutlery	0.51	PS
7	Glass bottles	0.46	–
8	Beverage cans	0.46	–
9	Straws stirrers	0.41	PE
10	Paper bags	0.332	–

Data from Ocean Conservancy. *CA* Cellulose acetate, *HDPE* High-density polyethylene

foams made from these plastics trap air. This is clearly the case with EPS foam used in floats, bait boxes and insulation that generally constitutes a highly visible and major fraction of persistent litter in the ocean environment.

The main items of debris are different plastic products (or their fragments) as illustrated in the global beach clean data compiled by Ocean Conservancy for 2009. The data in Table 3.1 summarize the beach cleanup efforts regularly sponsored by the organization: beach cleanup is carried out by volunteers who also count and tabulate the litter over an area assigned to each person. The data are aggregated and summarized by the Ocean Conservancy.

A second inefficiency in sampling of plastic debris at all marine sites is the minimum particle size isolated. The procedure of using plankton nets to sample water and separating particles visually after sieving or by floatation from sediment samples invariably fails to catch the micro-sized fragments of plastics. Commonly used nets have a mesh size of about 330 μm. While the meso-sized plastics are reasonably represented in these samples the micro-sized and nano-scale particles are grossly underestimated. Since a great majority of the floating litter is generated on land and transported to the ocean, one would expect the resin types in the litter to be consistent with the production volume shown in Table 3.2. As the mass fraction of the unsampled microplastics is likely miniscule by comparison to the macro-plastic debris, the statistics of plastics by resin type in water samples show PE and PP to be the most abundant, consistent with the production data in Table 3.2.

3.3 Fate of Plastics Entering the Oceans

Common plastics used in packaging and encountered in the marine environment are persistent recalcitrant materials. In common with other organic materials they do ultimately degrade but the rate at which environmental degradation proceeds is painstakingly slow for plastics. Several agencies can potentially bring about the

Table 3.2 The common classes of plastics found in ocean debris and those used in fishing gear along with their densities and the fraction of their global volume production. Items of lower specific gravity than seawater (~1.02 g cm⁻³) float

Plastic	Specific gravity ($g\ cm^{-3}$)	Percentage of production[a]	Main uses
Polyethylene (PE)	0.91–0.94	29.1	Packaging, fishing gear
Polypropylene (PP)	0.83–0.85	18.0	Packaging, fishing gear
Polystyrene (PS) and foam (EPS)	1.05 (variable)	7.8	Packaging, food service
Poly(vinyl chloride) (PVC)	1.38	15.3	Packaging
Poly(ethylene terephthalate) [PET]	1.37	20.0	Packaging
Nylon (PA)	1.13	~1	Fishing gear
Cellulose acetate (CA)	1.29	<1	Cigarette filter

[a]Percentage production is based on data taken from plastics news (accessed: December 2014): http://www.plasticsnews.com/article/20100305/FYI/303059995/global-thermoplastic-resin-capacity-2008

degradation (or chemical breakdown of the polymer molecules with consequent change in material properties) in the environment. These are primarily as follows:

(a) solar UV-induced photodegradation reactions
(b) thermal reactions including thermo-oxidation
(c) hydrolysis of the polymer
(d) microbial biodegradation

Of these, only the first or the light-induced oxidative degradation is particularly effective in the ocean environment and that only with plastics floating at the sea surface or littered on beaches (Cooper and Corcoran 2010). Slow thermal oxidation of plastics also proceeds in concert with photo-oxidation, especially on beaches. However, no hydrolysis or significant biodegradation of plastics is anticipated in the ocean.

Different measurable properties of a plastic might be altered as a result of weathering. Some of these are properties that are directly relevant to the performance of common products made from them (Singh and Sharma 2008). Others are changes at the molecular level that might be used to detect early stages of degradation. The popularly used characteristics of common plastics are as follows:

(a) decrease in average molecular weight of the plastic. This is conveniently measured using gel permeation chromatography (GPC) and also using solution (or melt) viscosity
(b) loss in bulk mechanical properties of the plastic, such as the tensile properties, compression properties or the impact properties
(c) loss in surface properties of the material including discoloration, micro-cracking or 'chalking' (release of white filler from filled plastic surfaces on weathering)
(d) changes in spectral characteristics that are markers for oxidative degradation or photodegradation. For polyolefins, the relative intensity of the carbonyl absorption band (in the Fourier transformed infra-red or the FTIR spectrum), which increases in percent crystallinity or level of unsaturation, might be monitored.

3.3.1 Photo-Oxidative Degradation

Photo-oxidation of polyolefin plastics is a free-radical reaction that is initiated by solar UV radiation. The sequence of oxidative chemical reactions involved, results in (a) incorporation of oxygen-containing functionalities into the polymer molecules, and (b) scission of long chain-like polymer molecules reducing the number-average molecular weight of the plastic material. Of these, it is the latter that drastically affects the useful properties of the polymer. Even at low levels of oxidation (often a percent or less) very substantial loss in mechanical strength can occur. High-energy UV-B (290–315 nm) and medium energy UV-A (315–400 nm) solar wavelengths are particularly efficient in facilitating photo-degradation of polymers (Andrady 1996). However, the fraction of longer wavelength radiation in sunlight is very much larger compared to that of the UV radiation and most of the light-induced damage occurs in the UV-A and/or the visible region of the spectrum.

The approximate region of the solar spectrum that accounts for the most degradation is represented by an activation spectrum. Activation spectra are generated in experiments where samples of a plastic are exposed to solar or solar-simulated radiation behind a series of cut-on filters that allow only wavelengths higher than a cut-on wavelength to be transmitted through them. The degradation rates for samples behind different filters can be used to construct the activation spectrum (for a discussion of the experimental procedures involved in generating such spectra see Singh and Sharma 2008). Figure 3.1 shows an activation spectrum for yellowing of polycarbonate exposed to solar radiation. It is clear from the figure that the UV-A region of sunlight (320–340 nm) causes the greatest damage, despite the shorter more energetic wavelengths <320 nm being present in the spectrum. The shorter wavelengths account for less than ~5 % of the solar radiation spectrum.

Rates of degra dation are markedly increased at higher ambient temperatures as the activation energies for oxidative degradation of common plastics are low

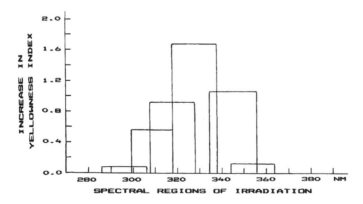

Fig. 3.1 Activation spectrum for yellowness index of un-stabilized lexan polycarbonate film (0.70 mm) exposed to natural sunlight facing 26°South in Miami, FL. Reproduced with permission from Andrady et al. (1992)

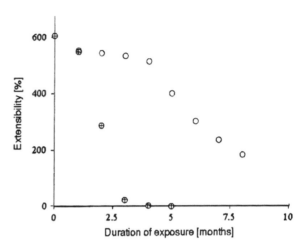

Fig. 3.2 Change in extensibility of polyethylene sheet samples after exposure to solar UV radiation in Dhahran, Saudi Arabia. The *open symbols* are for samples maintained at 25 °C. The *filled symbols* are for samples exposed at ambient temperatures of 26–36 °C. Reprinted with permission from Andrady et al. (1998)

(Hamid and Pritchard 1991; Tocháček and Vrátníčková 2014). Plastics lying on hot sand on beaches undergo faster photo-oxidation relative to those floating on water and being, therefore, maintained at a lower temperature. The same phenomenon is also responsible for differences in the rates of weathering of differently colored plastics. Darker shades of plastics exposed to sunlight tend to absorb more of the infrared energy in the solar spectrum, reaching higher sample temperatures. Consequently, they weather faster relative to lighter colored plastics. A particularly good measure of degradation in plastics is tensile extensibility. Figure 3.2 shows the effect of sample temperature on the loss in tensile extensibility of polyethylene film samples exposed in Dhahran, Saudi Arabia. One set of samples was exposed at ambient temperature of 26–36 °C. Another set of samples was maintained at a constant temperature of 25 °C. At different durations of exposure the samples (typically dumbbell-shaped pieces) were removed periodically for testing. In this test the dumbbell shaped sample (5–6 in. long) is held at its ends in a pair of grips and pulled along its long axis at a constant speed of 500 mm/min. The sample first extends and then snaps. The ratio of the grip separation at the point the sample snaps to that at the start of the extension, expressed as a percentage, is the extensibility or ultimate strain of the sample.

3.3.2 Mechanisms of Photo-Oxidation

The basic mechanism of light-induced degradation for the two plastics used in highest volume and therefore most numerous in marine debris, PE and PP, is well known. It is a free-radical reaction initiated by UV radiation or heat and propagated via hydrogen abstraction from the polymer. The polymer alkyl radicals formed react with oxygen to yield peroxy radicals, $ROO\bullet$, that are converted to a peroxide moiety by hydrogen abstraction. As peroxide products can themselves dissociate readily into radicals, the reaction sequence is autocatalytic. The main reactions involved in the sequence are as follows (François-Heude et al. 2015):

1. Initiation:

$$RH \rightarrow \text{Free radicals, ex., } R\bullet, H\bullet$$

2. Propagation:

$$R\bullet + O_2 \rightarrow ROO\bullet$$
$$ROO\bullet + RH \rightarrow ROOH + R\bullet$$

3. Termination:

$$ROO\bullet + ROO\bullet \rightarrow ROOR + O_2$$
$$R\bullet + R\bullet \rightarrow R-R$$
$$RO\bullet + H\bullet \rightarrow ROH$$
$$R\bullet + H\bullet \rightarrow RH$$

From a practical standpoint, it is the chain scission that accompanies this cyclic reaction sequence, which is of greater interest. The chain scission event is believed to be associated with one of the propagation reactions and is responsible for the loss in mechanical properties of the plastic material after exposure. Different mechanical properties (such as ultimate extensibility, the tensile modulus, or impact strength) having different functional dependence on the average molecular weight will change at different rates with the duration of exposure. There is, thus, no 'general' weathering curve for a given polymer but only for specific modes of damage of the polymer material under exposure to a specified light source such as sunlight or radiation from a xenon lamp. Chain scission is often directly estimated from gel permeation chromatography. Being associated with the number of propagation cycles it can also be correlated with the products of the chemical reactions, especially the accumulation of carbonyl compounds $\{>C=O\}$. This is often monitored using the relative intensity of the relevant bands in the FTIR spectrum of the polymer and has been demonstrated to correlate well with the ultimate extensibility of the sample (Andrady et al. 1993).

Other reactions that contribute to changes in the useful properties of plastics following exposure to solar radiation are also evident with common plastics. Yellowing discoloration of poly(vinyl chloride) (PVC) is an example of such a reaction. This is a light-induced de-hydrochlorination reaction that generates short sequences of conjugated unsaturation in the polymer (Andrady et al. 1989):

$$\sim CH_2 - CHCl - CH_2 - CHCl - CH_2 - CHCl \sim \rightarrow$$
$$\sim CH_2 - CH = CH - CH = CH - CHCl \sim + 2HCl$$

These absorb on the blue region and make the plastic appear yellow. However, polyolefins (both PE and PP) as well as PS also yellow on exposure to sunlight but the mechanism of such yellowing and the identity of the species involved are not well known. Polycarbonate (PC) plastic used in glazing applications is another example of a material that undergoes yellowing under exposure to sunlight. The main photodegradation reaction of PC, however, is a rearrangement reaction (Fries reaction) with no change in spectral qualities (Factor et al. 1987):

A second reaction that yields yellow oxidation products also occurs along with it, however, the mechanistic details of the second reaction are unknown.

3.3.3 Weathering Under Marine Conditions

While the main agencies involved and the mechanisms of weathering in the marine environment are the same as those on land environments, the rates at which weathering proceeds can be significantly slower in the former (Pegram and Andrady 1989). To better understand the differences, the marine environment must be regarded in separate zones: the beach environment, the surface water environment, and the deep water/benthic environment. The availability of weathering agencies in these are different as summarized in Table 3.3.

Availability of sunlight to initiate the degradation reactions is restricted in the case of floating plastics because of bio-fouling of their surface in seawater. Initial

Table 3.3 Comparison of the availability of weathering agents in the different zones within the marine environment

Weathering agent	Land[a]	Beach	Surface water	Deep water or sediment
Sunlight	Yes	Yes	Yes	No
Sample temperature	High	High	Moderate	Low
Oxygen levels	High	High	High/moderate	Low
Fouling (screens solar radiation)	No	No	Yes	Yes

[a]Land environment included for comparison

exposure of the plastic results in the formation of a surface biofilm (Lobelle and Cunliffe 2011) that is rapidly colonized by algae and other marine biota including encrusting organisms that increase the density of the plastic causing it to sink in seawater (Thangavelu et al. 2011). The plastic particles that sink due to this process may re-emerge at a later time once the foulants are foraged by marine consumers and the plastic decreases in density (Ye and Andrady 1991). Fouling shields the surface of plastic from exposure to sunlight interfering with the initiation of the oxidation process. This is a significant reason for the retardation of weathering degradation in plastics floating in seawater (Pegram and Andrady 1989). Also, attenuation of solar UV radiation in seawater is very rapid and light-induced initiation reactions cannot occur at depths beyond the photic zone.

The primary reason for the retardation of weathering degradation in floating samples is the relatively lower sample temperatures. In contact with a good heat sink (i.e. seawater), the samples do not undergo heat build up and reach high temperatures as in the case of samples exposed on land. The combined effect of these factors in retarding degradation is illustrated in Fig. 3.3 that compares the loss in extensibility of polypropylene exposed in Biscayne Bay, FL, floating in water and on land during the same period. This observation of retardation of the weathering at sea is generally true for all common plastic materials. With expanded polystyrene foam (EPS) plasticization by water and wave action result in the foam breaking up readily into individual beads of the polymer. However, the weathering degradation of these beads is a slow process.

Initial stages of oxidative breakdown of the plastic materials result in a marked decrease in their mechanical properties. However, the high-polymer nature persists even at extensive degradation where the mechanical integrity of the plastic material is fully compromised. Andrady (2011) as well as Klemchuk and Horng (1984) have demonstrated that for polyethylenes weathered even to the point of embrittlement with no extensibility of the material, the average molecular weights persisted in the 10s of thousands g/mole. These will likely not be further photodegraded so

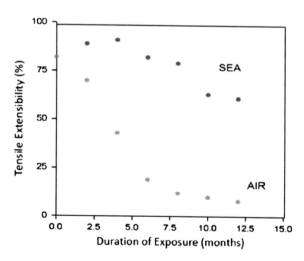

Fig. 3.3 Change in percent original tensile extensibility of polypropylene film exposed in air and floating in seawater at a beach location in Biscayne Bay, FL. Reproduced with permission from Andrady (2011)

that, being fouled or submerged in water, slow biodegradation is the only feasible mechanism for their removal from the environment.

Plastic debris in the ocean generally accumulates a biofilm that contains numerous diverse microorganisms (Ho et al. 1999). Such marine biota can secrete enzymes that can biodegrade common plastics such as polyethylenes as evidenced by surface depressions and pits caused by these on the plastic debris (Zettler et al. 2013). But, the relevant species are rare and the kinetics of biodegradation at sea is particularly slow. While strictly speaking, plastics do biodegrade at sea due to the action of marine organisms, however, the rate of the process is far too slow to either remove plastic debris from the environment or even to obtain obvious decreases in mechanical integrity attributable solely to this process. The exceptions are those plastics, such as aliphatic polyesters, that have structural features that allow facile biodegradation (Kita et al. 1997; Sudhakar et al. 2007) by a host of microorganisms present in the ocean. Biodegradation converts the carbon sequestered in the plastic to carbon dioxide (Narayan 2006). With a simple substrate such as glucose, the products depend on whether the process is aerobic or anaerobic (Tokiwa et al. 2009):

Aerobic biodegradation:

$$C_6H_{12}O_6 + 6O_2 \rightarrow 6CO_2 + 6H_2O \ \Delta G = -2870 \text{ kJ/mol}$$

Anaerobic biodegradation:

$$C_6H_{12}O_6 \rightarrow 3CO_2 + 3CH_4 \ \Delta G = -390 \text{ kJ/mol}$$

Most of the common plastics are hydrocarbons and the stoichiometry will be different from above (Shimao 2001).

3.4 Microplastics in the Oceans

An emerging pollutant of concern in the marine environment is microplastic material or plastic fragments of a size-range that allows their interaction with marine plankton (Cole et al. 2011). Their presence in surface water (Barnes et al. 2009; Song et al. 2014), beaches and sediment (Katsanevakis et al. 2007) has been reported from many parts of the world, including even the Arctic (Obbard et al. 2014). Additionally, microplastics have been reported in estuaries and freshwater bodies (Lima et al. 2014).

Many different definitions of the size scale that constitute 'microplastics' are reported in the research literature (Gregory and Andrady 2003; Betts 2008; Fendall and Sewell 2009). But there is growing consensus for categorizing microplastics as being <1 mm and >1 μm with the larger fragments that include virgin resin pellets being called 'mesoplastics'. Most of the studies that document the existence of plastic debris in the world's oceans focus almost exclusively on mesoplastics and larger pieces. Studies on true microplastics (<1 mm fraction) are rare because identification and quantification of the microscopic particles is challenging (Löder and Gerdts 2015). Plankton nets used to sample surface waters have a mesh size of ~330 microns and collect the mesoplastics. A majority of the literature, however, uses the term 'microplastics' loosely to mean both meso- and micro-scale particles. A clear definition of the particle sizes is important because it is the particle-size distribution that determines the set of marine organisms that are able to interact, particularly ingest, the microdebris. For instance, microplastics (as well as nanoplastics) are ingestible by zooplankton (Frias et al. 2014) at the bottom of the food pyramid while the mesoplastics including virgin plastic pellets are found in species such as dolphins (Di Beneditto and Ramos 2014).

While virgin plastics such as the prils used in manufacturing plastic products are generally non-toxic and not digestible by any marine organism, large fragments may cause distress due to physical obstruction of the gut or filter appendages (Kühn et al. 2015). The main concern, however, is that microplastics concentrate persistent organic pollutants (POPs) in seawater via partition. The distribution coefficients for organic compounds including POPs range in the 10^4–10^6. Their ingestion by marine organisms provides a credible pathway to transfer the environmental pollutants dissolved in water into the marine food web. Therefore, relatively low mass fractions of the microplastics can transport a disproportionately high dose of POPs into an ingesting organism. Where the organism is small as with zooplanktons (Frias et al. 2014; Lima et al. 2014), assuming high bioavailability, the body burden of the POPs that might be released into the organism can be significant. This is a particular concern as it involves the lower echelons of the marine food web, where any adverse impact may affect the entire food chain and potentially the global fish supply (Betts 2008). Others have suggested that this transfer pathway is likely of limited importance under equilibrium conditions (Gouin et al. 2011; Koelmans et al. 2013, 2014). At least in the lugworm *Arenicola marina*, conservative modeling suggests that the transfer of POPs (Bisphenol A and nonylphenol) from microplastics into the organism yields concentrations below the global environmental concentration of these chemicals (Koelmans et al. 2014).

The origins of meso-, micro- and nano-plastics in the oceans are attributed to either products that incorporate such particles (such as cosmetics, sandblasting media, virgin pellets) or to the weathering degradation of larger plastic debris in the marine environment (Thompson 2015). In the former instance they are referred to as primary microparticles being introduced into the ocean already as micro-debris while in the latter case they are generated in the ocean environment from macro-debris. As already pointed out (Table 3.3), where microplastics are derived from larger plastic litter, the process occurs particularly efficiently on beaches and least efficiently in deep water or sediment.

While weathering related oxidative mechanisms for polyolefins (PE and PP) are well known (Ojeda 2011), the concurrent embrittlement of the material has not been adequately studied. This is to be expected as material scientists have little interest in the weathering process beyond the point at which the material has lost its useful properties; embrittlement, however, occurs after this stage. It is the embrittlement phenomenon that is particularly interesting as it has the potential to generate microplastics. Associated with the oxidation reactions described in the previous section is an autocatalytic chain scission reaction. This is easily demonstrated by monitoring the change in average molecular weight of the plastic during weathering [for instance by gel permeation chromatography (GPC)] (Ojeda 2011). For instance, with PP exposed to UV radiation in an accelerated laboratory weathering experiment the molecular weight of the polymer at the surface of a test piece decreased by 51 % in six weeks of exposure (O'Donnell et al. 1994). At greater depths of a sample, the effect is less pronounced for two reasons: the attenuation of UV radiation with depth that restricts the initiation reaction and the limitation of the reaction due to slow diffusion of oxygen at greater depths.

Chain scission occurs exclusively in the amorphous fraction of semi-crystalline polymers and that, too, preferentially in the surface layer that is several hundred microns in thickness. This can, in theory, lead to two types of fracture: (a) the bulk fracture and (b) surface layer removal due to stresses on highly weathered samples. The former results in a sample such as virgin prils being fragmented gradually into several daughter particles. The latter results in a large number of particles

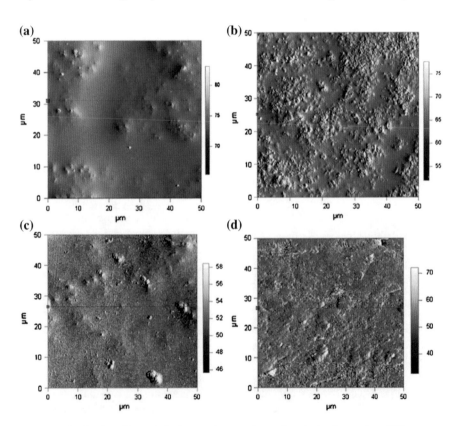

Fig. 3.4 AFM surface images of primer only-coated samples obtained at various UV exposure and salt fog tests: **a** 0 days, **b** 16 days of UV exposure, **c** 0 days of UV light after 80 days of salt fog, and **d** 16 days of UV exposure followed by 80 days of salt fog. Reproduced with permission from Asmatulu et al. (2011)

derived from the surface layer with particle size, at least in one dimension, equal to the thickness of that layer. Possibly both modes of fragmentation occur in natural weathering of plastics on beaches or in seawater.

Plastic samples collected from beach or surface water environments show surface patterns consistent with surface erosion and cracking due to weathering. The cracks and pits on the surface of PE and PP samples from the ocean environment are similar to those seen on samples exposed to weathering (or UV radiation) in the laboratory. It is reasonable to expect that it is this fragmentation process that yields derived microplastics in the ocean environment. The early evolution of surface damage from exposure to UV radiation can be easily discerned from atomic force microscopy (AFM) of the surface. Figure 3.4 shows the changes on an epoxy primer coating, exposed to UV radiation and/or salt fog. These micro-cracks propagate in time to form surface features that are easily visible under a low-power microscope. The cracks appear first on the edges and propagate towards the center of sample surface. The evolution of surface cracks under exposure to UV light has been reported for HDPE (Shimao 2001), LDPE (Cole et al. 2011) and PP (Yakimets et al. 2004). Some of the plastic samples collected from beaches as well as from surface waters in the ocean have extensive yellowing and cracking (Ogata et al. 2009; Cooper and Corcoran 2010). Figure 3.5 shows micrographs that illustrate this phenomenon.

Fig. 3.5 Development of visible cracks on exposure of LDPE samples to laboratory accelerated weathering. **a** Exposed to a xenon source (Atlas WeatherOmeter) for 1600 h at 63.5 °C and **b** exposed to a UV fluorescent lamp (QUV WeatherOmeter) for 800 h at 60.5 °C. Reproduced with permission from Küpper et al. (2004)

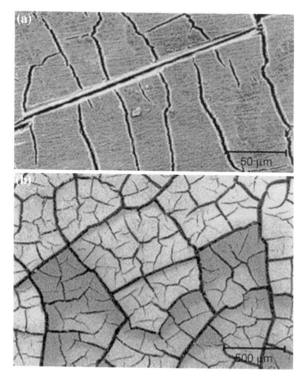

3.5 Conclusions

The degradation mechanisms, pathways and kinetic expressions are well-established in the literature. Detailed information is available particularly on the plastics used in high volume such as PE and PP. However, these studies either do not progress beyond the weakening of the plastic material to a point it cannot be used or the fragmentation process has not been investigated. Hitherto, there has been little interest in studying the fragmentation process or the changes in the ensuing particle size distribution of the plastics. With growing interest in microplastics in the ocean this aspect of polymer degradation will receive more attention.

References

Andrady, A. L. (1996). Wavelength sensitivity of common polymers: A review. *Advances in Polymer Science, 128*, 45–94.

Andrady, A. L. (2011). Microplastics in the marine environment. *Marine Pollution Bulletin, 62*, 1596–1605.

Andrady, A. L., Fueki, K., & Torikai, A. (1989). Photodegradation of rigid PVC formulations. I. Wavelength sensitivity of light induced yellowing by monochromatic light. *Journal of Applied Polymer Science, 37*, 935–946.

Andrady, A. L., Hamid, S. H., Hu, X., & Torikai, A. (1998). Effects of increased solar ultraviolet radiation on materials. *Journal of Photochemistry and Photobiology B: Biology, 46*, 96–103.

Andrady, A. L., & Neal, M. A. (2009). Applications and societal benefits of plastics. *Philosophical Transactions of the Royal Society of London B, 364*, 1977–1984.

Andrady, A. L., Pegram, J. E., & Tropsha, Y. (1993). Changes in carbonyl index and average molecular weight on embrittlement of enhanced photo-degradable polyethylene. *Journal of Environmental Degradation of Polymers, 1*, 171–179.

Andrady, A. L., Searle, N. D., & Crewdson, L. F. E. (1992). Wavelength sensitivity of unstabilized and UV stabilized polycarbonate to solar simulated radiation. *Polymer Degradation and Stability, 35*, 235–247.

Asmatulu, R., Mahmud, G. A., Hille, C., & Misak, H. E. (2011). Effects of UV degradation on surface hydrophobicity, crack, and thickness of MWCNT-based nanocomposite coatings. *Progress in Organic Coatings, 72*, 553–561.

Barnes, D. K. A., Galgani, F., Thompson, R. C., & Barlaz, M. (2009). Accumulation and fragmentation of plastic debris in global environments. *Philosophical Transactions of the Royal Society of London B, 364*, 1985–1998.

Betts, K. (2008). Why small plastic particles may pose a big problem in the oceans. *Environmental Science and Technology, 42*, 8995.

Browne, M. A., Crump, P., Niven, S. J., Teuten, E., Tonkin, A., & Galloway, T. S., et al. (2011). Accumulation of microplastic on shorelines worldwide: Sources and sinks. *Environmental Science and Technology, 45*, 9175–9179.

Cole, M., Lindeque, P., Halsband, C., & Galloway, T. S. (2011). Microplastics as contaminants in the marine environment: A review. *Marine Pollution Bulletin, 62*, 2588–2597.

Cooper, D. A., & Corcoran, P. L. (2010). Effects of mechanical and chemical processes on the degradation of plastic beach debris on the island of Kauai, Hawaii. *Marine Pollution Bulletin, 60*, 650–654.

Di Benedetto, A., & Ramos, R. (2014). Marine debris ingestion by coastal dolphins: what drives differences between sympatric species? *Marine Pollution Bulletin, 83*, 298–301.

Factor, A., Ligon, W. V., May, R. J., & Greenburg, F. H. (1987). Recent advances in polycarbonate photodegradation. In A. V. Patsis (Ed.), *Advances in stabilization and controlled degradation of polymers II* (pp. 45–58). Lancaster, PA: Technomic.

Fendall, L. S., & Sewell, M. A. (2009). Contributing to marine pollution by washing your face: Microplastics in facial cleansers. *Marine Pollution Bulletin, 58*, 1225–1228.

François-Heude, A., Richaud, E., Desnoux, E., & Colin, X. (2015). A general kinetic model for the photothermal oxidation of polypropylene. *Journal of Photochemistry and Photobiology A: Chemistry, 296*, 48–65.

Frias, P. G. L., Otero, V., & Sobral, P. (2014). Evidence of microplastics in samples of zooplankton from Portuguese coastal waters. *Marine Environmental Research, 95*, 89–95.

Galgani, F., Hanke, G. & Maes, T. (2015). Global distribution, composition and abundance of marine litter. In M. Bergmann, L. Gutow & M. Klages (Eds.), *Marine anthropogenic litter* (pp. 29–56). Berlin: Springer.

Gouin, T., Roche, N., Lohmann, R., & Hodges, G. (2011). A thermodynamic approach for assessing the environmental exposure of chemicals absorbed to microplastic. *Environmental Science and Technology, 45*, 1466–1472.

Gregory, M. R., & Andrady, A. L. (2003). Plastics in the marine environment. In A. L. Andrady (Ed.), *Plastics and the Environment* (pp. 379–401). Hoboken, NJ: Wiley.

Hamid, S. H., & Pritchard, W. (1991). Mathematical modelling of weather-induced degradation of polymer properties. *Journal of Applied Polymer Science, 43*, 651–678.

Hidalgo-Ruz, V., Gutow, L., Thompson, R. C., & Thiel, M. (2012). Microplastics in the marine environment: A review of the methods used for identification and quantification. *Environmental Science and Technology, 46*, 3060–3075.

Ho, K. L. G., Pometto, A. L., & Hinz, P. N. (1999). Effects of temperature and relative humidity on polylactic acid plastic degradation. *Journal of Environmental Polymer Degradation, 7*, 83–92.

Katsanevakis, S., Verriopoulos, G., Nicolaidou, A., & Thessalou-Legaki, M. (2007). Effect of marine litter on the benthic megafauna of coastal soft bottoms: A manipulative experiment. *Marine Pollution Bulletin, 54*, 771–778.

Kita, K., Mashiba, S., Nagita, M., Ishimaru, M., Okamoto, K., Yanase, H., et al. (1997). Cloning of poly(3-hydroxybutyrate) depolymerase from a marine bacterium, *Alcaligenes faecalis* AE122, and characterization of its gene product. *Biochimica et Biophysica Acta, 1352*, 113–122.

Klemchuk, P. P., & Horng, P. (1984). Perspectives on the stabilization of hydrocarbon polymers against thermo-oxidative degradation. *Polymer Degradation and Stability, 7*, 131–151.

Koelmans, A. A., Besseling, E., & Foekema, E. M. (2014). Leaching of plastic additives to marine organisms. *Environmental Pollution, 187*, 49–54.

Koelmans, A. A., Besseling, E., Wegner, A., & Foekema, E. M. (2013). Plastic as a carrier of POPs to aquatic organisms: a model analysis. *Environmental Science and Technology, 47*, 7812–7820.

Kühn, S., Bravo Rebolledo, E. L., & van Franeker, J. A. (2015). Deleterious effects of litter on marine life. In M. Bergmann, L. Gutow & M. Klages (Eds.), *Marine anthropogenic litter* (pp. 75–116). Berlin: Springer.

Küpper, L., Gulmine, J. V., Janissek, P. R., & Heise, H. M. (2004). Attenuated total reflection infrared spectroscopy for micro-domain analysis of polyethylene samples after accelerated ageing within weathering chambers. *Vibrational Spectroscopy, 34*, 63–72.

Lima, A. R. A., Costa, M. F., & Barletta, M. (2014). Distribution patterns of microplastics within the plankton of a tropical estuary. *Environmental Research, 132*, 146–155.

Lobelle, D., & Cunliffe, M. (2011). Early microbial biofilm formation on marine plastic debris. *Marine Pollution Bulletin, 62*, 197–200.

Löder, M. G. J. & Gerdts, G. (2015). Methodology used for the detection and identification of microplastics—A critical appraisal. In M. Bergmann, L. Gutow & M. Klages (Eds.), *Marine anthropogenic litter* (pp. 201–227). Berlin: Springer.

Narayan, R. (2006). Rationale, drivers, standards, and technology for biobased materials. In E. M. Graziani & P. Fornasiero (Eds.), *Renewable resources and renewable energy*. Boca Raton: CRC Press.

Ng, K. L., & Obbard, J. P. (2006). Prevalence of microplastics in Singapore's coastal marine environment. *Marine Pollution Bulletin, 52*, 761–767.

Obbard, R. W., Sadri, S., Wong, Y. Q., Khitun, A. A., Baker, I., & Thompson, R. C. (2014). Global warming releases microplastic legacy frozen in Arctic Sea ice. *Earth's Future, 2*, 2014EF000240.

O'Donnell, B., White, J. R., & Holding, S. R. (1994). Molecular weight measurement in weathered polymers. *Journal of Applied Polymer Science, 52*, 1607–1618.

Ogata, Y., Takada, H., Mizukawa, K., Hirai, H., Iwasa, S., Endo, S., et al. (2009). International pellet watch: Global monitoring of persistent organic pollutants (POPs) in coastal waters. Initial phase data on PCBs, DDTs, and HCHs1. *Marine Pollution Bulletin, 58*, 1437–1446.

Ojeda, T. (2011). Degradability of linear polyolefins under natural weathering. *Polymer Degradation and Stability, 96*, 703–707.

Pegram, J. E., & Andrady, A. L. (1989). Outdoor weathering of selected polymeric materials under marine exposure conditions. *Polymer Degradation and Stability, 26*, 333–345.

Shimao, M. (2001). Biodegradation of plastics. *Current Opinion in Biotechnology, 12*, 242–247.

Singh, B., & Sharma, N. (2008). Mechanistic implications of plastic degradation. *Polymer Degradation and Stability, 93*, 561–584.

Song, Y., Hong, S., Jang, M., Kang, J., Kwon, O., Han, G., et al. (2014). Large accumulation of micro-sized synthetic polymer particles in the sea surface microlayer. *Environmental Science and Technology, 48*, 9014–9021.

Spengler, A., & Costa, M. F. (2008). Methods applied in studies of benthic marine debris. *Marine Pollution Bulletin, 56*, 226–230.

Sudhakar, M., Trishul, A., Doble, M., Suresh Kumar, K., Syed Jahan, S., Inbakandan, D., et al. (2007). Biofouling and biodegradation of polyolefins in ocean waters. *Polymer Degradation and Stability, 92*, 1743–1752.

Thangavelu, M., Adithan, A., Karunamoorthy, L., Ramasamy, V., Loganathan, V., & Mukesh, D. (2011). Fouling and stability of polymers and composites in marine environment. *International Biodeterioration and Biodegradation, 65*, 276–284.

Thompson, R. C. (2015). Microplastics in the marine environment: Sources, consequences and solutions. In M. Bergmann, L. Gutow & M. Klages (Eds.), *Marine anthropogenic litter* (pp. 185–200). Berlin: Springer.

Tocháček, J., & Vrátníčková, Z. (2014). Polymer life-time prediction: The role of temperature in UV accelerated ageing of polypropylene and its copolymers. *Polymer Testing, 36*, 82–87.

Tokiwa, Y., Calabia, B. P., Ugwu, C. U., & Aiba, S. (2009). Biodegradability of plastics. *International Journal of Molecular Sciences, 10*, 3722–3742.

Watters, D. L., Yoklavich, M. M., Love, M. S., & Schroeder, D. M. (2010). Assessing marine debris in deep seafloor habitats off California. *Marine Pollution Bulletin, 60*, 131–138.

Yakimets, I., Lai, D., & Guigon, M. (2004). Effect of photooxidation cracks on behaviour of thick polypropylene samples. *Polymer Degradation and Stability, 86*, 59–67.

Ye, S., & Andrady, A. L. (1991). Fouling of floating plastic debris under Biscayne Bay exposure conditions. *Marine Pollution Bulletin, 22*, 608–613.

Zettler, E. R., Mincer, T. J., & Amaral-Zettler, L. A. (2013). Life in the "plastisphere": Microbial communities on plastic marine debris. *Environmental Science and Technology, 47*, 7137–7146.

Part II
Biological Implications
of Marine Litter

Chapter 4
Deleterious Effects of Litter on Marine Life

Susanne Kühn, Elisa L. Bravo Rebolledo and Jan A. van Franeker

Abstract In this review we report new findings concerning interaction between marine debris and wildlife. Deleterious effects and consequences of entanglement, consumption and smothering are highlighted and discussed. The number of species known to have been affected by either entanglement or ingestion of plastic debris has doubled since 1997, from 267 to 557 species among all groups of wildlife. For marine turtles the number of affected species increased from 86 to 100 % (now 7 of 7 species), for marine mammals from 43 to 66 % (now 81 of 123 species) and for seabirds from 44 to 50 % of species (now 203 of 406 species). Strong increases in records were also listed for fish and invertebrates, groups that were previously not considered in detail. In future records of interactions between marine debris and wildlife we recommend to focus on standardized data on frequency of occurrence and quantities of debris ingested. In combination with dedicated impact studies in the wild or experiments, this will allow more detailed assessments of the deleterious effects of marine debris on individuals and populations.

Keywords Marine debris · Plastic litter · Entanglement · Ingestion · Harm · Review

Electronic supplementary material The online version of this chapter (doi:10.1007/978-3-319-16510-3_4) contains supplementary material, which is available to authorized users.

S. Kühn (✉) · E.L. Bravo Rebolledo · J.A. van Franeker
IMARES Wageningen UR, P.O. Box 167, 1790 AD Den Burg, Texel, The Netherlands
e-mail: susanne.kuehn@wur.nl

4.1 Introduction

For several decades, it has been known that anthropogenic debris in the marine environment, in particular plastic, affects marine organisms (Shomura and Yoshida 1985; Laist 1997; Derraik 2002; Katsanevakis 2008). Plastic production grows at 5 % per year (Andrady and Neal 2009). Part of the material ends up as litter in the marine environment, to such an extent that the issue is considered to be of major global concern (UNEP 2011). Awareness has grown that plastics may become less visible but do not really disappear as they become fragmented into small persistent particles ('plastic soup') (Andrady 2015). Plastic fragmentation can be caused by abiotic factors (Andrady 2011) or through animal digestion processes (Van Franeker et al. 2011). The smaller the particle, the higher the availability to animals at the base of the food chain. The potential deleterious effects from ingestion, have heightened the urgency to evaluate the impact of plastics on the whole marine food chain and, ultimately, the consequences for humans as end consumers (Koch and Calafat 2009; UNEP 2011; Galloway 2015).

The most visible effect of plastic pollution on marine organisms concerns wildlife entanglement in marine debris, often in discarded or lost fishing gear and ropes (Laist 1997; Baulch and Perry 2014). Entangled biota are hindered in their ability to move, feed and breathe. In addition, many marine organisms mistake litter for food and ingest it (Day et al. 1985; Laist 1997). Indigestible debris such as plastics may accumulate in their stomachs and affect individual fitness, with consequences for reproduction and survival, even if not causing direct mortality (Van Franeker 1985; Bjorndal et al. 1994; McCauley and Bjorndal 1999). Marine birds, turtles and mammals have received most attention, but the consequences of entanglement and ingestion on other organism groups, e.g. fish and invertebrates, are becoming more evident. In addition to the issues of entanglement and ingestion, synthetic materials represent a long-lived substrate that may present the possibility of transporting hitch-hiking 'alien' species horizontally to ecosystems elsewhere (for more details see Kiessling et al. 2015) or vertically from the sea surface through the water column to the seafloor. Plastics may also smother water surfaces and sea bottoms where effects may range from suffocating organisms (e.g. Mordecai et al. 2011; Green et al. 2015) to offering new habitats for species that are otherwise unable to settle (e.g. Chapman and Clynick 2006).

Major reviews of the impacts of litter, in particular plastics, on marine life have been undertaken by Shomura and Yoshida (1985), Laist (1997), Derraik (2002) and Katsanevakis (2008). We used the species list of Laist (1997) as a basis for our work and conducted an extensive literature review to add not only birds and mammals, but also fish and invertebrates. Laist (1997) tabulated data on both entanglement and ingestion but focused discussions on the entanglement aspect. Therefore, we paid more attention to descriptions and discussion of the ingestion issue. This includes occurrence of smaller plastics in smaller organisms, including invertebrates but leaves the real microplastic issues to the dedicated chapter in this book (Lusher 2015). The table with species listings for ingestion and entanglement starts with marine birds and mammals because for these animal groups, literature coverage is far more complete than for lower taxonomic groups, and because this is directly comparable with Laist (1997). Further taxonomic groups are in traditional taxonomic sequence.

Tables 4.1, 4.2 and 4.3 summarise our findings on entanglement and ingestion for groups of species in comparison to the earlier review by Laist (1997). Table 4.4 gives a more specific overview of our findings, but all details for individual species and data sources are provided in our Online Supplement. Data in our tables only relate to observations on wild organisms. This excludes for example fisheries by-catch data for active fishing gear and laboratory experiments. Texts refer to these only where it does not overlap too much with the microplastics chapters in this book and the review by Cole et al. (2011). The main aim of our paper was to compile a factual overview of known records of interference of plastic debris with marine wildlife as a basis for current discussions and future work addressing the scale of impact and policies to be developed.

4.2 Entanglement

Entanglement of marine life occurs all over the world, from whales in the Arctic (Knowlton et al. 2012) and fur seals in the Southern Ocean (Waluda and Staniland 2013), to gannets in Spain (Rodríguez et al. 2013), octopuses in Japan (Matsuoka et al. 2005) and crabs in Virginia, USA (Bilkovic et al. 2014). One of the first entanglement records of marine debris was probably a shark, caught in a rubber automobile tyre in 1931 (Gudger and Hoffman 1931). Hundreds of thousands of marine birds and mammals are known to perish in active fishing gear (Read et al. 2006; Žydelis et al. 2013), but no estimates are available for the actual number of animals becoming entangled in synthetic fisheries debris and other litter. However, from species records in Table 4.1 and the Online Supplement, it appears that the problem is substantial. The percentage of species that have been recorded as entangled among various groups of marine organisms, is high: 100 % of marine turtles (7 of 7 species), 67 % of seals (22 of 33 species), 31 % of whales (25 of 80 species) and 25 % of seabirds (103 of 406). In comparison to the listings by Laist (1997) the number of bird + turtle + mammal species with known entanglement increased from 89 (21 %) to 161 (30 %) (Table 4.1). For other reptiles, fish and invertebrates the percentage of affected species is futile because there are many thousands of species which have not been properly investigated. Often it is considered less worthwhile to publish individual entanglement records for common fishes or invertebrates or inconspicuous small species, than, for example, for a large entangled whale washed ashore.

Temporal entanglement trends are difficult to establish, as they differ between species groups and population changes play an important role (Ryan et al. 2009). Fowler et al. (1992) found a decline in entanglement of northern fur seals (*Callorhinus ursinus*) from 1975 to 1992. In Antarctic fur seals (*Arctocephalus gazella*), Waluda and Staniland (2013) reported a peak in 1994 and then a decrease

Table 4.1 Number of species with documented records of entanglement in marine debris

Species group	Laist (1997)			This study		
	Spp. total	Entanglement		Spp. total	Entanglement	
	(n)	(n)	(%)	(n)	(n)	(%)
Seabirds	**312**	51	16	**406**	103	25.4
Anseriformes (marine ducks)	–	–	–	**13**	5	38.5
Gaviiformes (divers)	–	–	–	**5**	3	60.0
Sphenisciformes (penguins)	**16**	6	38	**18**	6	33.3
Procellariiformes (tubenoses)	**99**	10	10	**141**	24	17.0
Podicipediformes (grebes)	**19**	2	10	**23**	6	26.1
Pelecaniformes, suliformes, phaethontiformes (pelicans, gannets and boobies, tropicbirds)	**51**	11	22	**67**	20	29.9
Charadriiformes (gulls, skuas, terns and auks)	**122**	22	18	**139**	39	28.1
Marine mammals	**115**	32	28	**123**	51	41.5
Mysticeti (baleen whales)	**10**	6	60	**13**	9	69.2
Odontoceti (toothed whales)	**65**	5	8	**65**	16	24.6
Phocidae (true seals)	**19**	8	42	**19**	9	47.4
Otariidae (eared seals)	**14**	11	79	**13**	13	100.0
Sirenia (sea cows, dugongs)	**4**	1	25	**5**	2	40.0
Mustelidae (otters)	**1**	1	100	**2**	1	50.0
Ursidae (polar bears)	**0**	0	0	**1**	1	100.0
Turtles	**7**	6	86	**7**	7	100
Sea snakes	–	–	–	**62**	2	3.2
Fishes	–	34	–	**32,554**	89	0.27
Invertebrates	–	8	–	**159,000**	92	0.06
Marine birds, mammals and turtles	**434**	**89**	**20.5**	**536**	**161**	**30.0**
All species	–	**136**	–		**344**	

Comparative summary with the earlier major review by Laist (1997). Individual species and sources are documented in the Online Supplement. Observations only concern dead or living animals entangled in marine debris including derelict fishing gear. Between the two reviews, the number of species, in the groups considered, differ because of changes in accepted taxonomic status, and selection of which species groups should be considered to be 'marine'. For details see the Online Supplement

until 2012. In the same period of time (1978–2000), Cliff et al. (2002) found an increase in entanglement rates of dusky sharks (*Carcharhinus obscurus*).

4.2.1 Ways of Entanglement

The term "ghost fishing" has been established for lost or abandoned fishing gear (Breen 1990). Ghost nets may continue to trap and kill organisms and can damage benthic habitats (Pawson 2003; Good et al. 2010). Important factors, increasing the risks of entanglement, are the size and structure (Sancho et al. 2003) of the lost nets and their location. For example, nets that are stretched open by structures on the sea bed, tend to catch more organisms (Good et al. 2010). The estimated time, over which lost fishing gear continues to entangle and kill organisms varies substantially and is site and gear specific (Kaiser et al. 1996; Erzini 1997; Hébert et al. 2001; Humborstad et al. 2003; Revill and Dunlin 2003; Sancho et al. 2003; Tschernij and Larsson 2003; Matsuoka et al. 2005; Erzini et al. 2008; Newman et al. 2011). Matsuoka et al. (2005) estimated catch durations of derelict gill- and trammel-nets from different studies between 30 and 568 days. Ghost-fishing efficiency can sometimes decrease exponentially (Erzini 1997; Tschernij and Larsson 2003; Ayaz et al. 2006; Baeta et al. 2009). For example, Tschernij and Larsson (2003) found 80 % of the catch in bottom gill nets in the Baltic Sea during the first three months. Still, the nets continued fishing at a low rate until the end of the experiment after 27 months. Lost fishing gear can carry on trapping, until it is heavily colonised, altering weight, mesh size and visibility (Erzini 1997; Humborstad et al. 2003; Sancho et al. 2003). In deeper waters, ghost fishing seems to continue for longer periods of time, as fouling takes longer (Breen 1990; Humborstad et al. 2003; Large et al. 2009). A reduction of the duration of ghost fishing by using degradable materials unfortunately also affects the operational lifetime of equipment. However, easily replaced degradable escape cords in lobster traps may reduce ghost fishing of lost traps efficiently (Antonelis et al. 2011).

In addition to entanglement in derelict fishing gear, other anthropogenic material such as ropes, balloons, plastic bags, sheets and six-pack drink holders can cause entanglement (e.g. Plotkin and Amos 1990; Norman et al. 1995; Camphuysen 2001; Matsuoka et al. 2005; Gomerčić et al. 2009; Votier et al. 2011; Bond et al. 2012; Moore et al. 2009, 2013; Rodríguez et al. 2013).

Whales and dolphins tend to become entangled around their neck, flippers and flukes, often in several types of fishing gear (Moore et al. 2013; Van der Hoop et al. 2013). Seals become frequently entangled in synthetic fishing gear, packing straps or other loop-shaped items that encircle the neck at young age and create problems during growth (Fowler 1987; Lucas 1992; Allen et al. 2012) (see Fig. 4.2). Seabirds are well known to become entangled around the bill, wings and feet with rope-like materials, which constrains their ability to fly or forage properly (Camphuysen 2001; Rodríguez et al. 2013) (Fig. 4.1). In addition to entanglement in fishing gear and other debris (Bugoni et al. 2001) marine turtles face problems on beaches where hatchlings are prone to entanglement or entrapment in marine debris on their way to the sea (Kasparek 1995; Ozdilek et al. 2006; Triessing et al. 2012). Motile benthic organisms become primarily caught in derelict traps on the seafloor (Adey et al. 2008; Erzini et al. 2008; Antonelis et al. 2011; Anderson and Alford 2014; Bilkovic et al. 2014; Kim et al. 2014; Uhrin et al. 2014) (Fig. 4.3a) although sometimes escape has also been observed (Parrish and Kazama 1992; Godøy et al. 2003). If there is no possibility of escape, animals in these traps and pots die from starvation (Pecci et al. 1978) and serve as bait, which attracts new victims (Kaiser et al. 1996; Stevens et al. 2000; Hébert et al. 2001).

Fig. 4.1 Northern gannet entanglement. On a nest on Helgoland, Germany (*top*), on a beach on Texel, The Netherlands (*bottom left*) and with fishing nets wrapped around the neck (*bottom right*) (Photos: J.A. van Franeker (1, 2) and S. Kühn (3), IMARES)

Behavioural traits can be important factors in becoming entangled (Shaughnessy 1985; Woodley and Lavigne 1991). It has been suggested that sharks become entangled when investigating large floating items and when searching for food associated with clumps of lost fishing gear (Bird 1978). Prey fish, which use debris as a shelter, can increase entanglement risks for predators, such as sharks (Cliff et al. 2002) and fish (Tschernij and Larsson 2003). The 'playful' behaviour of marine mammals may increase the risk of entanglement (Mattlin and Cawthorn 1986; Laist 1987; Harcourt et al. 1994; Zavala-González and Mellink 1997; Hanni and Pyle 2000; Page et al. 2004). Zavala-González and Mellink (1997) and Hanni and Pyle (2000) explained a higher incidence of entanglement

Fig. 4.2 Marine Mammal entanglement and plastic ingestion. Stomach contents of Dutch harbour seals (*top*), entangled grey seal (*bottom left*) and harbor seal (Texel, The Netherlands, *bottom center*), Antarctic fur seal investigating a rope (Cape Shirreff, Antarctica, *bottom right*) (Photos: J.A. van Franeker (1, 2, 3) and E. Bravo Rebolledo (6) IMARES; S. de Wolf (4, 5), Ecomare)

in younger California sea lions (*Zalophus californianus*) by playful behaviour and curiosity in combination with lack of experience and a foraging habit closer to the water surface. Age plays a significant role in pinnipeds, as younger seals are more often entangled than adults (Lucas 1992; Henderson 2001; Hofmeyr et al. 2006).

Gannets and many other seabird species use seaweed to build their nests, but are known to frequently incorporate ropes, nets and other anthropogenic debris (Podolski and Kress 1989; Montevecchi 1991; Hartwig et al. 2007; Votier et al. 2011; Bond et al. 2012; Lavers et al. 2013; Verlis et al. 2014) (Fig. 4.1). Marine debris used in nest construction increases the risk of mortal entanglement for both adult birds and chicks (Fig. 4.1). In three of the six North American gannet populations, close to 75 % of the gannet nests contained fishing debris. Its frequency can be linked to the level of gillnet fishing effort in the waters around the colonies (Bond et al. 2012).

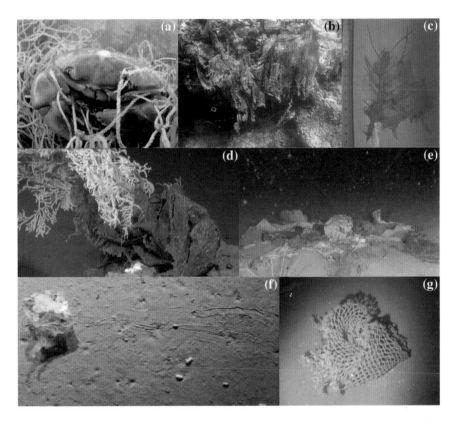

Fig. 4.3 Effects of litter on organisms on the seafloor. **a** Crab entangled in derelict net and **b** fishing net wrapped around coral, NW Hawaiian Islands (Photo: NOAA); **c** plastic fragment entangled in trawled sponge (*Cladorhiza gelida*) from HAUSGARTEN observatory (Arctic), 2,500 m depth (Photo: M. Bergmann, AWI); **d** rubbish bag wrapped around deep-sea gorgonian at 2,115 m depth in Astoria Canyon (Photo: © 2007, MBARI); **e** Mediterranean soft-sediment habitat at 450 m depth smothered with plastic litter (Photo: F. Galgani, AAMP); **f** evidence of plastic fragment causing disturbance and biogeochemical changes at the sediment-water interface by dragging along the seabed of the Molloy Deep, HAUSGARTEN IX, at 5,500 m depth (Photo: M. Bergmann, AWI); **g** cargo net entangled in a deep-water coral colony at 950 m in Darwin Mounds province with entrapped biota (Photo: V. Huvenne, National Oceanography Centre Southampton)

4.2.2 Effects of Entanglement

Entangled organisms may no longer be able to acquire food and avoid predators, or become so exhausted that they starve or drown (Laist 1997). Even if the organism does not die directly, wounds, restricted movements and reduced foraging ability will seriously affect the entangled animal (Arnould and Croxall 1995; Laist 1997; Moore et al. 2009; Allen et al. 2012). In turtles, entanglement is known to cause skin infections, amputations of legs and septic processes (Orós et al. 2005; Barreiros and Raykov 2014). Barreiros and Guerreiro (2014) reported a ring from a plastic bottle that became fixed around the operculum of a juvenile axillary sea bream (*Pagellus acarne*), which inflicted a deep cut in the anterior part of the fish and caused mortality. Discarded plastic lines and fishing gear, even if not directly drowning the animal, may cause complications in proper foraging or surfacing to breathe (Wabnitz and Nichols 2010). Illustrating the fact that such events may affect even unlikely species, the entanglement of a sea snake (*Hydrophis elegans*) in a ceramic ring caused starvation by restricting the passage of food (Udyawer et al. 2013).

In sharks, plastic entanglement reduced the mouth opening so as to impair foraging and gill ventilation (Sazima et al. 2002). A malformation of the backbone due to long-term entanglement of a shortfin mako shark (*Isurus oxyrinchus*) disturbed natural growth. In addition, biofouling on the rope probably reduced its swimming efficiency, maximum velocity and manoeuvrability (Wegner and Cartamil 2012). Lucas (1992) discovered a dead grey seal (*Halichoerus grypus*) with deformations. The size of the rubber trawl roller suggested that it had been entangled as a juvenile five years before.

Crabs, octopuses, fishes and a wide range of smaller marine biota are known to get caught in derelict traps on the seafloor and die from stress, injuries or starvation, as escape is difficult (Matsuoka 1999; Al-Masroori et al. 2004; Matsuoka et al. 2005; Erzini et al. 2008; Antonelis et al. 2011; Cho 2011). Derelict fishing lines and other gear are often covering structurally complex biota such as sponges, gorgonians (Fig. 4.3b) or (soft) corals (Pham et al. 2013; Smith and Edgar 2014) which suffer broken parts and may be more susceptible to infections and eventually die, as shown for shallow-water (soft) corals and gorgonians (Bavestrello et al. 1997; Schleyer and Tomalin 2000; Asoh et al. 2004; Yoshikawa and Asoh 2004; Chiappone et al. 2005). Contact with soft plastic litter also caused necrosis in the cold water coral *Lophelia pertusa* (Fabri et al. 2014).

Although examples of entanglement and various pathways of negative effects on individuals are abundant, it is rarely possible to assess the proportional damage to populations. However, Knowlton et al. (2012) reported that among a known number of 626 photo-identified individuals of the North Atlantic right whale (*Eubalaena glacialis*), 83 % showed evidence of entanglements in ropes and nets. On average, 26 % of adequately photographed animals acquired new wounds or scars every year. Allen et al. (2012) showed that entanglement reduces the longer-term survival of grey seals significantly. Studies like these, although not attributable with certainty to marine debris alone, do show that entanglement, although not directly obvious, can have a serious impact on wild populations.

4.3 Smothering

Marine debris on the seabed can have various effects on the resident flora and fauna that we do not consider to be 'entanglement' but rather describe as 'smothering'. Smith (2012) suggested that large quantities of litter may impede attempts to rehabilitate depleted mangrove forests in Papua New Guinea through smothering of seedlings. In the intertidal zone the weight and shading effects of debris may crush sensitive salt marsh vegetation or reduce light levels needed for growth, which can lead to denuded areas in these sensitive protected ecosystems (Uhrin and Schellinger 2011; Viehman et al. 2011). Two species of seagrass (*Thalassia testudinum, Syringodium filiforme*) had significantly decreased shoot densities after experimental deployment of traps on the sea bed (Uhrin et al. 2005). The weight of the traps caused blades to become abraded or crushed into the underlying *anoxic sediments*, likely suffocating the plants, reducing photosynthetic rates and leading to eventual senescence of above-ground biomass (Uhrin et al. 2005), which indicates long-lasting effects on ecosystem function and thus biodiversity of these vulnerable habitats.

Estimates on the impact of marine debris on local populations are available for corals: for example, Richards and Beger (2011) found that coral cover decreased significantly as macrodebris cover increased. Yoshikawa and Asoh (2004) reported that 65 % of coral colonies in Oahu, Hawaii were covered with fishing lines, and 80 % of colonies were either entirely or partially dead, which was, again, positively correlated with the percentage of colonies covered with fishing lines.

On one hand some debris may provide shelter for motile animals, and a habitat for sessile organisms, as was experimentally shown by Katsanevakis et al. (2007) and in the deep sea (Mordecai et al. 2011; Schlining et al. 2013). In the Majuro Lagoon, the coral *Porites rus* overgrew debris and appeared to thrive in locations of high debris cover (Richards and Beger 2011). On the other hand, derelict fishing gear, bags and large (agricultural) foils are known to cover parts of the seafloor at all depths (e.g. Galgani et al. 1996; Watters et al. 2010; Van Cauwenberghe et al. 2013; Pham et al. 2014) (Fig. 4.3e). Mordecai et al. (2011) reported anoxic sediments below a plastic bag on the deep seafloor of the Nazaré Canyon and suggested that this would alter the infaunal community underneath as it reduces the exchange of pore water with overlying water masses. Indeed, anoxic sediments, reduced primary productivity and organic matter and significantly lower abundances of infaunal invertebrates were recently recorded below plastic bags experimentally deployed on a beach.for 9 weeks (Green et al. 2015). Anoxic sediments below marine litter were also observed at two sites of a mangrove habitat from Papua New Guinea (Smith 2012). Dragged along the seafloor litter may cause further damage to fragile habitat engineers (coral, plants) and change biogeochemical seafloor properties (e.g. Fig. 4.3f). Macroplastics covering larger parts of corals, cannot only cause direct mechanical damage, but also diminish the capacity of phototrophic and heterotrophic nutrition (Richards and Beger 2011) (see also Fig. 4.3b, d). Also, a relationship between marine debris and coral diseases has been observed (Harrison et al. 2011). When corals die and release the debris, it moves on to a new spot and repeats the negative cycle (Donohue et al. 2001; Chiappone et al. 2005; Abu-Hilal and Al-Najjar 2009). In eastern Indonesian areas experimentally smothered by plastic, diatom densities were lower, probably due to the lack of light. However, meiofauna had a higher density beneath smothered test sites than on clean control sites, which was explained by the temporarily decomposing organic matter improving habitat quality for meiofauna (Uneputty and Evans 2009). Smothering may also limit the nutrition of filter feeders as it may restrict water circulation and thereby particles reaching the feeding apparatus (see Fig. 4.3b, c, d). In addition, marine debris on beaches can have negative effects on marine biota. Kasparek (1995) found marine turtle nesting sites on beaches in Syria, where the beaches were so polluted that females might not be able to dig a nest at an appropriate site. Litter may also lead to behavioural changes: for example, prolonged food searching time and increased self-burial in intertidal snails (*Nassarius pullus*) is strongly correlated with increased plastic cover, which was also reflected in low snail densities in areas of high litter cover (Aloy et al. 2011). Twenty-two taxa that are affected by smothering with litter are listed in Online Supplement 2 (provided by M. Bergmann) including four grasses, two types of sponges, 14 cnidarian species and one mollusc and crustacean.

4.4 Ingestion of Plastic

Ingestion of plastic by marine organisms is less visible than entanglement. Table 4.2 and Online Supplement 1 show that ingestion of plastic debris has currently been documented for 100 % of marine turtles (7 of 7 species), 59 % of whales (47 of 80), 36 % of seals (12 of 33), and 40 % of seabirds (164 of 406). In comparison to the review by Laist (1997) the number of bird + turtle + mammal species with known ingestion of plastics increased from 143 (33 %) to 233 (44 %). Studies on the ingestion of plastics by fish and invertebrates are largely a recent development. Currently, low proportions of fish and invertebrate species are presented in the tables, but a rapid increase in publications and species numbers are expected in this currently dynamic field of research. Records of plastic ingestion date back to the early days of plastic production in the 1960s. One of the first birds recorded to contain plastic was Leach's storm petrel *(Oceanodroma leucorhoa)* off New Foundland in 1962 (Rothstein 1973). The first report of a leatherback turtle *(Dermochelys coriacea)* with plastic dates back to 1968 (Mrosovsky et al. 2009). While the first record of anthropogenic debris in sperm whales *(Physeter macrocephalus)* was a fish hook found in a stomach in 1895, the first report of ingested plastic in sperm whales dates back to 1979 (de Stephanis et al. 2013). The first fish feeding on plastic was published in 1972 (Carpenter et al. 1972). The ingestion of plastics became a more commonly reported phenomenon from the 1970s onwards (Kenyon and Kridler 1969; Crockett and Reed 1976; Bourne and Imber 1982; Furness 1983; Day et al. 1985). A trend for birds ingesting plastic was probably first noted by Harper and Fowler (1987). Between 1958 and 1959 they found no plastic in prions *(Pachyptila* spp.) but from then on there was an upward trend in plastic consumption until 1977. A peak of plastic ingestion was detected in 1985 and 1995 in a number of long-term studies (Moser and Lee 1992; Robards et al. 1995; Spear et al. 1995; Mrosovsky et al. 2009; Van Franeker et al. 2011). In contrast to the continuing growth of global plastic use and increase in marine activities, the trend of plastic consumption decreased and stabilized from 2000 onwards approaching the 1980s level (Mrosovsky et al. 2009; Van Franeker et al.

Table 4.2 Number of species with documented records of ingestion of marine debris

Species group	Laist (1997)			This study		
	Spp. total	Ingestion		Spp. total	Ingestion	
	(n)	(n)	(%)	(n)	(n)	(%)
Seabirds	**312**	111	36	**406**	164	40.4
Anseriformes (marine ducks)	–	–	–	13	1	7.7
Gaviiformes (divers)	–	–	–	5	3	60.0
Sphenisciformes (penguins)	**16**	1	6	18	5	27.8
Procellariiformes (tubenoses)	**99**	62	63	141	84	59.6
Podicipediformes (grebes)	**19**	0	0	23	0	0.0
Pelecaniformes, suliformes, phaethontiformes (pelicans, gannets and boobies, tropicbirds)	**51**	8	16	67	16	23.9
Charadriiformes (gulls, skuas, terns and auks)	**122**	40	33	139	55	39.6
Marine mammals	**115**	26	23	123	62	50.4
Mysticeti (baleen whales)	**10**	2	20	13	7	53.8
Odontoceti (toothed whales)	**65**	21	32	65	40	61.5
Phocidae (true seals)	**19**	1	5	19	4	21.1
Otariidae (eared seals)	**14**	1	7	13	8	61.5
Sirenia (sea cows, dugongs)	**4**	1	25	5	3	60.0
Mustelidae (otters)	**1**	0	0	2	0	0.0
Ursidae (polar bears)	**0**	0	0	0	0	0.0
Turtles	7	6	86	7	7	100
Sea snakes	–	–	–	62	0	0.0
Fish	–	33	–	**32,554**	92	0.28
Invertebrates	–	1	–	**159,000**	6	0.004
Marine birds, mammals and turtles	**434**	143	32.9	536	233	**43.5**
All species		177			331	

Comparative summary with the earlier major review by Laist (1997). Individual species and sources are documented in Online Supplement. Observations only concern non-simulated dead or living wild animals found in their natural habitat. We thus exclude experimental studies showing the potential for ingestion by marine species. Between the two reviews, the number of species in the groups considered, differ because of changes in accepted taxonomic status, and selection of which species groups should be considered to be 'marine'. For details see the Online Supplement

2011; Bond et al. 2013). Figure 4.4 illustrates the ingestion of plastic by northern fulmars.

4.4.1 Ways of Plastic Ingestion

Plastics may be ingested intentionally or accidentally and both pathways deserve further discussion.

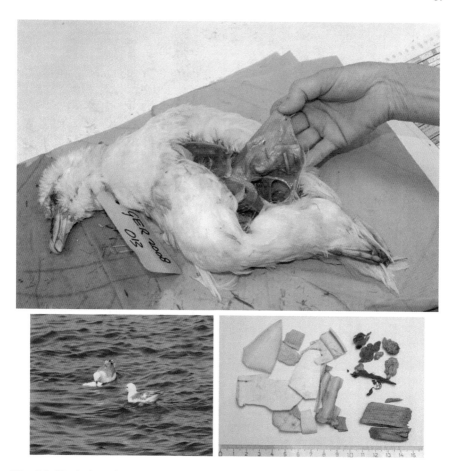

Fig. 4.4 Plastic ingestion by northern fulmars (*Fulmarus glacialis*). Unopened stomach with plastic inside (*top*), fulmars at sea chewing on a plastic fragment (*bottom left*), stomach content of a northern fulmar with fragments, foam, sheets and wood (*bottom right*) (Photos: J.A. van Franeker (1, 3) and S. Kühn (2), IMARES)

4.4.1.1 Intentional Ingestion

Why some animals intentionally ingest plastic debris may depend on a range of factors, and these may vary among different animal groups. Although many of these factors interact, it is useful to review at least some of them separately.

Foraging Strategy

In seabirds, plastic ingestion has been linked to foraging strategy by several authors (e.g. Day et al. 1985; Azzarello and Van Vleet 1987; Ryan 1987; Tourinho et al. 2010.) From their study on many different seabird species, Day et al. (1985)

concluded, that pursuit-diving birds have the highest frequency of plastic uptake, followed by surface-seizing and dipping seabirds. Provencher et al. (2010) reported that marine birds, feeding on crustaceans and cephalopods had ingested more plastic than piscivorous seabirds, and those omnivores are most likely to confuse prey and plastic. Seabirds with specialized diets are less likely to misidentify plastic, unless a particular type resembles their prey (Ryan 1987). Many gull species frequent rubbish bins and landfill areas, in addition to foraging in marine habitats and seem prone to ingest debris. However, ingested debris does not often show up in their stomachs during dissections because they clear them daily by regurgitating hard prey remains (Hays and Cormons 1974; Ryan and Fraser 1988; Lindborg et al. 2012). As regurgitation takes place regularly, plastics quantified from boluses reflect the ingestion of the very last period, rather than accumulated debris (Camphuysen et al. 2008; Ceccarelli 2009; Codina-García et al. 2013; Hong et al. 2013). Tubenosed seabirds mostly retain plastic and hard prey items (Mallory 2006) because they possess two stomachs with a constriction (*Isthmus gastris*) between the glandular proventriculus and the muscular gizzard (Furness 1985; Ryan and Jackson 1986). Even when spitting stomach oil to defend themselves or when feeding their chicks, only plastics from the proventriculus are regurgitated but items from the gizzard are retained (Rothstein 1973). Marine turtles frequently ingest plastic bags as they may mistake them for jellyfish, a common component of their diet (Carr 1987; Lutz 1990; Mrosovsky et al. 2009; Tourinho et al. 2010; Townsend 2011; Campani et al. 2013; Schuyler et al. 2014). While accidental plastic ingestion by filter-feeding baleen whales (Mysticeti) might be assumed to be common, Walker and Coe (1990) expected that toothed whales (Odontoceti) would have a low rate of plastic ingestion because they use echolocation or visual cues to locate their prey. However, Laist (1997), Simmonds (2012) and Baulch and Perry (2014) all made extensive descriptions of toothed whales that had ingested plastic. Indeed, our updated literature search showed that 54 and 62 % of the baleen and toothed whales, respectively, ingest plastics. It has also been suggested, that marine mammals could see plastic as a curiosity and while investigating it, they swallow it or become entrapped (Mattlin and Cawthorn 1986; Laist 1987). Large predatory fishes and birds are known to frequently inspect plastic debris and take bites out of larger plastic items. Cadée (2002) observed that 80 % of foamed plastic debris on the Dutch coast showed peckmarks of birds and suggested that the birds mistake polystyrene foam for cuttlebones or other food. Carson et al. (2013) observed bite marks of sharks or large predatory fishes on 16 % of plastic debris beached on Hawaii indicating 'testing' of materials. Choy and Drazen (2013) showed that among 595 individuals of seven such large predatory fish species, 19 % of individuals (range per species <1–58 %) had actually ingested plastic. Foraging strategies may vary under different conditions of food availability. Duguy et al. (2000) considered that decreased availability of jellyfish during winter could be the reason for the higher incidence of plastic bags during these months in the diet of turtles.

In conclusion it seems that although indiscriminate omnivorous predators or filter feeders appear most prone to plastic ingestion, there are many examples of ingestion among species with specialized foraging techniques and specific prey selection.

Color

One of the factors often considered to influence the consumption of marine debris is color as specific colors might attract predators when resembling the color of their prey. In seabirds, this has been suggested for e.g. greater shearwaters (*Puffinus gravis*) and red phalaropes *(Phalaropus fulicarius)* (Moser and Lee 1992). Parakeet auklets (*Aethia psittacula*) on the Alaskan coast, feeding naturally primarily on light-brown crustaceans, consumed mainly darker plastic granules, suggesting they were mistaken for food items (Day et al. 1985). In studies of marine turtles, the issue of color preference is controversial. Lutz (1990) indicated no preferential ingestion of different plastic colors; neither did Campani et al. (2013) in loggerhead turtles. However, others find light-colored and translucent plastics are most commonly ingested, suggesting similarity to their jellyfish prey (Bugoni et al. 2001; Tourinho et al. 2010). Schuyler et al. (2014) indicated such prey-similarity by the combination of translucency and flexibility of plastic bags and found that blue-colored items were less frequently eaten probably because of lower detection rates in open water. An additional visual factor could be shape as floating plastic bags resemble jellyfish. In fur seal scats, the colors of plastic were white, brown, blue, green and yellow (Eriksson and Burton 2003), however, no clear preference was evident.

White, clear, and blue plastics were primarily ingested by planktivorous fish from the North Pacific central gyre but similar color proportions were recorded from neuston samples (Boerger et al. 2010). By contrast, black particles were most prevalent in stomachs of fish from the English Channel but this study included both pelagic and demersal fish (Lusher et al. 2013). While two mesopelagic fish (*Lampris* spp.) species did not favour particular colors *Alepisaurus ferox* seemed to favour white and clear plastic pieces, which may resemble their gelatinous prey (Choy and Drazen 2013). The majority of strands reported from the intestines of Norway lobsters (*Nephrops norvegicus*) were also transparent (Murray and Cowie 2011).

Studies on the color-specific uptake often do not take into account that color may change in the gastrointestinal tract (e.g. Eriksson and Burton 2003). Also, there are rarely quantitative data on the abundance of various color categories in the foraging ranges of the species studied. In general, light colors seem to be most common in floating marine debris ranging from 94 % of the abundance in the Sargasso Sea (Carpenter et al. 1972) and 82–89 % in the South Atlantic (Ryan 1987) to 72 % in the North Pacific (Day et al. 1985). The frequently observed prevalence of translucent or brightly colored objects in stomachs may thus reflect the availability of such items the ambient environment rather than color selectivity.

Age

Among seabirds, it has been well-established that younger northern fulmars have more plastic in their stomachs than adults (Day et al. 1985; Van Franeker et al. 2011). The same has been shown for flesh-footed and short-tailed shearwater (*Puffinus carneipes* and *P. tenuirostris*, respectively, Hutton et al. 2008; Acampora et al. 2014). The chicks of Laysan albatrosses (*Phoebastria immutabilis*) at colonies (Auman et al. 1997) have a much higher load of plastic than adults at sea (e.g. Gray et al. 2012). In marine turtles, Plotkin and Amos (1990) found a decreasing trend in plastic consumption with age and attributed this to the fact that young turtles linger along drift-lines, where plastic accumulates. However, in the Adriatic Sea no clear age or size-related differences were apparent in loggerhead sea turtles (*Caretta caretta*) (Lazar and Gracan 2011; Campani et al. 2013). Schuyler et al. (2013) concludes that turtles ingest most debris during their younger oceanic life stages. Significantly higher levels of plastics were recorded in younger franciscana dolphins (*Pontoporia blainvillei*) off the Argentinian coast (Denuncio et al. 2011). Younger harbour seals (*Phoca vitulina*) in the Netherlands had significantly more plastic in their stomach than older ones (Bravo Rebolledo et al. 2013) (illustrated by Fig. 4.2). There were no differences in the plastic consumption of different age classes of cat fishes (Ariidae) from a Brazilian estuary (Possatto et al. 2011). Similarly, there was no relationship between ingested litter mass and sex, maturity and body length in deep-water blackmouth catsharks (*Galeus melastomus*, Anastasopoulou et al. 2013). By contrast, the mean number of plastic items ingested by planktivorous fish from the North Pacific gyre increased as the size of fish increased, reaching a maximum of seven pieces per fish for the 7-cm size class (Boerger et al. 2010). However, this may also be explained by higher plastic uptake of larger individuals during the capture process in the codend (Davison and Asch 2011). Larger individuals of the Norway lobster had fewer plastic threads in their intestines indicating higher ingestion rates of smaller/younger animals (Murray and Cowie 2011) that also have higher incidence of infaunal prey such as polychaetes (Wieczorek et al. 1999).

In summary, it seems that where age differences were shown, younger animals are most affected. The reasons for this are not clear. In seabirds, this could partly be explained by parental delivery of food by regurgitation to chicks at the nest. In such chicks, elevated loads of plastic could be the consequence of being fed by two parents, each transferring much of its own plastic load, which has accumulated in the proventricular stomach over an extended period of time before breeding. In addition, a less developed grinding action in the gizzards of young birds could slow the mechanical break-down of plastic and removal through the intestines. Some species of albatross and shearwater chicks may lose an excess load of plastic by regurgitating proventricular stomach contents prior to fledging (Auman et al. 1997; Hutton et al. 2008). However, in fulmars the high level of plastic persists in immature birds and only gradually disappears after several years (Jensen 2012) and thus cannot be completely explained by parental feeding and stomach functioning. Perhaps, young animals are less efficient at foraging, and therefore less specific in their prey selection (Day et al. 1985; Baird and Hooker 2000; Denuncio et al. 2011). One important open question therefore is whether higher loads of plastic in younger animals reflect a learning process or mortality of those individuals that ingested too much plastic. Both explanations are speculative, but the latter suggests serious deleterious effects at the population level.

Sex

To date, there is no evidence that sex affects plastic ingestion. Studies that specifically evaluated male and female ingestion, found no significant differences in the plastic load (e.g. Day et al. 1985; Van Franeker and Meijboom 2002; Lazar and Gracan 2011; Murray and Cowie 2011; Anastasopoulou et al. 2013; Bravo Rebolledo et al. 2013). However, species showing strong sexual dimorphy or sex-dependent foraging ranges or winter distributions may show sex-specific uptake rates.

4.4.1.2 Accidental and Secondary Ingestion

Filter-feeding marine organisms, ranging in size from small crustaceans, to shell-fish, fish, some seabirds (prions, *Pachyptila* spp.) and ultimately large baleen whales may be prone to plastic ingestion. These species obtain their nutrition by filtering large volumes of water, which may contain debris in addition to the targeted food source. Although non-food items can be ejected before passage into the digestive system, this is not always the case. In their natural habitat, ingested plastics have been found in filter-feeding crustaceans such as goose barnacles (*Lepas* spp.; Goldstein and Goodwin 2013) and mussels (*Mytilus edulis*, Van Cauwenberghe et al. 2012; Leslie et al. 2013; Van Cauwenberghe and Janssen 2014). Large baleen whales have been long known to occasionally ingest debris (Laist 1997; Baulch and Perry 2014). In France, a young minke whale (*Balaenoptera acutorostrata*) beached with various plastic bags completely filling its stomachs (De Pierrepont et al. 2005). Curiously, we have found no record of plastic ingestion by obligate filter-feeding large fish such as basking shark (*Cetorhinus maximus*) or manta ray (*Manta birostris*). Some bony fish species partially use filter-feeding, but also directional feeding making it difficult to assign the pathway of debris ingestion. Uptake of plastic by filter-feeding fish has been reported for herring (*Clupea harengus*) and horse mackerel (*Trachurus trachurus*) from the North Sea and English Channel (Foekema et al. 2013; Lusher et al. 2013).

Accidental ingestion of a mixture of food and debris is not restricted to filter feeders. In the Clyde Sea, 83 % of Norway lobsters (*Nephrops norvegicus)* had plastic in their stomach, which was attributed either to passive ingestion of sediment while feeding or to secondary ingestion (Murray and Cowie 2011), although it could be argued that the fibres ingested may resemble benthic polychaete prey. Plastics and other non-food items found in stomachs of harbour seals in the Netherlands were considered to have been accidentally ingested when catching prey fishes (Bravo Rebolledo et al. 2013). A similar route for plastic ingestion was proposed by Di Beneditto and Ramos (2014), who showed that plastic in franciscana dolphins was related to benthic feeding habits, in which disturbance of sediment probably induced accidental intake of plastic debris. Florida manatees (*Trichechus manatus latirostris*) may take up plastic accidentally during foraging on plants (Beck and Barros 1991). Pelagic loggerhead sea turtles may ingest plastic because they feed indiscriminately or graze on organisms settled on floating plastic (McCauley and Bjorndal 1999; Tomas et al. 2002). A special case of such accidental ingestion is known for the Laysan albatross who take up plastic particles in combination with eggs strings of flying fish. The fishes attach their eggs to floating items: previously seaweed, bits of wood or pumice, but nowadays often plastic objects (Pettit et al. 1981). This phenomenon has also been observed in loggerhead turtles. The plastic in their stomachs was sometimes covered by the eggs of the insect *Halobates micans* (Frick et al. 2009).

A final case of unintentional plastic ingestion is that of secondary ingestion, which occurs when animals feed on prey, which had already ingested debris. This may concern both prey swallowed as a whole or scavenging. In seabirds, skuas are known to forage on smaller seabirds that consume plastic (Ryan 1987). Great skuas (*Stercorarius skua*) from the South Atlantic Ocean predate several seabird species, and their regurgitated boluses showed a link with the amount of secondarily ingested plastic and their main prey species (Bourne and Imber 1982; Ryan and Fraser 1988). In the monitoring study on northern fulmars (*Fulmarus glacialis*) in the North Sea intact stomachs from scavenged fulmars or black-legged kittiwakes (*Rissa tridactyla*) were occasionally found, which contained plastic (Van Franeker et al. 2011). A spectacular example of secondary ingestions was provided by Perry et al. (2013) who reported a ball of nylon fishing line in the stomach of a little auk (*Alle alle*), that was found in the stomach of a goose fish (*Lophius americanus*). The presence of small plastic particles in the faeces of fur seals on Macquarie Island was attributed to secondary ingestion through the consumption of myctophid fishes (Eriksson and Burton 2003). High abundance of small plastics in myctophid fishes (Boerger et al. 2010; Davison and Asch 2011), in combination with the fact that this type of fish is a common prey for many larger marine predators, suggest that secondary ingestion may be more common than reported.

4.4.2 Impacts of Plastic Ingestion

Plastic ingestion may directly cause mortality or can affect animals by slower sublethal physical and chemical effects which are best considered separately.

4.4.2.1 Direct Mortality Caused by Plastic Ingestion

When the gastrointestinal tract becomes completely blocked or severely damaged ingested plastic may lead to rapid death. Even small pieces can cause the blockage of the intestines of animals, if orientated in the wrong way (Bjorndal et al. 1994). An ingested straw led to the death of a Magellanic penguin (*Spheniscus magellanicus*) by perforation of the stomach wall (Brandao et al. 2011). Other examples of lethal impacts in seabirds were provided, for example, by Kenyon and Kridler (1969), Pettit et al. (1981) and Colabuono et al. (2009). Cases of mortality among marine turtles have been reported by e.g. Bjorndal et al. (1994), Bugoni et al. (2001), Mrosovsky et al. (2009) and Tourinho et al. (2010). Unlike most birds, turtles seem to pass plastic debris easily into the gut, and therefore most plastics have been found in the intestines rather than the stomach (e.g. Bjorndal et al. 1994; Bugoni et al. 2001; Tourinho et al. 2010, Campani et al. 2013). As a consequence, physical impact in turtles may often be related to gut functioning or damage. In the Mediterranean Sea, the death of a sperm whale of 4.5 t, was attributed to 7.6 kg of plastic debris in its stomach, which was ruptured probably due to the large plastic load (de Stephanis et al. 2013). Often, it is difficult to produce evidence for causal links between ingested debris and mortality, and as a consequence, documented cases of death through plastic ingestion are rare (Sievert and Sileo 1993; Colabuono et al. 2009). A direct lethal result from ingestion probably does not occur at a frequency relevant at the population level. Indirect, sub-lethal effects are probably more relevant.

4.4.3 Indirect Physical Effects of Plastic Ingestion

Impacts that are deleterious for the individual but not directly lethal become relevant to populations if many individuals are affected. Partial blockage or moderate damage of the digestive tract in Laysan albatross chicks was not a major cause of direct mortality, but may contribute to poor nutrition or dehydration (Auman et al. 1997). Since virtually every chick in this population (frequency of occurrence: 97.6 %) had a considerable quantity of plastic in the stomach, debris ingestion must be considered a relevant factor in overall fledging success of the population. Major proportions of tubenosed seabird species and marine turtles ingest plastic on a very regular basis. This raises urgent questions concerning the cumulative physical and chemical impacts at the population level. Sub-lethal physical impacts may have various consequences.

Firstly, stomach volume occupied by debris may limit optimal food intake. For example, tubenosed seabirds have large proventricular stomachs because they depend on irregular patchy food availability. Reduced storage capacity affects optimal foraging at times when this should be possible. Partial blockage of food passage through the digestive tract may cause gradual deterioration of body condition of fish (Hoss and Settle 1990). Efficiency of digestive processes may be reduced when sheet-like plastics or fragments cover parts of the intestinal wall. Sometimes ulcerations are found on stomach walls of organisms that ingested plastic (Pettit et al. 1981; Hoss and Settle 1990). A potentially important physical impact from ingested plastics may be a feeling of satiation as receptors signal satiety to the brain and reduce the feeling of hunger (Day et al. 1985), which may reduce the drive to search for food (Hoss and Settle 1990). High volumes of plastic can reduce proventricular contraction, responsible for the stimulation of appetite (Sturkie 1976).

All these factors may lead to a deterioration of the body condition of animals. In young loggerhead turtles, McCauley and Bjorndal (1999) found experimental evidence, that volume reduction in stomachs by non-food material caused lower nutrient and energy uptake. Similarly Lutz (1990) found a negative correlation between plastic consumption and nutritional condition in experiments with green turtles (*Chelonia mydas*) and loggerhead turtles. Ryan (1988) provided evidence for a negative effect on uptake of food and growth rate among chickens (*Gallus gallus domesticus*) that had been fed plastic pellets under controlled laboratory conditions, compared to control chickens.

In many non-experimental studies, researchers have looked for correlations between plastic loads and body condition. Some seabird studies indicate negative correlations between ingested plastics and body condition (e.g. Connors and Smith 1982; Harper and Fowler 1987; Donnely-Greenan et al. 2014; Lavers et al. 2014). However, no such correlation was found by Day et al. (1985), Furness (1985), Sileo et al. (1990), Moser and Lee (1992), Van Franeker and Meijboom (2002) and Vliestra and Parga (2002). In these non-experimental studies, it is always problematic to distinguish cause and consequence: do animals increase ingestion of abnormal items such as plastics when in poor condition, or do they loose condition because of the plastic debris in their stomach? This is even more complicated because many studies are based on corpses of beached animals that often starved before being washed ashore with potentially aberrant foraging activity.

We conclude that the estimated impact from plastic ingestion on body condition is difficult to document in wild populations. However, as mentioned above, experimental studies clearly indicate that eating plastic reduces an individual's body condition. This may not be directly lethal but will translate into negative effects on average survival and reproductive success in populations in which plastic ingestion is a common phenomenon.

4.4.3.1 Chemical Effects from Plastic Ingestion

The chemical substances added during manufacture or adsorbed to plastics at sea are an additional source of concern in terms of sublethal effects. Potential chemical impacts from the ingestion of plastic are not exhaustively discussed in this chapter, as chemical transfer and impacts are discussed in more detail in the contributions by Koelmans (2015) and Rochman (2015). We would like to stress, however, that in larger organisms, plastics often have a long residence time, during which objects may be fragmented to smaller sizes due to mechanical or enzymatic digestive processes. In such conditions, the chemical additives may play a more prominent role than chemicals adsorbed to the surface. We conclude that although research to quantify body burden and consequences of plastic-derived chemicals in marine organisms is still in its infancy, there is a risk to species frequently ingesting synthetic debris. This will remain a complicated issue due to the widespread presence of many chemicals and their accumulation in marine foodwebs along routes other than plastics alone.

4.4.3.2 Chain of Impacts Related to Plastic Ingestion

By ingesting plastics, marine biota, and in particular seabirds, accidentally facilitate and catalyse the global distribution of plastic through bio-transportation. Studies of polar tubenosed seabirds returning to clean breeding areas after overwintering in more polluted regions are a good example. Similarly, Van Franeker and Bell (1988) found that cape petrels (*Daption capense*) process and excrete some 75 % of their initial plastic load by grinding particles in the gizzard during one month in Antarctica. Plastics are thus excreted as smaller particles in other places than where they were taken up and become available to other trophic levels in marine and terrestrial habitats. Similar data were obtained for northern fulmars and thick-billed murres (*Uria lomvia*) in the Canadian high Arctic (Mallory 2008; Provencher et al. 2010, Van Franeker et al. 2011). In the Antarctic, Van Franeker and Bell (1988) also found that 75 % of Wilsons storm petrel (*Oceanites oceanicus*) chicks that died before fledging had plastics in their stomachs, fed to them by their parents and now permanently deposited around Antarctic breeding colonies. Transport of materials may be considerable. Van Franeker (2011) calculated that northern fulmars in the North Sea area (plastic incidence 95 %, average number 35 plastic items, average mass 0.31 g per bird) annually reshape and redistribute ca. 630 million pieces or 6 t of plastic. As fulmars range over large areas, widespread secondary distribution of plastics will occur. Chemicals may be brought to other environments by seabirds (Blais et al. 2005)—potentially partly linked to plastics. From an average plastic mass of 10 g in healthy Laysan albatross chicks on Midway Atoll to about 20 g in chicks that died (Auman et al. 1997) it may be conservatively estimated, that this species with locally ca. 600,000 breeding pairs, annually brings ashore some 6 t of marine plastic debris. Also, some crustaceans reshape and redistribute plastics: Davidson (2012) showed that boring crustacean *Sphaeroma* sp. could release into the environment thousands of small particles per burrow. One of the open questions is how plastic items reach the deep sea despite their low density and therefore low sinking rates. Along with increased density by fouling processes (Ye and Andrady 1991) plastic may also be transported to the deep sea either through sinking of carcasses containing plastics, in marine snow (Van Cauwenberghe et al. 2013) or repackaged in the faeces of zooplankton (Cole et al. 2013) or other pelagic organisms. Vertical export may also be facilitated by migratory behaviour of mesopelagic fish in the water column, which had fed on plastic items (Choy and Drazen 2013). Thus, marine life is as a significant factor in the environmental production and redistribution of secondary microplastics.

4.4.4 Impacts from Species Dispersal

One of the potentially deleterious effects of marine debris is that it offers opportunities for the dispersal, or 'hitch hiking' of species around the world. Organisms can colonise non-degradable material and be transported by the currents and winds. Once settled in a new habitat, this can lead to massive population growth of 'alien species' that can outcompete original ecosystem components (Kiessling et al. 2015). Oceanic plastics can also provide new or increased habitat opportunities for specialized species such as ocean skaters (Goldstein et al. 2012; Majer et al. 2012) or whole pelagic or benthic communities (Goldberg 1997; Bauer et al. 2008; Zettler et al. 2013; Goldstein et al. 2014). For more details on hitch-hiking species see Kiessling et al. (2015).

4.5 Discussion

The total number of marine species with documented records of either entanglement and/or ingestion has doubled with an increase from 267 species in Laist (1997) to 557 species in this new review (Table 4.3 and Online Supplements). The increase in number of affected species is substantial in all groups. The documented impact for marine turtles increased from 86 to 100 % of species (now 7 of 7 species), for marine mammals from 43 to 66 % of species (now 81 of 123 species) and for seabirds from 44 to 50 % of species (now 203 of 406 species). Among marine mammals the percentage of affected whales increased from 37 to 68 % of species (now 54 of 80 species) and seals from 58 to 67 % of species (now 22 of 32 species) (see Table 4.3).

Laist (1997) addressed groups such as fish and invertebrates only marginally, so comparative figures in such groups (Tables 4.1, 4.2, 4.3) are currently of less use. We may have missed sources, and recently publications have been published at such high frequency that we cannot guarantee completeness as given in full in the online supplement, with derived data in Table 4.4.

We have stopped our additions to the online supplement and thus to derived tables on the 9th of December 2014. We welcome documentation on missed or new records of entanglement or ingestion for future updates. It remains important to continue such documentation of species affected by marine debris. However, given sufficient time and research effort, all species of marine organisms will get documented examples of interaction with marine debris. Any species can become the victim of entanglement. Furthermore, the filter-feeding habits of many lower trophic levels, and secondary ingestion by higher trophic levels, make it almost unavoidable that any species in the marine food web will at some stage pass at least some plastic debris through the intestinal tract.

As a consequence, to improve on current knowledge, future assessments of deleterious effects of debris on marine life require comparable standardized data on frequency of occurrence, ingestion quantification and categorisation of ingested debris. It is only through study of the various impacts (including frequency and quantity) on

Table 4.3 Number of species with documented records of entanglement in, and/or ingestion of marine debris

Species group	Laist (1997)			This study		
	Spp. total	Species affected		Spp. total	Species affected	
	(n)	(n)	(%)	(n)	(n)	(%)
Seabirds	314	138	43.9	406	203	50.0
Anseriformes (marine ducks)	–	1	–	**13**	5	38.5
Gaviiformes (divers)	–	–	–	**5**	4	80.0
Sphenisciformes (penguins)	**16**	6	38.0	**18**	9	50.0
Procellariiformes (tubenoses)	**99**	63	64.0	**141**	85	60.3
Podicipediformes (grebes)	**19**	2	10.0	**23**	6	26.1
Pelecaniformes, suliformes, phaethontiformes (pelicans, gannets and boobies, tropicbirds)	**51**	17	33.3	**67**	27	40.3
Charadriiformes (gulls, skuas, terns, auks)	**122**	50	41.0	**139**	67	48.2
Marine mammals	115	49	43	123	81	65.9
Mysticeti (baleen whales)	**10**	6	60.0	**13**	10	76.9
Odontoceti (toothed whales)	**65**	22	34.0	**65**	44	66.2
Phocidae (true seals)	19	8	42.0	**19**	9	42.1
Otariidae (eared seals)	14	11	79.0	**13**	13	100.0
Sirenia (sea cows, dugongs)	**4**	1	25.0	**5**	3	60.0
Mustelidae (otters)	**1**	1	100.0	**2**	1	50.0
Ursidae (polar bears)	–	–	–	**1**	1	100.0
Turtles	**7**	6	85.7	**7**	7	100.0
Sea snakes	–	0	–	**62**	2	3.2
Fish	–	60	–	**32,554**	166	0.6
Invertebrates	–	9	–	**159,000**	98	0.1
All species	–	267	–	–	557	–
Marine birds, mammals and turtles	436	193	44.3	**536**	291	54.3
Species associated with smothering	–	–	–	–	22	–

Comparative summary with the earlier major review by Laist (1997). See notes in captions of Tables 4.1 and 4.2. Numbers of species affected and group percentages are not a simple sum of Tables 4.1 and 4.2 because many species suffer from entanglement as well as ingestion. For details, see the Online Supplement

different species and their interactions, combined with dedicated observational or experimental studies, that we can ultimately gain areal understanding of the many deleterious impacts of marine plastic debris on wild populations. A number of recommendations can be made to assist collection of comparable high-quality data sets:

- Accurate data on frequency of occurrence of entanglement or ingestion of debris require a proper a priori protocol, staff that has experience with identifying (symptoms of) marine debris and adequate samples sizes.
- Concerning frequency of entanglement in debris, protocols for assessment are complicated by the distinction between interaction with active fishing gear and

Table 4.4 Number of species of major groups of marine organisms with documented records of marine debris impacts in natural habitats, separately for entanglement or ingestion and in combination (search closed 9th December 2014)

Species group	Species (n)	Entanglement (n)	Entanglement (%)	Ingestion (n)	Ingestion (%)	Total species affected (n)	Total species affected (%)
Seabirds	**406**	**103**	**25.4**	**164**	**40.4**	**203**	**50.0**
Anseriformes	**13**	**5**	**38.5**	**1**	**7.7**	**5**	**38.5**
Anatidae (marine ducks)	13	5	38.5	1	7.7	5	38.5
Gaviiformes	**5**	**3**	**60.0**	**3**	**60.0**	**4**	**80.0**
Gaviidae (divers, loons)	5	3	60.0	3	60.0	4	80.0
Sphenisciformes	**18**	**6**	**33.3**	**5**	**27.8**	**9**	**50.0**
Spheniscidae (penguins)	18	6	33.3	5	27.8	9	50.0
Procellariiformes	**141**	**24**	**17.0**	**84**	**59.6**	**85**	**60.3**
Diomedeidae (albatrosses)	21	12	57.1	17	81.0	17	81.0
Procellariidae (petrels, shear-waters, prions)	92	10	10.9	55	59.8	56	60.9
Hydrobatidae (storm petrels)	24	2	8.3	10	41.7	10	41.7
Pelecanoididae (diving petrels)	4	0	0.0	2	50.0	2	50.0
Podicipediformes	**23**	**6**	**26.1**	**0**	**0.0**	**6**	**26.1**
Podicipedidae (grebes)	23	6	26.1	0	0.0	6	26.1
Phaethontiformes	**3**	**0**	**0.0**	**2**	**66.7**	**2**	**66.7**
Phaethontidae (tropicbirds)	3	0	0.0	2	66.7	2	66.7
Pelecaniformes	**8**	**4**	**50.0**	**2**	**25.0**	**5**	**62.5**
Pelecanidae (pelicans)	8	4	50.0	2	25.0	5	62.5
Suliformes	**56**	**16**	**28.6**	**12**	**21.4**	**20**	**35.7**
Fregatidae (frigatebirds)	5	0	0.0	1	20.0	1	20.0
Sulidae (gannets, boobies)	10	6	60.0	5	50.0	8	80.0

(continued)

Table 4.4 (continued)

Species group	Species (n)	Entanglement (n)	Entanglement (%)	Ingestion (n)	Ingestion (%)	Total species affected (n)	Total species affected (%)
Phalacrocoracidae (cormorants, shags)	41	10	24.4	6	14.6	11	26.8
Charadriiformes	**139**	**39**	**28.1**	**55**	**39.6**	**67**	**48.2**
Chionidae (sheathbills)	2	0	0.0	1	50.0	1	50.0
Scolopacidae (phalaropes)	3	0	0.0	2	66.7	2	66.7
Laridae (gulls, noddies, skimmers, terns)	102	28	27.5	32	31.4	42	41.2
Stercorariidae (skuas)	7	2	28.6	6	85.7	6	85.7
Alcidae (murres, guillemots, murrelets, auks, auklets, puffins)	25	9	36.0	14	56.0	16	64.0
Marine mammals	**123**	**51**	**41.5**	**62**	**50.4**	**81**	**65.9**
Mysticeti	**13**	**9**	**69.2**	**7**	**53.8**	**10**	**76.9**
Balaenidae (right whales)	4	3	75	2	50	3	75.0
Neobalaenidae (pygmy right whales)	1	1	100	1	100	1	100.0
Eschrichtiidae (gray whales)	1	1	100	0	0	1	100.0
Balaenopteridae (rorquals)	7	4	57.1	4	57.1	5	71.4
Odontoceti	**67**	**16**	**23.9**	**40**	**59.7**	**44**	**65.7**
Physeteridae (sperm whales)	1	1	100	1	100	1	100.0
Kogiidae (dwarf and pygmy sperm whales)	2	1	50	2	100	2	100.0
Pontoporiidae (La Plata river dolphins)	1	0	0	1	100	1	100.0
Monodontidae (narwhals, belugas)	2	1	50	0	0	1	50.0

(continued)

Table 4.4 (continued)

Species group	Species (n)	Entanglement (n)	Entanglement (%)	Ingestion (n)	Ingestion (%)	Total species affected (n)	Total species affected (%)
Phocoenidae (porpoises)	6	2	33.3	4	66.7	4	66.7
Delphinidae (oceanic dolphins)	34	10	29.4	19	55.9	**22**	**64.7**
Ziphiidae (beaked whales)	21	1	4.8	13	61.9	13	61.9
Pinniped	**33**	**22**	**66.7**	**12**	**36.4**	22	66.7
Phocidae (true seals)	19	9	47.4	4	21.1	9	47.4
Otariidae (eared seals)	13	13	100	8	61.5	13	100.0
Sirenia	**5**	**2**	**40**	**3**	**60**	**3**	**60.0**
Trichechidae (manatees)	2	1	50.0	2	100.0	2	100.0
Dugongidae (dugongs)	2	1	50	1	50	1	50.0
Carnivora	**3**	**2**	**66.7**	**0**	**0.0**	**2**	**66.7**
Mustelidae (otters)	2	1	50	0	0	1	50.0
Ursidae (polar bears)	1	1	100	0	0	1	100.0
Turtles	**7**	**7**	**100.0**	**7**	**100.0**	**7**	**100.0**
Carettinae	3	3	100.0	3	3.0	3	100.0
Cheloniidae	3	3	100.0	3	3.0	3	100.0
Dermochelyidae	1	1	100.0	1	1.0	1	100.0
Sea snakes	**62**	**2**	**3.2**	**0**	**0.0**	**2**	**3.2**
Hydrophiidae (sea snakes)	62	2	3.2	0	0.0	2	3.2
Fish	**32,554**	**89**	**0.27**	**92**	**0.28**	**166**	**0.51**
Elasmobranchii	**692**	**21**	**3.03**	**18**	**2.60**	**30**	**4.34**
Hexanchiformes (frill and cow sharks)	6	1	16.67	0	0.00	1	16.67
Orectolobiformes (carpet sharks)	44	0	0.00	1	2.27	1	2.27

(continued)

Table 4.4 (continued)

Species group	Species (n)	Entanglement (n)	Entanglement (%)	Ingestion (n)	Ingestion (%)	Total species affected (n)	Total species affected (%)
Lamniformes (mackerel sharks)	16	6	37.50	5	31.25	7	43.75
Charcharhiniformes (ground sharks)	282	11	3.90	8	2.84	14	4.96
Squaliformes (bramble, sleeper, dogfish sharks)	129	1	0.78	3	2.33	4	3.10
Myliobatiformes (stingrays)	215	2	0.93	1	0.47	3	1.40
Holocephali	**50**	**1**	**2.00**	**0**	**0.00**	**1**	**2.0**
Chimaeriformes (chimaeras)	50	1	2.00	0	0.00	1	2.0
Actinopterygii	**22,916**	**67**	**0.29**	**74**	**0.32**	**135**	**0.6**
Amiiformes (bowfins)	1	1	100.00	0	0.00	1	100.0
Anguilliformes (eels, morays)	906	1	0.11	0	0.00	1	0.11
Clupeiformes (herrings)	390	2	0.51	2	0.51	3	0.77
Siluriformes (cat fish)	3589	2	0.06	3	0.08	4	0.11
Osmeriformes (smelts)	319	1	0.31	0	0.00	1	0.31
Salmoniformes (salmons)	215	4	1.86	1	0.47	5	2.33
Stomiiformes (light fish, dragon fish)	413	0	0.00	3	0.73	3	0.73
Aulopiformes (grinners)	255	0	0.00	1	0.39	1	0.39
Myctophiformes (lantern fish)	215	0	0.00	12	5.58	12	5.58
Lampriformes (velifers, tube-eyes, ribbon fish)	24	0	0.00	3	12.50	3	12.50
Gadiformes (cod-like)	614	2	0.33	7	1.14	8	1.30
Batrachoidiformes (toad fish)	82	2	2.44	0	0.00	2	2.44
Lophiiformes (angler fish)	353	0	0.00	1	0.28	1	0.28

(continued)

Table 4.4 (continued)

Species group	Species (n)	Entanglement (n)	Entanglement (%)	Ingestion (n)	Ingestion (%)	Total species affected (n)	Total species affected (%)
Atheriniformes (silversides)	338	0	0.00	1	0.30	1	0.30
Cyprinodontiformes (rivulines, killi fish, live bearers)	1249	1	0.08	0	0.00	1	0.08
Beloniformes (needle fish)	254	0	0.00	1	0.39	1	0.39
Zeiformes (dories)	33	0	0.00	1	3.03	1	3.03
Scorpaeniformes (scorpion fish, flatheads)	1622	18	1.11	5	0.31	23	1.42
Perciformes (perch-like)	10,837	20	0.18	29	0.27	47	0.43
Pleuronectiformes (flatfish)	778	11	1.41	4	0.51	14	1.80
Tetraodontiformes (puffers, file fish)	429	2	0.47	0	0.00	2	0.47
Invertebrates	**159,000**	**92**	**0.06**	**6**	**0.004**	**99**	**0.06**
Crustacea	67,000	46	0.07	3	0.00	49	0.07
Echinodermata	7000	21	0.30	0	0.00	21	0.30
Mollusca	85,000	25	0.03	3	0.00	29	0.03
*All species**		**344**		**331**		**557**	

*Exclusive 22 species associated with smothering, see online supplement 2

interaction with marine debris. For example, even for experts using standard protocols, it is difficult to distinguish whether wounds are caused by entanglement in active or derelict fishing gear, even when remains of nets or similar are found on the body. Some suggestions are being developed concerning entanglement rates in ghost nets or for bird entanglement in synthetic materials used for nest construction (MSFD-TSGML 2013).

- For ingestion, in addition to frequency of occurrence ('incidence') it is recommended to collect data on quantities of ingested debris not only on the basis of numbers of items but also by mass of categories.

- In such ingestion records, as a minimum it is recommended to separate industrial plastics (pellets) from consumer-waste plastics (see Table 4.5). The latter if possible can be further specified following the categorisation recommended for ingestion by birds, mammals and fishes according to the EU Marine Strategy Directive (MSFD-TSGML 2013), that is into categories of sheetlike, threadlike, foamed, hard fragmented, and other synthetic items, plus categories of non-plastic rubbish.

- For averaged data, information should be provided as 'population averages' with standard error of the mean. Population averages are calculated with the inclusion of individuals without ingested plastics. Additional data can be maximum levels observed, or proportions of animals exceeding a particular limit [such as the 0.1-g critical limit in the Ecological Quality Objective for plastic ingestion by northern fulmars (Van Franeker et al. 2011)] (see Table 4.5). We emphasize this explicit use of population averages because in quite a few of the publications checked for this review averages had been calculated just over those individuals that had plastic, often not specifying that zero values had been omitted.

- Negative species results (e.g. Avery-Gomm et al. 2013; Provencher et al. 2014) are also relevant but again should be based on an adequate sample size of animals studied according to a proper protocol. Thus, records of absence of debris for an individual sample should be as firm as those on presence. From experience in our own research group, we know of claims on absence or near absence of plastics in stomachs or guts of several species of which diets were studied, but without dedicated methods or data recording for marine debris (including zeros). Once proper methods were established for laboratory procedures and data recording, each of those species was found to contain debris regularly (e.g. Bravo-Rebolledo et al. 2013).

- Examples of protocols for ingested debris in intestinal tracts of larger organisms can be found in e.g. MSFD-TSGML (2013), with further information for ingestion by marine birds in Van Franeker et al. (2011) and marine turtles in Camedda et al. (2014). Standard protocols for marine mammals, invertebrates have not yet been established in detail but may largely follow those for seabirds and turtles. In general, these studies consider debris of ≥ 1 mm by using sieves with such mesh size.

- Only when using the above approaches on frequency of occurrence (proportion of animals in populations affected) and gravity of interaction (quantity of ingested material; damage level from entanglement), it becomes possible to design experimental or other dedicated studies that allow estimates of the true impact of plastic ingestion on wildlife populations. This relates to both the physical and chemical types of impacts, and will ultimately require model predictions using demographic characteristics of the species involved (Criddle et al. 2009).

Table 4.5 Recommended mode of data presentation for ingested plastic debris, using the example of plastics ingested by northern fulmars in different sub-regions of the North Sea (modified from Van Franeker and the 'Save the North Sea Fulmar' study group, 2013)

Regions	Sample (n)	Industrial granules			User plastics			Total plastics				
2007–2011 period		Incidence (%)	Average number n ± se	Average mass g ± se	Incidence (%)	Average number n ± se	Average mass g ± se	Incidence (%)	Average number n ± se	Average mass g ± se	Geometric mean mass	EcoQO (%) (over 0.1 g)
Scottish Islands	121	47	1.3 ± 0.2	0.03 ± 0.00	90	21.2 ± 3.0	0.32 ± 0.06	90	22.5 ± 3.0	0.35 ± 0.06	0.091	58
East England	51	75	4.2 ± 0.9	0.09 ± 0.02	98	42.2 ± 6.2	0.26 ± 0.06	98	46.4 ± 6.7	0.35 ± 0.07	0.154	76
Channel area	72	82	7.1 ± 1.6	0.15 ± 0.03	99	44.6 ± 8.0	0.39 ± 0.06	99	51.7 ± 9.3	0.54 ± 0.08	0.278	86
SE North Sea	493	57	3.1 ± 0.6	0.07 ± 0.01	94	24.9 ± 1.8	0.29 ± 0.04	95	28.0 ± 2.1	0.36 ± 0.04	0.105	60
Skagerrak	79	53	3.4 ± 0.8	0.07 ± 0.02	94	49.2 ± 14.9	0.24 ± 0.04	94	52.6 ± 15.5	0.31 ± 0.05	0.105	56
North Sea total	816	58	3.3 ± 0.4	0.07 ± 0.01	94	29.5 ± 2.0	0.30 ± 0.03	95	32.8 ± 2.2	0.37 ± 0.03	0.115	62

Given are sample size, percentage of individuals with ingested material (incidence or frequency of occurrence), and population averages (including zero values) with standard error (se) for both number of items (n) and mass. These are specified for industrial plastics and consumer waste separately, and in total. Added to total plastics in this example are geometric mean mass and EcoQO performance that is the percentage of fulmars that had more than the critical level of 0.1 g of total plastic in the stomach

It will take considerable time and effort to collect these data and conduct dedicated studies before firm conclusions can be drawn on the level of detrimental impact of marine plastic debris on wildlife. However, in our opinion the suffering and death of individuals, in combination with the likelihood of higher-level population effects, indicates the need for a rapid reduction of input of plastic debris into the marine environment. If wildlife problems are not convincing: recent studies show that chemical and physical impacts are likely to occur in marine food webs (e.g. Van Cauwenberghe and Janssen 2014; Rochman et al. 2013, 2014), which implies potential impacts on human end consumers (Galloway 2015).

Long-term studies on seabirds have shown that measures to reduce loss of plastics to the environment do have relatively rapid effects. After considerable attention to the massive loss of industrial pellets to the marine environment in the early 1980s, improvements in production and transport methods were reflected in a visible result in the marine environment within one to two decades: several studies from around the globe showed that by the early 2000s the number of industrial granules in seabird stomachs had approximately halved from levels observed in the 1980s (Van Franeker and Meijboom 2002; Vlietstra and Parga 2002; Ryan 2008; Van Franeker et al. 2011; Van Franeker and Law 2015). These examples indicate that it is possible to reduce deleterious impacts from marine plastic debris on marine wildlife in shorter time frames than the longevity of the material might suggest.

Acknowledgments We are very grateful for the language corrections made by Dan Turner. We thank two anonymous reviewers and the editorial team for their constructive comments and suggestions, which contributed to considerable improvements in the manuscript.

References

Abu-Hilal, A., & Al-Najjar, T. (2009). Marine litter in coral reef areas along the Jordan Gulf of Aqaba, Red Sea. *Journal of Environmental Management, 90,* 1043–1049.

Acampora, H., Schuyler, Q. A., Townsend, K. A., & Hardesty, B. D. (2014). Comparing plastic ingestion in juvenile and adult stranded short-tailed shearwaters (*Puffinus tenuirostris*) in eastern Australia. *Marine Pollution Bulletin, 78,* 63–68.

Adey, J., Smith, I., Atkinson, R. J. A., Tuck, I., & Taylor, A. (2008). Ghost fishing' of target and non-target species by Norway lobster, *Nephrops norvegicus*, creels. *Marine Ecology Progress Series, 366,* 119–127.

Allen, R., Jarvis, D., Sayer, S., & Mills, C. (2012). Entanglement of grey seals, *Halichoerus grypus*, at a haul out site in Cornwall, UK. *Marine Pollution Bulletin, 64,* 2815–2819.

Al-Masroori, H., Al-Oufi, H., McIlwain, J., & McLean, E. (2004). Catches of lost fish traps (ghost fishing) from fishing grounds near Muscat, Sultanate of Oman. *Fisheries Research, 69,* 407–414.

Aloy, A. B., Vallejo, B. M, Jr, & Juinio-Meñez, M. A. (2011). Increased plastic litter cover affects the foraging activity of the sandy intertidal gastropod *Nassarius pullus*. *Marine Pollution Bulletin, 62,* 1772–1779.

Anastasopoulou, A., Mytilineou, C., Smith, C. J., & Papadopoulou, K. N. (2013). Plastic debris ingested by deep-water fish of the Ionian Sea (eastern Mediterranean). *Deep-Sea Research I, 74*, 11–13.

Anderson, J. A., & Alford, A. B. (2014). Ghost fishing activity in derelict blue crab traps in Louisiana. *Marine Pollution Bulletin, 79*, 261–267.

Andrady, A. L. (2015). Persistence of plastic litter in the oceans. In M. Bergmann, L. Gutow & M. Klages (Eds.), *Marine anthropogenic litter* (pp. 57–72). Berlin: Springer.

Andrady, A. L. (2011). Microplastics in the marine environment. *Marine Pollution Bulletin, 62*, 1596–1605.

Andrady, A. L., & Neal, M. A. (2009). Applications and societal benefits of plastics. Philosophical transactions of the Royal Society of London *B, 364*, 1977–1984.

Antonelis, K., Huppert, D., Velasquez, D., & June, J. (2011). Dungeness crab mortality due to lost traps and a cost-benefit analysis of trap removal in Washington state waters of the Salish Sea. *North American Journal of Fisheries Management, 31*, 880–893.

Arnould, J. P. Y., & Croxall, J. P. (1995). Trends in entanglement of Antarctic fur seals (*Arctocephalus gazella*) in man-made debris at South Georgia. *Marine Pollution Bulletin, 30*, 707–712.

Asoh, K., Yoshikawa, T., Kosaki, R., & Marschall, E. A. (2004). Damage to cauliflower coral by monofilament fishing lines in Hawaii. *Conservation Biology, 18*, 1645–1650.

Auman, H. J., Ludwig, J. P., Giesy, J. P., & Colborn, T. (1997). Plastic ingestion by Laysan Albatross chicks on Sand Island, midway atoll, in 1994 and 1995. In G. Robinson & R. Gales (Eds.), *Albatross biology and conservation*. Surrey Beatty and Sons: Chipping Norton.

Avery-Gomm, S., Provencher, J. F., Morgan, K. H., & Bertram, D. F. (2013). Plastic ingestion in marine-associated bird species from the eastern North Pacific. *Marine Pollution Bulletin, 72*, 257–259.

Ayaz, A., Acarli, D., Altinagac, U., Ozekinci, U., Kara, A., & Ozen, O. (2006). Ghost fishing by monofilament and multifilament gillnets in Izmir Bay, Turkey. *Fisheries Research, 79*, 267–271.

Azzarello, M. Y., & Van Vleet, E. S. (1987). Marine birds and plastic pollution. *Marine Ecology Progress Series, 37*, 295–303.

Baeta, F., Costa, M. J., & Cabral, H. (2009). Trammel nets' ghost fishing off the Portuguese central coast. *Fisheries Research, 98*, 33–39.

Baird, R. W., & Hooker, S. K. (2000). Ingestion of plastic and unusual prey by a juvenile harbour porpoise. *Marine Pollution Bulletin, 40*, 719–720.

Barreiros, J. P., & Guerreiro, O. (2014). Notes on a plastic debris collar on a juvenile *Pagellus acarne* (Perciformes: Sparidae) from Terceira Island, Azores, NE Atlantic. *Bothalia-Pretoria, 44*, 2–5.

Barreiros, J. P., & Raykov, V. S. (2014). Lethal lesions and amputation caused by plastic debris and fishing gear on the loggerhead turtle *Caretta caretta* (Linnaeus 1758). Three case reports from Terceira Island, Azores (NE Atlantic). *Marine Pollution Bulletin, 86*, 518–522.

Bauer, L. J., Kendall, M. S., & Jeffrey, C. F. (2008). Incidence of marine debris and its relationships with benthic features in Gray's Reef National Marine Sanctuary, Southeast USA. *Marine Pollution Bulletin, 56*, 402–413.

Baulch, S., & Perry, C. (2014). Evaluating the impacts of marine debris on cetaceans. *Marine Pollution Bulletin, 80*, 210–221.

Bavestrello, G., Cerrano, C., Zanzi, D., & Cattaneo-Vietti, R. (1997). Damage by fishing activities to the gorgonian coral *Paramuricea clavata* in the Ligurian Sea. *Aquatic Conservation: Marine and Freshwater Ecosystems, 7*, 253–262.

Beck, C. A., & Barros, N. B. (1991). The impact of debris on the Florida manatee. *Marine Pollution Bulletin, 22*, 508–510.

Bilkovic, D. M., Havens, K., Stanhope, D., & Angstadt, K. (2014). Derelict fishing gear in Chesapeake Bay, Virginia: Spatial patterns and implications for marine fauna. *Marine Pollution Bulletin, 80*, 114–123.

Bird, P. M. (1978). Tissue regeneration in three carcharhinid sharks encircled by embedded straps. *Copeia, 1978*, 345–349.

Bjorndal, K. A., Bolten, A. B., & Lagueux, C. J. (1994). Ingestion of marine debris by Juvenile Sea turtles in coastal Florida habitats. *Marine Pollution Bulletin, 28*, 154–158.

Blais, J. M., Kimpe, L. E., McMahon, D., Keatley, B. E., Mallory, M. L., Douglas, M. S., et al. (2005). Arctic seabirds transport marine-derived contaminants. *Science, 309*, 445–445.

Boerger, C. M., Lattin, G. L., Moore, S. L., & Moore, C. J. (2010). Plastic ingestion by plank-tivorous fishes in the North Pacific central gyre. *Marine Pollution Bulletin, 60*, 2275–2278.

Bond, A. L., Montevecchi, W. A., Guse, N., Regular, P. M., Garthe, S., & Rail, J. F. (2012). Prevalence and composition of fishing gear debris in the nests of northern gannets (*Morus bassanus*) are related to fishing effort. *Marine Pollution Bulletin, 64*, 907–911.

Bond, A. L., Provencher, J. F., Elliot, R. D., Ryan, P. C., Rowe, S., Jones, I. L., et al. (2013). Ingestion of plastic marine debris by common and thick-billed murres in the northwestern Atlantic from 1985 to 2012. *Marine Pollution Bulletin, 77*, 192–195.

Bourne, W. R. P., & Imber, M. J. (1982). Plastic pellets collected by a prion on Gough Island, Central South Atlantic Ocean. *Marine Pollution Bulletin, 13*, 20–21.

Brandao, M. L., Braga, K. M., & Luque, J. L. (2011). Marine debris ingestion by magellanic penguins, *Spheniscus magellanicus* (Aves: Sphenisciformes), from the Brazilian Coastal Zone. *Marine Pollution Bulletin, 62*, 2246–2249.

Bravo Rebolledo, E. L., Van Franeker, J. A., Jansen, O. E., & Brasseur, S. M. (2013). Plastic ingestion by harbour seals (*Phoca vitulina*) in the Netherlands. *Marine Pollution Bulletin, 67*, 200–202.

Breen, P. A. (1990). A review of ghost fishing by traps and gillnets. In R. S. Shomura & M. L. Godfrey (Eds.), *Proceedings of the Second International Conference of Marine Debris* (pp. 571–599). Honolulu, Hawaii: U.S. Department of Commerce, NOAA Tech Memo, NMFS.

Bugoni, L., Krause, L., & Petry, M. V. (2001). Marine debris and human impact on sea turtles in southern Brazil. *Marine Pollution Bulletin, 42*(12), 1330–1334.

Cadée, G. C. (2002). Seabirds and floating plastic debris. *Marine Pollution Bulletin, 44*, 1294–1295.

Camedda, A., Marra, S., Matiddi, M., Massaro, G., Coppa, S., Perilli, A., et al. (2014). Interaction between loggerhead sea turtles (*Caretta caretta*) and marine litter in Sardinia (western Mediterranean Sea). *Marine Environmental Research, 100*, 25–32.

Campani, T., Baini, M., Giannetti, M., Cancelli, F., Mancusi, C., Serena, F., et al. (2013). Presence of plastic debris in loggerhead turtle stranded along the tuscany coasts of the Pelagos Sanctuary for Mediterranean Marine Mammals (Italy). *Marine Pollution Bulletin, 74*, 225–230.

Camphuysen, C. J. (2001). Northern gannets morus bassanus found dead in the Netherlands 1970–2000. *Atlantic Seabirds, 3*, 15–30.

Camphuysen, C. J., Boekhout, S., Gronert, A., Hunt, V., Van Nus, T., & Ouwehand, J. (2008). Bizarre prey items: Odd food choices in herring gulls and lesser black-backed gulls at texel. *Sula, 21*, 29–61.

Carpenter, E. J., Anderson, S. J., Harvey, G. R., Miklas, H. P., & Peck, B. B. (1972). Polystyrene spherules in coastal waters. *Science, 178*, 749–750.

Carr, A. (1987). Impact of nondegradable marine debris on the ecology and survival outlook of sea turtles. *Marine Pollution Bulletin, 18*, 352–356.

Carson, H. S., Lamson, M. R., Nakashima, D., Toloumu, D., Hafner, J., Maximenko, N. et al. (2013). Tracking the sources and sinks of local marine debris in Hawai'i. *Marine Environmental Research, 84*, 76–83.

Ceccarelli, D. M. (2009). Impacts of plastic debris on Australian marine wildlife. C&R Consulting (Ed.), *Report by C&R consulting for the department of the environment, water, Heritage and the arts* (p. 83).

Chapman, M. G., & Clynick, B. G. (2006). Experiments testing the use of waste material in estuaries as habitat for subtidal organisms. *Journal of Experimental Marine Biology and Ecology, 338*, 164–178.

Chiappone, M., Dienes, H., Swanson, D. W., & Miller, S. L. (2005). Impacts of lost fishing gear on coral reef sessile invertebrates in the Florida Keys National Marine Sanctuary. *Biological Conservation, 121,* 221–230.

Cho, D. (2011). Removing derelict fishing gear from the deep seabed of the East Sea. *Marine Policy, 35,* 610–614.

Choy, C. A., & Drazen, J. C. (2013). Plastic for dinner? Observations of frequent debris ingestion by pelagic predatory fishes from the Central North Pacific. *Marine Ecology Progress Series, 485,* 155–163.

Cliff, G., Dudley, S. F. J., Ryan, P. G., & Singleton, N. (2002). Large sharks and plastic debris in Kwazulu-Natal, South Africa. *Marine and Freshwater Research, 53,* 575–581.

Codina-García, M., Militão, T., Moreno, J., & González-Solís, J. (2013). Plastic debris in mediterranean seabirds. *Marine Pollution Bulletin, 77,* 220–226.

Colabuono, F. I., Barquete, V., Domingues, B. S., & Montone, R. C. (2009). Plastic ingestion by procellariiformes in southern Brazil. *Marine Pollution Bulletin, 58,* 93–96.

Cole, M., Lindeque, P., Halsband, C., & Galloway, T. S. (2011). Microplastics as contaminants in the marine environment: A review. *Marine Pollution Bulletin, 62,* 2588–2597.

Cole, M., Lindeque, P., Fileman, E., Halsband, C., Goodhead, R., Moger, J. et al. (2013). Microplastic ingestion by zooplankton. *Environmental Science and Technology, 47,* 6646–6655.

Connors, P. G., & Smith, K. G. (1982). Oceanic plastic particle pollution: Suspected effect on fat deposition in Red Phalaropes. *Marine Pollution Bulletin, 13,* 18–20.

Criddle, K., Amos, A., Carroll, P., Coe, J., Donohue, M., Harris, J. et al. (2009). *Tackling marine debris in the 21st century* (206 p). Washington, DC, USA: The National Academies Press.

Crockett, D. E., & Reed, S. M. (1976). Phenomenal antarctic fulmar wreck. *Notornis, 23,* 250–262.

Davidson, T. M. (2012). Boring crustaceans damage polystyrene floats under docks polluting marine waters with microplastic. *Marine Pollution Bulletin, 64,* 1821–1828.

Davison, P., & Asch, R. G. (2011). Plastic ingestion by mesopelagic fishes in the North Pacific subtropical gyre. *Marine Ecology Progress Series, 432,* 173–180.

Day, R. H., Wehle, D. H. S., & Coleman, F. C. (1985). Ingestion of plastic pollutants by marine birds. In R. S. Shomura & H. O. Yoshida (Eds.), *Proceedings of the Workshop on the Fate and Impact of Marine Debris* (pp. 344–386). Honolulu, Hawaii: U.S. Dep. Commer., NOAA Tech. Memo. NMFS.

De Pierrepont, J. F., Dubois, B., Desormonts, S. M., Santos, B. O., & Robin, J. P. (2005). Stomach contents of english channel cetaceans stranded on the Coast of Normandy. *Journal of the Marine Biological Association of the U.K., 85,* 1539–1546.

De Stephanis, R., Gimenez, J., Carpinelli, E., Gutierrez-Exposito, C., & Canadas, A. (2013). As main meal for sperm whales: Plastics debris. *Marine Pollution Bulletin, 69,* 206–214.

Denuncio, P., Bastida, R., Dassis, M., Giardino, G., Gerpe, M., & Rodriguez, D. (2011). Plastic ingestion in franciscana dolphins, *Pontoporia blainvillei* (Gervais and D'Orbigny, 1844), from Argentina. *Marine Pollution Bulletin, 62,* 1836–1841.

Derraik, J. G. B. (2002). The pollution of the marine environment by plastic debris: A review. *Marine Pollution Bulletin, 44,* 842–852.

Di Beneditto, A. P. M., & Ramos, R. M. A. (2014). Marine debris ingestion by coastal dolphins: What drives differences between sympatric species? *Marine Pollution Bulletin, 83,* 298–301.

Donnelly-Greenan, E. L., Harvey, J. T., Nevins, H. M., Hester, M. M., & Walker, W. A. (2014). Prey and plastic ingestion of pacific northern fulmars (*Fulmarus glacialis rogersii*) from Monterey Bay. *California. Marine Pollution Bulletin, 85*(1), 214–224.

Donohue, M. J., Boland, R. C., Sramek, C. M., & Antonelis, G. A. (2001). Derelict fishing gear in the northwestern Hawaiian Islands: Diving surveys and debris removal in 1999 confirm threat to coral reef ecosystems. *Marine Pollution Bulletin, 42,* 1301–1312.

Duguy, R., Moriniere, P., & Meunier, A. (2000). L'ingestion Des Déchets Flottants Par La Tortue Luth *Dermochelys coriacea* (Vandelli, 1761) Dans Le Golfe De Gascogne. *Annales de la Société de Sciences Naturelles de la Charente-Maritimes, 8,* 1035–1038.

Eriksson, C., & Burton, H. (2003). Origins and biological accumulation of small plastic particles in fur seals from Macquarie Island. *AMBIO: A Journal of the Human Environment, 32*, 380–384.

Erzini, K. (1997). An experimental study of gill net and trammel net 'ghost fishing' off the algarve (southern Portugal). *Marine Ecology Progress Series, 158*, 257–265.

Erzini, K., Bentes, L., Coelho, R., Lino, P. G., Monteiro, P., Ribeiro, J., et al. (2008). Catches in ghost-fishing octopus and fish traps in the northeastern Atlantic Ocean (Algarve, Portugal). *Fishery Bulletin, 106*, 321–327.

Fabri, M. C., Pedel, L., Beuck, L., Galgani, F., Hebbeln, D., & Freiwald, A. (2014). Megafauna of vulnerable marine ecosystems in French mediterranean submarine canyons: Spatial distribution and anthropogenic impacts. *Deep-Sea Research II, 104*, 184–207.

Foekema, E. M., De Gruijter, C., Mergia, M. T., van Franeker, J. A., Murk, A. J., & Koelmans, A. A. (2013). Plastic in North Sea fish. *Environmental Science and Technology, 47*, 8818–8824.

Fowler, C. W. (1987). Marine debris and northern fur seals: A case study. *Marine Pollution Bulletin, 18*, 326–335.

Fowler, C. W., Ream, R., Robson, B., & Kiyota, M., (1992). *Entanglement studies on juvenile male northern fur seals, St. Paul Island, 1991* (p. 42). Seattle: U.S. Department of Commerce, Alaska Fisheries Science Center.

Frick, M. G., Williams, K. L., Bolten, A. B., Bjorndal, K. A., & Martins, H. R. (2009). Foraging ecology of oceanic-stage loggerhead turtles *Caretta caretta. Endangered Species Research, 9*, 91–97.

Furness, R. W. (1983). Ingestion of plastic particles by seabirds at Gough Island, South Atlantic Ocean. *Environmental Pollution, 38*, 261–272.

Furness, R. W. (1985). Plastic particle pollution: Accumulation by procellariiform seabirds at Scottish colonies. *Marine Pollution Bulletin, 16*, 103–106.

Galgani, F., Souplet, A., & Cadiou, Y. (1996). Accumulation of debris on the deep sea floor off the French mediterranean coast. *Marine Ecology Progress Series, 142*, 225–234.

Galloway T. S. (2015). Micro- and nano-plastics and human health. In M. Bergmann, L. Gutow & M. Klages (Eds.), *Marine anthropogenic litter* (pp. 347–370). Berlin: Springer.

Gill, F., & Donsker, D. (2013). IOC world bird list (Version 3.3). Available at http://www.Worldbirdnames.Org. Accessed 5 Feb 2013.

Godøy, H., Furevik, D. M., & Stiansen, S. (2003). Unaccounted mortality of red king crab (*Paralithodes camtschaticus*) in deliberately lost pots off northern Norway. *Fisheries Research, 64*, 171–177.

Goldberg, E. (1997). Plasticizing the seafloor: An overview. *Environmental Technology, 18*, 195–201.

Goldstein, M. C., Rosenberg, M., & Cheng, L. (2012). Increased oceanic microplastic debris enhances oviposition in an endemic pelagic insect. *Royal Society Biology Letters, 8*, 817–820.

Goldstein, M. C., & Goodwin, D. S. (2013). Gooseneck barnacles (*Lepas* spp.) ingest microplastic debris in the North Pacific subtropical gyre. *PeerJ, 1*(e184), 17.

Goldstein, M. C., Carson, H. S., & Eriksen, M. (2014). Relationship of diversity and habitat area in north pacific plastic-associated rafting communities. *Marine Biology, 161*, 1–13.

Gomerčić, M. D., Galov, A., Gomerčić, T., Škrtić, D., Ćurković, S., Lucić, H., et al. (2009). Bottlenose dolphin (*Tursiops truncates*) depredation resulting in larynx strangulation with gill-net parts. *Marine Mammal Science, 25*, 392–401.

Good, T. P., June, J. A., Etnier, M. A., & Broadhurst, G. (2010). Derelict fishing nets in puget sound and the northwest straits: Patterns and threats to marine fauna. *Marine Pollution Bulletin, 60*, 39–50.

Gray, H., Lattin, G. L., & Moore, C. J. (2012). Incidence, mass and variety of plastics ingested by laysan (*Phoebastria immutabilis*) and black-footed albatrosses (*P. nigripes*) recovered as by-catch in the North Pacific Ocean. *Marine Pollution Bulletin, 64*, 2190–2192.

Green, D. S., Boots, B., Blockley, D. J., Rocha, C., & Thompson, R. C. (2015). Impacts of discarded plastic bags on marine assemblages and ecosystem functioning. *Environmental Science and Technology, 49*, 5380–5389.

Gudger, E. W., & Hoffman, W. H. (1931). A shark encircled with a rubber automobile tire. *Scientific Monthly, 33*, 275–277.

Hanni, K. D., & Pyle, P. (2000). Entanglement of pinnipeds in synthetic materials at South-East Farallon Island, California, 1976–1998. *Marine Pollution Bulletin, 40*, 1076–1081.

Harcourt, R., Aurioles, D., & Sanchez, J. (1994). Entanglement of California sea lions at Los Islotes, Baja California Sur, Mexico. *Marine Mammal Science, 10*, 122–125.

Harper, P. C., & Fowler, J. A. (1987). Plastic pellets in New Zealand storm-killed prions. *Notornis, 34*, 65–70.

Harrison, J. P., Sapp, M., Schratzberger, M., & Osborn, A. M. (2011). Interactions between microorganisms and marine microplastics: A call for research. *Marine Technology Society Journal, 45*, 12–20.

Hartwig, E., Clemens, T., & Heckroth, M. (2007). Plastic debris as nesting material in a kittiwake (*Rissa tridactyla*) colony at the Jammerbugt, Northwest Denmark. *Marine Pollution Bulletin, 54*, 595–597.

Hays, H., & Cormons, G. (1974). Plastic particles found in tern pellets, on coastal beaches and at factory sites. *Marine Pollution Bulletin, 5*, 44–46.

Hébert, M., Miron, G., Moriyasu, M., Vienneau, R., & DeGrâce, P. (2001). Efficiency and ghost fishing of snow crab (*Chionoecetes opilio*) traps in the Gulf of St Lawrence. *Fisheries Research, 52*, 143–153.

Henderson, J. R. (2001). A pre-and post-marpol annex v summary of Hawaiian monk seal entanglements and marine debris accumulation in the Northwestern Hawaiian Islands, 1982–1998. *Marine Pollution Bulletin, 42*, 584–589.

Hofmeyr, G. J. G., Bester, M. N., Kirkman, S. P., Lydersen, C., & Kovacs, K. M. (2006). Entanglement of antarctic fur seals at Bouvetøya, Southern Ocean. *Marine Pollution Bulletin, 52*, 1077–1080.

Hong, S., Lee, J., Jang, Y. C., Kim, Y. J., Kim, H. J., Han, D., et al. (2013). Impacts of marine debris on wild animals in the coastal area of Korea. *Marine Pollution Bulletin, 66*, 117–124.

Hoss, D. E., & Settle, L. R. (1990). Ingestion of plastics by teleost fishes. In R. S. Shomura & M. L. Godfrey (Eds.), *Proceedings of the Second International Conference of Marine Debris* (pp. 693–709). Honolulu, Hawaii: U.S. Department of Commerce, NOAA Tech Memo, NMFS.

Humborstad, O.-B., Løkkeborg, S., Hareide, N.-R., & Furevik, D. M. (2003). Catches of greenland halibut (*Reinhardtius hippoglossoides*) in deepwater ghost-fishing gillnets on the norwegian continental slope. *Fisheries Research, 64*, 163–170.

Hutton, I., Carlile, N., & Priddel, D. (2008). Plastic ingestion by flesh-footed shearwaters, *Puffinus carneipes*, and wedge-tailed shearwaters, *Puffinus pacificus. Papers and Proceedings of the Royal Society of Tasmania, 142*, 67–72.

Jensen, J.-K. (2012). Mallemukken På Færøerne/the fulmar on the Faroe Islands, Nolsoy.

Kaiser, M. J., Bullimore, B., Newman, P., Lock, K., & Gilbert, S. (1996). Catches in 'ghost fishing' set nets. *Marine Ecology Progress Series, 145*, 11–16.

Kasparek, M. (1995). The nesting of marine turtles on the coast of Syria. *Zoology in the Middle East, 11*, 51–62.

Katsanevakis, S., Verriopoulos, G., Nicolaidou, A., & Thessalou-Legaki, M. (2007). Effect of marine litter on the benthic megafauna of coastal soft bottoms: A manipulative field experiment. *Marine Pollution Bulletin, 54*, 771–778.

Katsanevakis, S. (2008). *Marine debris, a growing problem: Sources, distribution, composition, and impacts. Marine pollution: New research* (pp. 53–100). New York: Nova Science Publishers.

Kenyon, K. W., & Kridler, E. (1969). Laysan albatrosses swallow indigestible matter. *The Auk, 86*, 339–343.

Kiessling T., Gutow L., & Thiel M. (2015). Marine litter as a habitat and dispersal vector. In M. Bergmann, L. Gutow & M. Klages (Eds.), *Marine anthropogenic litter* (pp. 141–181). Berlin: Springer.

Kim, S.-G., Lee, W.-I. L., & Yuseok, M. (2014). The estimation of derelict fishing gear in the coastal waters of South Korea: Trap and gill-net fisheries. *Marine Policy, 46*, 119–122.

Knowlton, A. R., Hamilton, P. K., Marx, M. K., Pettis, H. M., & Kraus, S. D. (2012). Monitoring north Atlantic right whale *Eubalaena glacialis* entanglement rates: A 30 year retrospective. *Marine Ecology Progress Series, 466*, 293–302.

Koch, H. M., & Calafat, A. M. (2009). Human body burdens of chemicals used in plastic manufacture. *Philosophical transactions of the Royal Society of London B, 364*, 2063–2078.

Koelmans A. A. (2015). Modeling the role of microplastics in bioaccumulation of organic chemicals to marine aquatic organisms. Critical review. In M. Bergmann, L. Gutow & M. Klages (Eds.), *Marine anthropogenic litter* (pp. 313–328). Berlin: Springer.

Laist, D. W. (1987). Overview of the biological effects of lost and discarded plastic debris in the marine environment. *Marine Pollution Bulletin, 18*, 319–326.

Laist, D. W. (1997). *Impacts of marine debris: Entanglement of marine life in marine debris including a comprehensive list of species with entanglement and ingestion records. Springer Series on Environmental Management.* New York: Springer.

Large, P. A., Graham, N. G., Hareide, N.-R., Misund, R., Rihan, D. J., Mulligan, M. C., et al. (2009). Lost and abandoned nets in deep-water gillnet fisheries in the Northeast Atlantic: Retrieval exercises and outcomes. *ICES Journal of Marine Science, 66*, 323–333.

Lavers, J. L., Hodgson, J. C., & Clarke, R. H. (2013). Prevalence and composition of marine debris in brown booby (*Sula leucogaster*) nests at ashmore reef. *Marine Pollution Bulletin, 77*, 320–324.

Lavers, J. L., Bond, A. L., & Hutton, I. (2014). Plastic ingestion by flesh-footed shearwaters (*Puffinus carneipes*): Implications for fledgling body condition and the accumulation of plastic-derived chemicals. *Environmental Pollution, 187*, 124–129.

Lazar, B., & Gracan, R. (2011). Ingestion of marine debris by loggerhead sea turtles, *Caretta caretta*, in the Adriatic Sea. *Marine Pollution Bulletin, 62*, 43–47.

Leslie, H. A., Van Velzen, M. J. M., & Vethaak, A. D. (2013). Microplastic survey of the Dutch environment—Novel data set of microplastics in North Sea sediments, treated wastewater effluents and marine biota. *IVM Institute for Environmental Studies Final Report R-13/11*, Free University, Amsterdam.

Lindborg, V. A., Ledbetter, J. F., Walat, J. M., & Moffett, C. (2012). Plastic consumption and diet of glaucous-winged gulls (*Larus glaucescens*). *Marine Pollution Bulletin, 64*, 2351–2356.

Lucas, Z. (1992). Monitoring persistent litter in the marine environment on Sable Island, Nova Scotia. *Marine Pollution Bulletin, 24*, 192–199.

Lusher A. (2015). Microplastics in the marine environment: distribution, interactions and effects. In M. Bergmann, L. Gutow & M. Klages (Eds.), *Marine anthropogenic litter*. Berlin: Springer.

Lusher, A. L., McHugh, M., & Thompson, R. C. (2013). Occurrence of microplastics in the gastrointestinal tract of pelagic and demersal fish from the English Channel. *Marine Pollution Bulletin, 67*, 94–99.

Lutz, P. L. (1990). Studies on ingestion of plastic and latex by sea turtles. In R. S. Shomura & M. L. Godfrey (Eds.), *Proceedings of the Second International Conference of Marine Debris* (pp. 719–735). Honolulu, Hawaii: U.S. Department of Commerce, NOAA Tech Memo, NMFS.

Majer, A. P., Vedolin, M. C., & Turra, A. (2012). Plastic pellets as oviposition site and means of dispersal for the ocean-skater insect *Halobates*. *Marine Pollution Bulletin, 64*, 1143–1147.

Mallory, M. L. (2006). The northern fulmar (*Fulmarus glacialis*) in Arctic Canada: Ecology, threats, and what it tells us about marine environmental conditions. *Environmental Reviews, 14*, 187–216.

Mallory, M. L. (2008). Marine plastic debris in northern fulmars from the Canadian high Arctic. *Marine Pollution Bulletin, 56*, 1501–1504.

Marine Mammal Commission. (2013). http://www.Mmc.Gov/Species/Speciesglobal2.Shtml. Accessed 25 Apr 2013.

Matsuoka, T. (1999). Ghost fishing by lost fish-traps in Azuma-cho water. *Mini Review Data File Fisheries Research, 8*, 64–69.

Matsuoka, K., Nakashima, T., & Nagasawa, N. (2005). A review of ghost fishing: Scientific approaches to evaluation and solutions. *Fisheries Science, 71*, 691–702.

Mattlin, R. H., & Cawthorn, M. W. (1986). Marine debris—an international problem. *New Zealand Environment, 51*, 3–6.

McCauley, S. J., & Bjorndal, K. A. (1999). Conservation implications of dietary dilution from debris ingestion: Sublethal effects in post-hatchling Loggerhead Sea turtles. *Conservation Biology: The Journal of the Society for Conservation Biology, 13*, 925–929.

Montevecchi, W. A. (1991). Incidence and types of plastic in gannet's nets in the Northwest Atlantic. *Canadian Journal of Zoology, 69*, 295–297.

Moore, E., Lyday, S., Roletto, J., Litle, K., Parrish, J. K., Nevins, H., et al. (2009). Entanglement of marine mammals and seabirds in Central California and the North–West coast of the United States 2001–2005. *Marine Pollution Bulletin, 58*, 1045–1051.

Moore, M., Andrews, R., Austin, T., Bailey, J., Costidis, A., George, C., et al. (2013). Rope trauma, sedation, disentanglement, and monitoring-tag associated lesions in a terminally entangled North Atlantic right whale (*Eubalaena glacialis*). *Marine Mammal Science, 29*, E98–E113.

Mordecai, G., Tyler, P. A., Masson, D. G., & Huvenne, V. A. I. (2011). Litter in submarine canyons off the west coast of Portugal. *Deep-Sea Research II, 58*, 2489–2496.

Moser, M. L., & Lee, D. S. (1992). A fourteen-year survey of plastic ingestion by western North Atlantic seabirds. *Colonial Waterbirds, 15*, 83–94.

Mrosovsky, N., Ryan, G. D., & James, M. C. (2009). Leatherback turtles: The menace of plastic. *Marine Pollution Bulletin, 58*, 287–289.

MSFD-TSGML (2013). *Guidance on monitoring of marine litter in European Seas—a guidance document within the common implementation strategy for the marine strategy framework directive. EUR-26113 EN.* JRC Scientific and Policy Reports JRC83985. 128 p. http://dx.doi.org/10.2788/99475.

Murray, F., & Cowie, P. R. (2011). Plastic Contamination in the Decapod Crustacean, *Nephrops norvegicus* (Linnaeus 1758). *Marine Pollution Bulletin, 62*, 1207–1217.

Newman, S. J., Skepper, C. L., Mitsopoulos, G. E. A., Wakefield, C. B., Meeuwig, J. J., & Harvey, E. S. (2011). Assessment of the potential impacts of trap usage and ghost fishing on the northern demersal scalefish fishery. *Reviews in Fisheries Science, 19*, 74–84.

Norman, F. I., Menkhorst, P. W., & Hurley, V. G. (1995). Plastics in nests of Australasian gannets, *Morus serrator*, in Victoria, Australia. *Emu, 95*, 129–133.

Orós, J., Torrent, A., Calabuig, P., & Déniz, S. (2005). Diseases and causes of mortality among sea turtles stranded in the Canary Islands, Spain (1998–2001). *Diseases of Aquatic Organisms, 63*, 13–24.

Ozdilek, H. G., Yalcin-Ozdilek, S., Ozaner, F. S., & Sonmez, B. (2006). Impact of accumulated beach litter on *Chelonia mydas* L. 1758 (green turtle) hatchlings of the Samandag Coast, Hatay, Turkey. *Fresenius Environmental Bulletin, 15*, 95–103.

Page, B., McKenzie, J., McIntosh, R., Baylis, A., Morrissey, A., Calvert, N., et al. (2004). Entanglement of Australian sea lions and New Zealand fur seals in lost fishing gear and other marine debris before and after government and industry attempts to reduce the problem. *Marine Pollution Bulletin, 49*, 33–42.

Parrish, F. A., & Kazama, T. (1992). Evaluation of ghost fishing in the Hawaiian lobster fishery. *Fishery Bulletin, 90*, 720–725.

Pawson, M. G. (2003). The catching capacity of lost static fishing gears: Introduction. *Fisheries Research, 64*, 101–105.

Pecci, K. J., Cooper, R. A., Newell, C. D., Clifford, R. A., & Smolowitz, R. J. (1978). Ghost fishing of vented and unvented lobster, *Homarus americanus*, traps. *Marine Fisheries Review, 40*, 9–43.

Perry, M. C., Olsen, G. H., Richards, R. A., & Osenton, P. C. (2013). Predation on dovekies by goosefish over deep water in the Northwest Atlantic Ocean. *Northeastern Naturalist, 20*, 148–154.

Pettit, T. N., Grant, G. S., & Whittow, G. C. (1981). Ingestion of plastics by Laysan albatross. *Auk, 98*, 839–841.

Pham, C. K., Gomes-Pereira, J. N., Isidro, E. J., Santos, R. S., & Morato, T. (2013). Abundance of litter in the condor seamount. *Deep-Sea Research II, 98*, 204–208.

Pham, C. K., Ramirez-Llodra, E., Alt, C. H., Amaro, T., Bergmann, M., Canals, M., et al. (2014). Marine litter distribution and density in European seas, from the shelves to deep basins. *PLoS ONE, 9*, e95839.

Plotkin, P., & Amos, A. F. (1990). Effects of anthropogenic debris on sea turtles in the north-western Gulf of Mexico. In R. S. Shomura & M. L. Godfrey (Eds.), *Proceedings of the Second International Conference of Marine Debris* (pp. 736–743). Honolulu, Hawaii: U.S. Department of Commerce, NOAA Tech Memo, NMFS.

Podolski, R. H., & Kress, S. W. (1989). Plastic debris incorporated into double-crested cormorant nests in the Gulf of Maine. *Journal of Field Ornithology, 60,* 248–250.

Possatto, F. E., Barletta, M., Costa, M. F., Ivar do Sul, J. A., & Dantas, D. V. (2011). Plastic debris ingestion by marine catfish: An unexpected fisheries impact. *Marine Pollution Bulletin, 62,* 1098–1102.

Provencher, J. F., Gaston, A. J., Mallory, M. L., O'Hara, P. D., & Gilchrist, H. G. (2010). Ingested plastic in a diving seabird, the thick-billed murre (*Uria lomvia*), in the eastern Canadian Arctic. *Marine Pollution Bulletin, 60,* 1406–1411.

Provencher, J. F., Bond, A. L., Hedd, A., Montevecchi, W. A., Muzaffar, S. B., Courchesne, S. J., et al. (2014). Prevalence of marine debris in marine birds from the North Atlantic. *Marine Pollution Bulletin, 84,* 411–417.

Read, A. J., Drinker, P., & Northridge, S. (2006). Bycatch of marine mammals in U.S. and global fisheries. *Conservation Biology, 20,* 163–169.

Revill, A. S., & Dunlin, G. (2003). The fishing capacity of gillnets lost on wrecks and on open ground in UK coastal waters. *Fisheries Research, 64,* 107–113.

Richards, Z. T., & Beger, M. (2011). A quantification of the standing stock of macro-debris in majuro lagoon and its effect on hard coral communities. *Marine Pollution Bulletin, 62,* 1693–1701.

Robards, M. D., Piatt, J. F., & Wohl, K. D. (1995). Increasing frequency of plastic particles ingested by seabirds in the subarctic North Pacific. *Marine Pollution Bulletin, 30,* 151–157.

Rochman C. M. (2015). The complex mixture, fate and toxicity of chemicals associated with plastic debris in the marine environment. In M. Bergmann, L. Gutow & M. Klages (Eds.), *Marine anthropogenic litter* (pp. 117–140). Berlin: Springer.

Rochman, C. M., Hoh, E., Kurobe, T., & Teh, S. J. (2013). Ingested plastic transfers hazardous chemicals to fish and induces hepatic stress. *Scientific Reports, 3* (3263), 7.

Rochman, C. M., Kurobe, T., Flores, I., & Teh, S. J. (2014). Early warning signs of endocrine disruption in adult fish from the ingestion of polyethylene with and without sorbed chemical pollutants from the marine environment. *Science of the Total Environment, 493,* 656–661.

Rodríguez, B., Bécares, J., Rodríguez, A., & Arcos, J. M. (2013). Incidence of entanglement with marine debris by northern gannets (*Morus bassanus*) in the non-breeding grounds. *Marine Pollution Bulletin, 75,* 259–263.

Rothstein, S. I. (1973). Plastic particle pollution of the surface of the Atlantic Ocean: Evidence from a seabird. *Condor, 75,* 344–345.

Ryan, P. G. (1987). The incidence and characteristics of plastic particles ingested by seabirds. *Marine Environmental Research, 23,* 175–206.

Ryan, P. G. (1988). Effects of ingested plastic on seabird feeding: Evidence from chickens. *Marine Pollution Bulletin, 19,* 125–128.

Ryan, P. G. (2008). Seabirds indicate changes in the composition of plastic litter in the Atlantic and South-Western Indian Oceans. *Marine Pollution Bulletin, 56,* 1406–1409.

Ryan, P. G., & Jackson, S. (1986). Stomach pumping: Is killing birds necessary? *The Auk, 103,* 427–428.

Ryan, P. G., & Fraser, M. W. (1988). The use of great skua pellets as indicators of plastic pollution in seabirds. *Emu, 88,* 16–19.

Ryan, P. G., Moore, C. J., van Franeker, J. A., & Moloney, C. L. (2009). Monitoring the abundance of plastic debris in the marine environment. *Philosophical transactions of the Royal Society of London B, 364,* 1999–2012.

Sancho, G., Puente, E., Bilbao, A., Gomez, E., & Arregi, L. (2003). Catch rates of monkfish (*Lophius* spp.) by lost tangle nets in the Cantabrian Sea (northern Spain). *Fisheries Research, 64,* 129–139.

Sazima, I., Gadig, O. B. F., Namora, R. C., & Motta, F. S. (2002). Plastic debris collars on juvenile carcharhinid sharks (*Rhizoprionodon lalandii*) in Southwest Atlantic. *Marine Pollution Bulletin, 44*, 1149–1151.

Schleyer, M. H., & Tomalin, B. J. (2000). Damage on South African coral reefs and an assessment of their sustainable diving capacity using a fisheries approach. *Bulletin of Marine Science, 67*, 1025–1042.

Schlining, K., von Thun, S., Kuhnz, L., Schlining, B., Lundsten, L., Jacobsen Stout, N., et al. (2013). Debris in the deep: Using a 22-year video annotation database to survey marine litter in Monterey Canyon, Central California, USA. *Deep-Sea Research I, 79*, 96–105.

Schuyler, Q. A., Hardesty, B. D., Wilcox, C., & Townsend, K. (2013). Global analysis of anthropogenic debris ingestion by sea turtles. *Conservation Biology, 28*, 129–139.

Schuyler, Q. A., Wilcox, C., Townsend, K., Hardesty, B. D., & Marshall, N. (2014). Mistaken identity? Visual similarities of marine debris to natural prey items of sea turtles. *BMC Ecology, 14*(14), 7.

Shaughnessy, P. D. (1985). Entanglement of grey seals *Halichoerus grypus* at a haul out site in Cornwall, UK. *Marine Pollution Bulletin, 64*, 2815–2819.

Shomura, R. S., & Yoshida, H. O. (1985). *Proceedings of the Workshop on the Fate and Impact of Marine Debris*. Honolulu, Hawaii.

Sievert, P. R., & Sileo, L. (1993). *The effects of ingested plastic on growth and survival of albatross chicks*. Otawa: Canadian Wildlife Service Special Publication.

Sileo, L., Sievert, P. R., Samuel, M. D., & Fefer, S. I. (1990). Prevalence and characteristics of plastic ingested by Hawaiian seabirds. In R. S. Shomura & M. L. Godfrey (Eds.), *Proceedings of the Second International Conference of Marine Debris* (pp. 665–681). Honolulu, Hawaii: U.S. Department of Commerce, NOAA Tech Memo, NMFS.

Simmonds, M. P. (2012). Cetaceans and marine debris: The great unknown. *Journal of Marine Biology, 2012*, 1–8.

Smith, S. D. A. (2012). Marine debris: A proximate threat to marine sustainability in Bootless Bay, Papua New Guinea. *Marine Pollution Bulletin, 64*, 1880–1883.

Smith, S. D. A., & Edgar, R. J. (2014). Documenting the density of subtidal marine debris across multiple marine and coastal habitats. *PLoS One, 9*, e94593.

Spear, L. B., Ainley, D. G., & Ribic, C. A. (1995). Incidence of plastic in seabirds from the tropical pacific 1984–91: Relation with distribution of species, sex, age, season, year and body weight. *Marine Environmental Research, 40*, 123–146.

Stevens, B. G., Vining, I., Byersdorfer, S., & Donaldson, W. (2000). Ghost fishing by tanner crab (*Chionoecetes bairdi*) pots off Kodiak, Alaska: Pot density and catch per trap as determined from sidescan sonar and pot recovery data. *Fisheries Bulletin, 98*, 389–399.

Sturkie, P. D. (1976). Avian Physiology. New York.

Tomas, J., Guitart, R., Mateo, R., & Raga, J. A. (2002). Marine debris ingestion in loggerhead sea turtles, *Caretta caretta*, from the western Mediterranean. *Marine Pollution Bulletin, 44*, 211–216.

Tourinho, P. S., Ivar do Sul, J. A., & Fillmann, G. (2010). Is Marine debris ingestion still a problem for the coastal marine biota of southern Brazil? *Marine Pollution Bulletin, 60*, 396–401.

Townsend, S.E. (2011). Impact of ingested marine debris on sea turtles of eastern Australia: Life history stage susceptibility, pathological implications and plastic bag preference. In *Fifth International Marine Debris Conference* (pp. 136–140). Honolulu, Hawaii.

Triessing, P., Roetzer, A., & Stachowitsch, M. (2012). Beach condition and marine debris: New hurdles for sea turtle hatchling survival. *Chelonian Conservation and Biology, 11*, 68–77.

Tschernij, V., & Larsson, P. O. (2003). Ghost fishing by lost cod gill nets in the Baltic Sea. *Fisheries Research, 64*, 151–162.

Udyawer, V., Read, M. A., Hamann, M., Simpfendorfer, C. A., & Heupel, M. R. (2013). First record of sea snake (*Hydrophis elegans*, Hydrophiinae) entrapped in marine debris. *Marine Pollution Bulletin, 73*, 336–338.

Uhrin, A. V., Fonseca, M. S., & DiDomenico, G. P. (2005). Effect of caribbean spiny lobster traps on seagrass beds of the Florida Keys National Marine Sanctuary: Damage assessment and evaluation of recovery. American Fisheries Society Symposium (pp. 579–588). American Fisheries Society.

Uhrin, A. V., & Schellinger, J. (2011). Marine debris impacts to a tidal fringing-marsh in North Carolina. *Marine Pollution Bulletin, 62*, 2605–2610.

Uhrin, A. V., Matthews, T., & Lewis, C. (2014). Lobster trap debris in the Florida Keys National Marine Sanctuary: Distribution, abundance, density and patterns of accumulation. *Management and Ecosystem Science, 6*, 20–32.

UNEP (2011). *UNEP year book: Emerging issues in our global environment* (79 p). Nairobi: United Nations Environmental Programme.

Uneputty, P., & Evans, S. M. (2009). The impact of platic debris on the biota of tidal flats in Ambon Bay (eastern Indonesia). *Marine Environmental Research, 44*, 233–242.

Van Cauwenberghe, L., & Janssen, C. R. (2014). Microplastics in bivalves cultured for human consumption. *Environmental Pollution, 193*, 65–70.

Van Cauwenberghe, L., Claessens, M., Vandegehuchte, M., & Janssen, C. R. (2012). Occurrence of microplastics in mussels (*Mytilus edulis*) and lugworms (*Arenicola marina*) collected along the French-Belgian-Dutch Coast. p. 88. In J. Mees & J. Seys (Eds.), *Book of Abstracts—VLIZ Young Scientists Day. Brugge, Belgium, 24 February 2012* (pp. xi + 150). Oostende, Belgium: VLIZ Special Publication 55. Flanders Marine Institute (VLIZ).

Van Cauwenberghe, L., Vanreusel, A., Mees, J., & Janssen, C. R. (2013). Microplastic pollution in deep-sea sediments. *Environmental Pollution, 182*, 495–499.

Van der Hoop, J., Moore, M., Fahlman, A., Bocconcelli, A., Gearge, C., et al. (2013). Behavioral impact of disentanglement of a right whale under sedation and the energetic costs of entanglement. *Marine Mammal Science, 30*, 282–307.

Van Franeker, J. A. (1985). Plastic ingestion in the North Atlantic fulmar. *Marine Pollution Bulletin, 16*, 367–369.

Van Franeker, J. A. (2011). Reshape and relocate: seabirds as transformers and transporters of microplastics. In *Fifth International Marine Debris Conference*, Honolulu, Hawaii.

Van Franeker, J. A., & Bell, P. J. (1988). Plastic ingestion by petrels breeding in Antarctica. *Marine Pollution Bulletin, 19*, 672–674.

Van Franeker, J. A., & Meijboom, A. (2002). Litter NSV—marine litter monitoring by northern fulmars. A pilot study, Alterra-rapport 401. Alterra, Wageningen.

Van Franeker, J.A. & Law, K.L. (2015). Seabirds, gyres and global trends in plastic pollution. *Environmental Pollution, 203*, 89–96.

Van Franeker, J. A., Blaize, C., Danielsen, J., Fairclough, K., Gollan, J., Guse, N., et al. (2011). Monitoring plastic ingestion by the northern fulmar *Fulmarus glacialis* in the North Sea. *Environmental Pollution, 159*, 2609–2615.

Verlis, K. M., Campbell, M. L., & Wilson, S. P. (2014). Marine debris is selected as nesting material by the brown booby (*Sula leucogaster*) within the Swain Reefs, Great Barrier Reef, Australia. *Marine Pollution Bulletin, 87*(1–2), 180–190.

Viehman, S., Vander Pluym, J. L., & Schellinger, J. (2011). Characterization of marine debris in North Carolina salt marshes. *Marine Pollution Bulletin, 62*, 2771–2779.

Vliestra, L. S., & Parga, J. A. (2002). Long-term changes in the type, but not amount, of ingested plastic particles in short-tailed shearwaters in the Southeastern Berin Sea. *Marine Pollution Bulletin, 44*, 945–955.

Votier, S. C., Archibald, K., Morgan, G., & Morgan, L. (2011). The use of plastic debris as nesting material by a colonial nesting seabird and associated entanglement mortality. *Marine Pollution Bulletin, 62*, 168–172.

Wabnitz, C., & Nichols, W. J. (2010). Plastic pollution: An ocean emergency. *Marine Turtle Newsletter, 129*, 1–4.

Walker, W. A., & Coe, J. M. (1990). Survey of marine debris ingestion by odontocete cetaceans. In R. S. Shomura & M. L. Godfrey (Eds.), *Proceedings of the Second International Conference of Marine Debris* (pp. 747–774). Honolulu, Hawaii: U.S. Department of Commerce, NOAA Tech Memo, NMFS.

Waluda, C. M., & Staniland, I. J. (2013). Entanglement of Antarctic fur seals at Bird Island, South Georgia. *Marine Pollution Bulletin, 74*, 244–252.

116

S. Kühn et al.

Watters, D. L., Yoklavich, M. M., Love, M. S., & Schroeder, D. M. (2010). Assessing marine debris in deep seafloor habitats off California. *Marine Pollution Bulletin, 60*, 131–138.

Wegner, N. C., & Cartamil, D. P. (2012). Effects of prolonged entanglement in discarded fishing gear with substantive biofouling on the health and behavior of an adult shortfin mako shark, *Isurus oxyrinchus*. *Marine Pollution Bulletin, 64*, 391–394.

Wieczorek, S. K., Campagnuolo, S., Moore, P. G., Froglia, C., Atkinson, R. J. A., Gramitto, E. M., et al. (1999). The composition and fate of discards from *Nephrops* trawling in Scottish and Italian waters. EC 96/092, Millport, U.K.

Woodley, T. H., & Lavigne, D. M. (1991). *Incidental capture of pinnipeds in commercial fishing gear*. ICS. International Marine Mammal Association (Ed.), Technical Report No. 91–01.

Ye, S., & Andrady, A. L. (1991). Fouling of floating plastic debris under Biscayne Bay exposure conditions. *Marine Pollution Bulletin, 22*, 608–613.

Yoshikawa, T., & Asoh, K. (2004). Entanglement of monofilament fishing lines and coral death. *Biological Conservation, 117*, 557–560.

Zavala-González, A., & Mellink, E. (1997). Entanglement of California sea lions, *Zalophus californianuscalifornianus*, in fishing gear in the central-northern part of the Gulf of California, Mexico. *Fishery Bulletin, 95*, 180–184.

Zettler, E. R., Mincer, T. J., & Amaral-Zettler, L. A. (2013). Life in the "plastisphere": Microbial communities on plastic marine debris. *Environmental Science and Technology, 47*, 7137–7146.

Żydelis, R., Small, C., & French, G. (2013). The incidental catch of seabirds in gillnet fisheries: A global review. *Biological Conservation, 162*, 76–88.

Chapter 5
The Complex Mixture, Fate and Toxicity of Chemicals Associated with Plastic Debris in the Marine Environment

Chelsea M. Rochman

Abstract For decades we have learned about the physical hazards associated with plastic debris in the marine environment, but recently we are beginning to realize the chemical hazards. Assessing hazards associated with plastic in aquatic habitats is not simple, and requires knowledge regarding organisms that may be exposed, the exposure concentrations, the types of polymers comprising the debris, the length of time the debris was present in the aquatic environment (affecting the size, shape and fouling) and the locations and transport of the debris during that time period. Marine plastic debris is associated with a 'cocktail of chemicals', including chemicals added or produced during manufacturing and those present in the marine environment that accumulate onto the debris from surrounding seawater. This raises concerns regarding: (i) the complex mixture of chemical substances associated with marine plastic debris, (ii) the environmental fate of these chemicals to and from plastics in our oceans and (iii) how this mixture affects wildlife, as hundreds of species ingest this material in nature. The focus of this chapter is on the mixture of chemicals associated with marine plastic debris. Specifically, this chapter discusses the diversity of chemical ingredients, byproducts of manufacturing and sorbed chemical contaminants from the marine environment among plastic types, the role of marine plastic debris as a novel medium for environmental partitioning of chemical contaminants in the ocean and the toxic effects that may result from plastic debris in marine animals.

Keywords Monomers and additives · Persistent organic pollutants · Metals · Cocktail of contaminants · Toxicity

C.M. Rochman (✉)
Aquatic Health Program, School of Veterinary Medicine,
University of California, Davis, CA, 95616, USA
e-mail: cmrochman@ucdavis.edu

© The Author(s) 2015
M. Bergmann et al. (eds.), *Marine Anthropogenic Litter*,
DOI 10.1007/978-3-319-16510-3_5

5.1 Introduction

Since the Industrial Revolution, there has been an exponential increase in the production and use of chemical substances, such that now the amount of chemicals produced annually is more than 400× greater than the amount produced annually four decades ago (Binetti et al. 2008). Among these chemical substances are several of the ingredients used in the manufacturing of plastics (Lithner et al. 2011). This increasing production and use is inevitably accompanied by an increase in waste, creating a challenge for waste management. Several mechanisms have recently been developed for managing waste, including landfill, wastewater treatment and recycling. Still, these mechanisms are not 100 % efficient and/or do not yet exist in several locations worldwide. The marine environment, residing at the end of most watersheds, is thus often the ultimate sink for many of these substances, including plastic, when not properly managed. As a consequence, plastic debris and many chemical contaminants are detected in our oceans globally.

In parallel with chemicals, the production rate of plastics has increased exponentially, from 0.5 million tons produced annually in 1950 to greater than 299 million tons produced annually today (Thompson et al. 2009; PlasticsEurope 2013). Of this material, less than 50 % was accounted for in the waste stream in 2012 (Rochman et al. 2013a). While some of these products may be still in use, others become litter. Today, marine plastic pollution has become ubiquitous and is reported globally from the ocean surface (Thompson et al. 2004; Goldstein et al. 2013; Eriksen et al. 2014; Law et al. 2014; Desforges et al. 2014) to the deep sea (Goldberg 1997; Galgani et al. 2000).

For decades we have learned about the physical hazards associated with this pollution in the marine environment (Laist 1987), but recently we are beginning to realize the chemical hazards. Marine plastic debris is associated with a 'cocktail of chemicals', including chemicals added or produced during manufacturing (Lithner et al. 2011) and those present in the marine environment that accumulate onto the debris from surrounding seawater (Mato et al. 2001; Ogata et al. 2009). This begs several questions regarding: (i) the complex mixture of chemical substances associated with marine plastic debris, (ii) the environmental fate of these chemicals to and from plastics in our oceans and (iii) how this mixture affects wildlife, as hundreds of species ingest this material in nature (CBD 2012). The focus of this chapter is on the mixture of chemicals associated with marine plastic debris. Specifically, this chapter discusses the diversity of chemical ingredients, byproducts of manufacturing and sorbed chemical contaminants from the marine environment, the role of marine plastic debris as a novel medium for environmental partitioning of chemical contaminants in the ocean and the toxic effects that may result from plastic debris in marine animals.

5.1.1 Plastic Marine Debris: A Complex Mixture of Chemicals

Marine plastic debris is associated with a complex mixture of chemicals, including those that are ingredients of the plastic material (e.g. monomers and additives),

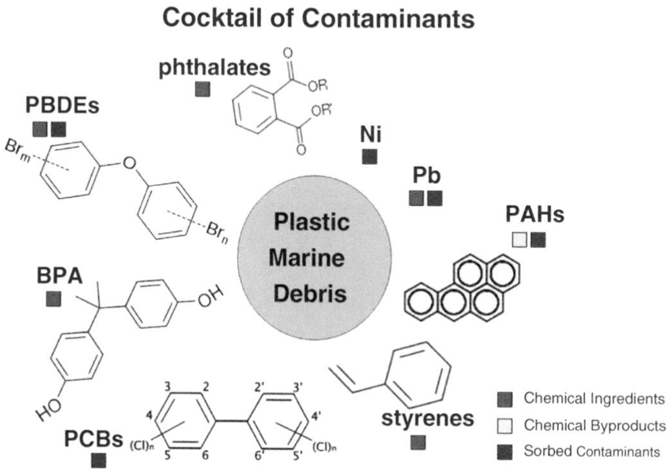

Fig. 5.1 Cocktail of contaminants associated with marine plastic debris. Contaminants associated with marine debris include chemical ingredients (*red squares*), byproducts of manufacturing (*yellow squares*) and those that accumulate from surrounding ocean water in the marine environment (*blue squares*)

byproducts of manufacturing (e.g. chemicals composed during the combustion of the raw material petroleum) and chemical contaminants in the ocean that accumulate on plastic when it becomes marine debris (e.g. persistent organic pollutants (POPs) and metals). There is evidence that this mixture, or 'cocktail of contaminants' (Fig. 5.1; Rochman 2013), can be bioavailable to whales (Fossi et al. 2012, 2014), basking sharks (Fossi et al. 2014), seabirds (Teuten et al. 2009; Tanaka et al. 2013; Lavers et al. 2014), amphipods (Chua et al. 2014), crickets (Gaylor et al. 2012), lugworms (*Arenicola marina*) (Besseling et al. 2013; Browne et al. 2013) and fish (Rochman et al. 2013b) upon ingestion. This is a cause for concern, as the (USEPA 2013) and the European Union (European Commission 2014) list several of these chemicals as priority pollutants because they are persistent, bioacummulative and/or toxic. In fact, of the chemicals listed as priority pollutants by the US EPA, 78 % are associated with marine plastic debris (Rochman et al. 2013a). This section discusses the complex mixture of chemicals associated with marine plastic debris, including those that originate from manufacturing, that accumulate from surrounding ocean water and how this mixture may vary according to the location where plastic is discarded and the plastic type.

5.1.2 Plastics and Their Chemical Ingredients

There are several different types of plastics manufactured into a diversity of products. Each is produced by polymerizing individual monomers, forming the backbone of the polymer. These are made using solvents and other chemicals that may be used as initiators and catalysts. Next, several additives (e.g. flame retardants, stabilizers, pigments and fillers) are included to give the plastic certain character-istics (e.g. flexibility, strength and color; OECD 2004; Lithner et al. 2011). Such chemicals, in addition to byproducts, may be released during production, use and disposal of the product, several of which can be harmful (Oehlmann et al. 2009; Teuten et al. 2009; Halden et al. 2010; Lithner et al. 2011; Papaleo et al. 2011). According to United Nations and European Union frameworks, >50 % of the plas-tics that are produced are hazardous based upon their constituent monomers, addi-tives and byproducts (Lithner et al. 2011).

The backbone structure, derived from long chains of monomers, are thought to be biochemically inert due to their large molecular size (Teuten et al. 2009; Lithner et al. 2011). Still, several of these are shown to have harmful effects (Xu et al. 2004; Halden et al. 2010; Lithner et al. 2011). Bisphenol A, used in the production of polycarbonate, can have endocrine disrupting effects (Oehlmann et al. 2009; Crain et al. 2007; Halden et al. 2010) and the styrene and polyvi-nyl chloride monomer, used in the production of polystyrene and polyvinyl chloride (PVC), can be carcinogenic and/or mutagenic (Papaleo et al. 2011; Xu et al. 2004; Lithner et al. 2011) and are listed as toxic substances by the USEPA, ATSDR and OSPAR.

Several of the chemicals used in the production process, including solvents, suspension aids, surfactants, initiators, catalysts and byproducts can also be harm-ful upon exposure (Lithner et al. 2011). For example, tributyltin (shown to cause endocrine disruption in molluscs; Oehlmann et al. 1996) and copper chloride (shown to have developmental effects on fish; Anderson et al. 1991) are added as catalysts during production (Lithner et al. 2011) and several solvents (shown to be carcinogenic; Lynge et al. 1997) are used in the production process (e.g. methanol, cyclohexane and 1,2-dichlorobenzene; Braun et al. 2005; Gowariker et al. 2003; Lithner et al. 2011).

The additive ingredients include plasticizers, antioxidants, flame-retardants and UV-stabilizers. In some cases, the ingredients make up a large proportion of the plastic product. Phthalates may constitute up to 50 % of the total weight of PVC plastics (Bauer and Herrmann 1997). The use of additives is also not equally distributed across plastic types—PVC requires the most additives accounting for 73 % of the world production of additives by volume, followed by polyethylene and polypropylene (10 % by volume) and styrenics (5 % by vol-ume) (Lithner et al. 2011). Several of these have been recognized or suggested to be hazardous, including the brominated flame retardants (PBDEs), phthalate plasticizers and lead heat stabilizers (Oehlmann et al. 2009; Halden et al. 2010; Lithner et al. 2011).

Finally, some hazardous chemicals may be produced as byproducts during manufacturing. PAH formation occurs during the production cycle of polysty-rene (Zabaniotou and Kassidi 2003; Kwon and Castaldi 2008). Residuals of these chemicals may be difficult to remove, therefore carry over into the plastic product and become one of the many chemicals in the cocktail of contaminants associated with marine plastic debris. Thus, when considering the hazards associated with plastic debris, it is important to consider polymer type.

5.1.3 The Accumulation of Chemicals on Plastic Debris in the Marine Environment

Because of their physical and chemical properties, plastics accumulate a complex mixture of chemical contaminants present in the surrounding seawater (Mato et al. 2001; Teuten et al. 2007, 2009; Rochman et al. 2013c; Holmes et al. 2012; Engler 2012), adding to the cocktail of chemicals already present from manufacturing. As a result of widespread global contamination of chemical contaminants (Ogata et al. 2009; Ross and Birnbaum 2010) and plastic debris (Thompson et al. 2004; Barnes et al. 2009; Browne et al. 2011), marine plastic debris is recovered globally with measurable amounts of POPs (e.g. polychlorinated biphenyls (PCBs), PAHs and PBDEs) and other persistent bioaccumulative and toxic substances (PBTs) (e.g. halogenated flame retardants, pesticides and nonylphenol; Mato et al. 2001; Endo et al. 2005; Ogata et al. 2009; Hirai et al. 2011; Heskett et al. 2012; Rios et al. 2010) and metals (e.g. lead, copper and cadmium; Ashton et al. 2010; Holmes et al. 2012; Rochman et al. 2014a).

PBTs, which include those listed as POPs by the Stockholm Convention, generally have a low water-solubility (i.e. are hydrophobic) and tend to partition out of the water column and onto another environmental matrix with similar hydrophobic properties (e.g. sediment, organic matter); thus, when PBTs encounter plastic debris they tend to sorb to this material (Engler 2012). Thus, it is not surprising that an early study reported PCBs on marine plastic debris (Carpenter and Smith 1972) or that plastics are used as passive samplers to quantify PBTs in aquatic environments (Huckins et al. 1993; Lohmann 2012).

Today, the accumulation of PBTs on plastic debris is unequivocal. Global samples show the presence of PBTs on plastic debris collected from coastal beaches (Van et al. 2011; Heskett et al. 2012; Fries et al. 2012; Fisner et al. 2013; Antunes et al. 2013) all the way to the remote open-ocean (Rios et al. 2007, 2010; Hirai et al. 2011). As such, plastic pre-production pellets, a recognizable component of marine debris, are now used to examine the global pattern of PBTs (Ogata et al. 2009; Takada et al. 2006), acting as passive samplers and providing baseline information regarding PBT contamination in the ocean. International Pellet Watch leads this effort, collecting plastic pellets globally and measuring the concentrations of various PBTs sorbed to plastic debris (Takada et al. 2006; Ogata et al. 2009; see also Fig. 5.3 in Hidalgo-Ruz and Thiel 2015 in this volume).

The presence of organic chemicals on plastic debris may be established globally, but the presence of a complex mixture of metals on plastic debris has only been recently demonstrated (Ashton et al. 2010; Holmes et al. 2012; Nakashima et al. 2011, 2012; Rochman et al. 2014a). Similar to organic chemicals, several metals have long been additive ingredients of plastics (e.g. lead added to PVC; Lithner et al. 2011; Nakashima et al. 2011, 2012), but now we have evidence that plastic debris accumulates metals from ocean water (Ashton et al. 2010; Holmes et al. 2012; Rochman et al. 2014a). Environmental accumulation of metals onto plastics may have been expected, as the surfaces of plastic containers are known to accumulate metals from water samples (Fischer et al. 2007; Weijuan et al. 2001; Robertson 1968). The accumulation of metals on marine plastic debris may be explained by both the chemical ingredients of the plastic (e.g. catalysts, fillers, plasticizers; Robertson 1968) and the degradation and fouling of aquatic plastic debris via microbial biofilms and colonization by algae and invertebrates (Holmes et al. 2012; Tien and Chen 2013) that may generate active sites for the sorption and/or bioaccumulation of metals. As such, similar to organic chemicals, plastic pellets may also serve as a passive sampler for metal contamination in the marine environment.

5.1.3.1 Spatial Variability

It should be noted that, similar to passive samplers, the types and concentrations of sorbed chemicals associated with marine plastic debris reflect the types and concentrations of chemical contaminants in ambient seawater. As such, the type and concentrations of chemicals sorbed to plastic debris will vary based upon the location that the debris is recovered on large (Ogata et al. 2009; Hirai et al. 2011; Fig. 5.2) and small spatial scales (Rochman et al. 2013c). Data from International Pellet Watch show that PCBs on plastic pellets collected from beaches spanning over 1000 km of the California coastline vary by one order of magnitude (23–605 ng/g; www. pelletwatch.org). Similarly, pellets recovered from eight beaches throughout a single Californian county, San Diego (3.8–42 ng/g total PCBs; Van et al. 2011), and pellets deployed at different locations within a single bay, San Diego Bay (3.4–35 ng/g; Rochman et al. 2013c), had concentrations of PCBs that also varied by one order of magnitude. Moreover, variations in concentrations of metal contaminants recovered from four beaches along a stretch of coastline in Devon, England (Ashton et al. 2010) and deployed at several locations within the San Diego Bay, CA (Rochman et al. 2014a) varied similarly to organic contaminants, also by one order of magnitude. Thus, variability in contaminant burden

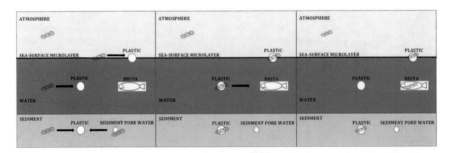

Fig. 5.2 Environmental fate diagram including plastic debris. The diagram represents how plastic debris may mediate the fate of some contaminants (e.g. PAHs) among different environmental compartments, including the sea-surface microlayer, water, sediment, sediment pore water and biota, in the marine environment

reflects local sources and global chemical contamination. Such spatial variation has implications for the management, as discarded plastics may pose a greater hazard and thus a greater management priority in locations with greater point sources and greater concentrations of chemical pollutants.

5.1.3.2 Variability by Plastic Type and Size

Also of implication for management is how the accumulation of chemical contaminants may vary among type and size of plastic debris. If some plastic types tend to carry a smaller burden of contaminants due to their physical and chemical properties, items that often become marine debris (e.g. fishing and aquaculture gear; Andrady 2011) could be produced out of potentially safer plastic types.

There are several reasons why we might expect the behavior of contaminants to vary according to plastic type. The physical and chemical properties of each type of plastic [e.g. surface area (Teuten et al. 2007), diffusivity (Karapanagiot and Klontza 2008; Pascall et al. 2005; Rusina et al. 2007; Mato et al. 2001) and crystallinity (Karapanagioti and Klontza 2008; Mato et al. 2001)] influence the accumulation of chemicals to plastic debris (Pascall et al. 2005; Rusina et al. 2007) and accumulation patterns will be compound-specific (e.g. increasing in affinity to the polymer with greater hydrophobicity; Smedes et al. 2009). For organic chemicals, several studies show that polyethylene, polypropylene and polystyrene sorb greater concentrations of organic contaminants than PVC and polyethylene terephthalate (PET; Pascall et al. 2005; Teuten et al. 2007; Karapagioti and Klontza 2008; Rochman et al. 2013c, d). For some chemicals (i.e. PAHs), polyethylene and polystyrene sorb greater concentrations than polypropylene (Lee et al. 2014; Rochman et al. 2013c), whereas for others (i.e. PCBs) there is no detectable difference (Rochman et al. 2013c). Rubbery polymers, such as polyethylene and polypropylene are expected to demonstrate greater diffusion than the glassy polymers, PET and PVC, which may explain their greater sorptive capacity (Pascall et al. 2005). Polyethylene has a greater sorptive capacity than polypropylene (Teuten et al. 2007), probably due to its greater surface area (Teuten et al. 2007) and free volume (Pascall et al. 2005). Moreover, diffusion into the polymer has been observed in polyethylene pellets, but not in polypropylene (Karapanagioti and Klontza 2008). This is likely the reason that polyethylene has a large affinity for a wide range of organic contaminants varying in hydrophobicity (Müller et al. 2001) and is often used as a passive-sampling device (Lohmann 2012; Pascall et al. 2005). For polystyrene, the presence of benzene increases the distance between adjacent polymeric chains, which can make it easier for a chemical to diffuse into the polymer (Pascall et al. 2005), and may explain why contaminants sorb similarly to polystyrene as they do to polyethylene, despite being a glassy polymer (Pascall et al. 2005).

These trends may not extend to all classes of contaminants. Compound-specific interactions in the polymer phase are also important (Smedes et al. 2009). For example, PVC has a greater affinity for alkylbenzenes than does polyethylene

(Wu et al. 2001). Moreover, for metal contaminants, there does not appear to be a large difference in sorption concentration among polymer types (Rochman et al. 2014a). A possible explanation is that the accumulation of metals to plastic may be immaterial, and the process may be mediated by a biofilm (Rochman et al. 2014a). In the aquatic environment, including marine systems, it is well established that biofilms have sorptive properties and accumulate metals and other contaminants (Decho 2000; Tien et al. 2009) and it has been suggested that the composition of biofilm does not vary significantly among plastic types (Ye and Andrady 1991; Zettler et al. 2013).

The sorption behavior of chemicals to plastic will also vary by size. Because of the difference in surface area, plastic of differing sizes will sorb contaminants accordingly (Koelmans et al. 2013; Velzeboer et al. 2014). Size will affect both the sorptive capacity and the rate at which chemicals are sorbed (Teuten et al. 2009). For example, nano- and micrometre-sized plastic debris may exchange organic chemicals faster than millimetre-sized plastic debris due to its larger surface area and short diffusion path lengths (Koelmans et al. 2013). Plastic may enter the ocean as nano- and micro-sized debris (e.g. as microscrubbers or laundry lint; Browne et al. 2011) or it may become smaller over time via photodegradation. Photodegradation of the polymer surface accelerates and increases the sorption capacity by altering surface properties and increasing surface area (Mato et al. 2001; Holmes et al. 2012; Rochman et al. 2013). The combination of increased sorption rate and capacity in smaller plastic debris may constitute increased risk in marine organisms (Velzeboer et al. 2014).

Thus, when assessing the hazard associated with plastic debris it is important to think holistically. Patterns are not simple or straightforward. While several lines of evidence show that polyethylene, polypropylene and polystyrene accumulate relatively large concentrations of some contaminants (e.g. POPs), polyethylene and polypropylene are made from the least hazardous monomers (Lithner et al. 2011) while polystyrene is made from the styrene monomer, which is a priority pollutant. Thus, the complex mixture of chemicals associated with plastic debris will be dependent on the type and size of the plastic and the location where it becomes marine debris.

5.1.4 Plastic Debris, Environmental Chemical Contaminants and Environmental Fate

The long-range transport, persistence and global dynamics of plastic debris are key aspects to understanding the ultimate fate of this material and any potential impacts of plastic debris on marine ecosystems. Because it is now globally accepted that plastic debris accumulates chemical contaminants (Ogata et al. 2009; Teuten et al. 2009), it is also important to understand how plastic debris mediates these same key aspects for environmental chemical contaminants. This then begs questions regarding: (1) how plastic debris fits into environmental fate models for chemical contaminant distribution and (2) how important plastic debris is relative to other media (e.g. water, sediment, biota) in driving processes of chemical distribution in the global oceans.

Chemical contaminants partition onto various environmental media, a process dependent upon the physical and chemical properties of each chemical and the physical and chemical properties of the environmental medium (e.g. sediment, water, organic matter, living biota). These processes, along with the chemical, physical and biological degradation of each chemical contaminant (Sinkonnen et al. 2000), help to determine their environmental fate globally. The addition of plastic to the marine environment adds a novel medium for chemical contaminants to interact with, and thus it is important to understand how plastic debris should be considered in future environmental fate models (Fig. 5.3). This section will discuss plastics as a novel environmental matrix and its potential role in helping to mediate the fate and distribution of chemical contaminants globally. Specifically, this section will discuss plastic debris as a sink and a source for chemical contaminants in the marine environment and how plastic may facilitate the global transport of chemicals in the marine environment and the transport of chemicals into marine foodwebs.

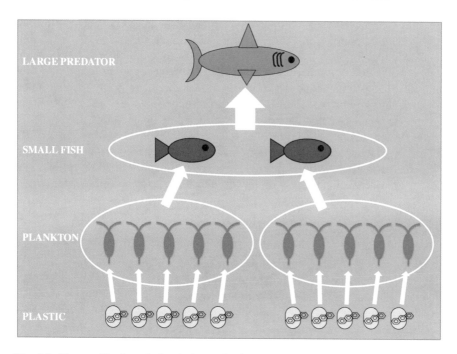

Fig. 5.3 Biomagnification of chemicals up the food chain. The diagram depicts a scenario whereby organic chemicals (e.g. PAHs) from plastic may transfer into lower trophic level organisms (e.g. zooplankton) via ingestion and accumulate at much greater concentrations via biomagnification in higher trophic level organisms (e.g. small fish and sharks), which may ultimately lead to contaminated seafood for humans as a result of plastic contamination in marine foodwebs. The size of the arrows depicts how the body burden (i.e. bioaccumulation of chemicals) may magnify in predators as compared to their prey

5.1.5 Plastic Debris as a Sink for Environmental Contaminants

There is no doubt that plastic debris acts as a sink for chemical contaminants in the marine environment. What is less understood is the process by which this occurs, how this varies over space and time and by polymer and chemical, and how important plastic is as a sink for chemicals relative to other environmental media (e.g. sediment, water and biota). The primary focus of this section will be on processes of chemical accumulation, temporal trends and comparisons among other environmental media.

5.1.5.1 Process of Accumulation

As noted above, the concentration of chemicals that sorb to plastic debris varies according to polymer type and chemical substance. One reason for these differences is the mechanism by which chemicals accumulate on the plastic. For some plastics (i.e. polyethylene and polyoxymethylene), organic chemicals absorb into the polymeric matrix (providing greater surface area for chemicals to accumulate), whereas for other plastics the chemicals adsorb to the surface (Karapanagioti and Klontza 2008). As discussed above, these differences are related to the structure of each polymer type. Despite this, observations are not consistent among contaminant groups. As such, the polymer structure may be less important and instead accumulation may be primarily facilitated by other factors. In this case, it is important to note that changes that occur to the plastic when it becomes marine debris (e.g. fouling and degradation) can alter the structure of the plastic debris (e.g. increasing their surface area and/or charge; Artham et al. 2009; Holmes et al. 2012, 2014), changing the way of how chemicals accumulate on the material.

5.1.5.2 Rate of Accumulation

Alterations of the plastic material when it becomes marine debris also impact the rate at which chemicals accumulate, also varying by polymer type and chemical of interest. For example, chemicals with less hydrophobicity and a lighter molecular weight reach saturation faster than those with greater hydrophobicity and a heavier molecular weight (Müller et al. 2001; Rochman et al. 2013c). Moreover, the mechanism by which chemicals accumulate will affect the rate of accumulation. For example, we observe faster saturation of POPs on PET and PVC whose glassy structure allow for adsorption only, than for polyethylene, where diffusion into the polymeric matrix facilitates a rapid adsorption to the surface followed by a slower increase via absorption (Rochman et al. 2013c).

Of concern for management is the slower rate of accumulation that occurs in the marine environment when compared to a laboratory setting. In a laboratory, chemicals (e.g. PAHs and metals) reach equilibrium on plastic in less than 72 h (Teuten et al. 2007; Holmes et al. 2012), whereas in the marine environment equilibrium occurs much slower (Mato et al. 2001; Rochman et al. 2013c). For example, on plastic pellets of various types (PET, PVC, polyethylene and polypropylene) that were deployed in a contaminated bay for up to 1 year, neither PCBs, PAHs or metals reached equilibrium for at least 3 months and in several cases did not reach equilibrium within the 1-year time period (Rochman et al. 2013c, 2014a). In the marine environment, the surface properties of the plastic debris consistently change. As plastic debris weathers it gains surface area, generates oxygen groups (increasing polarity; Mato et al. 2001; Fotopoulou and Karapanagioti 2012) and fouls (increasing their charge, roughness and porosity) (Artham et al. 2009)—all allowing plastic debris to accumulate increasingly larger concentrations of chemical contaminants (Holmes et al. 2012; Fotopoulou and Karapanagioti 2012; Rochman et al. 2013c, 2014a). Thus, in general, the longer the plastic is in the water, the greater concentrations of chemical contaminants it will accumulate (Engler et al. 2012), suggesting that plastic debris may become more hazardous the longer it remains at sea.

5.1.5.3 Comparisons with Other Environmental Media

Of greatest concern for management appears to be how the concentrations of hazardous chemicals accumulating on plastic debris compared to other environmental media. For several chemical groups it has been shown that pollutants can partition to plastic at greater than or equal concentrations than other environmental media. For example, POPs can accumulate on plastic debris at concentrations up to six orders of magnitude greater than ambient water (Ogata et al. 2009). Floating plastics are associated with the sea-surface microlayer, at the interface between the ocean and atmosphere, where large concentrations of chemical contaminants (e.g. PBTs and metals) accumulate due to its unique chemical composition (i.e. of lipids, fatty acids and proteins). Here, concentrations of organic contaminants are found at concentrations up to 500 × greater than underlying waters (Wurl and Obbard 2004). As such, floating plastic debris could easily accumulate relatively large concentrations of chemical contaminants from this sea-surface microlayer (Mato et al. 2001), and may be one reason why we find large concentrations of chemicals on floating plastic debris recovered globally, including in remote regions (Heskett et al. 2012; Hirai et al. 2011). When comparing plastic debris to other solid matrices, concentrations of POPs on plastics have been found to accumulate at concentrations up to two orders of magnitude greater than on sediment and suspended particulates (Mato et al. 2001; Teuten et al. 2007) and concentrations of metals on plastics have been found at similar concentrations (Holmes et al. 2012) to those on nearby sediment. Still, a thermodynamically-based model, assuming equilibrium, predicts that with the current concentrations of plastic debris in the oceans the total fraction of POPs sorbed to plastic debris is negligible in relation to all other media globally (i.e. <1 %; Gouin et al. 2011).

The accumulation of chemicals on plastic debris has several potential implications for management. It could be positive if plastic debris aids in the removal of some chemicals from the environment. For example, they may act as a permanent sink when plastic debris transports vertically to the bottom of the ocean in the same way that sinking natural particles (e.g. phytoplankton cells and fecal pellets) are considered a final sink when they sequester POPs (Dachs et al. 2002). However, it may also be considered negative if it aids in the transport of hazardous chemicals to remote regions of the world and/or to marine food webs. As such, it is important to understand how plastic debris acts as a sink, and also as a source.

5.1.6 Plastic Debris as a Source of Environmental Contaminants

Plastic, like other PBTs (e.g. POPs; Sinkonnen et al. 2000; Dachs et al. 2002), are persistent and bioaccumulative, and thus can be transported long distances via ocean currents (Law et al. 2010; Maximenko et al. 2012) or by the migration of ocean life. As such, plastics debris may play a role in the transport of sorbed chemical contaminants and chemical ingredients globally (Engler et al. 2012; Cheng et al. 2013; Endo et al. 2013; Kwon et al. 2014).

5.1.7 Global Transport

While sorbed onto floating plastic debris, chemical contaminants may be transported long distances, including across or even to adjacent oceans (Zarfl and Matthies 2010; Engler et al. 2012). Negatively buoyant plastics, or plastic debris that becomes negatively buoyant upon fouling, will sink to the seafloor transporting any sorbed contaminants to the benthos where sediment-dwelling organisms reside. If these chemicals are released upon degradation of the material, plastic debris may be a source of chemical contaminants into pelagic and benthic marine habitats (Teuten et al. 2007; Hirai et al. 2011). While some contaminants may be lost due to biological or physical degradation (Sinkonnen et al. 2000; Rochman et al. 2013c), leaching of chemicals back to the environment may be of concern in remote and more pristine regions where sources of chemical contaminants are sparse (Teuten et al. 2007; Hirai et al. 2011; Heskett et al. 2012). Laboratory studies have found that plastics with sorbed POPs release a considerable amount of these chemicals upon being placed in clean water (Teuten et al. 2007; Endo et al. 2013).

The behavior of chemicals from plastic debris will likely be dependent upon location-specific considerations that include temperature, salinity, the intensity of solar radiation, biodegradation rates, and the presence of co-contaminants (Sinkonnen et al. 2000; Dachs et al. 2002; Bakir et al. 2012, 2014; Holmes et al. 2014). This process will also vary according to the hydrophobicity of the chemicals, such that chemicals with a greater hydrophobicity desorb much slower and may take years or even centuries to fully attain equilibrium (Endo et al. 2013). This has implications for management, as certain chemicals may transfer long distances holding onto the plastic as it migrates from a contaminated region to one that is remote and/or more pristine. Plastic debris sampled from remote regions with sporadic large concentrations of chemical support this theory (Hirai et al. 2011; Heskett et al. 2012).

A further consideration is how the transport of chemicals hitchhiking on plastic debris compares to other transport mechanisms, such as atmospheric or ocean currents. A group of researchers used thermodynamically-based model calculations (assuming sorptive equilibrium) to determine the relative importance of plastic debris as a source of PBTs to the remote Arctic Ocean (Zarfl and Matthies 2010). Their models conclude that transport via atmospheric and ocean currents are orders of magnitude larger than via plastic particles, determining that the contribution of PBTs from plastic debris may be negligible compared with annual PBT flux from other global-transport mechanisms (Zarfl and Matthies 2010). The authors warn that their model estimations include considerable uncertainty and suggest that future studies test the importance of plastic-mediated transport for chemicals with greater hydrophobicity and that are not generally transported via air or ocean currents (Zarfl and Matthies 2010). Moreover, there is a need to better understand the influence from different types of polymers and chemical contaminants (Gouin et al. 2011). The physical and chemical properties (e.g. boiling point, vapor pressure, water solubility and octanol-water partitioning) of the monomers and additive ingredients in addition to properties of the polymer (e.g. the size of the plastic and its pore size) are important when assessing the environmental fate of associated chemicals (Teuten et al. 2009; Lithner et al. 2011). For example, glassy polymers, like PVC, have a slower desorption rate than rubbery polymers, such as polyethylene (Teuten et al. 2009). Moreover, one should consider how desorption may differ in the presence of the microbial biofouling that populates plastic debris in the marine environment (Zettler et al. 2013) and may provide greater surface area for sorption, biodegrade and/or transform the chemical contaminants, or facilitate chemical leaching or transport into other environmental media, including the biota (Gouin et al. 2011).

5.1.7.1 Food Web Transport

Several researchers have tried to understand the role of plastic debris as a source of chemical contaminants into the foodweb, raising several questions regarding: (i) whether contaminants transfer from the plastic to animals upon ingestion, (ii) how important this may be relative to other sources of contaminants in foodwebs and (iii) if contaminants from plastic debris biomagnify in top predators. These questions have been explored using computer modeling (Teuten et al. 2007; Gouin et al. 2011; Koelmans et al. 2013, 2014; Koelmans 2015), assessing correlations between plastic ingestion and chemical body burdens (Ryan et al. 1988;

Teuten et al. 2009; Yamashita et al. 2011; Tanaka et al. 2013; Lavers et al. 2014) and/or using experimental techniques to measure the bioaccumulation of chemicals from plastic in laboratory animals (Gaylor et al. 2012; Besseling et al. 2013; Browne et al. 2013; Rochman et al. 2013b; Chua et al. 2014). Modeling approaches are useful for interpreting experimental and observation data, as well as for risk assessment of the hazards caused by plastic ingestion in wildlife (Koelmans 2015). Because this is discussed in detail by Koelmans (2015), the discussion here will be limited to observational data in the field and experimental data in the laboratory.

There are several lines of evidence suggesting that chemical contaminants do transfer from plastic debris to marine animals. Correlative evidence in the field and laboratory shows that the concentrations of PCBs (Ryan et al. 1988; Teuten et al. 2009; Yamashita et al. 2011) and trace metals (Lavers et al. 2014) in seabirds are positively correlated with the mass of ingested plastic. Moreover, seabirds collected from the North Pacific were found with similar congener patterns of PBDEs in their tissues as those found on the ingested plastic in their gut content (Tanaka et al. 2013) and myctophid fish collected from the South Atlantic were found with similar congener patterns of PBDEs in their tissues as those found on the plastic debris in the region (Rochman et al. 2014b). These observational data suggest that plastic-associated chemicals from plastic do transfer to wildlife upon ingestion.

This hypothesis has been further investigated in controlled laboratory studies, providing a stronger weight of evidence. Two studies demonstrated the bioaccumulation of additive PBDEs in crickets (*Acheta domesticus*; Gaylor et al. 2012) and amphipods (*Allorchestes compressa*; Chua et al. 2014) as a result of the ingestion of plastic. Another study showed greater concentrations of PCBs in lugworms exposed to contaminated sediment with polystyrene as opposed to contaminated sediment without plastic, suggesting that the existence of the plastic in the experiment facilitated the transfer of chemicals to lugworms (*Arenicola marina*; Besseling et al. 2013). Another laboratory study demonstrated that both additive chemicals and chemicals that accumulate in nature (nonylphenol, phenanthrene, PBDE-47 and triclosan) desorb from PVC and can transfer into the tissues of lugworms upon ingestion (*A. marina*; Browne et al. 2013). Lastly, a study measuring the bioaccumulation of POPs sorbed to plastics demonstrated the transfer of chrysene, PCB 28 and several congeners of PBDEs to fish from the ingestion of polyethylene pellets (Rochman et al. 2013b). Thus, there is strong evidence showing that chemical contaminants can bioaccumulate in marine life when plastic debris is ingested. What remains less understood, is whether these plastic-associated contaminants biomagnify in higher trophic level animals as a direct result of plastic ingestion (potentially leading to bioaccumulation of plastic-derived chemicals in seafood; Fig. 5.3) and how important bioaccumulation from plastic is relative to bioaccumulation from other sources of chemical contamination in the environment (e.g. chemical contamination that is ubiquitous in water, sediments and food webs globally; Ross and Birnbaum 2010). Still, the fact that chemicals from plastic debris can transfer to marine animals begs the question, how do these chemicals associated with plastic debris impact marine organisms?

5.2 Toxicity of Plastic Debris to Marine Life

Hazardous substances may be emitted during all phases of the life cycle of a plastic product (Lithner et al. 2011; Galloway 2015). When the ultimate fate of a plastic product is the marine environment, plastic debris carries a cocktail of contaminants, including those that accumulate on the material from the ocean water. If these chemicals become bioavailable, they can penetrate cells and chemically interact with biologically important molecules, and may cause adverse effects including changes in behavior (Browne et al. 2013), liver toxicity (Rochman et al. 2013b) and endocrine disruption (Teuten et al. 2009; Rochman et al. 2014c). This section will discuss what is currently understood regarding the potential for chemicals associated with marine plastic debris to impact marine organisms.

5.2.1 Hazards Associated with Plastic Ingredients

Several of the ingredients associated with plastics are considered hazardous by regulatory agencies (Lithner et al. 2011; Rochman 2013; Browne et al. 2013; USEPA 2013; European Commision 2014). Polymerization reactions are rarely complete and unpolymerized residual monomers can migrate off the plastic (Lithner et al. 2011). Moreover, additive ingredients are not usually bound to the polymer matrix, and often account for the major leaching and emissions of chemical substances from plastic materials (Lithner et al. 2011; Engler et al. 2012). Release of hazardous substances, including phthalates, brominated flame retardants, bisphenol A, formaldehyde, acetaldehyde, 4-nonylphenol and many volatile organic compounds, from plastic products has been shown (Crain et al. 2007; Lithner et al. 2012). As such, these chemicals may be bioavailable to marine life and thus there is potential for organisms to be impacted by the chemical ingredients associated with plastic debris.

For some plastics, the monomer that makes up the polymer itself is classified as hazardous. For example, polyurethane foam, PVC, polycarbonate and high-impact polystyrene, are composed of monomers that are considered carcinogenic, mutagenic or toxic for reproduction (Lithner et al. 2011). Other monomers that have been described as the most environmentally hazardous are m-phenylenediamine, p-phenylenediamine, 1,4-dichlorobenzene and the phthalate plasticizer BBP (used as a monomer in some PVC), all of which have been found to be acutely toxic to aquatic life (Lithner et al. 2011).

Several plastics are composed of monomers considered to be non-hazardous (e.g. polyethylene and polypropylene), but contain harmful additives. Some of the most hazardous additives include brominated flame retardants, polyfluoronated compounds, triclosan, phthalate plasticizers and lead heat stabilizers (Halden et al. 2010; Lithner et al. 2011). Phthalates, for example, have been shown to target nuclear hormone receptor signaling pathways (Grün and Blumberg 2007) and cause endocrine disrupting effects in fish (Kim et al. 2002). Adverse effects related to the brominated flame-retardants include neurobehavioral development disorders, thyroid hormone alterations, teratogenicity and reduction in spawning success (Darnerud 2003; de Wit 2002). Other additives have hazardous degradation products. For example, nonylphenol is a degradation product of nonylphenol ethoxylates, a surfactant, and can cause endocrine disruption in fish (Gray and Metcalfe 1997; Seki et al. 2003; Kawahata et al. 2004).

When trying to understand how plastic debris may impact ocean organisms, it is critical to measure effects at environmentally relevant concentrations and under environmentally relevant exposure conditions (Rochman and Boxall 2014). While some evidence of toxicity for these substances occurs at levels greater than those found in the environment, for several chemical ingredients (e.g. phthalates and bisphenol A) adverse effects have been demonstrated at environmentally relevant concentrations (Crain et al. 2007; Oehlmann et al. 2009). Moreover, organisms are rarely exposed to one chemical in isolation, and the interaction of several chemicals may induce synergistic effects. As such, when considering the impacts to organisms from plastic debris one must consider the complex mixture of chemicals associated with this material in the marine environment.

5.2.2 Hazards Associated with the Complex Mixture of Plastic and Sorbed Pollutants

Here, we will not focus on any physical adverse effects from the material itself, although it is worth noting that plastic debris may act as a multiple stressor to marine organisms as a result of the combination of both physical and chemical stressors (Rochman 2013). In this chapter, we will discuss the existing evidence of adverse chemical effects from the complex mixture of chemicals associated with plastic products and plastic marine debris. See Kühn et al. (2015) and Lusher (2015) for information regarding any adverse effects from plastic debris not related to the chemical impacts.

Some studies have assessed the toxicity of the leachates from plastic products. These incorporate adverse effects from the complex mixture of chemical ingredients associated with the material. One researcher exposed *Daphnia magna* to leachates from several plastic products and found that all leachates from PVC, polyurethane and epoxy products were acutely toxic (48-h EC50 s) at concentrations ranging from 2 to 235 grams of plastic per liter of water (Lithner et al. 2009, 2012). Another study found that most of the 500 plastic products sampled leached chemicals that had estrogenic activity, detected by an E-screen assay (Yang et al. 2011). Similarly, Wagner and Oehlmann (2009, 2011) detected estrogenic contamination in PET water bottles, concluding that PET packaging materials are a source of estrogen-like compounds.

Moreover, when plastic becomes marine debris, it accumulates several other priority pollutants from the surrounding seawater, including several organic pollutants and metals. Ecotoxicological work has shown that priority pollutants such as these can degrade the structure and functions of ecosystems. Key physiological processes of organisms (e.g. cell-division, immunity, secretion of hormones) can be disrupted, causing disease (e.g. cancer; Zhuang et al. 2009; Vasseur and Cossu-Leguille 2006; Oehlmann et al. 2009) and reducing the ability to escape predators (Cartwright et al. 2006) and reproduce (Brown et al. 2004). Recently, some evidence has emerged regarding the impacts associated with the complex mixture of plastic and sorbed contaminants to organisms. One laboratory study found that the ingestion of PVC with sorbed triclosan altered feeding behavior and caused mortality in lugworms (*A. marina*; Browne et al. 2013). Another study fed fish polyethylene that had been deployed in the San Diego Bay, CA (i.e. allowing the plastic to accumulate environmentally relevant concentrations of priority pollutants). After a two-month dietary exposure to plastic with a complex mixture of sorbed priority pollutants (POPs and metals), fish suffered from liver toxicity, including glycogen depletion, lipidosis, cellular death and tumor promotion (Rochman et al. 2013b) and showed signs of endocrine disruption via changes in gene expression and abnormal growth of germ cells in the gonads (Rochman et al. 2014c). In both studies, adverse effects were demonstrated from the plastic alone, but organisms suffered greater effects when exposed to the mixture of plastic with sorbed chemical contaminants (Browne et al. 2013; Rochman et al. 2013b), further supporting the idea that when assessing the hazards of plastic debris it is important to consider the complex mixture of plastic debris, chemical ingredients and any sorbed chemical contaminants.

5.3 Conclusion

There are many different plastic polymers and several thousand plastic additives. The combination of these makes a large variation in chemical composition of plastic products. The unique combination of chemical ingredients may render some types of plastic more hazardous than others when their chemical constituents are bioavailable to organisms. Moreover, the chemical properties of plastic facilitate the accumulation of relatively large concentrations of contaminants, producing a complex mixture of chemical contaminants on marine plastic debris. The mixture and concentrations of hazardous chemicals will vary based upon the plastic type, the location where the material is discarded and the time it is left in the aquatic environment. Research has shown that marine plastic debris may act as both a sink and a source for contaminants in the marine environment, including their transfer into marine foodwebs, and thus need to be considered in models assessing the environmental fate of contaminants in the ocean. Moreover, this complex mixture of contaminants associated with marine plastic debris should be considered under risk assessment for marine animals exposed to this debris. Assessing such hazards associated with plastic in aquatic habitats is not simple, and requires knowledge regarding organisms that may be exposed, the exposure concentrations, the types of polymers comprising the debris, the length of time the debris was present in the aquatic environment (affecting the size, shape and fouling) and the locations and transport of the debris during that time period.

Although the scientific understanding regarding the fate and consequences of this material in the environment is growing, there remain several gaps in our understanding regarding the cocktail of chemicals associated with marine plastic debris. To design effective management strategies for mitigating any impacts, policy-makers will benefit from a greater understanding regarding the importance of plastic debris as a sink and source for global contaminants, its role in the global transport of chemicals substances, the bioaccumulation of plastic ingredients and accumulated chemical contaminants in wildlife, the importance of plastic as a mechanism for foodweb contamination relative to other sources of priority pollutants and whether or not these chemicals biomagnify in top predators (including humans) as a consequence of plastic debris entering marine foodwebs. Today, while researchers continue to expand our knowledge base, policy-makers can begin to act with the current information available, as there are no signs that the amount of plastic debris entering the marine environment is decreasing (Law et al. 2010; Goldstein et al. 2012) and if we continue business-as-usual the planet will hold another 33 billion tons of plastic by the year 2050 (Rochman 2013; Browne et al. 2013).

References

Artham, T., Sudhakar, M., Venkatesan, R., Madhavan Nair, C., Murty, K. V. G. K., & Doble, M. (2009). Biofouling and stability of synthetic polymers in sea water. *International Biodeterioration and Biodegradation, 63*(7), 884–890.

Anderson, B. S., Middaugh, D. P., Hunt, J. W., & Turpen, S. L. (1991). Copper toxicity to sperm, embryos and larvae of topsmelt *Atherinops affinis*, with notes on induced spawning. *Marine Environmental Research, 31*(1), 17–35.

Andrady, A. L. (2011). Microplastics in the marine environment. *Marine Pollution Bulletin, 62*(8), 1596–1605.

Antunes, J. C., Frias, J. G. L., Micaelo, A. C., & Sobral, P. (2013). Resin pellets from beaches of the Portuguese coast and adsorbed persistent organic pollutants. *Estuarine Coastal and Shelf Science, 130*, 62–69.

Ashton, K., Holmes, L., & Turner, A. (2010). Association of metals with plastic production pellets in the marine environment. *Marine Pollution Bulletin, 60*, 2050–2055.

Bakir, A., Rowland, S. J., & Thompson, R. C. (2012). Competitive sorption of persistent organic pollutants onto microplastics in the marine environment. *Marine Pollution Bulletin, 64*, 2782–2789.

Bakir, A., Rowland, S. J., & Thompson, R. C. (2014). Transport of persistent organic pollutants by microplastics in estuarine conditions. *Estuarine Coastal and Shelf Science, 140*, 14–21.

Barnes, D. K., Galgani, F., Thompson, R. C., & Barlaz, M. (2009). Accumulation and fragmentation of plastic debris in global environments. *Philosophical Transactions of the Royal Society B, 364*(1526), 1985–1998.

Bauer, M. J., & Herrmann, R. (1997). Estimation of the environmental contamination by phthalic acid esters leaching from household wastes. *Science of the Total Environment, 208*, 49–57.

Besseling, E., Wegner, A., Foekema, E., Van Den Heuvel-Greve, M., & Koelmans, A. A. (2013). Effects of microplastic on fitness and PCB bioaccumulation by the lugworm *Arenicola marina* (L.). *Environmental Science and Technology, 47*, 593–600.

Binetti, R., Costamagna, F. M., & Marcello, I. (2008). Exponential growth of new chemicals and evolution of information relevant to risk control. *Annali dell'Istituto Superiore di Sanita, 44*, 13–15.

Braun, D., Cherdron, H., Rehahn, M., Ritter, H., & Voit, B. (2005). *Polymer synthesis: Theory and practice—fundamentals, methods, experiments* (4th ed.). Berlin: Springer.

Brown, R. J., Galloway, T. S., Lowe, D., Browne, M. A., Dissanayake, A., et al. (2004). Differential sensitivity of three marine invertebrates to copper assessed using multiple biomarkers. *Aquatic Toxicology, 66*, 267–278.

Browne, M. A., Crump, P., Niven, S. J., Teuten, E. L., Tonkin, A., Galloway, T., et al. (2011). Accumulations of microplastic on shorelines worldwide: Sources and sinks. *Environmental Science and Technology, 45*, 9175–9179.

Browne, M. A., Niven, S. J., Galloway, T. S., Rowland, S. J., & Thompson, R. C. (2013). Microplastic moves pollutants and additives to worms, reducing functions linked to health and biodiversity. *Current Biology, 23*(23), 2388–2392.

Carpenter, E., & Smith, K. (1972). Plastics on the Sargasso Sea surface. *Science, 175*, 1240–1241.

Cartwright, S. R., Coleman, R. A., & Browne, M. A. (2006). Ecologically relevant effects of pulse application of copper on the limpet. *Patella vulgata. Marine Ecology Progress Series, 326*, 187–194.

Cheng, W., Xie, Z., Blais, J. M., Zhang, P., Li, M., Yang, C., et al. (2013). Organophosphorus esters in the oceans and possible relation with ocean gyres. *Environmental Pollution, 180*, 159–164.

Chua, E., Shimeta, J., Nugegoda, D., Morrison, P. D., & Clarke, B. O. (2014). Assimilation of Polybrominated diphenyl ethers from microplastics by the marine amphipod, *Allorchestes compressa. Environmental Science and Technology, 48*(14), 8127–8134.

Crain, D. A., Eriksen, M., Iguchi, T., Jobling, S., Laufer, H., LeBlanc, G. A., et al. (2007). An ecological assessment of bisphenol-A: Evidence from comparative biology. *Reproductive Toxicology, 24*(2), 225–239.

Dachs, J., Lohmann, R., Ockenden, W. A., Méjanelle, L., Eisenreich, S. J., & Jones, K. C. (2002). Oceanic biogeochemical controls on global dynamics of persistent organic pollutants. *Environmental Science and Technology, 36*(20), 4229–4237.

Darnerud, P. O. (2003). Toxic effects of brominated flame retardants in man and in wildlife. *Environment International, 29*(6), 841–853.

Decho, A. W. (2000). Microbial biofilms in intertidal systems: An overview. *Continental Shelf Research, 20*, 1257–1273.

Desforges, J. P. W., Galbraith, M., Dangerfield, N., & Ross, P. S. (2014). Widespread distribution of microplastics in subsurface seawater in the NE Pacific Ocean. *Marine Pollution Bulletin, 79*(1), 94–99.

de Wit, C. A. (2002). An overview of brominated flame retardants in the environment. *Chemosphere, 46*(5), 583–624.

Endo, S., Takizawa, R., Okuda, K., Takada, H., Chiba, K., et al. (2005). Concentration of polychlorinated biphenyls (PCBs) in beached resin pellets: Variability among individual particles and regional differences. *Marine Pollution Bulletin, 50*(10), 1103–1114.

Endo, S., Yuyama, M., Takada, H. (2013). Desorption kinetics of hydrophobic organic contaminants from marine plastic pellets. *Marine Pollution Bulletin, 74*(1), 125–131.

Engler, R. E. (2012). The complex interaction between marine debris and toxic chemicals in the ocean. *Environmental Science and Technology, 46*(22), 12302–12315.

Eriksen, M., Lebreton, L. C. M., Carson, H. S., Thiel, M., Moore, C. J., Borerro, J. C., et al. (2014). Plastic pollution in the world's oceans: More than 5 trillion plastic pieces weighing over 250,000 tons afloat at sea. *PLoS ONE, 9*, e111913.

European Commission. (2014). Priority substances and certain other pollutants according to Annex II of directive 2008/105/EC. http://ec.europa.eu/environment/water/water-framework/priority_substances.htm.

Fischer, A. C., Kroon, J. J., Verburg, T. G., Teunissen, T., & Wolterbeer, H. T. (2007). On the relevance of iron adsorption to container materials in small-volume experiments on iron marine chemistry: ^{55}Fe-aided assessment of capacity, affinity and kinetics. *Marine Chemistry, 107*, 533–546.

Fisner, M., Taniguchi, S., Moreira, F., Bícego, M. C., & Turra, A. (2013). Polycyclic aromatic hydrocarbons (PAHs) in plastic pellets: Variability in the concentration and composition at different sediment depths in a sandy beach. *Marine Pollution Bulletin, 70*, 219–226.

Fossi, M. C., Panti, C., Guerranti, C., Coppola, D., Giannetti, M., Marsili, L., et al. (2012). Are baleen whales exposed to the threat of microplastics? A case study of the Mediterranean fin whale (*Balaenoptera physalus*). *Marine Pollution Bulletin, 64*(11), 2374–2379.

Fossi, M. C., Coppola, D., Baini, M., Giannetti, M., Guerranti, C., Marsili, L., et al. (2014). Large filter feeding marine organisms as indicators of microplastic in the pelagic environment: The case studies of the Mediterranean basking shark (*Cetorhinus maximus*) and fin whale (*Balaenoptera physalus*). *Marine Environmental Research 100*, 17–24.

Fotopoulou, K. N., & Karapanagioti, H. K. (2012). Surface properties of beached plastic pellets. *Marine Environmental Research, 81*, 70–77.

Fries, E., & Zarfl, C. (2012). Sorption of polycyclic aromatic hydrocarbons (PAHs) to low and high density polyethylene (PE). *Environmental Science and Pollution Research, 19*(4), 1296–1304.

Galgani, F., Leaute, J. P., Moguedet, P., Souplet, A., Verin, Y., Carpentier, A. et al. (2000). Litter on the sea floor along European coasts. *Marine Pollution Bulletin, 40*(6), 516–527.

Galloway, T. S. (2015). Micro- and nano-plastics and human health. In M. Bergmann, L. Gutow & M. Klages (Eds.), *Marine anthropogenic litter* (pp. 347–370). Berlin: Springer.

Gaylor, M. O., Harvey, E., & Hale, R. C. (2012). House crickets can accumulate polybrominated diphenyl ethers (PBDEs) directly from polyurethane foam common in consumer products. *Chemosphere, 86*(5), 500–505.

Goldberg, E. D. (1997). Plasticizing the seafloor: an overview. *Environmental Technology, 18*(2), 195–201.

Goldstein, M. C., Rosenberg, M., & Cheng, L. (2012). Increased oceanic microplastic debris enhances oviposition in an endemic pelagic insect. *Biology Letters, 8*(5), 817–820.

Goldstein, M. C., Titmus, A. J., & Ford, M. (2013). Scales of spatial heterogeneity of plastic marine debris in the northeast Pacific ocean. *PLoS One, 8*(11), e80020.

Gouin, T., Roche, N., Lohmann, R., & Hodges, G. (2011). A thermodynamic approach for assessing the environmental exposure of chemicals absorbed to microplastic. *Environmental Science and Technology, 45*(4), 1466–1472.

Gowariker, V. R., Viswanathan, N. V., & Sreedhar, J. (2003). *Polymer science*. New Dehli: New Age International.

Gray, M. A., & Metcalfe, C. D. (1997). Induction of testis-ova in Japanese medaka (*Oryzias latipes*) exposed to p-nonylphenol. *Environmental Toxicology and Chemistry, 16*(5), 1082–1086.

Grün, F., & Blumberg, B. (2007). Perturbed nuclear receptor signaling by environmental obesogens as emerging factors in the obesity crisis. *Reviews in Endocrine and Metabolic Disorders, 8*(2), 161–171.

Halden, R. U. (2010). Plastics and health risks. *Annual Review of Public Health, 31*, 179–194.

Heskett, M., Takada, H., Yamashita, R., Yuyama, M., Ito, M., et al. (2012). Measurement of persistent organic pollutants (POPs) in plastic resin pellets from remote islands: Toward establishment of background concentrations for International Pellet Watch. *Marine Pollution Bulletin, 64*, 445–448.

Hidalgo-Ruz, V., & Thiel, M. (2015). The contribution of citizen scientists to the monitoring of marine litter. In M. Bergmann, L. Gutow, M. Klages (Eds.), *Marine anthropogenic litter* (pp. 433–451). Berlin: Springer.

Hirai, H., Takada, H., Ogata, Y., Yamashita, R., Mizukawa, K., et al. (2011). Organic micropollutants in marine plastic debris from the open ocean and remote and urban beaches. *Marine Pollution Bulletin, 62*, 1683–1692.

Holmes, L. A., Turner, A., & Thompson, R. C. (2012). Adsorption of trace metals to plastic resin pellets in the marine environment. *Environmental Pollution, 160*, 42–48.

Holmes, L. A., Turner, A., & Thompson, R. C. (2014). Interactions between trace metals and plastic production pellets under estuarine conditions. *Marine Chemistry, 167*, 25–32.

Huckins, J., Manuweera, G., Petty, J., Mackay, D., & Lebo, J. (1993). Lipid-containing semi-permeable membrane devices for monitoring organic contaminants in water. *Environmental Science and Technology, 27*, 2489–2496.

Karapanagioti, H. K., & Klontza, I. (2008). Testing phenanthrene distribution properties of virgin plastic pellets and plastic eroded pellets found on Lesvos island beaches (Greece). *Marine Environmental Research, 65*, 283–290.

Kawahata, H., Ohta, H., Inoue, M., & Suzuki, A. (2004). Endocrine disrupter nonylphenol and bisphenol A contamination in Okinawa and Ishigaki Islands, Japan—within coral reefs and adjacent river mouths. *Chemosphere, 55*(11), 1519–1527.

Kim, E. J., Kim, J. W., & Lee, S. K. (2002). Inhibition of oocyte development in Japanese medaka (*Oryzias latipes*) exposed to di-2-ethylhexyl phthalate. *Environment International, 28*(5), 359–365.

Koelmans, A. A. (2015). Modeling the role of microplastics in bioaccumulation of organic chemicals to marine aquatic organisms. Critical review. In M. Bergmann, L. Gutow & M. Klages (Eds.), *Marine anthropogenic litter* (pp. 313–328). Berlin: Springer.

Koelmans, A. A., Besseling, E., Wegner, A., & Foekema, E. M. (2013). Plastic as a carrier of POPs to aquatic organisms: A model analysis. *Environmental Science and Technology, 47*, 7812–7820.

Koelmans, A. A., Besseling, E., & Foekema, E. M. (2014). Leaching of plastic additives to marine organisms. *Environmental Pollution, 187*, 49–54.

Kühn, S., Bravo Rebolledo, E. L., & van Franeker, J. A. (2015). Deleterious effects of litter on marine life. In M. Bergmann, L. Gutow & M. Klages (Eds.), *Marine anthropogenic litter* (pp. 75–116). Berlin: Springer.

Kwon, E., & Castaldi, M. J. (2008). Investigation of mechanisms of polycyclic aromatic hydrocarbons (PAHs) initiated from the thermal degradation of styrene butadiene rubber (SBR) in N_2 atmosphere. *Environmental Science and Technology, 42*, 2175–2180.

Kwon, B. G., Saido, K., Koizumi, K., Sato, H., Ogawa, N., Chung, S. Y., et al. (2014). Regional distribution of styrene analogues generated from polystyrene degradation along the coastlines of the North-East Pacific Ocean and Hawaii. *Environmental Pollution, 188*, 45–49.

Laist, D. W. (1987). Overview of the biological effects of lost and discarded plastic debris in the marine environment. *Marine Pollution Bulletin, 18*(6), 319–326.

Lavers, J. L., Bond, A. L., & Hutton, I. (2014). Plastic ingestion by flesh-footed Shearwaters (*Puffinus carneipes*): Implications for fledgling body condition and the accumulation of plastic-derived chemicals. *Environmental Pollution, 187*, 124–129.

Law, K. L., Morét-Ferguson, S., Maximenko, N. A., Proskurowski, G., Peacock, E. E., et al. (2010). Plastic accumulation in the North Atlantic subtropical gyre. *Science, 329*(5996), 1185–1188.

Law, K. L., Morét-Ferguson, S. E., Goodwin, D. S., Zettler, E. R., DeForce, E., Kukulka, T., er al. (2014). Distribution of surface plastic debris in the Eastern Pacific Ocean from an 11-year data set. *Environmental Science and Technology, 48*(9), 4732–4738.

Lee, H., Shim, W. J., & Kwon, J. H. (2014). Sorption capacity of plastic debris for hydrophobic organic chemicals. *Science of the Total Environment, 470*, 1545–1552.

Lithner, D., Damberg, J., Dave, G., & Larsson, Å. (2009). Leachates from plastic consumer products–screening for toxicity with *Daphnia magna*. *Chemosphere, 74*(9), 1195–1200.

Lithner, D., Larsson, A., & Dave, G. (2011). Environmental and health hazard ranking and assessment of plastic polymers based on chemical composition. *Science of the Total Environment, 409*, 3309–3324.

Lithner, D., Nordensvan, I., & Dave, G. (2012). Comparative acute toxicity of leachates from plastic products made of polypropylene, polyethylene, PVC, acrylonitrile–butadiene–styrene, and epoxy to *Daphnia magna*. *Environmental Science and Pollution Research, 19*(5), 1763–1772.

Lohmann, R. (2012). Critical review of low-density polyethylene's partitioning and diffusion coefficients for trace organic contaminants and implications for its use as a passive sampler. *Environmental Science and Technology, 46*, 606–618.

Lusher, A. (2015). Microplastics in the marine environment: Distribution, interactions and effects. In M. Bergmann, L. Gutow & M. Klages (Eds.), *Marine anthropogenic litter* (pp. 245–308). Berlin: Springer.

Lynge, E., Anttila, A., & Hemminki, K. (1997). Organic solvents and cancer. *Cancer Causes and Control, 8*(3), 406–419.

Mato, Y., et al. (2001). Plastic resin pellets as a transport medium for toxic chemicals in the marine environment. *Environmental Science and Technology, 35,* 318–324.

Maximenko, N., Hafner, J., & Niiler, P. (2012). Pathways of marine debris derived from trajectories of Lagrangian drifters. *Marine Pollution Bulletin, 65*(1), 51–62.

Müller, J. F., Manomanii, K., Mortimer, M. R., & McLachlan, M. S. (2001). Partitioning of polycyclic aromatic hydrocarbons in the polyethylene/water system. *Fresenius Journal of Analytical Chemistry, 371,* 816–822.

Nakashima, E., Isobe, A., Magome, S., Kako, S. I., & Deki, N. (2011). Using aerial photography and *in situ* measurements to estimate the quantity of macro-litter on beaches. *Marine Pollution Bulletin, 62*(4), 762–769.

Nakashima, E., Isobe, A., Kako, S. I., Itai, T., & Takahashi, S. (2012). Quantification of toxic metals derived from macroplastic litter on Ookushi Beach, Japan. *Environmental Science and Technology, 46*(18), 10099–10105.

OECD. (2004). *Emission scenario document on plastic additives. Series on emission scenario documents, no. 3.* Paris: Environmental Directorate, OECD Environmental Health and Safety Publications.

Oehlmann, J., Fioroni, P., Stroben, E., & Markert, B. (1996). Tributyltin (TBT) effects on *Ocinebrina aciculate* (Gastropoda: Muricidae): Imposex development, sterilization, sex change and population decline. *Science of the Total Environment, 188*(2), 205–223.

Oehlmann, J., Schulte-Oehlmann, U., Kloas, W., Jagnytsch, O., Lutz, I., et al. (2009). A critical analysis of the biological impacts of plasticizers on wildlife. *Philosophical Transactions of the Royal Society B, 364*(1526), 2047–2062.

Ogata, Y., Takada, H., Mizukawa, K., Hirai, H., Iwasa, S., Endo, S., et al. (2009). International pellet watch: Global monitoring of persistent organic pollutants (POPs) in coastal waters. 1. Initial phase data on PCBs, DDTs, and HCHs. *Marine Pollution Bulletin, 58,* 1437–1446.

Papaleo, B., Caporossi, L., Bernardini, F., Cristadoro, L., Bastianini, L., De Rosa, M., et al. (2011). Exposure to styrene in fiberglass-reinforced plastic manufacture: Still a problem. *Journal of Occupational and Environmental Medicine, 53*(11), 1273–1278.

Pascall, M. A., Zabik, M. E., Zabik, M. J., & Hernandez, R. J. (2005). Uptake of Polychlorinated biphenyls (PCBs) from an aqueous medium by polyethylene, polyvinyl chloride, and polystyrene films. *Journal of Agricultural and Food Chemistry, 53,* 164–169.

PlasticsEurope, Plastics—the facts 2013: An analysis of European plastics production, demand and waste data. http://www.plasticseurope.org/Document/plastics-the-facts-2013.aspx?FolID=2.

Rios, L. M., Moore, C., & Jones, P. R. (2007). Persistent organic pollutants carried by synthetic polymers in the ocean environment. *Marine Pollution Bulletin, 54*(8), 1230–1237.

Rios, L. M., Jones, P. R., Moore, C., & Narayan, U. V. (2010). Quantitation of persistent organic pollutants adsorbed on plastic debris from the Northern Pacific Gyre's "eastern garbage patch". *Journal of Environmental Monitoring, 12*(12), 2226–2236.

Robertson, D. E. (1968). The adsorption of trace chemicals in sea water on various container surfaces. *Analytica Chimica Acta, 42,* 533–536.

Rochman, C. M. (2013). Plastics and priority pollutants: A multiple stressor in aquatic habitats. *Environmental Science and Technology, 47,* 2439–2440.

Rochman, C. M., Browne, M. A., Halpern, B. S., Hentschel, B. T., Hoh, E., Karapanagioti, H., et al. (2013a). Classify plastic waste as hazardous. *Nature, 494,* 169–171.

Rochman, C. M., Hoh, E., Kurobe, T., & Teh, S. J. (2013b). Ingested plastic transfers hazardous chemicals to fish and induces hepatic stress. *Scientific Reports, 3,* 3263.

Rochman, C. M., Hoh, E., Hentschel, B. T., & Kaye, S. (2013c). Long-term field measurement of sorption of organic contaminants to five types of plastic pellets: Implications for plastic marine debris. *Environmental Science and Technology, 47,* 1646–1654.

Rochman, C. M., Manzano, C., Hentschel, B., Massey, L., Simonich, S., & Hoh, E. (2013d). Polystyrene plastic: A source and sink for polycyclic aromatic hydrocarbons in the marine environment. *Environmental Science and Technology, 47*, 13976–13984.

Rochman, C. M., Hentschel, B. T., & Teh, S. J. (2014a). Long-term sorption of metals is similar among plastic types: Implications for plastic debris in aquatic environments. *PLOS One, 9*, e85433.

Rochman, C. M., Lewison, R. L., Eriksen, M., Allen, H., Cook, A. M., & Teh, S. J. (2014b). Polybrominated diphenyl ethers (PBDEs) in fish tissue may be an indicator of plastic contamination in marine habitats. *Science of the Total Environment, 476*, 622–633.

Rochman, C. M., Kurobe, T., Flores, I., & Teh, S. J. (2014c). Early warning signs of endocrine disruption in adult fish from the ingestion of polyethylene with and without sorbed chemical pollutants from the marine environment. *Science of the Total Environment, 493*, 656—661.

Rochman, C. M., & Boxall, A. B. A. (2014). Environmental relevance: A necessary component of experimental design to answer the question, "So what?" *Integrated Environmental Assessment and Management, 10*, 311–312.

Ross, P. S., & Birnbaum, L. S. (2010). Integrated human and ecological risk assessment: A case study of persistent organic pollutants (POPs) in humans and wildlife. *Human and Ecological Risk Assessment, 9*, 303–324.

Rusina, T., Smedes, F., Klanova, J., Booij, K., & Holoubek, I. (2007). Polymer selection for passive sampling: A comparison of critical properties. *Chemosphere, 68*, 1344–1351.

Ryan, P. G., Connell, A. D., & Gardner, B. D. (1988). Plastic ingestion and PCBs in seabirds: Is there a relationship? *Marine Pollution Bulletin, 19*(4), 174–176.

Secretariat of the Convention on Biological Diversity. (2012). Impacts of marine debris on biodiversity. CBD technical series no. 67. http://www.thegef.org/gef/sites/thegef.org/files/publication/cbd-ts-67-en.pdf.

Seki, M., Yokota, H., Maeda, M., Tadokoro, H., & Kobayashi, K. (2003). Effects of 4-nonylphenol and 4-tert-octylphenol on sex differentiation and vitellogenin induction in medaka (*Oryzias latipes*). *Environmental Toxicology and Chemistry, 22*(7), 1507–1516.

Sinkonnen, S., & Paasivirta, J. (2000). Degradation half-life times of PCDDs, PCDFs and PCBs for environmental fate modeling. *Chemosphere, 40*, 943–949.

Smedes, F., Geertsma, R. W., Van der Zande, T., & Booij, K. (2009). Polymer-water partition coefficients of hydrophobic compounds for passive samplilng: Application for cosolvent models for validation. *Environmental Science and Technology, 43*, 7047–7054.

Takada, H. (2006). Call for pellets! international pellet watch global monitoring of POPs using beached plastic resin pellets. *Marine Pollution Bulletin, 52*(12), 1547–1548.

Tanaka, K., Takada, H., Yamashita, R., Mizukawa, K., Fukuwaka, M. A., & Watanuki, Y. (2013). Accumulation of plastic-derived chemicals in tissues of seabirds ingesting marine plastics. *Marine Pollution Bulletin, 69*, 219–222.

Teuten, E. L., Rowland, S. J., Galloway, T. S., & Thompson, R. C. (2007). Potential for plastics to transport hydrophobic contaminants. *Environmental Science and Technology, 41*(22), 7759–7764.

Teuten, E. L., Saquing, J. M., Knappe, D. R., Barlaz, M. A., Jonsson, S., Björn, A., et al. (2009). Transport and release of chemicals from plastics to the environment and to wildlife. *Philisophical Transactions of the Royal Society B, 364*, 2027–2045.

Tien, C., Wu, W., Chuang, T., & Chen, C. S. (2009). Development of river biofilms on artificial substrates and their potential for biomonitoring water quality. *Chemosphere, 76*, 1288–1295.

Tien, C., & Chen, C. S. (2013). Patterns of metal accumulation by natural river biofilms during their growth and seasonal succession. *Archives of Environmental Contamination and Toxicology, 64*, 605–616.

Thompson, R. C., Olsen, Y., Mitchell, R. P., Davis, A., Rowland, S. J., John, A. W. G., et al. (2004). Lost at sea: Where is all the plastic? *Science, 304*, 838.

Thompson, R. C., Swan, S. H., Moore, C. J., & vom Saal, F. S. (2009). Our plastic age. *Philosophical Transactions of the Royal Society B, 364*(1526), 1973–1976.

USEPA. (2013). *Water: CWA methods—priority pollutants*. http://water.epa.gov/scitech/methods/cwa/pollutants.cfm.

Van, A., Rochman, C. M., Flores, E. M., Hill, K. L., Vargas, E., Vargas, S. A., et al. (2011). Persistent organic pollutants in plastic marine debris found on beaches in San Diego, California. *Chemosphere, 86,* 258–263.

Vasseur, P., & Cossu-Leguille, C. (2006). Linking molecular interactions to consequent effects of persistent organic pollutants (POPs) upon populations. *Chemosphere, 63,* 1033–1042.

Velzeboer, I., Kwadijk, C. J. A. F., & Koelmans, A. A. (2014). Strong sorption of PCBs to nanoplastics, microplastics, carbon nanotubes, and fullerenes. *Environmental Science and Technology, 48*(9), 4869–4876.

Wagner, M., & Oehlmann, J. (2009). Endocrine disruptors in bottled mineral water: Total estrogenic burden and migration from plastic bottles. *Environmental Science and Pollution Research, 16*(3), 278–286.

Wagner, M., & Oehlmann, J. (2011). Endocrine disruptors in bottled mineral water: Estrogenic activity in the E-Screen. *The Journal of Steroid Biochemistry and Molecular Biology, 127*(1), 128–135.

Weijuan, L., Yougian, D., & Zuyi, T. (2001). Americium(III) adsorption on polyethylene from very dilute aqueous solutions. *Journal of Radioanalytical and Nuclear Chemistry, 250,* 497–500.

Wu, B., Taylor, C. M., Knappe, D. R. U., Nanny, M. A., & Barlaz, M. A. (2001). Factors controlling alkylbenzene sorption to municipal solid waste. *Environmental Science and Technology, 35,* 4569–4576.

Wurl, O., & Obbard, J. P. (2004). A review of pollutants in the sea-surface microlayer (SML): A unique habitat for marine organisms. *Marine Pollution Bulletin, 48*(11), 1016–1030.

Xu, H., Vanhooren, H. M., Verbeken, E., Yu, L., Lin, Y., Nemery, B., et al. (2004). Pulmonary toxicity of polyvinyl chloride particles after repeated intratracheal instillations in rats. Elevated CD4/CD8 lymphocyte ratio in bronchoalveolar lavage. *Toxicology and Applied Pharmacology, 194*(2), 122–131.

Yamashita, R., Takada, H., Fukuwaka, M. A., & Watanuki, Y. (2011). Physical and chemical effects of ingested plastic debris on short-tailed shearwaters, *Puffinus tenuirostris*, in the North Pacific Ocean. *Marine Pollution Bulletin, 62*(12), 2845–2849.

Yang, C. Z., Yaniger, S. I., Jordan, V. C., Klein, D. J., & Bittner, G. D. (2011). Most plastic products release estrogenic chemicals: A potential health problem that can be solved. *Environmental Health Perspectives, 119*(7), 989.

Ye, S., & Andrady, A. L. (1991). Fouling of floating plastic debris under Biscayne Bay exposure conditions. *Marine Pollution Bulletin, 22,* 608–613.

Zabaniotou, A., & Kassidi, E. (2003). Life cycle assessment applied to egg packaging made from polystyrene and recycled paper. *Journal Cleaner Production, 11*(5), 549–559.

Zarfl, C., & Matthies, M. (2010). Are marine plastic particles transport vectors for organic pollutants to the Arctic? *Marine Pollution Bulletin, 60*(10), 1810–1814.

Zettler, E. R., Mincer, T. J., & Amaral-Zettler, L. A. (2013). Life in the "Plastisphere": Microbial communities on plastic marine debris. *Environmental Science and Technology, 47,* 7137–7146.

Zhuang, P., McBride, M. B., Xia, H., Li, N., & Li, Z. (2009). Health risk from heavy metals via consumption of food crops in the vicinity of Dabaoshan mine, South China. *Science of the Total Environment, 407*(5), 1551–1561.

Chapter 6
Marine Litter as Habitat and Dispersal Vector

Tim Kiessling, Lars Gutow and Martin Thiel

Abstract Floating anthropogenic litter provides habitat for a diverse community of marine organisms. A total of 387 taxa, including pro- and eukaryotic micro-organisms, seaweeds and invertebrates, have been found rafting on floating litter in all major oceanic regions. Among the invertebrates, species of bryozoans, crustaceans, molluscs and cnidarians are most frequently reported as rafters on marine litter. Micro-organisms are also ubiquitous on marine litter although the composition of the microbial community seems to depend on specific substratum characteristics such as the polymer type of floating plastic items. Sessile suspension feeders are particularly well-adapted to the limited autochthonous food resources on artificial floating substrata and an extended planktonic larval development seems to facilitate colonization of floating litter at sea. Properties of floating litter, such as size and surface rugosity, are crucial for colonization by marine organisms and the subsequent succession of the rafting community. The rafters themselves affect substratum characteristics such as floating stability, buoyancy, and degradation. Under the influence of currents and winds marine litter can transport associated organisms over extensive distances. Because of the great persistence (especially of plastics) and the vast quantities of litter in the world's oceans, rafting dispersal has become more prevalent in the marine environment, potentially facilitating the spread of invasive species.

T. Kiessling · M. Thiel
Facultad Ciencias del Mar, Universidad Católica del Norte, Larrondo 1281, Coquimbo, Chile

L. Gutow
Biosciences | Functional Ecology, Alfred-Wegener-Institut Helmholtz-Zentrum für Polar- und Meeresforschung, Bremerhaven, Germany

M. Thiel (✉)
Centro de Estudios Avanzados en Zonas Áridas (CEAZA), Coquimbo, Chile
e-mail: thiel@ucn.cl

M. Thiel
Nucleus Ecology and Sustainable Management of Oceanic Island (ESMOI), Coquimbo, Chile

Keywords Anthropogenic flotsam · Rafting community · Succession · Biogeography · Biological invasions · Plastic pollution

6.1 Introduction

Litter in the marine environment poses a hazard for a great variety of animals. Various species of marine vertebrates including fish, seabirds, turtles and marine mammals become easily entangled in floating marine litter, resulting in reduced mobility, strangulation and drowning (Derraik 2002; Kühn et al. 2015). Additionally, ingested litter can damage or block intestines, thereby affecting nutrition with often lethal effects (reviewed by Derraik 2002; Kühn et al. 2015). On the seafloor, marine litter can smother the substratum and thus cause hypoxia in benthic organisms (Moore 2008; Gregory 2009). In addition to these immediate hazardous effects on marine biota, marine litter has been suggested to facilitate the spread of non-indigenous species (Lewis et al. 2005). Biological invasions are considered a major threat to coastal ecosystems (Molnar et al. 2008).

Like any other submerged substrata, marine litter provides a habitat for organisms that are able to settle and persist on artificial surfaces. Once colonized by marine biota, litter items floating at the sea surface can facilitate dispersal of the associated rafters at different spatial scales. Previous studies have reported over 1200 taxa that are associated with natural and anthropogenic flotsam (Thiel and Gutow 2005a) and the extreme localities that rafting organisms can reach when transported over large distances by currents and wind (Barnes and Fraser 2003; Barnes and Milner 2005). While floating macroalgae, wood and volcanic pumice have been part of the natural flotsam assemblage of the oceans for millions of years, marine litter adds a new dimension to the dispersal opportunities of potential rafters (Barnes 2002). Marine litter is diverse (e.g. domestic waste, derelict fishing gear, detached buoys), persistent (afloat for longer than many natural substrata-Thiel and Gutow 2005b; Bravo et al. 2011), widespread (Barnes et al. 2009; Eriksen et al. 2014) and abounds in oceanic regions where natural floating substrata, such as macroalgae, occur less frequently (Rothäusler et al. 2012).

Unlike biotic substrata, anthropogenic litter is of no nutritional value to most organisms. Additionally, marine litter items differ from natural substrata in their physical and chemical characteristics such as surface rugosity and floating behavior. Accordingly, rafters need to overcome specific challenges with regard to food acquisition and attachment in order to persist for extended time periods on artificial floating substrata. The specific properties of marine litter are likely to influence colonization and succession processes, and thus the composition of the associated rafting community (Bravo et al. 2011).

In this chapter, we compiled information from peer-reviewed scientific literature on the biota associated with marine floating litter and on characteristics of litter items that affect the composition of the rafting community. Information on the biological traits of species associated with floating marine litter was used to characterize the rafting assemblage's functionally and to identify specific conditions that rafters on floating marine litter have to cope with. Finally, the environmental implications of litter rafting will be discussed, including the dispersal and invasion potential of non-indigenous species.

6.2 Floating Litter as a Habitat

Marine flotsam can be classified according to its nature (abiotic or biotic) and its origin (natural or anthropogenic). Biotic flotsam comprises macroalgae, animal remains/carcasses, wood and other parts of terrestrial plants such as seeds and leaf litter. Abiotic flotsam of natural origin consists mostly of volcanic pumice and ice. Flotsam of anthropogenic origin includes every kind of discarded material: biotic anthropogenic flotsam consists mainly of manufactured wood, discarded food (e.g. fruits) and oil/tar lumps, but the great majority of anthropogenic flotsam is abiotic and comprises any artificial object at sea.

Floating marine litter consists of consumer and household articles, industrial waste products or objects that had previously served maritime and fishery purposes (Fig. 6.1).

Fig. 6.1 Taxa floating on different marine litter items, **a** the tropical coral *Favia fragum* on a metal cylinder found in The Netherlands (Reprinted with permission from Hoeksema et al. 2012), **b** *Lepas* and a bryozoan colony growing on a toothbrush handle (Reprinted with permission from Goldstein et al. 2014), **c** extensive *Lepas* cover on a floating buoy (Reprinted with permission from Goldstein et al. 2014)

Discarded or lost consumer articles usually start their floating journey in a "clean" state, i.e. free of fouling biota. Floating litter from maritime activities comprises detached buoys, discarded fishing gear and chunks of piers and harbor infrastructure. These objects usually have spent long time periods in the marine environment, and therefore often host an extensive and reproductively active fouling biota, before they become part of marine floating litter, e.g. after detachment from anchorings. For example, Astudillo et al. (2009) found diverse rafting communities in advanced successional stages on lost aquaculture buoys floating off the Chilean coast. Detached buoys might carry with them anchoring lines, which extend into greater depths, thereby offering a habitat less influenced by harsh surface conditions. Highly buoyant items, such as Styrofoam, often have low floating stability and tip over more easily, a process, which suppresses colonization by fouling organisms (Bravo et al. 2011). However, colonization by fouling organisms may stabilize the floating item, equivalent to the "biological keel" of attached organisms on floating pumice described by Bryan et al. (2012). Accordingly, the degree of colonization has substantial impact on the floating behavior of the substratum at sea and therefore on the succession of the rafting community.

The rafting community on litter is described as being similar to but less species rich than that of floating macroalgae (Stevens et al. 1996; Winston et al. 1997; Gregory 2009). Winston et al. (1997) attribute this partly to the higher structural complexity and the soft mechanical properties of macroalgae compared to smooth and hard plastic particles. In contrast, Barnes and Milner (2005) report a significantly higher amount of encrusting organisms on floating wood and plastic compared to floating kelp. Only few studies allow for a comparison of the rafting communities on different marine litter substrata, probably because the vast majority of the floating litter is composed of plastics. Wong et al. (1974) found similar organisms colonizing larger plastic items and tar lumps of the same size. In a colonization experiment, organisms settled rapidly on floating substrata regardless of its type (plastic, Styrofoam or pumice—Bravo et al. 2011). However, in an early stage of colonization fewer species were found on plastic surfaces than on Styrofoam and pumice, indicating that surface rugosity of the substratum facilitates initial colonization of floating objects (Bravo et al. 2011—Fig. 6.2). Similarly, Carson et al. (2013) observed more diatoms, though not bacteria, on rough surfaces.

Only few studies have considered the material differences between types of plastic. Though there is no evidence that the polymer type is relevant for the composition of the rafting macrobiota, it was shown that it influences the composition of micro-organisms: Carson et al. (2013) found significantly more bacteria on polystyrene than on polyethylene and polypropylene, probably because of the surface characteristics of the material. Zettler et al. (2013) found distinct bacterial assemblages on polypropylene and polyethylene with a compositional overlap of less than 50 %.

Biotic flotsam occurs in a wide size range with floating macroalgae and tree trunks often reaching several metres in diameter or length. The majority of abiotic flotsam is generally smaller and rarely reaches a size of 1 m (Thiel and Gutow 2005b). Marine litter of any size, ranging from fragments in the order of millimetres (Gregory 1978; Minchin 1996) to larger items, such as lost buoys (Astudillo et al. 2009) and even refrigerators (Dellinger et al. 1997) are colonized by organisms. Carson et al. (2013) found that a larger surface area of plastic fragments is

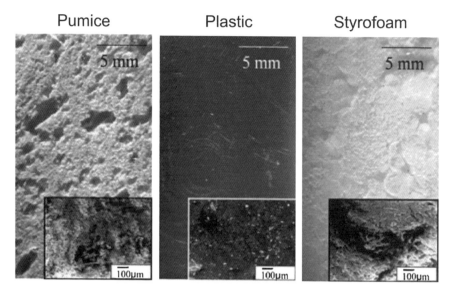

Fig. 6.2 Macro-photographs of the surface of pumice, plastic and Styrofoam, illustrating the different rugosities of the materials (Reprinted with permission from Bravo et al. 2011)

Fig. 6.3 Number of taxa in relation to the surface area of floating litter items (modified after Goldstein et al. 2014)

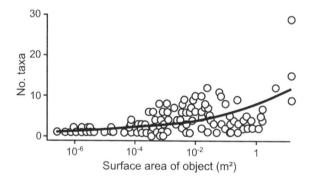

associated with a higher taxonomic richness, though not necessarily abundance, of microbiota. Similarly, Goldstein et al. (2014) recorded a positive correlation between the surface area of floating litter items in the North Pacific and species richness of the rafting community (Fig. 6.3). Most of these larger litter items consisted of fishing gear, which are more likely to harbor a diverse biota before being discarded or lost than are smaller domestic litter items. Other possible explanations involve stochastic effects (a random distribution of organisms on marine flotsam leads to a higher quantity on larger items), biased sampling efforts (small items sink already when colonized by only few organisms) or other raft characteristics, e.g. stability (Goldstein et al. 2014). A floating experiment conducted by Ye and Andrady (1991) revealed that larger surfaces are more quickly colonized by macrobiota than smaller surfaces. Wong et al. (1974) did not find algae and invertebrates on plastic fragments, which were significantly smaller than floating pumice in the same region. Lepadid barnacles seem to have species-specific preferences for litter of certain size, and some species (*Lepas pectinata* and *Dosima fascicularis*) associated with smaller litter items develop morphological adaptations, such as a small body size and light-weight valves, that minimize the risk of sinking of colonized flotsam (Whitehead et al. 2011). A size-specific selection of floating substrata has previously been shown for lepadid barnacles rafting on tar pellets (Minchin 1996).

Abiotic and biotic flotsam differ in their expected longevity. The persistence of biotic flotsam, such as floating seaweeds, is clearly limited by physical factors such as temperature and biological processes such as consumption and decomposition (Vandendriessche et al. 2007; Rothäusler et al. 2009). Therefore, the longevity of floating macroalgae is in the range of a few weeks up to six months (Thiel and Gutow 2005b). Floating litter is of no nutritional value for metazoans, and so far only few microorganisms have been shown capable of plastic digestion (Zettler et al. 2013). Accordingly, biological degradation is slow and marine litter, especially plastic, is expected to persist for years or even centuries in the marine environment (Derraik 2002; O'Brine and Thompson 2010). Plastics are particularly persistent at sea because lower temperatures and oxygen levels decelerate decomposition processes (Andrady 2011). Attached biota may protect the raft from degradation through solar radiation (Winston et al. 1997), thereby further extending its lifetime.

Estimating the time a floating item has spent in the marine environment is complicated and at present no reliable method exists. Age estimations for floating litter are inferred from (a) drift trajectories and velocities based on the supposed origin of the items (Ebbesmeyer and Ingraham 1992; Rees and Southward 2009; Hoeksema et al. 2012), (b) the successional stage of the rafting community (Cundell 1974), (c) the size of rafting organisms of known growth rates, e.g. bryozoans or lepadid barnacles (Stevens 1992 cited by Winston et al. 1997; Barnes and Fraser 2003; Tsikhon-Lukanina et al. 2001), or (d) the degradation of the substratum, for example by measuring the tensile extensibility of the material (Andrady 2011). However, all these methods have drawbacks, introducing a high degree of uncertainty to age estimates for floating litter. The sources of litter items are often unknown and floating velocities can be highly variable due to seasonal variations in wind and current conditions. Additionally, the composition and the successional stage of the rafting community may change the floating behavior of a litter item. Biological interactions such as predation and competition may influence the composition and the age structure of a rafting community rendering the size of specific rafting organisms an unreliable predictor of the duration of the floating period. Moreover, unlike floating macroalgae, abiotic flotsam may repeatedly return to the sea even after extended periods on the shore, which likely influences the state of degradation of the raft as well as the composition of the associated biota. Bravo et al. (2011) discussed that degradation of marine litter may either facilitate colonization by producing more rugose surfaces or alternatively impede it by abrasion processes. Overall, degradation and fragmentation of litter items into smaller pieces reduces the size of individual rafts, thereby changing settlement opportunities for species of a certain size range.

Removal of floating litter rafts from the sea surface occurs through stranding, sinking or ingestion by aquatic animals. Sinking of litter rafts mostly occurs because of high epibiont biomass that increases the weight of a floating object (Barnes et al. 2009; Bravo et al. 2011). Depending on environmental conditions, a critical accumulation of biomass that forces a substratum to sink can develop within 8–10 weeks on smaller household plastic items and plastic bags (Ye and Andrady 1991). Sinking flotsam may facilitate the transport of associated organisms to the seafloor. However, subsequent establishment of rafters in the benthic environment is unlikely, especially in the deep sea. The loss of buoyancy is reversible if epibionts die at greater water depth and fall off their substratum (Ye and Andrady 1991). Consequentially, the item may resurface, initiating a new cycle of colonization. Rafting organisms likely benefit from neutral buoyancy of a litter item because they are less exposed to desiccation and solar radiation on a substratum that barely emerges above the sea surface (Bravo et al. 2011; Carson et al. 2013). Vertical export of litter into deeper waters may be facilitated by wind-driven mixing or eddies (Kukulka et al. 2012).

6.3 Composition of Rafting Assemblages on Floating Litter

6.3.1 Taxonomic Overview

A review of 82 publications revealed a total of 387 marine litter rafting taxa, of which 244 were identified to the species, and 143 to the genus level (for complete species list see Appendix 1). In this review we included publications that report on organisms associated with floating litter in the field as well as experimental studies on the colonization of anthropogenic flotsam. We did not consider the many experimental studies on the succession of fouling communities on rigidly fixed artificial substrata because these items do not display the specific floating behavior, which probably affects the colonization by marine biota. To avoid potential overlaps, taxa identified at genus level were excluded if a species-level identification existed for the same genus. The identification of some micro-organisms was vague despite the use of advanced analytical methods such as electron microscopy and RNA analysis. Most taxa (335) were associated with plastic substrata (domestic waste, plastic fragments or buoys made of plastic), which constitute the large majority of anthropogenic floating litter in the oceans (Galgani et al. 2015). Accordingly, only few taxa (17) were recorded from other floating litter items consisting of metal, glass and paper. For 83 taxa, the floating substrata were of unknown composition or were composed of various materials. The given numbers exceed the total number of 387 taxa because some species have been found on more than just one substratum type. 132 taxa were recorded from items, which previously served maritime purposes (mainly buoys and fishing gear). A large proportion (60 %) of the rafting taxa was sampled in situ, associated with their floating substrata, whereas 35 % of the taxa are only known from beached litter. For 2 %, the ability to raft on floating litter was inferred from floating experiments (Bravo et al. 2011) and the remaining 3 % consist of taxa that could not be reliably identified but were assigned to a certain genus or species by the respective authors.

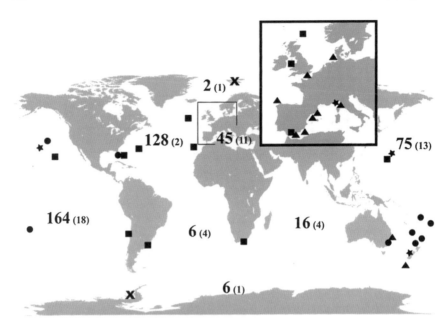

Fig. 6.4 Number of observed rafting taxa on floating marine litter (number of studies in brackets) in major oceanic regions (from *top left* Arctic, North Atlantic, Mediterranean, North Pacific, South Pacific, South Atlantic, Indian Ocean, Southern Ocean). The symbols represent reports of frequently observed rafting species on marine litter: *Circles* = *Jellyella tuberculata*, *squares* = *Lepas anatifera*, *triangles* = *Idotea metallica*, *stars* = *Fiona pinnata*. The two *crosses* represent the northern- and southernmost observations of rafters on marine litter

The highest numbers of rafting taxa on floating litter were found in the Pacific and North Atlantic, which might be explained by the overall high research effort undertaken in these regions (Fig. 6.4). A considerable number of rafters were also found in the Mediterranean while only few taxa were reported from the South Atlantic and from the Indian Ocean. Some rafters have even been found in the Arctic at 79°N (Barnes and Milner 2005) and in Antarctica at approximately 67°S (Barnes and Fraser 2003). The percentage of anthropogenic litter items colonized varied significantly with latitude. Barnes and Milner (2005) found that at low latitudes (0–15°) about 50 % of all beached litter items were colonized by marine biota while at higher latitudes (15–40°) only 25 % of the litter items had attached organisms. This rate decreased further to 5–10 % at 40–60° latitude and beyond 60° colonization of marine litter was rarely observed (Fig. 6.5). This geographic pattern was evident for remote sites as well as for sites close to the continental shore (Barnes 2002). A similar latitudinal decrease of the colonization rate was evident on a smaller spatial scale for the Indian Ocean (Barnes 2004).

Numerous taxa of bacteria, protists and algae (most prominently diatoms and Rhodophyta) form part of the rafting community on marine floating litter (Table 6.1). Four studies examined the microbiota associated with marine microplastics (i.e. plastic particles in the size range of millimetres and a few centimetres—Fortuño et al. 2010; Carson et al. 2013; Zettler et al. 2013; Reisser et al. 2014) and found a total

Fig. 6.5 Proportion of marine litter colonized according to latitude (modified after Barnes and Milner 2005)

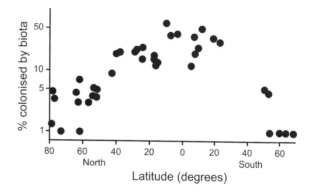

of 44 bacteria and 56 Chromista taxa. Micro-organisms seem to be ubiquitous on marine litter as Carson et al. (2013) found microbes on each plastic item sampled in the North Pacific gyre. Plastic litter offers a habitat for various functional microbial groups including autotrophs, symbionts, heterotrophs (including phagotrophs) and predators (Zettler et al. 2013). Harmful micro-organisms were also found on floating litter, including potential human and animal pathogens of the genus *Vibrio* (Zettler et al. 2013), the ciliate *Halofolliculina* sp., which causes skeletal eroding band disease in corals (Goldstein et al. 2014) and the dinoflagellates *Ostreopsis* sp., *Coolia* sp. and *Alexandrium taylori*, known to form harmful algal blooms under favorable conditions (Masó et al. 2003). The composition of the microbial community clearly differs from the surrounding seawater suggesting that plastic litter forms a novel habitat for micro-biota (termed 'microbial reef' by Zettler et al. 2013). Some organisms found on plastic samples are otherwise strictly associated with open seawater and their presence was probably the result of entanglement (Zettler et al. 2013). Carson et al. (2013) characterized the encountered microbial community in the North Pacific gyre as dominated by rod-shaped bacteria and pennate diatoms, each at densities of roughly 1,000 cells m^{-2}. Less frequent microbiota on plastic samples comprised coccoid bacteria, centric diatoms, dinoflagellates, coccolithophores, and radiolarians. A surprisingly low morphological diversity among the abundant diatoms was mentioned.

Macroalgae have occasionally been found attached to floating marine litter, among them red (11 taxa), brown (6 taxa) and some green algae (4 taxa). However, rarely was a single taxon encountered more than once. Diatoms (29 taxa), dinoflagellates (5 taxa) and foraminiferans (7 taxa) seem to be more common, although likewise, very few taxa were reported more than once, probably owing to the low number of studies focusing on micro-organisms.

The most common invertebrate groups on marine litter are crustaceans, bryozoans, molluscs and cnidarians (Table 6.1). The composition of taxa retrieved from beached litter tends to be biased towards sessile organisms with hard (calcified) structures such as bryozoans, foraminiferans, tubeworms and barnacles (Stevens et al. 1996; Winston et al. 1997; Gregory 2009). Mobile organisms such as crustaceans and annelids are more frequently observed on rafts collected while afloat (Astudillo et al. 2009; Goldstein et al. 2014). Some taxa have repeatedly been observed associated with floating litter (Fig. 6.4) and thus, may not just be accidental rafters.

Table 6.1 Taxonomic overview of marine litter rafters (for complete taxonomic list see Appendix 1)

Kingdom	Phylum	Class	Order	Number of taxa
Bacteria				44
Chromista				
	Ciliophora			2
	Foraminifera			7
	Myzozoa			
		Dinophyceae		5
	Haptophyta			7
	Ochrophyta			
		Bacillariophyceae		29
		Phaeophyceae		6
Plantae				
	Charophyta			1
	Chlorophyta			3
	Rhodophyta			11
Animalia				
	Porifera			2
	Cnidaria			
		Anthozoa		10
		Hydrozoa		26
	Nemertea			1
	Annelida			
		Polychaeta		27
	Arthropoda			
		Pycnogonida		1
		Insecta		3
		Ostracoda		1
		Maxillopoda		
			Kentrogonida	1
			Lepadiformes	11
			Sessilia	15
		Malacostraca		
			Decapoda	22
			Amphipoda	21
			Isopoda	8
			Tanaidacea	1
	Mollusca			
		Gastropoda		18
		Bivalvia		21
	Echinodermata			3
	Bryozoa			

(continued)

Table 6.1 (continued)

Kingdom	Phylum	Class	Order	Number of taxa
		Gymnolaemata		66
		Stenolaemata		10
	Chordata			
		Ascidiacea		4
Total				**387**

Stalked barnacles of the genus *Lepas* are by far the most frequently encountered hitchhikers in all major oceanic regions except for the Arctic and Southern Ocean. Seven *Lepas* species have been found rafting on litter, the most frequently observed and widespread being *L. anatifera* and *L. pectinata*. *Lepas* are prominent fouling species and readily colonize a variety of floating objects, a process likely facilitated by their extended planktonic larval stage (Southward 1987).

Isopods of the genus *Idotea* are frequently found on marine litter in the Atlantic, Pacific and Mediterranean. While *I. metallica* and *I. baltica* have repeatedly been reported on floating litter items other species such as *I. emarginata* are less common. *Idotea metallica* is an obligate rafter without benthic populations, and the constant replenishment of an otherwise not self-sustaining population in the North Sea illustrates its conformity with the rafting environment (Gutow and Franke 2001). *Idotea metallica* shows specific adaptations to the rafting life-style such as reduced "locomotive activity and a tight association to the substratum" and low food requirements compared to its congener *I. baltica* (Gutow et al. 2006, 2007). The latter species predominantly colonizes algal rafts, which are rapidly consumed by this voracious herbivore (Gutow 2003; Vandendriessche et al. 2007).

Other frequently encountered crustaceans include the three pelagic species of crab, *Planes major*, *P. marinus* and *P. minutus*, found in the Atlantic, Pacific and Indian Ocean; and five species of the diverse amphipod genus *Caprella*, whose members show morphological adaptations in the form of reduced abdominal appendages enabling them to cling to flotsam (Takeuchi and Sawamoto 1998).

Bryozoans from the closely related genera *Membranipora* and *Jellyella* were found rafting on marine litter in the Atlantic, Pacific, Mediterranean and even in Arctic waters. *Jellyella tuberculata* was the most frequently encountered species in the Atlantic and Pacific and is known to colonize a wide range of substrata including plastic litter and macroalgae (Winston et al. 1997). The species typically occurs at tropical and subtropical latitudes (Gregory 1978), however, sightings on marine litter are reported from all major oceanic regions with the exception of polar seas (Fig. 6.4). The most common gastropod on floating litter, *Fiona pinnata*, was sighted in the Pacific and Mediterranean. According to Willan (1979), *F. pinnata* has a cosmopolitan distribution and commonly inhabits floating wood and macroalgae where it can exploit its *Lepas* prey, growing on the same substratum.

6.3.2 Biological Traits of Rafting Invertebrates on Floating Litter

Given the specific habitat conditions on floating marine litter, it can be expected that certain biological traits will predominate among the assemblage of rafting organisms. Of the 215 invertebrate species considered for this analysis, 25 (12 %) have been classified as obligate rafters that live exclusively on floating objects. 165 species (77 %) are facultative rafters that occupy benthic habitats as well. For 25 species (12 %) the available information was not sufficient to determine their raft status.

6.3.2.1 Mobility

Fifty-nine percent of the rafting species on floating litter are fully sessile whereas 5 % of the species can be classified as semi-sessile (with the ability to detach and re-attach). Only 27 % of the reported species are mobile, for the remaining species the information was insufficient. In contrast to these numbers, Astudillo et al. (2009) and Goldstein et al. (2014) found more mobile than sessile taxa on floating litter, indicating that the inclusion of studies from beached litter is likely leading to an underestimation of mobile taxa. Nevertheless, the high proportion of sessile and semi-sessile species highlights the necessity for a firm attachment of rafting species to the often smooth and solid abiotic surfaces of floating litter items. It further illustrates the often low structural complexity of litter items compared to, for example, floating macroalgae which host a much higher proportion of mobile species that can efficiently cling to the often complex algal thalli with numerous branches and highly structured holdfasts (Thiel and Gutow 2005a). Disadvantages for sessile organisms arise when unstable rafts change positions and expose organisms to surface conditions (Bravo et al. 2011), or if the raft sinks or strands (Winston 2012).

6.3.2.2 Feeding Biology

The great majority (72 %) of the rafting taxa on marine floating litter are suspension feeders whereas only 7 % of the species feed as grazers and borers, and 9 % as predators and scavengers (for the remaining 12 % no feeding mode could be identified). The high proportion of suspension feeders on marine litter is not surprising. Abiotic floating substrata are of no nutritional value for associated rafters, making them dependent on food from the surrounding environment. On floating seaweeds, which are consumed by associated herbivores, the proportion of suspension feeders is substantially lower (approx. 40 %) and the proportion of grazers and borers higher (approx. 20 %—Thiel and Gutow 2005a). Rafting suspension feeders benefit from the concentration of their rafts and suspended organic material in surface fronts generated by the convergence of surface waters, wind-induced Langmuir cells and other surface features (Woodcock 1993; Marmorino et al. 2011). The accumulation of suspended matter and nutrients in these convergence zones apparently fuels diverse rafting communities on floating abiotic substrata, which also encompass primary producers, herbivores, and predators.

6.3.2.3 Reproductive Traits

Forty-eight percent of the rafting invertebrate species on marine floating litter reproduce sexually (of which 42 % are hermaphroditic and 58 % are gonochoric) and 38 % have, at least theoretically, the ability to reproduce both sexually and asexually while for 14 % of the species no information on the reproductive mode is available. Bryozoans, constituting most of the species that are capable of asexual and sexual reproduction, reproduce primarily asexually. This facilitates establishment and rapid local spread. However, encrusting bryozoans seem to reproduce exclusively sexually (Thomsen and Hakansson 1995). Bryozoans also perform "spermcast mating" where sperm is accumulated from the surrounding water and stored prior to fertilization (Bishop and Pemberton 2006), a strategy which appears particularly beneficial for rafting organisms because there may be no (or only few) conspecifics nearby. If bryozoans grow in isolation many have the ability to self-fertilize rather than to rely on neighbouring colonies (Maturo 1991 cited by Winston et al. 1997).

About 9 % of the rafting species on marine litter have benthic larvae or larvae with a short pelagic development of less than two days and 12 % release fully developed individuals. Thirty percent of the species have pelagic larvae with an extended planktonic phase of up to several weeks. For 49 % of the invertebrate species no details on larval biology were available. Winston et al. (1997) suggest that long-lived larvae may be beneficial for settlement on litter floating in the open ocean, although upwelling events and storms may facilitate the colonization of litter items by species with short larval development. Astudillo et al. (2009) found mainly rafters with short larval development or direct development on floating buoys in the south-eastern Pacific, a region under influence of upwelling regimes. Stevens et al. (1996) also reported many bryozoans with short larval development on beached litter in northern New Zealand. Given the long distances floating litter can travel, some stranded items may have been under the influence of upwelling regions as described for the South Taranaki Bight (summarized by Foster and Battaerd 1985), approximately 500 km to the south of the sampled location.

6.3.3 Other Species Attracted to Marine Litter

Fishes and other marine vertebrates and invertebrates are known to aggregate around floating objects at sea (for example Hunter and Mitchell 1967; Taquet et al.

2007). Aliani and Molcard (2003) observed dolphins, sea turtles and fish below larger items (mostly plastics) in the Mediterranean. Fish that aggregate below rafts (of natural or anthropogenic origin) may also become dispersed over long oceanic distances, occasionally even crossing oceanic barriers (Luiz et al. 2012). Possibly, the increasing number of observations of raft-associated fish species near oceanic islands (e.g. Afonso et al. 2013) is due to increasing densities of floating litter in these regions (e.g. Law et al. 2010). It is still not well known why fish aggregate around floating objects, especially because they are rarely observed feeding on organisms living on flotsam (e.g. Ibrahim et al. 1996). On the other hand, fish and shark bite marks in plastic litter might indicate that fishes prey actively on the biota on floating litter (Winston et al. 1997; Carson 2013). A review by Castro et al. (2002) concludes that the reasons why fish aggregate around floating objects, and especially macroalgae assemblages, may be manifold, including serving as a refuge, a source for food, and a meeting point for solitary fish. Seabirds may accidentally ingest litter items if they confuse artificial flotsam such as Styrofoam with food (e.g. van Franeker 1985; Kühn et al. 2015). Some species may also ingest litter while feeding on the organisms growing on small litter items.

6.3.4 Succession of the Rafting Community

The colonization of artificial floating substrata follows a general pattern that has been investigated experimentally in several studies (Ye and Andrady 1991; Artham et al. 2009; Bravo et al. 2011; Lobelle and Cunliffe 2011): first, a biofilm consisting of bacteria and biopolymers develops within hours after submergence. This first phase is primarily controlled by the physico-chemical properties of the substratum (such as rugosity and hydrophobicity) whereas biological processes seem less important at this stage (Artham et al. 2009). The exact development and composition of the biofilm is highly variable, even on similar substrata at the same site (Ye and Andrady 1991) and probably influenced by seasonal (Artham et al. 2009) and other environmental variables (temperature, salinity—Carson et al. 2013). The composition of the initial colonizer assemblage affects the further succession of the fouling community (Ye and Andrady 1991; Bravo et al. 2011), although bryozoans readily colonize clean substrata without a biofilm (Maki et al. 1989; Zardus et al. 2008). In general, invertebrates and macroalgae may colonize submerged substrata within three to four weeks (Ye and Andrady 1991; Bravo et al. 2011). Results from a fouling experiment conducted by Dean and Hurd (1980) suggest that initial colonization of organisms on artificial substrata may facilitate some later arrivers but inhibit others.

The settlement of invertebrates seems to depend mainly on the availability of propagules (larvae and juveniles) in the surrounding environment (Stevens et al. 1992 cited by Winston et al. 1997; Barnes 2002) but less on the distance from the coast (Barnes 2002). Further information on later successional stages of rafting communities on floating litter has been collected from floating and stranded

Fig. 6.6 Succession of a rafting community on floating objects, among them marine litter. The y-axis gives the share of the respective taxa in terms of abundance. Higher invertebrates are mainly represented by amphipods. Modified after Tsikhon-Lukanina et al. (2001)

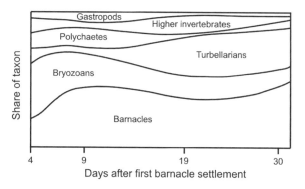

substrata and from experiments: during an experimental exposure of different plastic items for 13–19 weeks, an initial biofilm with green algae was replaced after seven weeks by hydroid colonies followed by bryozoans and ascidians (Ye and Andrady 1991). Bravo et al. (2011) found a peak in taxonomic richness on abiotic substrata (plastics, Styrofoam and pumice) that had been submerged for eight weeks. The community was initially dominated by diatoms, whereas later successional stages were characterized by hydrozoans (mainly *Obelia* sp.), barnacles (*Austromegabalanus psittacus*) and an ascidian (*Diplosoma* sp.). Tsikhon-Lukanina et al. (2001), studying natural and anthropogenic flotsam in the western North Pacific, recognized a bryozoan-dominated phase with a higher abundance of polychaetes and gastropods, followed by a lepadid barnacle phase with a higher incidence of malacostracan crustaceans, especially amphipods (Fig. 6.6). Turbellarians increased in abundance and biomass throughout the experimental duration. Winston et al. (1997) found no signs of succession on beached litter in Florida and Bermuda, which may have been obscured by the state of desiccated animals. In contrast to the initial biofilm formation, later successional stages are much more controlled by biological processes. For example, the bryozoan *Electra tenella* occurs exclusively on plastic items (floating off the U.S. Atlantic coast), thereby avoiding competition, mainly with *Membranipora tuberculata*, which frequently overgrows *E. tenella* on natural substrata (Winston 1982).

6.4 Floating Litter as Dispersal Vector

Floating litter can facilitate the dispersal of associated organisms when moved across the ocean surface by winds and currents. The efficiency of rafting dispersal depends on the availability and the persistence of floating substrata in the oceans. Already established populations may disperse regionally with the help of marine litter, as was observed by Whitehead et al. (2011) for lepadid barnacles in South Africa, by Serrano et al. (2013) for a Mediterranean population of the coral *Oculina patagonica* and also by Davidson (2012) for the isopod *Sphaeroma quoianum*, which "manufactures" its own raft by causing fragmentation of Styrofoam/polystyrene dock floats.

Several taxa, including potential invaders, were found on marine litter far beyond their natural dispersal range: stranded barnacles (of the genera *Dosima*, *Lepas* and *Perforatus*) were observed in Ireland and Wales (having spent considerable time rafting in the North Atlantic), though individuals were not found alive (Minchin 1996; Rees and Southward 2009). Studies from the Netherlands report the reef coral *Favia fragum*, also dead and having rafted from the Caribbean (Hoeksema et al. 2012, Fig. 6.1a) and shell parts of the bivalve *Pinctada imbricata* (Cadée 2003). Barnes and Milner (2005) recorded *Austrominius modestus* (as *Elminius modestus*), an exotic invader, on drift plastic on the Shetland Islands (Scotland, UK), although this was not the first record of that barnacle there. By far the biggest piece of long-distance-rafting flotsam is described by Choong and Calder (2013): A 188-ton piece of a former dock, dislodged during a tsunami in Japan in 2011, stranded in Oregon and offered a rafting opportunity for over 100 species, non-native to the U.S. coast. Several other large pieces of tsunami debris of the same origin transported further species to the North Pacific east coast (Calder et al. 2014).

To successfully establish a founding population rafting organisms not only have to survive the journey but be able to reproduce upon reaching a potential habitat. In general, colonial organisms have the highest potential to successfully establish in new habitats as every individual "represents a potential founder population" (Winston 2012). Reproductively active organisms have been observed on numerous occasions, including bryozoans, as far south as Adelaide Island, Antarctica (Barnes and Fraser 2003), and egg-bearing crustaceans in many different regions (e.g. Spivak and Bas 1999; Gutow and Franke 2003; Poore 2012; Cabezas et al. 2013). Resting cysts of dinoflagellates attached to plastic have been observed (Masó et al. 2003) as well as egg masses of gastropods, even though no adult specimens were present (Winston et al. 1997; Bravo et al. 2011). The pelagic insects *Halobates sericeus* (Goldstein et al. 2012) and *H. micans* (Majer et al. 2012) are known to deposit eggs on marine plastics, and the ubiquity of this substratum helps these species to overcome limitations of suitable oviposition sites.

On numerous occasions, rafting taxa have been reported for the very first time on marine litter in a given region (Jara and Jaramillo 1979; Stevens et al. 1996; Winston et al. 1997; Cadée 2003), a mentionable feat considering the stochastic nature of rafting events. Like other floating substrata marine litter is under the influence of winds and currents, but due to high buoyancy some litter items may be pushed along different trajectories than other flotsam, such as mostly submerged macroalgae. However, unlike other potential dispersal vectors for invasive species, especially transport by ship (ballast water and hull fouling), it is not expected that marine litter opens up novel pathways that are not available for other floating substrata (Lewis et al. 2005).

Given the high persistence of marine litter and the enormous abundances in the world's oceans (Eriksen et al. 2014) it becomes evident that the littering of the oceans with plastics over the past decades has substantially enhanced rafting opportunities for marine organisms, and it is estimated that floating marine litter doubles or even triples the dispersal of marine organisms (Barnes 2002, however doubted by Lewis et al. 2005). The implications of the increasing amounts of long-lived floating substrata in the oceans are pointed out by Goldstein et al. (2012) who suggest that the populations of the ocean skater *H. sericeus* are no longer limited by the availability of floating objects, used for egg attachment. Similar effects may be responsible for the reported population expansion of other common rafters (e.g. Winston 1982 for *Electra tenella*).

More importantly, floating litter is not only more abundant than natural floating substrata in many parts of the world's oceans, but its abundances are chronically high, throughout all seasons and across years. This continuous presence of large amounts of floating litter contrasts strongly with the highly episodic appearance of pumice rafting opportunities (e.g. Bryan et al. 2012) and few natural rafting opportunities in tropical waters (Rothäusler et al. 2012). It is likely that this change in the temporal and spatial availability of abiotic rafts dramatically affects the dynamics of rafting transport and colonization by associated organisms.

6.5 Summary and Outlook

In an earlier global compilation Thiel and Gutow (2005a) listed 108 invertebrate species that have been found rafting on plastics in the ocean. Since then the list of rafting invertebrates on marine litter (including plastics and other anthropogenic litter) has almost doubled to 215 species. Additionally, some recent studies revealed the ubiquity of micro-organisms on marine litter. Sessile suspension feeders seem to be particularly well adapted to life on solid artificial substrata with specific surface characteristics and limited autochthonous food supply. The colonization of floating litter items is apparently facilitated by larvae with an extended planktonic development. Sexual and asexual reproduction is equally common among rafting species on marine litter with asexual reproduction probably allowing for rapid monopolization, especially of colonial species (e.g. bryozoans) on isolated floating substrata. Physical characteristics of the raft, such as surface rugosity and floating behavior, are crucial for colonization processes and subsequent succession of the rafting invertebrate community. The associated organisms themselves can influence the persistence and stability of their raft indicating complex interaction between the rafting substratum and the associated biota.

Abundant floating marine litter has been suggested to facilitate the spread of invasive species and, in fact, some species have been observed rafting on marine litter beyond their natural distributional limits. Marine litter has probably not opened new rafting routes in the oceans. However, the permanent availability of high densities of persistent floating litter items, especially in regions where natural flotsam occurs in low densities or only episodically, has substantially increased rafting opportunities for species that are able to persist on abiotic flotsam.

Accordingly, the continuous supply of individuals from distant up-current regions probably facilitates the establishment of species in new regions.

Recent studies have not only enhanced our understanding of the role of marine litter as a habitat and dispersal vector for marine biota but also revealed open questions that clearly deserve more research effort. Ocean current models have been used to identify drift trajectories and major accumulation zones of floating marine litter in the Atlantic, Pacific and Indian Ocean (Lebreton et al. 2012; Maximenko et al. 2012), which could be confirmed by field surveys (see for example Law et al. 2010; Goldstein et al. 2013). These models are primarily based on drift trajectories of surface buoys equipped with drogues extending several metres below the sea surface and are thus suitable for identifying broad distributional patterns and large-scale accumulation zones of litter in the oceans. In coastal waters, currents are much more variable and complex and litter objects floating at the sea surface are more strongly influenced by wind than common drifter buoys (e.g. Astudillo et al. 2009). However, our knowledge on how wind and currents influence the floating behavior of different litter items is limited (Neumann et al. 2014). Experimental studies on the floating speed and direction of different categories of floating litter under the influence of variable wind and current conditions would improve our abilities to model floating trajectories of marine litter, predict potential rafting routes, and identify sources of marine floating litter.

Persistence of a litter item in the sea is crucial for its suitability as a habitat and dispersal vector for marine biota. However, the dynamics of degradation of the various litter types under variable marine environmental conditions are poorly understood. Likewise, more research is required to understand how marine biota can accelerate or decelerate degradation processes of marine litter. Investigations on the degradation processes should combine in situ monitoring of litter items in the marine environment and biochemical laboratory studies, e.g. on the enzymatic decomposition of plastic polymers.

The degradation of plastics may induce the release of chemicals, some of which are known to affect the health of marine organisms (Rochman 2015). The role of ingested microplastics for the transport of contaminants to marine biota may be limited also because of the rapid gut passage of the small particles (Koelmans 2015). However, the firm attachment of a sessile organism to an artificial surface is permanent and it is yet unknown whether this form of chronic exposure might allow for a slow but continuous transfer of contaminants from plastics to animals via epithelia or with chemically enriched water from the micro-layer on the plastic surface. These studies would require laboratory measurements on the chemical load and the health status of litter rafters, but should also involve organisms collected from litter at sea.

Combined, new and sound information on floating trajectories, raft persistence, and performance of associated organisms will help to estimate the potential of marine litter for the transport of invasive species or entire rafting communities, and therefore add to our understanding of the hazardous character of marine litter beyond the immediate effects of ingestion and entanglement.

Acknowledgments We thank Miriam Goldstein and Emmett Clarkin for valuable comments on an earlier draft of the manuscript. This is publication no. 37793 of the Alfred-Wegener-Institut Helmholtz-Zentrum für Polar- und Meeresforschung.

References

Aliani, S., & Molcard, A. (2003). Hitch-hiking on floating marine debris: Macrobenthic species in the western Mediterranean Sea. *Hydrobiologia, 503*, 59–67.

Afonso, P., Porteiro, F. M., Fontes, J., Tempera, F., Morato, T., Cardigos, F., et al. (2013). New and rare coastal fishes in the Azores islands: Occasional events or tropicalization process? *Journal of Fish Biology, 83*, 272–294.

Andrady, A. L. (2011). Microplastics in the marine environment. *Marine Pollution Bulletin, 62*, 1596–1605.

Artham, T., Sudhakar, M., Venkatesan, R., Nair, C. M., Murty, K. V. G. K., & Doble, M. (2009). Biofouling and stability of synthetic polymers in sea water. *International Biodeterioration and Biodegradation, 63*, 884–890.

Astudillo, J. C., Bravo, M., Dumont, C. P., & Thiel, M. (2009). Detached aquaculture buoys in the SE Pacific: Potential dispersal vehicles for associated organisms. *Aquatic Biology, 5*, 219–231.

Barnes, D. K. A. (2004). Natural and plastic flotsam stranding in the Indian Ocean. In D. Davenport & J. Davenport (Eds.), *The effects of human transport on ecosystems: Cars and planes, boats and trains* (pp. 193–205). Dublin: Royal Irish Academy.

Barnes, D. K. A. (2002). Invasions by marine life on plastic debris. *Nature, 416*, 808–809.

Barnes, D. K. A., & Milner, P. (2005). Drifting plastic and its consequences for sessile organism dispersal in the Atlantic Ocean. *Marine Biology, 146*, 815–825.

Barnes, D. K. A., & Fraser, K. P. P. (2003). Rafting by five phyla on man-made flotsam in the Southern Ocean. *Marine Ecology Progress Series, 262*, 289–291.

Barnes, D. K. A., Galgani, F., Thompson, R. C., & Barlaz, M. (2009). Accumulation and fragmentation of plastic debris in global environments. *Philosophical Transactions of the Royal Society B, 364*, 1985–1998.

Bishop, J. D. D., & Pemberton, A. J. (2006). The third way: Spermcast mating in sessile marine invertebrates. *Integrative and Comparative Biology, 46*, 398–406.

Bravo, M., Astudillo, J. C., Lancellotti, D., Luna-Jorquera, G., Valdivia, N., & Thiel, M. (2011). Rafting on abiotic substrata: Properties of floating items and their influence on community succession. *Marine Ecology Progress Series, 439*, 1–17.

Bryan, S. E., Cook, A. G., Evans, J. P., Hebden, K., Hurrey, L., Colls, P., et al. (2012). Rapid, long-distance dispersal by pumice rafting. *PLoS ONE, 7*, e40583.

Cabezas, M. P., Navarro-Barranco, C., Ros, M., & Guerra-García, J. M. (2013). Long-distance dispersal, low connectivity and molecular evidence of a new cryptic species in the obligate rafter *Caprella andreae* Mayer, 1890 (Crustacea: Amphipoda: Caprellidae). *Helgoland Marine Research, 67*, 483–497.

Cadée, M. (2003). Een vondst van de Atlantische Pareloester *Pinctada imbracata* (Röding, 1789) in een plastic fles op het Noordwijkse strand. *Het Zeepard, 63*, 76–78.

Calder, D. R., Choong, H. H., Carlton, J. T., Chapman, J. W., Miller, J. A., & Geller, J. (2014). Hydroids (Cnidaria: Hydrozoa) from Japanese tsunami marine debris washing ashore in the northwestern United States. *Aquatic Invasions, 9*, 425–440.

Carson, H. S. (2013). The incidence of plastic ingestion by fishes: from the prey's perspective. *Marine Pollution Bulletin, 74*, 170–174.

Carson, H. S., Nerheim, M. S., Carroll, K. A., & Eriksen, M. (2013). The plastic-associated microorganisms of the North Pacific Gyre. *Marine Pollution Bulletin, 75*, 126–132.

Castro, J. J., Santiago, J. A., & Santana-Ortega, A. T. (2002). A general theory on fish aggregation to floating objects: an alternative to the meeting point hypothesis. *Reviews in Fish Biology and Fisheries, 11*, 255–277.

Choong, H. H. C., & Calder, D. R. (2013). *Sertularella mutsuensis* Stechow, 1931 (Cnidaria: Hydrozoa: Sertulariidae) from Japanese tsunami debris: systematics and evidence for transoceanic dispersal. *BioInvasions Records, 2*, 33–38.

Cundell, A. (1974). Plastics in the marine environment. *Environmental Conservation, 1*, 63–68.

Davidson, T. M. (2012). Boring crustaceans damage polystyrene floats under docks polluting marine waters with microplastic. *Marine Pollution Bulletin, 64*, 1821–1828.

Dean, T. A., & Hurd, L. E. (1980). Development in an estuarine fouling community: the influence of early colonists on later arrivals. *Oecologia, 46*, 295–301.

Dellinger, T., Davenport, J., & Wirtz, P. (1997). Comparisons of social structure of Columbus crabs living on loggerhead sea turtles and inanimate flotsam. *Journal of the Marine Biological Association of the UK, 77*, 185–194.

Derraik, J. G. (2002). The pollution of the marine environment by plastic debris: A review. *Marine Pollution Bulletin, 44*, 842–852.

Ebbesmeyer, C. C., & Ingraham, W. J. (1992). Shoe spill in the North Pacific. *EOS, Transactions, American Geophysical Union, 73*, 361–368.

Eriksen, M., Lebreton, L. C. M., Carson, H. S., Thiel, M., Moore, C. J., Borerro, J. C., et al. (2014). Plastic pollution in the world's oceans: More than 5 trillion plastic pieces weighing over 250,000 Tons afloat at sea. *PLoS ONE, 9*, e111913.

Fortuño, J., Masó, M., Sáez, R., De Juan, S., & Demestre, M. (2010). SEM microphotographs of biofouling organisms on floating and benthic plastic debris. *Rapport Commission International Mer Mediterranée, 39*, 358.

Foster, B. A., & Battaerd, W. R. (1985). Distribution of zooplankton in a coastal upwelling in New Zealand. *New Zealand Journal of Marine and Freshwater Research, 19*, 213–226.

Galgani, F., Hanke, G., & Maes, T. (2015) Global distribution, composition and abundance of marine litter. In M. Bergmann, L. Gutow, M. Klages (Eds.), *Marine anthropogenic litter* (pp. 29–56). Berlin: Springer

Goldstein, M. C., Rosenberg, M., & Cheng, L. (2012). Increased oceanic microplastic debris enhances oviposition in an endemic pelagic insect. *Biology Letters, 8*, 817–820.

Goldstein, M. C., Titmus, A. J., & Ford, M. (2013). Scales of spatial heterogeneity of plastic marine debris in the Northeast Pacific Ocean. *PLoS ONE, 8*, e80020.

Goldstein, M. C., Carson, H. S., & Eriksen, M. (2014). Relationship of diversity and habitat area in North Pacific plastic-associated rafting communities. *Marine Biology, 161*, 1441–1453.

Gregory, M. R. (1978). Accumulation and distribution of virgin plastic granules on New Zealand beaches. *New Zealand Journal of Marine and Freshwater Research, 12*, 399–414.

Gregory, M. R. (2009). Environmental implications of plastic debris in marine settings—entanglement, ingestion, smothering, hangers-on, hitch-hiking and alien invasions. *Philosophical Transactions of the Royal Society B, 364*, 2013–2025.

Gutow, L. (2003). Local population persistence as a pre-condition for large-scale dispersal of *Idotea metallica* (Crustacea, Isopoda) on drifting habitat patches. *Hydrobiologia, 503*, 45–48.

Gutow, L., & Franke, H. D. (2001). On the current and possible future status of the neustonic isopod *Idotea metallica* Bosc in the North Sea: A laboratory study. *Journal of Sea Research, 45*, 37–44.

Gutow, L., & Franke, H. D. (2003). Metapopulation structure of the marine isopod *Idotea metallica*, a species associated with drifting habitat patches. *Helgoland Marine Research, 56*, 259–264.

Gutow, L., Strahl, J., Wiencke, C., Franke, H. D., & Saborowski, R. (2006). Behavioural and metabolic adaptations of marine isopods to the rafting life style. *Marine Biology, 149*, 821–828.

Gutow, L., Leidenberger, S., Boos, K., & Franke, H. D. (2007). Differential life history responses of two *Idotea* species (Crustacea: Isopoda) to food limitation. *Marine Ecology Progress Series, 344,* 159–172.

Hoeksema, B. W., Roos, P. J., & Cadée, G. C. (2012). Trans-Atlantic rafting by the brooding reef coral *Favia fragum* on man-made flotsam. *Marine Ecology Progress Series, 445,* 209–218.

Hunter, J. R., & Mitchell, C. T. (1967). Association of fishes with flotsam in the offshore waters of Central America. *Fisheries Bulletin, 66,* 13–29.

Ibrahim, S., Ambak, M. A., Shamsudin, L., & Samsudin, M. Z. (1996). Importance of fish aggregating devices (FADs) as substrates for food organisms of fish. *Fisheries Research, 27,* 265–273.

Jara, C., & Jaramillo, E. (1979). Hallazgo de *Planes marinus* Rathbun, 1914, sobre boya a la deriva de Maiquillahue, Chile (Crustacea, Decapoda, Grapsidae). *Medio Ambiente, 4,* 108–113.

Koelmans, A. A. (2015). Modeling the role of microplastics in bioaccumulation of organic chemicals to marine aquatic organisms. Critical review. In M. Bergmann, L. Gutow, M. Klages (Eds.). *Marine Anthropogenic Litter* (pp. 313–328). Berlin: Springer

Kühn, S., Bravo Rebolledo, E. L., & van Franeker, J. A. (2015). Deleterious effects of litter on marine life. In M. Bergmann, L. Gutow, M. Klages (Eds.). *Marine Anthropogenic Litter* (pp. 75–116). Berlin: Springer

Kukulka, T., Proskurowski, G., Morét-Ferguson, S., Meyer, D. W., & Law, K. L. (2012). The effect of wind mixing on the vertical distribution of buoyant plastic debris. *Geophysical Research Letters, 39,* L07601.

Law, K. L., Morét-Ferguson, S., Maximenko, N. A., Proskurowski, G., Peacock, E. E., Hafner, J., et al. (2010). Plastic accumulation in the North Atlantic Subtropical Gyre. *Science, 329,* 1185.

Lebreton, L. C. M., Greer, S. D., & Borrero, J. C. (2012). Numerical modelling of floating debris in the world's oceans. *Marine Pollution Bulletin, 64,* 653–661.

Lewis, P. N., Riddle, M. J., & Smith, S. D. A. (2005). Assisted passage or passive drift: A comparison of alternative transport mechanisms for non-indigenous coastal species into the Southern Ocean. *Antarctic Science, 17,* 183–191.

Lobelle, D., & Cunliffe, M. (2011). Early microbial biofilm formation on marine plastic debris. *Marine Pollution Bulletin, 62,* 197–200.

Luiz, O. J., Madin, J. S., Robertson, D. R., Rocha, L. A., Wirtz, P., & Floeter, S. R. (2012). Ecological traits influencing range expansion across large oceanic dispersal barriers: Insights from tropical Atlantic reef fishes. *Proceedings of the Royal Society B, 279,* 1033–1040.

Majer, A. P., Vedolin, M. C., & Turra, A. (2012). Plastic pellets as oviposition site and means of dispersal for the ocean-skater insect *Halobates. Marine Pollution Bulletin, 64,* 1143–1147.

Maki, J. S., Rittschof, D., Schmidt, A. R., Snyder, A. G., & Mitchell, R. (1989). Factors controlling attachment of bryozoan larvae: A comparison of bacterial films and unfilmed surfaces. *Biological Bulletin, 177,* 295–302.

Marmorino, G. O., Miller, W. D., Smith, G. B., & Bowles, J. H. (2011). Airborne imagery of a disintegrating *Sargassum* drift line. *Deep-Sea Research, 158,* 316–321.

Masó, M., Garcés, E., Pagès, F., & Camp, J. (2003). Drifting plastic debris as a potential vector for dispersing Harmful Algal Bloom (HAB) species. *Scientia Marina, 67,* 107–111.

Maturo, F. J. (1991). Self-fertilisation in gymnolaemate Bryozoa. *Bulletin de la Société des Sciences Naturelles de l'Ouest de la France, 1,* 572.

Maximeno, N., Hafner, J., & Niiler, P. (2012). Pathways of marine debris derived from trajectories of Lagrangian drifters. *Marine Pollution Bulletin, 65,* 51–62.

Minchin, D. (1996). Tar pellets and plastics as attachment surfaces for lepadid cirripedes in the North Atlantic Ocean. *Marine Pollution Bulletin, 32,* 855–859.

Molnar, J. L., Gamboa, R. L., Revenga, C., & Spalding, M. D. (2008). Assessing the global threat of invasive species to marine biodiversity. *Frontiers in Ecology and the Environment, 6,* 485–492.

Moore, C. J. (2008). Synthetic polymers in the marine environment: a rapidly increasing, long-term threat. *Environmental Research, 108,* 131–139.

Neumann, D., Callies, U., & Matthies, M. (2014). Marine litter ensemble transport simulations in the southern North Sea. *Marine Pollution Bulletin, 86,* 219–228.

O'Brine, T., & Thompson, R. C. (2010). Degradation of plastic carrier bags in the marine environment. *Marine Pollution Bulletin, 60*, 2279–2283.

Poore, G. C. B. (2012). Four new valviferan isopods from diverse tropical Australian habitats (Crustacea: Isopoda: Holognathidae and Idoteidae). *Memoirs of Museum Victoria, 69*, 327–340.

Rees, E. I. S., & Southward, A. J. (2009). Plastic flotsam as an agent for dispersal of *Perforatus perforatus* (Cirripedia: Balanidae). *Marine Biodiversity Records, 2*, e25.

Reisser, J., Shaw, J., Hallegraeff, G., Proietti, M., Barnes, D. K. A., Thums, M., et al. (2014). Millimeter-sized marine plastics: a new pelagic habitat for microorganisms and invertebrates. *PLoS ONE, 9*, e100289.

Rochman, C. M. (2015). The complex mixture, fate and toxicity of chemicals associated with plastic debris in the marine environment. In M. Bergmann, L. Gutow, M. Klages (Eds.) *Marine anthropogenic litter* (pp. 117–140). Berlin: Springer

Rothäusler, E., Gómez, I., Hinojosa, I. A., Karsten, U., Tala, F., & Thiel, M. (2009). Effect of temperature and grazing on growth and reproduction of floating *Macrocystis* spp. (Phaeophyceae) along a latitudinal gradient. *Journal of Phycology, 45*, 547–559.

Rothäusler, E., Gutow, L., & Thiel, M. (2012). Floating seaweeds and their communities. In C. Wiencke & K. Bischof (Eds.), *Seaweed Biology* (pp. 359–380). Berlin Heidelberg: Springer.

Serrano, E., Coma, R., Ribes, M., Weitzmann, B., García, M., & Ballesteros, E. (2013). Rapid northward spread of a zooxanthellate coral enhanced by artificial structures and sea warming in the western Mediterranean. *PLoS ONE, 8*, e52739.

Southward, A. J. (1987). *Barnacle biology*. Rotterdam: Balkema.

Spivak, E. D., & Bas, C. C. (1999). First finding of the pelagic crab *Planes marinus* (Decapoda: Grapsidae) in the southwestern Atlantic. *Journal of Crustacean Biology, 19*, 72–76.

Stevens, L. M. (1992). *Marine plastic debris: Fouling and degradation*. Unpublished M.Sc. thesis, University of Auckland.

Stevens, L. M., Gregory, M. R., & Foster, B. A. (1996). Fouling bryozoans on pelagic and moored plastics from northern New Zealand. In D. P. Gordon, A. M. Smith, & J. A. Grant-Mackie (Eds.), *Bryozoans in Space and Time* (pp. 321–340). Wellington: NIWA.

Takeuchi, I., & Sawamoto, S. (1998). Distribution of caprellid amphipods (Crustacea) in the western North Pacific based on the CSK International Zooplankton Collection. *Plankton Biology and Ecology, 45*, 225–230.

Taquet, M., Sancho, G., Dagorn, L., Gaertner, J. C., Itano, D., Aumeeruddy, R., et al. (2007). Characterizing fish communities associated with drifting fish aggregating devices (FADs) in the Western Indian Ocean using underwater visual surveys. *Aquatic Living Resources, 20*, 331–341.

Thiel, M., & Gutow, L. (2005a). The ecology of rafting in the marine environment. II. The rafting organisms and community. *Oceanography and Marine Biology: An Annual Review, 43*, 279–418.

Thiel, M., & Gutow, L. (2005b). The ecology of rafting in the marine environment. I. The floating substrata. *Oceanography and Marine Biology: An Annual Review, 42*, 181–264.

Thomsen, E., & Hakansson, E. (1995). Sexual versus asexual dispersal in clonal animals: Examples from cheilostome bryozoans. *Paleobiology, 21*, 496–508.

Tsikhon-Lukanina, E. A., Reznichenko, O. G., & Nikolaeva, G. G. (2001). Ecology of invertebrates on the oceanic floating substrata in the northwest Pacific ocean. *Russian Academy of Sciences. Oceanology, 41*, 525–530.

van Franeker, J. A. (1985). Plastic ingestion in the North Atlantic fulmar. *Marine Pollution Bulletin, 16*, 367–369.

Vandendriessche, S., Vincx, M., & Degraer, S. (2007). Floating seaweeds and the influences of temperature, grazing and clump size on raft longevity—a microcosm study. *Journal of Experimental Marine Biology and Ecology, 343*, 64–73.

Whitehead, T. O., Biccard, A., & Griffiths, C. L. (2011). South African pelagic goose barnacles (Cirripedia, Thoracica): substratum preferences and influence of plastic debris on abundance and distribution. *Crustaceana, 84*, 635–649.

Willan, R. C. (1979). New Zealand locality records for the aeolid nudibranch *Fiona pinnata* (Eschscholtz). *Tane, 25*, 141–147.

Winston, J. E. (1982). Drift plastic—an expanding niche for a marine invertebrate? *Marine Pollution Bulletin, 13*, 348–351.

Winston, J. E. (2012). Dispersal in marine organisms without a pelagic larval phase. *Integrative and Comparative Biology, 52*, 447–457.

Winston, J. E., Gregory, M. R., & Stevens, L. M. (1997). Encrusters, epibionts, and other biota associated with pelagic plastics: A review of biogeographical, environmental, and conservation issues. In J. M. Coe & D. B. Rogers (Eds.), *Marine Debris* (pp. 81–97). New York: Springer.

Wong, C. S., Green, D. R., & Cretney, W. J. (1974). Quantitative tar and plastic waste distributions in the Pacific Ocean. *Nature, 247*, 30–32.

Woodcock, A. H. (1993). Winds subsurface pelagic *Sargassum* and Langmuir circulations. *Journal of Experimental Marine Biology and Ecology, 170*, 117–125.

Ye, S., & Andrady, A. L. (1991). Fouling of floating plastic debris under Biscayne Bay exposure conditions. *Marine Pollution Bulletin, 22*, 608–613.

Zardus, J. D., Nedved, B. T., Huang, Y., Tran, C., & Hadfield, M. G. (2008). Microbial biofilms facilitate adhesion in biofouling invertebrates. *Biological Bulletin, 214*, 91–98.

Zettler, E. R., Mincer, T. J., & Amaral-Zettler, L. A. (2013). Life in the "Plastisphere": Microbial communities on plastic marine debris. *Environmental Science and Technology, 47*, 7137–7146.

Appendix 1

Table of marine floating litter rafters. Raft substrata: P = Plastic, G = Glass, M = Metal, Pa = Paper, U = Unknown composition of multiple materials. Ocean: Arctic = Arctic Ocean, A(N) = North Atlantic, A(S) = South Atlantic, P(N) = North Pacific, P(S) = South Pacific, Ind = Indian Ocean, Southern = Southern Ocean, Med = Mediterranean. Inference (rafting evidence): fl = floating, in situ, str = collected from stranded items, exp = inferred from floating experiments, spec = speculative because of uncertain identification (only 'strongest' rafting evidence is listed fl > str > exp > spec). The taxonomic classification (and taxa names) follows the World Register of Marine Species (WoRMS)

Taxon	Substratum	Region	Inference	Reference
Bacteria				
Acinetobacter sp.	P	A(N)	fl	Zettler et al. (2013)
Albidovulum sp.	P	A(N)	fl	Zettler et al. (2013)
Alteromonas sp.	P	A(N)	fl	Zettler et al. (2013)
Bacteriovorax sp.	P	A(N)	fl	Zettler et al. (2013)
Bdellovibrio sp.	P	A(N)	fl	Zettler et al. (2013)
Blastopirellula sp.	P	A(N)	fl	Zettler et al. (2013)
Devosia sp.	P	A(N)	fl	Zettler et al. (2013)
Erythrobacter sp.	P	A(N)	fl	Zettler et al. (2013)
Filomicrobium sp.	P	A(N)	fl	Zettler et al. (2013)
Fulvivirga sp.	P	A(N)	fl	Zettler et al. (2013)
Haliscomenobacter sp.	P	A(N)	fl	Zettler et al. (2013)

(continued)

(continued)

Taxon	Substratum	Region	Inference	Reference
Hellea sp.	P	A(N)	fl	Zettler et al. (2013)
Henriciella sp.	P	A(N)	fl	Zettler et al. (2013)
Hyphomonas sp.	P	A(N)	fl	Zettler et al. (2013)
Iamia sp.	P	A(N)	fl	Zettler et al. (2013)
Idiomarina sp.	P	A(N)	fl	Zettler et al. (2013)
Labrenzia sp.	P	A(N)	fl	Zettler et al. (2013)
Lewinella sp.	P	A(N)	fl	Zettler et al. (2013)
Marinoscillum sp.	P	A(N)	fl	Zettler et al. (2013)
Microscilla sp.	P	A(N)	fl	Zettler et al. (2013)
Muricauda sp.	P	A(N)	fl	Zettler et al. (2013)
Nitratireductor sp.	P	A(N)	fl	Zettler et al. (2013)
Oceaniserpentilla sp.	P	A(N)	fl	Zettler et al. (2013)
Parvularcula lutaonensis	P	A(N)	fl	Zettler et al. (2013)
Parvularcula sp.	P	A(N)	fl	Zettler et al. (2013)
Pelagibacter sp.	P	A(N)	fl	Zettler et al. (2013)
Phormidium sp.	P	A(N)	fl	Zettler et al. (2013)
Phycisphaera sp.	P	A(N)	fl	Zettler et al. (2013)
Plectonema sp.	P	A(N)	fl	Zettler et al. (2013)
Pleurocapsa sp.	P	A(N)	fl	Zettler et al. (2013)
Prochlorococcus sp.	P	A(N)	fl	Zettler et al. (2013)
Pseudoalteromonas sp.	P	A(N)	fl	Zettler et al. (2013)
Pseudomonas sp.	P	A(N)	fl	Zettler et al. (2013)
Psychrobacter sp.	P	A(N)	fl	Zettler et al. (2013)
Rhodovulum sp.	P	A(N)	fl	Zettler et al. (2013)
Rivularia sp.	P	A(N)	fl	Zettler et al. (2013)
Roseovarius sp.	P	A(N)	fl	Zettler et al. (2013)
Rubrimonas sp.	P	A(N)	fl	Zettler et al. (2013)
Rubritalea sp.	P	A(N)	fl	Zettler et al. (2013)
Saprospira sp.	P	A(N)	fl	Zettler et al. (2013)
Synechococcus sp.	P	A(N)	fl	Zettler et al. (2013)
Tenacibaculum sp.	P	A(N)	fl	Zettler et al. (2013)
Thalassobius sp.	P	A(N)	fl	Zettler et al. (2013)
Thiobios sp.	P	A(N)	fl	Zettler et al. (2013)
Vibrio sp.	P	A(N)	fl	Zettler et al. (2013)
Chromista–Ciliophora				
Ephelota sp.	P	A(N)	fl	Zettler et al. (2013)
Halofolliculina sp.	P	P(N)	fl	Goldstein et al. (2014)
Chromista–Foraminifera				
Acervulina sp.	P	A(N)	str	Winston et al. (1997)
Cibicides sp.	P	A(N)	str	Winston et al. (1997)
Discorbis sp.	P	A(N)	str	Gregory (1983)

(continued)

(continued)

Taxon	Substratum	Region	Inference	Reference
Homotrema rubra	P	P(S)	str	Gregory (1990), Winston et al. (1997)
Planogypsina acervalis	P	A(N)	spec	Winston (2012)
Planulina ornata	P	P(N)	fl	Goldstein et al. (2014)
Rosalina sp.	P	A(N)	str	Winston et al. (1997)
Chromista–Myzozoa–Dinophyceae				
Alexandrium taylori	P	Med	str	Masó et al. (2003)
Alexandrium sp.	P	A(N)	fl	Zettler et al. (2013)
Ceratium macroceros	P	Ind	fl	Reisser et al. (2014)
Ceratium sp.	P	Ind or P(S)	fl	Reisser et al. (2014)
Coolia sp.	P	Med	str	Masó et al. (2003)
Ostreopsis sp.	P	Med	str	Masó et al. (2003)
Prorocentrum sp.	P	Med	str	Masó et al. (2003)
Chromista–Haptophyta				
Calcidiscus leptoporus	P	Ind	fl	Reisser et al. (2014)
Calciosolenia sp.	P	Ind	fl	Reisser et al. (2014)
Coccolithus pelagicus	P	Ind	fl	Reisser et al. (2014)
Emiliania huxleyi	P	Ind	fl	Reisser et al. (2014)
Gephyrocapsa oceanica	P	Ind	fl	Reisser et al. (2014)
Umbellosphaera tenuis	P	Ind	fl	Reisser et al. (2014)
Umbilicosphaera hulburtiana	P	Ind	fl	Reisser et al. (2014)
Chromista–Ochrophyta–Bacillariophyceae				
Achnanthes sp.	P	Ind or P(S), Med	fl	Fortuño et al. (2010), Reisser et al. (2014)
Amphora sp.	P	Ind or P(S)	fl	Reisser et al. (2014)
Ardissonea sp.	P	P(N)	spec	Carson et al. (2013)
Chaetoceros sp.	P	A(N)	fl	Zettler et al. (2013)
Cocconeis sp.	P	Ind or P(S)	fl	Reisser et al. (2014)
Cyclotella meneghiniana	P	A(N)	fl	Carpenter and Smith (1972)
Cylindrotheca sp.	P	Med	fl	Fortuño et al. (2010)
Cymbella sp.	P	Ind or P(S)	fl	Reisser et al. (2014)
Diploneis sp.	P	P(N)	spec	Carson et al. (2013)
Fragilaria sp.	P	P(N)	spec	Carson et al. (2013)
Frustulia sp.	P	P(N)	spec	Carson et al. (2013)
Grammatophora sp.	P	Ind or P(S)	fl	Reisser et al. (2014)
Haslea sp.	P	P(N), Ind or P(S)	fl	Carson et al. (2013), Reisser et al. (2014)
Licmophora sp.	P	Ind or P(S)	fl	Reisser et al. (2014)
Mastogloia angulata	P	A(N)	fl	Carpenter and Smith (1972)

(continued)

(continued)

Taxon	Substratum	Region	Inference	Reference
Mastogloia hulburti	P	A(N)	fl	Carpenter and Smith (1972)
Mastogloia pusilla	P	A(N)	fl	Carpenter and Smith (1972)
Mastogloia sp.	P	P(N), Ind or P(S)	fl	Carson et al. (2013), Reisser et al. (2014)
Microtabella sp.	P	Ind or P(S)	fl	Reisser et al. (2014)
Minidiscus trioculatus	P	Ind or P(S)	fl	Reisser et al. (2014)
Navicula sp.	P	A(N), Ind or P(S),Med	fl	Fortuño et al. (2010), Zettler et al. (2013), Reisser et al. (2014)
Nitzschia longissima	P	Ind or P(S)	fl	Reisser et al. (2014)
Nitzschia sp.	P	A(N), Ind or P(S)	fl	Zettler et al. (2013), Reisser et al. (2014)
Pleurosigma sp.	P	A(N)	fl	Carpenter and Smith (1972)
Protoraphis sp.	P	P(N)	spec	Carson et al. (2013)
Sellaphora sp.	P	A(N)	fl	Zettler et al. (2013)
Stauroneis sp.	P	A(N)	fl	Zettler et al. (2013)
Tabularia sp.	P	Med	fl	Fortuño et al. (2010)
Thalassionema nitzschioides	P	Ind or P(S)	fl	Reisser et al. (2014)
Thalassionema sp.	P	P(N), Med	fl	Fortuño et al. (2010), Carson et al. (2013)
Thalassiosira sp.	P	Ind or P(S), Med	fl	Fortuño et al. (2010), Reisser et al. (2014)
Chromista–Ochrophyta–Phaeophyceae				
Cystoseira sp.	P	Med	fl	Aliani and Molcard (2003)
Ectocarpus acutus	P	P(S)	fl	Astudillo et al. (2009)
Hincksia granulosa	P	P(S)	fl	Astudillo et al. (2009)
Petalonia sp.	P	P(S)	exp	Bravo et al. (2011)
Sargassum sp.	P	A(N)	str	Winston et al. (1997)
Scytosiphon lomentaria	P	P(S)	fl	Astudillo et al. (2009), Bravo et al. (2011)
Plantae–Charophyta				
Closterium sp.	P	Med	fl	Fortuño et al. (2010)
Plantae–Chlorophyta				
Bryopsis rhizophora	P	P(S)	fl	Astudillo et al. (2009)
Codium fragile	P	P(S)	fl	Astudillo et al. (2009)
Ulva rigida	P		spec	Morton and Britton (2000a, b)
Ulva sp.	P, U	P(S)	fl	Thiel et al. (2003), Astudillo et al. (2009)

(continued)

(continued)

Taxon	Substratum	Region	Inference	Reference
Plantae–Rhodophyta				
Amphiroa sp.	P	A(N)	str	Winston et al. (1997)
Antithamnion densum	P	P(S)	fl	Astudillo et al. (2009)
Antithamnion sp.	P	P(S)	exp	Bravo et al. (2011)
Corallina officinalis	P	P(S)	fl	Astudillo et al. (2009)
Fosliella sp.	P	A(N)	str	Gregory (1983), Winston et al. (1997)
Gelidium sp.	P	P(S)	fl	Astudillo et al. (2009)
Hydrolithon farinosum	P	Med	fl	Aliani and Molcard (2003)
Jania sp.	P	A(N)	str	Winston et al. (1997)
Lithophyllum sp.	P	A(N)	str	Winston et al. (1997)
Mesophyllum sp.	P	A(N)	str	Winston et al. (1997)
Polysiphonia mollis	P	P(S)	fl	Astudillo et al. (2009)
Polysiphonia sp.	P	P(S)	exp	Bravo et al. (2011)
Rhodymenia sp.	P	P(S)	fl	Astudillo et al. (2009)
Animalia–Porifera				
Halichondria panicea	P	P(N)	fl	Goldstein et al. (2014)
Sycon sp.	U	P(N)	fl	Goldstein et al. (2014)
Animalia–Cnidaria–Anthozoa				
Actinia sp.	U	P(N)	fl	Goldstein et al. (2014)
Anthopleura dixoniana	P	P(N)	fl	Goldstein et al. (2014)
Anthopleura sp.	P	P(N)	fl	Goldstein et al. (2014)
Anthothoe chilensis	P	P(S)	fl	Astudillo et al. (2009)
Calliactis sp.	U	P(N)	fl	Goldstein et al. (2014)
Diadumene lineata	P, U	P(N)	str	Zabin et al. (2004)
Favia fragum	M	A(N)	str	Hoeksema et al. (2012)
Metridium sp.	P	P(N)	fl	Goldstein et al. (2014)
Oculina patagonica	P, M	Med	str	Fine et al. (2001)
Phyllangia americana	P	A(N)	str	Winston et al. (1997)
Pocillopora sp.	G	P(N)	str	Jokiel (1984)
Animalia–Cnidaria–Hydrozoa				
Aglaophenia latecarinata	P, U	A(N)	str	Calder (1993) (cited by Calder 1995)
Amphisbetia furcata	U	P(S)	str	Calder et al. (2014)
Bougainvillia muscus	U	P(S)	str	Calder et al. (2014)
Clytia gracilis	P	A(N)		Carpenter and Smith (1972)
Clytia gregaria	P	P(N)	fl	Goldstein et al. (2014)
Clytia hemisphaerica	P, U	A(N), Med	fl	Calder (1993) (cited by Calder 1995), Aliani and Molcard (2003)

(continued)

(continued)

Taxon	Substratum	Region	Inference	Reference
Clytia sp.	P	A(N)	str	Winston et al. (1997)
Eudendrium sp.	P	Med	fl	Aliani and Molcard (2003)
Eutima japonica	U	P(S)	str	Calder et al. (2014)
Gonothyraea loveni	P	A(N), Med	fl	Carpenter and Smith (1972), Aliani and Molcard (2003)
Halecium sp.	P	A(N)	str	Winston et al. (1997)
Halecium tenellum	U	P(S)	str	Calder et al. (2014)
Hydrodendron gracilis	U	P(S)	str	Calder et al. (2014)
Laomedea angulata	P	Med	fl	Aliani and Molcard (2003)
Millepora sp.	P	A(N)	str	Winston et al. (1997)
Obelia dichotoma	P, U	A(N) Med	fl	Calder (1993) (cited by Calder 1995), Aliani and Molcard (2003)
Obelia griffini	U	P(S)	str	Calder et al. (2014)
Obelia longissima	U	P(S)	str	Calder et al. (2014)
Obelia sp.	P, U	A(N), P(N), P(S)	fl	Winston et al. (1997), Astudillo et al. (2009), Bravo et al. (2011), Goldstein et al. (2014)
Orthopyxis integra	U	P(S)	str	Calder et al. (2014)
Phialella sp.	U	P(S)	str	Calder et al. (2014)
Plumularia margaretta	P, U	A(N)	str	Calder (1993) (cited by Calder 1995)
Plumularia setacea	P, U	P(N), P(S)	fl	Astudillo et al. (2009), Calder et al. (2014), Goldstein et al. (2014)
Plumularia sp.	P, U	P(S)	str	Bravo et al. (2011), Calder et al. (2014)
Plumularia strictocarpa	P, U	A(N)	str	Calder (1993) (cited by Calder 1995)
Sertularella mutsuensis	U	P(N)	str	Choong and Calder (2013)
Sertularella sp.	U	P(S)	str	Calder et al. (2014)
Sertularia sp.	P	A(N)	str	Winston et al. (1997)
Stylactaria sp.	U	P(S)	str	Calder et al. (2014)
Tubularia sp.	P	P(S)	exp	Bravo et al. (2011)
Zanclea alba	P, U	A(N)	str	Calder (1993) (cited by Calder 1995)
Animalia–Nemertea				
Oerstedia dorsalis	U	P(S)	str	Calder et al. (2014)

(continued)

(continued)

Taxon	Substratum	Region	Inference	Reference
Animalia–Annelida–Polychaeta				
Amaeana sp.	P	P(S)	fl	Astudillo et al. (2009)
Amphinome rostrata	P, U	P(N)	fl	Inatsuchi et al. (2010), Goldstein et al. (2014)
Branchiomma sp.	P	P(S)	fl	Astudillo et al. (2009)
Circeis spirillum	P	A(N)	str	Winston et al. (1997)
Cirratulus sp.	P	P(S)	fl	Astudillo et al. (2009)
Dodecaceria opulens	P	P(S)	fl	Astudillo et al. (2009)
Eunice sp.	U	P(N)	fl	Goldstein et al. (2014)
Halosydna patagonica	P	P(S)	fl	Astudillo et al. (2009)
Halosydna sp.	P	P(N)	fl	Goldstein et al. (2014)
Hipponoe gaudichaudi	P	P(N)	fl	Goldstein et al. (2014)
Hydroides dianthus	P	A(N)	str	Winston et al. (1997)
Hydroides elegans	P		str	Winston et al. (1997)
Hydroides sanctaecrucis	U		spec	Stafford and Willan (2007)
Hydroides sp.	P	A(N)	str	Gregory (1983), Winston et al. (1997)
Myrianida simplex	P	P(S)	fl	Astudillo et al. (2009)
Myrianida sp.	P	P(S)	fl	Astudillo et al. (2009)
Nereis falsa	P	Med	fl	Aliani and Molcard (2003)
Nereis grubei	P	P(S)	fl	Astudillo et al. (2009)
Nereis sp.	U	P(N)	fl	Goldstein et al. (2014)
Odontosyllis sp.	P	P(S)	fl	Astudillo et al. (2009)
Paleanotus sp.	P	P(S)	fl	Astudillo et al. (2009)
Platynereis australis	P	P(S)	fl	Astudillo et al. (2009)
Polycirrus sp.	P	P(S)	fl	Astudillo et al. (2009)
Romanchella pustulata	P	P(S)	fl	Astudillo et al. (2009)
Salmacina sp.	U	P(N)	fl	Goldstein et al. (2014)
Spirobranchus polytrema	P	Med	fl	Aliani and Molcard (2003)
Spirobranchus triqueter	P	A(N)	str	Southward et al. (2004)
Spirorbis corrugatus	P	A(N)	str	Winston et al. (1997)
Spirorbis spirorbis	P	A(N)	str	Winston et al. (1997)
Spirorbis sp.	P	A(N), P(N), P(S)	fl	Gregory (1983, 1990), Goldstein et al. (2014)
Steggoa magalaensis	P	P(S)	fl	Astudillo et al. (2009)
Typosyllis magdalena	P	P(S)	fl	Astudillo et al. (2009)
Animalia–Arthropoda–Pycnogonida				
Phoxichilidium quadradentatum	P	P(N)	fl	Goldstein et al. (2014)

(continued)

(continued)

Taxon	Substratum	Region	Inference	Reference
Animalia–Arthropoda–Insecta				
Halobates micans	P	A(S)	str	Majer et al. (2012)
Halobates sericeus	P	P(N)	fl	Goldstein et al. (2012)
Halobates sp.	P	Ind or P(S)	fl	Reisser et al. (2014)
Halocladius variabilis	P	A(N)	exp	Ingólfsson (1998)
Animalia–Arthropoda–Ostracoda				
Cypris sp.	P	P(S)	fl	Astudillo et al. (2009)
Animalia–Arthropoda–Maxillopoda–Kentrogonida				
Heterosaccus sp.	U	P(N)	fl	Goldstein et al. (2014)
Animalia–Arthropoda–Maxillopoda–Lepadiformes				
Conchoderma auritum	P	A(N)	spec	Gittings et al. (1986)
Conchoderma virgatum	G, U	P(N), P(S)	fl	MacIntyre (1966), Newman (1972)
Dosima fascicularis	P, U	A(N), P(N), Ind	str	Cheng and Lewin (1976), Zevina and Memmi (1981), Minchin (1996), Whitehead et al. (2011)
Dosima sp.	P, G, M, U	Ind	str	Whitehead et al. (2011)
Lepas anatifera	P, M, U	A(N), A(S), P(N),P(S), Ind, Med	fl	Patel (1959), MacIntyre (1966), Green et al. (1994), Minchin (1996), Dellinger et al. (1997), Winston et al. (1997), Spivak and Bas (1999), Barnes and Milner (2005), Astudillo et al. (2009), Whitehead et al. (2011), Cabezas et al. (2013), Goldstein and Goodwin (2013), Goldstein et al. (2014)
Lepas anserifera	P, G, U	P(N), Ind	fl	Newman (1972), Celis et al. (2007), Inatsuchi et al. (2010), Whitehead et al. (2011)
Lepas australis	P	A(S), P(S), Ind	fl	Barnes and Milner (2005), Astudillo et al. (2009), Whitehead et al. (2011)
Lepas hillii	G	P(N)	str	Newman (1972)
Lepas pacifica	P, U	P(N)	fl	Cheng and Lewin (1976), Goldstein and Goodwin (2013), Goldstein et al. (2014)

(continued)

(continued)

Taxon	Substratum	Region	Inference	Reference
Lepas pectinata	P, U	A(N), P(N), P(S),Ind, Med	fl	Minchin (1996), Winston et al. (1997), Tsikhon-Lukanina et al. (2001), Aliani and-Molcard (2003), Wirtz et al. (2006), Astudillo et al. (2009), Bravo et al. (2011), Whitehead et al. (2011), Ryan and Branch (2012)
Lepas testudinata	P, U	Ind	str	Whitehead et al. (2011), Ryan and Branch (2012)
Lepas sp.	P, G, M, U	P(N), P(S), Ind,Med	fl	Woods Hole Oceanographic Institution (1952), Dell (1964), Willan (1979), Holdway and Maddock (1983b), Frazier and Margaritoulis (1990), Gregory (1990), Astudillo et al. (2009), Whitehead et al. (2011), Calder et al. (2014), Goldstein et al. (2014), Reisser et al. (2014)
Animalia–Arthropoda–Maxillopoda–Sessilia				
Amphibalanus amphitrite	P, U	A(N), P(N)	fl	Winston et al. (1997), Stafford and Willan (2007), Goldstein et al. (2014)
Amphibalanus eburneus	P	A(N)	str	Winston et al. (1997)
Austromegabalanus psittacus	P	P(S)	fl	Astudillo et al. (2009), Bravo et al. (2011)
Austrominius modestus	P	A(N), Med	str	Southward et al. (2004), Barnes and Milner (2005)
Balanus flosculus	P	P(S)	fl	Astudillo et al. (2009)
Balanus laevis	P	P(S)	fl	Astudillo et al. (2009)
Balanus trigonus	M	A(N)	str	Hoeksema et al. (2012)
Balanus sp.	P	P(S)	exp	Bravo et al. (2011)
Chelonibia patula	P	Med	str	Frazier and Margaritoulis (1990)
Chthamalus sp.	U	P(N)	fl	Goldstein et al. (2014)
Hesperibalanus fallax	P, U	A(N), Med	str	Kerckhof (1997) (cited by Kerckhof 2002), Southward et al. (2004)

(continued)

(continued)

Taxon	Substratum	Region	Inference	Reference
Megabalanus rosa	P, U	P(N), P(S)	fl	Calder et al. (2014), Goldstein et al. (2014)
Megabalanus tulipiformis	P, U	A(N), Med	str	Southward et al. (2004)
Perforatus perforatus	P	A(N), Med	str	Southward et al. (2004), Rees and Southward (2009)
Semibalanus balanoides	P	Arctic	str	Barnes and Milner (2005)
Semibalanus cariosus	U	P(N)	str	Choong and Calder (2013)
Animalia–Arthropoda–Malacostraca–Decapoda				
Acanthocyclus sp.	P	P(S)	fl	Astudillo et al. (2009)
Allopetrolisthes spinifrons	P	P(S)	fl	Astudillo et al. (2009)
Cancer setosus	P	P(S)	fl	Astudillo et al. (2009)
Chorilia sp.	U	P(N)	fl	Goldstein et al. (2014)
Halicarcinus planatus	P	P(S)	fl	Astudillo et al. (2009)
Herbstia sp.	U	P(N)	fl	Goldstein et al. (2014)
Hippolyte sp.	P	P(S)	fl	Astudillo et al. (2009)
Latreutes antiborealis	P	P(S)	fl	Astudillo et al. (2009)
Liopetrolisthes mitra	P	P(S)	fl	Astudillo et al. (2009)
Lysmata sp.	P	P(S)	fl	Astudillo et al. (2009)
Pachycheles sp.	P	P(S)	fl	Astudillo et al. (2009)
Palaemon affinis	U	P(N)	fl	Goldstein et al. (2014)
Petrolisthes tuberculosus	P	P(S)	fl	Astudillo et al. (2009)
Pilumnoides perlatus	P	P(S)	fl	Astudillo et al. (2009)
Pilumnus sp.	U	P(N)	fl	Goldstein et al. (2014)
Pisoides edwardsii	P	P(S)	fl	Astudillo et al. (2009)
Plagusia immaculata	P		spec	Donlan and Nelson (2003)
Plagusia sp.	U	P(N)	fl	Goldstein et al. (2014)
Planes major	P, U	P(N), P(S)	fl	Chace (1951), Goldstein et al. (2014)
Planes minutus	P, U	A(N), P(N), Ind	fl	Dellinger et al. (1997), Winston et al. (1997), Ryan and Branch (2012), Goldstein et al. (2014)
Planes sp.	P, U	P(N)	fl	Goldstein et al. (2014)
Synalpheus spinifrons	P	P(S)	fl	Astudillo et al. (2009)
Taliepus dentatus	P	P(S)	fl	Astudillo et al. (2009)

(continued)

(continued)

Taxon	Substratum	Region	Inference	Reference
Animalia–Arthropoda–Malacostraca–Amphipoda				
Aora sp.	P	P(S)	fl	Astudillo et al. (2009)
Calliopius laeviusculus	P	A(N)	exp	Ingólfsson (1998)
Caprella andreae	U	A(S), Med	fl	Spivak and Bas (1999), Cabezas et al. (2013)
Caprella equilibra	P, U	P(S)	fl	Thiel et al. (2003), Astudillo et al. (2009)
Caprella hirsuta	U	Med	fl	Cabezas et al. (2013)
Caprella mutica	U	P(S)	str	Calder et al. (2014)
Caprella scaura	P, U	P(S)	fl	Thiel et al. (2003), Astudillo et al. (2009)
Caprella verrucosa	P, U	P(S)	fl	Thiel et al. (2003), Astudillo et al. (2009)
Caprella sp.	P, U	P(N)	fl	Goldstein et al. (2014)
Deutella venenosa	P, U	P(S)	fl	Thiel et al. (2003), Astudillo et al. (2009)
Dexamine thea	P	A(N)	exp	Ingólfsson (1998)
Elasmopus brasiliensis	U	Med	fl	Cabezas et al. (2013)
Ericthonius sp.	P	P(S)	fl	Astudillo et al. (2009)
Gammarus locusta	P, Pa	A(N)	spec	Vandendriessche et al. (2006)
Hyale grimaldii	U	Med	fl	Cabezas et al. (2013)
Jassa cadetta	U	Med	fl	Cabezas et al. (2013)
Jassa marmorata	P	P(S)	fl	Astudillo et al. (2009)
Jassa slatteryi	P	P(S)	fl	Astudillo et al. (2009)
Jassa sp.	U	A(N)	fl	LeCroy (2007)
Paracaprella pusilla	P	P(S)	fl	Astudillo et al. (2009)
Paradexamine pacifica	P	P(S)	fl	Astudillo et al. (2009)
Phtisica marina	P	Med	fl	Aliani and Molcard (2003)
Stenothoe sp.	P	P(S)	fl	Astudillo et al. (2009)
Animalia–Arthropoda–Malacostraca–Isopoda				
Sphaeroma quoianum	P	P(N)	fl	Davidson (2008)
Ianiropsis serricaudis	U	P(S)	str	Calder et al. 2014
Idotea balthica	P, Pa, U	A(N), Med	fl	Holdway and Maddock (1983a, b), Franke et al. (1999), Gutow and Franke (2003), Vandendriessche et al. (2006)
Idotea emarginata	U	A(N)	fl	Gutow and Franke (2003)

(continued)

(continued)

Taxon	Substratum	Region	Inference	Reference
Idotea metallica	P, U	A(N), P(S), Med	fl	Holdway and Maddock (1983a, b), Davenport and Rees (1993), Poore and Lew-Ton (1993), Franke et al. (1999), Aliani and Molcard (2003), Gutow and Franke (2003), Abelló et al. (2004), Cabezas et al. (2013)
Idotea sp.	P, U	P(N)	fl	Goldstein et al. (2014)
Sphaeroma terebrans	P	A(N), P(N)	fl	Davidson (2012)
Synidotea innatans	U	P(S)	fl	Poore (2012)
Synidotea marplatensis	P		spec	Masunari et al. (2000) (cited by Loyola-Silva and Melo 2008)
Animalia–Arthropoda–Malacostraca–Tanaidacea				
Zeuxo marmoratus	P	P(S)	fl	Astudillo et al. (2009)
Animalia–Mollusca–Gastropoda				
Berthella sp.	U	P(N)	fl	Goldstein et al. (2014)
Crepidula fornicata	P	A(N)	str	Cadée (2003)
Crepidula sp.	P	P(S)	fl	Astudillo et al. (2009)
Crucibulum sp.	P	P(S)	fl	Astudillo et al. (2009)
Doto uva	P	P(S)	fl	Astudillo et al. (2009)
Doto sp.	P	Med	fl	Aliani and Molcard (2003)
Erronea sp.	U	P(N)	fl	Goldstein et al. (2014)
Evalea tenuisculpta	P	P(N)	fl	Goldstein et al. (2014)
Fiona pinnata	P, U	P(N), P(S), Med	fl	Willan (1979), Aliani and Molcard (2003), Inatsuchi et al. (2010), Goldstein et al. (2014)
Fissurella cumingi	P	P(S)	fl	Astudillo et al. (2009)
Fissurella latimarginata	P	P(S)	fl	Astudillo et al. (2009)
Fissurella sp.	P	P(S)	fl	Astudillo et al. (2009)
Laevilitorina antarctica	P	Southern	str	Barnes and Fraser (2003)
Litiopa melanostoma	P, U	P(N)	fl	Goldstein et al. (2014)
Mitrella sp.	P	P(S)	fl	Astudillo et al. (2009)
Nassarius sp.	P	P(S)	fl	Astudillo et al. (2009)
Petaloconchus varians	U	A(S)	str	Breves and Skinner (2014)
Phidiana lottini	P	P(S)	fl	Astudillo et al. (2009)

(continued)

(continued)

Taxon	Substratum	Region	Inference	Reference
Prisogaster sp.	P	P(S)	fl	Astudillo et al. (2009)
Scurria viridula	P	P(S)	fl	Astudillo et al. (2009)
Thecacera darwini	P	P(S)	fl	Astudillo et al. (2009)
Animalia–Mollusca–Bivalvia				
Aequipecten opercularis	P	A(N)	str	Cadée (2003)
Anomia ephippium	P	A(N)	str	Southward et al. (2004)
Anomia sp.	P	A(N)	str	Winston et al. (1997)
Argopecten purpuratus	P	P(S)	fl	Astudillo et al. (2009)
Brachidontes granulatus	P	P(S)	fl	Astudillo et al. (2009)
Chama congregata	M	A(N)	str	Hoeksema et al. (2012)
Chama sp.	P	A(N)	str	Winston et al. (1997)
Chioneryx grus	M	A(N)	str	Hoeksema et al. 2012
Chlamys sp.	U	P(N)	fl	Goldstein et al. (2014)
Crassostrea gigas	P, U	P(N)	fl	Goldstein et al. (2014)
Crassostrea sp.	P	A(N)	str	Winston et al. (1997)
Hiatella arctica	M	A(N)	str	Hoeksema et al. (2012)
Isognomon sp.	P	A(N)	str	Winston et al. (1997)
Lopha cristagalli	P	P(S)	fl	Gardner (1971) (cited by Gregory 2009),Winston et al. (1997)
Musculus cupreus	U	P(S)	str	Calder et al. (2014)
Mytilus edulis	P, U	A(N)	exp	Ingólfsson (1998), Cardigos et al. (2006)
Mytilus galloprovincialis	P, U	P(N), P(S)	fl	Calder et al. (2014), Goldstein et al. (2014)
Mytilus sp.	P	Med	str	Frazier and Margaritoulis (1990)
Ostrea edulis	P	Med	str	Frazier and Margaritoulis (1990)
Ostrea equestris	M	A(N)	str	Hoeksema et al. (2012)
Pinctada imbricata	P	A(N)	str	Ávila et al. (2000) (cited by Cardigos 2006), Cadée (2003)
Pinctada sp.	P, U	A(N), P(N)	fl	Winston et al. (1997), Gregory (2009), Goldstein et al. (2014)
Pteria sp.	P	A(N)	str	Winston et al. (1997)
Rocellaria dubia	M	A(N)	str	Hoeksema et al. (2012)
Semimytilus algosus	P	P(S)	fl	Astudillo et al. (2009)
Zirfaea sp.	P	P(N)	fl	Goldstein et al. (2014)

(continued)

(continued)

Taxon	Substratum	Region	Inference	Reference
Animalia–Echinodermata				
Arbacia lixula	P	Med	fl	Aliani and Molcard (2003)
Patiria chilensis	P	P(S)	fl	Astudillo et al. (2009)
Tetrapygus niger	P	P(S)	fl	Astudillo et al. (2009)
Animalia–Bryozoa–Gymnolaemata				
Aetea sp.	P	A(N)	str	Winston et al. (1997)
Aimulosia antarctica	P	Southern	str	Barnes and Fraser (2003)
Aimulosia marsupium	P	P(S)	str	Stevens et al. (1996)
Amphiblestrum contentum	P	P(S)	str	Stevens et al. (1996)
Arachnopusia inchoata	P	Southern	str	Barnes and Fraser (2003)
Arachnopusia unicornis	P	P(S)	str	Stevens et al. (1996)
Beania inermis	P	P(S)	str	Stevens et al. (1996)
Beania plurispinosa	P	P(S)	str	Stevens et al. (1996)
Biflustra arborescens	P		str	Winston et al. (1997)
Biflustra savartii	P	A(N)	str	Key et al. (1996), Winston et al. (1997)
Bitectipora cincta	P	P(S)	str	Stevens et al. (1996)
Bowerbankia gracilis	P	Med	fl	Aliani and Molcard (2003)
Bowerbankia sp.	P, U	P(N)	fl	Goldstein et al. (2014)
Bugula flabellata	P	P(S)	fl	Stevens et al. (1996), Astudillo et al. (2009), Bravo et al. (2011)
Bugula minima	P	A(N)	str	Winston et al. (1997)
Bugula neritina	P, U	A(N), P(S)	fl	Southward et al. (2004), Stafford and Willan (2007), Astudillo et al. (2009), Bravo et al. (2011)
Bugula sp.	P, U	P(N)	fl	Goldstein et al. (2014)
Caberea rostrata	P	P(S)	str	Stevens et al. (1996)
Caberea zelandica	P	P(S)	str	Stevens et al. (1996)
Callopora lineata	P	Med	fl	Aliani and Molcard (2003)
Calloporina angustipora	P	P(S)	str	Stevens et al. (1996)
Calyptotheca immersa	P	P(S)	str	Stevens et al. (1996)
Celleporaria agglutinans	P	P(S)	str	Stevens et al. (1996)
Celleporella cancer	P	P(S)	str	Stevens et al. (1996)

(continued)

(continued)

Taxon	Substratum	Region	Inference	Reference
Celleporella tongima	P	P(S)	str	Stevens et al. (1996)
Celleporina hemiperistomata	P	P(S)	str	Stevens et al. (1996), Winston et al. (1997)
Celleporina sp.	P	P(S)	str	Stevens et al. (1996)
Chaperia acanthina	P	P(S)	str	Stevens et al. (1996)
Chaperiopsis sp.	P	P(S)	str	Stevens et al. (1996)
Chiastosella sp.	P	P(S)	str	Stevens et al. (1996)
Crepidacantha crinispina	P	P(S)	str	Stevens et al. (1996)
Cryptosula pallasiana	P	P(S)	fl	Stevens et al. (1996), Winston et al. (1997), Astudillo et al. (2009)
Electra angulata	P		str	Key et al. (1996), Winston et al. (1997)
Electra posidoniae	P	Med	fl	Aliani and Molcard (2003)
Electra tenella	P	A(N), P(S)	str	Winston 1982, Gordon and Mawatari (1992), Stevens et al. (1996), Winston et al. (1997)
Ellisina antarctica	P	Southern	str	Barnes and Fraser (2003)
Escharoides angela	P	P(S)	str	Stevens et al. (1996)
Escharoides excavata	P	P(S)	str	Stevens et al. (1996)
Eurystomella foraminigera	P	P(S)	str	Stevens et al. (1996)
Exochella armata	P	P(S)	str	Stevens et al. (1996)
Exochella tricuspis	P	P(S)	str	Stevens et al. (1996)
Fenestrulina disjuncta	P	P(S)	str	Stevens et al. (1996)
Fenestrulina rugula	P	Southern	str	Barnes and Fraser (2003)
Foveolaria cyclops	P	P(S)	str	Stevens et al. (1996)
Galeopsis polyporus	P	P(S)	str	Stevens et al. (1996)
Galeopsis porcellanicus	P	P(S)	str	Stevens et al. (1996)
Inversiula fertilis	P	P(S)	str	Stevens et al. (1996)
Jellyella eburnea	P, U	P(N), P(S)	fl	Stevens et al. (1996), Goldstein et al. (2014)
Jellyella tuberculata	P, U	A(N), P(N), P(S)	fl	Gregory (1978, 1990, 2009), Stevens et al. (1996), Winston et al. (1997), Goldstein et al. (2014)
Jellyella sp.	P	P(N)	fl	Goldstein et al. (2014)
Macropora grandis	P	P(S)	str	Stevens et al. (1996)

(continued)

(continued)

Taxon	Substratum	Region	Inference	Reference
Membranipora isabelleana	P	P(S)	fl	Astudillo et al. (2009), Bravo et al. (2011)
Membranipora membranacea	P	Arctic, Med	fl	Aliani and Molcard (2003), Barnes and Milner (2005)
Membranipora tenella	P, U	P(N)	fl	Goldstein et al. (2014)
Membranipora sp.	P	A(N), P(N)	fl	Winston et al. (1997), Goldstein et al. (2014)
Micropora brevissima	P	Southern	str	Barnes and Fraser (2003)
Micropora mortenseni	P	P(S)	str	Stevens et al. (1996)
Microporella agonistes	P	P(S)	str	Stevens et al. (1996)
Microporella speculum	P	P(S)	str	Stevens et al. (1996)
Opaeophora lepida	P	P(S)	str	Stevens et al. (1996)
Parasmittina sp.	P	P(S)	str	Stevens et al. (1996)
Rhynchozoon larreyi	P	P(S)	str	Stevens et al. (1996)
Schizoporella pungens	P	A(N)	str	Winston (2012)
Schizosmittina cinctipora	P	P(S)	str	Stevens et al. (1996)
Schizosmittina sp.	P	P(S)	str	Stevens et al. (1996)
Smittina torques	P	P(S)	str	Stevens et al. (1996)
Smittoidea maunganuiensis	P	P(S)	str	Stevens et al. (1996)
Smittoidea sp.	P	P(S)	str	Stevens et al. (1996)
Steginoporella magnifica	P	P(S)	str	Stevens et al. (1996)
Thalamoporella evelinae	P	A(N)	str	Winston et al. (1997)
Tricellaria inopinata	U	P(S)	str	Calder et al. (2014)
Victorella sp.	P	P(N)	fl	Goldstein et al. (2014)
Watersipora subtorquata	P, U	P(S)	str	Stevens et al. (1996), Winston et al. (1997), Stafford and Willan (2007)
Animalia–Bryozoa–Stenolaemata				
Diastopora sp.	P	P(S)	str	Stevens et al. (1996)
Disporella sibogae	P	P(S)	str	Stevens et al. (1996)
Disporella sp.	P	P(S)	str	Stevens et al. (1996)
Eurystrotos ridleyi	P	P(S)	str	Stevens et al. (1996)
Favosipora sp.	P	P(S)	str	Stevens et al. (1996)
Filicrisia sp.	P	P(N)	fl	Goldstein et al. (2014)
Hastingsia sp.	P	P(S)	str	Stevens et al. (1996)

(continued)

(continued)

Taxon	Substratum	Region	Inference	Reference
Lichenopora novaezelandiae	P	P(S)	str	Stevens et al. (1996)
Platonea sp.	P	P(S)	str	Stevens et al. (1996)
Stomatopora sp.	P	P(N)	fl	Goldstein et al. (2014)
Tubulipora sp.	P	P(N), P(S)	fl	Stevens et al. (1996), Goldstein et al. (2014)
Animalia–Chordata–Ascidiacea				
Diplosoma sp.	P	P(S)	fl	Astudillo et al. (2009), Bravo et al. (2011)
Ascidia sp.	U	P(S)	fl	Thiel et al. (2003)
Ciona intestinalis	P	P(S)	fl	Astudillo et al. (2009), Bravo et al. (2011)
Pyura chilensis	P, U	P(S)	fl	Thiel et al. (2003), Astudillo et al. (2009), Bravo et al. (2011)

Abelló, P., Guerao, G., & Codina, M., (2004). Distribution of the neustonic isopod *Idotea metallica* in relation to shelf-slope frontal structures. *Journal of Crustacean Biology, 24,* 558–566.

Ávila, S. P., (2000). Shallow-water marine molluscs of the Azores: biogeographical relationships. *Arquipélago—Life and Marine Sciences,* Supp. 2(Part A): 99–131.

Breves, A., & Skinner, L. F., (2014). First record of the vermetid *Petaloconchus varians* (d'Orbigny, 1841) on floating marine debris at Ilha Grande, Rio de Janeiro, Brazil. *Journal of Integrated Coastal Zone Management, 14,* 159–161.

Calder, D. R., (1993). Local distribution and biogeography of the hydroids (Cnidaria) of Bermuda. *Caribbean Journal of Science, 29,* 61–74.

Calder, D. R., (1995). Hydroid assemblages on holopelagic *Sargassum* from the Sargasso Sea at Bermuda. *Bulletin of Marine Science, 56,* 537–546.

Cardigos, F., Tempera, F., Ávila, S., Gonçalves, J., Colaço, A., & Santos, R. S., (2006). Non-indigenous marine species of the Azores. *Helgoland Marine Research, 60,* 160–169.

Carpenter, E. J., & Smith, K. L., (1972). Plastics on the Sargasso Sea surface. *Science, 175,* 1240–1241.

Celis, A., Rodríguez-Almaráz, G., & Álvarez, F., (2007). The shallow-water thoracican barnacles (Crustacea) of Tamaulipas, Mexico. *Revista Mexicana de Biodiversidad, 78,* 325–337.

Chace, F. A., (1951). The oceanic crabs of the genera *Planes* and *Pachygrapsus. Proceedings of the United States National Museum, 101,* 3272.

Cheng, L., & Lewin, R. A., (1976). Goose barnacles (Cirripedia: Thoracica) on flotsam beached at La Jolla, California. *Fishery Bulletin, 74,* 212–217.

Davenport, J., & Rees, E. I. S., (1993). Observations on neuston and floating weed patches in the Irish Sea. Estuarine, *Coastal and Shelf Science, 36,* 395–411.

Davidson, T. M., (2008). Prevalence and distribution of the introduced burrowing isopod, *Sphaeroma quoianum*, in the intertidal zone of a temperate northeast Pacific estuary (Isopoda, Flabellifera). *Crustaceana, 81,* 155–167.

Dell, R. K., (1964). The oceanic crab, *Pachygrapsus marinus* (Rathbun) in the South-West Pacific. *Crustaceana, 7,* 79–80.

Donlan, C. J., & Nelson, P. A., (2003). Observations on invertebrate colonized flotsam in the eastern tropical Pacific with a discussion of rafting. *Bulletin of Marine Science, 72,* 231–240.

Fine, M., Zibrowius, H., Loya, Y., (2001). *Oculina patagonica*: a non-lessepsian scleractinian coral invading the Mediterranean Sea. *Marine Biology, 138,* 1195–1203.

Frazier, J. G., & Margaritoulis, D., (1990). The occurrence of the barnacle, *Chelonibia patula* (Ranzani, 1818), on an inanimate substratum (Cirripedia, Thoracica). *Crustaceana, 59,* 213–218.

Gardner, N. N., (1971). *Lopha cristagali* (Linne). *Poirieria, 5*, 104.

Gittings, S. R., Dennis, G. D., & Harry, H. W., (1986). *Annotated guide to the barnacles of the northern Gulf of Mexico.* Texas A&M University.

Goldstein, M. C., & Goodwin, D. S., (2013). Gooseneck barnacles (*Lepas* spp.) ingest microplastic debris in the North Pacific Subtropical gyre. *PeerJ, 1*, e184.

Gordon DP, Mawatari SF (1992). *Atlas of marine-fouling Bryozoa of New Zealand ports and harbours* (Vol. 107). Miscellaneous Publications, N.Z. Oceanographic Institute, 1–52.

Green, A., Tyler, P. A., Angel, M. V., & Gage, J. D., (1994). Gametogenesis in deep- and surface-dwelling oceanic stalked barnacles from the NE Atlantic Ocean. *Journal of Experimental Marine Biology and Ecology, 184*, 143–158.

Gregory, M. R., (1983). Virgin plastic granules on some beaches of eastern Canada and Bermuda. *Marine Environmental Research,* 10, 73–92.

Gregory, M. R., (1990). Plastics: accumulation, distribution, and environmental effects of meso-, macro-, and megalitter in surface waters and on shores of the Southwest Pacific. In R. S. Shomura, & M. L.Godfrey, (Eds.), *Proceedings of the Second International Conference on Marine Debris, Honolulu,* (pp. 55–84). Hawaii.

Holdway, P., & Maddock, L., (1983a). A comparative survey of neuston: geographical and temporal distribution patterns. *Marine Biology,* 76, 263–270.

Holdway, P., & Maddock, L., (1983b). Neustonic distributions. *Marine Biology,* 77, 207–214.

Inatsuchi, A., Yamato, S., & Yusa, Y., (2010). Effects of temperature and food availability on growth and reproduction in the neustonic pedunculate barnacle *Lepas anserifera. Marine Biology, 157*, 899–905.

Ingólfsson, A., (1998). Dynamics of macrofaunal communities of floating seaweed clumps off western Iceland: a study of patches on the surface of the sea. *Journal of Experimental Marine Biology and Ecology, 231*, 119–137.

Jokiel, P. L., (1984). Long distance dispersal of reef corals by rafting. *Coral Reefs, 3,* 113–116.

Kerckhof, F., (1997). Waarnemingen van de Afrikaanse zeepok *Solidobalanus fallax* langs de Franse Atlantische kust en op drijvende voorwerpen aangespoeld op het Belgische strand. *De Strandvlo, 17,* 34–44.

Kerckhof, F., (2002). Barnacles (Cirripedia, Balanomorpha) in Belgian waters, an overview of the species and recent evolutions, with emphasis on exotic species. *Bulletin Van Het Koninklijk Belgisch Instituut voor Natuurwetenschappen—Biologie, 72,* 93–104.

Key, M. M., Jeries, W. B., Voris, H. K., & Yang, C. M., (1996). Epizoic bryozoans and mobile ephemeral host substrata. In D. P. Gordon, A. M. Smith, J. A. Grant-Mackie, (Eds.), *Bryozoans in space and time. National Institute of Water and Atmospheric Research, Wellington,* 157–165.

LeCroy, S. E., (2007). *An illustrated identification guide to the nearshore marine and estuarine gammaridean amphipoda of Florida* (Vol. 4) University of Southern Mississippi.

Loyola-Silva, J., & Melo, S. G., (2008). Population structure, descriptive complements as a new occurrence of *Synidotea marplatensis* in Paraná State, Brazil. *Acta Biológica Paranaense, 37,* 217–227.

MacIntyre, R. J., (1966). Rapid growth in stalked barnacles. *Nature, 212,* 637–638.

Masunari, S., Loyola-Silva, J., Alcantara, G. B., & Almeida, R., (2000). Ocorrência de *Synidotea marplatensis* Giambiagi, 1922. (Crustacea: Isopoda: Idoteidae) em substrato artificial na região de Itapoá, litoral do Estado de Santa Catarina. *Scientific Journal,* 4(Supp1):16.

Morton, B., & Britton, J. C., (2000a). The origins of the coastal and marine flora and fauna of the Azores. In Barnes H, A. D. Ansell, R. N. Gibson, (Eds.), *Oceanography and Marine Biology: An Annual Review* (Vol. 38), University College London Press.

Morton, B., & Britton, J. C., (2000b). Origins of the Azorean intertidal biota: the significance of introduced species, survivors of chance events. *Arquipélago—Life and Marine Sciences,* Supp. 2(Part A): 29–51.

Newman, W.A. (1972). Lepadids from the Caroline Islands (Cirripedia Thoracica). *Crustaceana,* 31–38.

Patel, B., (1959). The influence of temperature on the reproduction and moulting of *Lepas anatifera* L. under laboratory conditions. *Journal of the Marine Biological Association of the UK,* 38, 589–597.

Poore, G. C. B., & Lew-Ton, H. M., (1993). Idoteidae of Australia and New Zealand (Crustacea: Isopoda: Valvifera). *Invertebrate Taxonomy, 7,* 197–278.

Ryan, P. G., & Branch, G. M., (2012). The November 2011 irruption of buoy barnacles *Dosima fascicularis* in the Western Cape, South Africa. *African Journal of Marine Science, 34,* 157–162.

Southward, A. J., Hiscock, K., Kerckhof, F., Moyse, J., & Elfimov, A. S., (2004). Habitat and distribution of the warm-water barnacle *Solidobalanus fallax* (Crustacea: Cirripedia). *Journal of the Marine Biological Association of the UK, 84,* 1169–1177.

Stafford, H., & Willan, R. C., (2007). *Is it a pest? Introduced and naturalised marine animal species of Torres Strait Northern Australia.* Queensland Department of Primary Industries and Fisheries, Cairns.

Thiel, M., Guerra-García, J. M., Lancellotti, D. A., & Vásquez, N., (2003). The distribution of littoral caprellids (Crustacea: Amphipoda: Caprellidea) along the Pacific coast of continental Chile. *Revista Chilena de Historia Natural, 76,* 297–312.

Vandendriessche, S., Vincx, M., & Degraer, S., (2006). Floating seaweed in the neustonic environment: a case study from Belgian coastal waters. *Journal of Sea Research, 55,* 103–112.

WoRMS Editorial Board. (2014). World Register of Marine Species. http://www.marinespecies.org, VLIZ. (Accessed on July 2014).

Part III
Microplastics

Chapter 7
Microplastics in the Marine Environment: Sources, Consequences and Solutions

Richard C. Thompson

Abstract Microplastics are small fragments of plastic debris that have accumulated in the environment on a global scale. They originate from the direct release of particles of plastic and as a consequence of the fragmentation of larger items. Microplastics are widespread in marine habitats from the poles to the equator; from the sea surface and shoreline to the deep sea. They are ingested by a range of organisms including commercially important fish and shellfish and in some populations the incidence of ingestion is extensive. Laboratory studies indicate that ingestion could cause harmful toxicological and/or physical effects. However, our understanding of the relative importance of these effects in natural populations is very limited. Looking to the future it seems inevitable that the quantity of microplastic will increase in the environment, since even if we could stop new items of debris entering the ocean, fragmentation of the items already present would continue for years to come. The term microplastics has only been in popular usage for a decade and while many questions remain about the extent to which they could have harmful effects, the solutions to reducing this contamination are at hand. There are considerable synergies to be achieved by designing plastic items for both their lifetime in service and their efficient end-of-life recyclability, since capturing waste via recycling will reduce usage of non-renewable oil and gas used in the production of new plastics and at the same time reduce the accumulation of waste in managed facilities such as land fill as well as in the natural environment.

Keywords Microplastic · Microbeads · Accumulation · Impact · Toxicology · Solution

R.C. Thompson (✉)
Marine Biology and Ecology Research Centre, School of Marine Science and Engineering, Plymouth University, Plymouth PL4 8AA, UK
e-mail: R.C.Thompson@plymouth.ac.uk

7.1 Introduction

Microplastics is used as a collective term to describe a truly heterogeneous mixture of particles ranging in size form a few microns to several millimetres in diameter; including particles of various shapes from completely spherical to elongated fibres. Microplastics have been reported in a range of colors. However, pieces that differ in appearance according to their shape size or color to ambient natural particulates are most commonly reported, for example blue or red fibres (Hidalgo-Ruz et al. 2012). The term microplastics has been widely used in relation to anthropogenic debris since 2004 when Thompson et al. used the term to illustrate and describe the accumulation of truly microscopic pieces of plastic in marine sediments and in the water column in European waters (Fig. 7.1). Microplastic contamination has since been reported on a global scale from the poles to the equator (Barnes et al. 2009; Browne et al. 2011; Hidalgo-Ruz et al. 2012) and contaminates the water surface of the open ocean (Law et al. 2010; Collignon et al. 2012; Goldstein et al. 2012; Ivar do Sul et al. 2013), estuaries (Sadri and Thompson 2014) and lakes (Eriksen et al. 2013) together with marine (Browne et al. 2011; Santos et al. 2009) and freshwater shorelines (Imhof et al. 2013) and subtidal sediments (Browne et al. 2011) down to the deep sea (Van Cauwenberghe et al. 2013; Woodall et al. 2014). Microplastics have also been reported in considerable concentrations in Arctic sea ice (Obbard et al. 2014; Fig. 7.2). Over the past decade, interest in the topic has grown immensely and there are now well over 100 publications on microplastic (Fig. 7.3) and numerous reviews (Browne et al. 2007; Arthur et al. 2009; Andrady 2011; Cole et al. 2011; Zarfl et al. 2011; Wright et al. 2013b; Ivar do Sul and Costa 2014; Law and Thompson 2014) spanning sources, occurrence, abundance, ingestion by biota and consequences. Alongside this scientific research there has been growing interest from the media, the public and policy makers. The first policy centered workshop on the topic was hosted by NOAA in the USA during 2008 (Arthur et al. 2009). Specific reference to microplastics was later made within EU legislation via the Marine Strategy Framework Directive in 2010 (Galgani et al. 2010), and more recently there has been legislation and voluntary actions by industry to reduce the use of microplastics in cosmetics. However, even in the unlikely event that inputs of larger items of debris were to cease immediately, it is likely that the quantities of microplastics would continue to increase in the environment due to the fragmentation of legacy items of larger debris. Hence, it is essential to gain further understanding about the sources, consequences and fate of microplastics in the ocean.

Microplastics originate from a variety of sources, but these can be broadly categorized as primary: the direct release of small particles, for example, as a result of release of pellets or powders, or secondary, which results from fragmentation of larger items (Andrady 2011; Cole et al. 2011; Hidalgo-Ruz et al. 2012). Microplastic-sized particles are directly used in a wide range of applications. Plastic pellets (around 5 mm diameter) and powders (less than 0.5 mm) are used as a feedstock for the production of larger items and the presence of these pellets (also

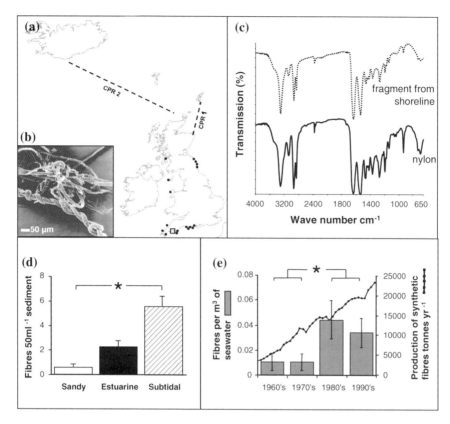

Fig. 7.1 **a** Sampling locations in the northeast Atlantic: six sites near Plymouth used to compare the abundance of microplastic among habitats, (*open square*) (see Fig. 7.1d). Other shores where similar fragments were found (*black solid circles*). *Dashed lines* show routes sampled by Continuous Plankton Recorder (CPR 1 and 2) and used to assess changes in microplastic abundance since 1960. **b** One of numerous fragments found among marine sediments and identified as plastic using FT-IR spectroscopy. **c** FT-IR spectra of a microscopic fragment matched that of nylon. **d** Microplastics were more abundant in subtidal habitats than in sandy beaches (* = $F_{2,3} = 13.26$, $P < 0.05$), but abundance was consistent among sites within habitat types. **e** Microscopic plastic in CPR samples revealed a significant increase in abundance when comparing the 1960s and 1970s to the 1980s and 1990s (* = $F_{3,3} = 14.42$, $P < 0.05$). Approximate global production of synthetic fibres overlain for comparison. Microplastics were also less abundant along oceanic route CPR 2 than CPR 1 ($F_{1, 24} = 5.18$, $P < 0.05$). Reproduced from Thompson et al. (2004) with permission

known as nurdles or mermaids tears) has been widely reported as a consequence of industrial spillage (Hays and Cormons 1974; Bourne and Imber 1982; Harper and Fowler 1987; Shiber 1987; Blight and Burger 1997). Small plastic particles typically around 0.25 mm are also widely used as abrasive in cosmetic products (Fig. 7.4) and as an industrial shot-blasting abrasive. Microplastics from cosmetics and cleaning agents (also known as microbeads) will be carried with waste water via sewers and are unlikely to be effectively removed by sewage treatment, and hence are accumulating in the environment (Zitko and Hanlon 1991; Gregory

Fig. 7.2 Sea ice core being collected during the NASA ICESCAPE expedition in July 2010 (Photo: D. Perovich, U.S. Army Corps of Engineers Cold Regions Research & Engineering)

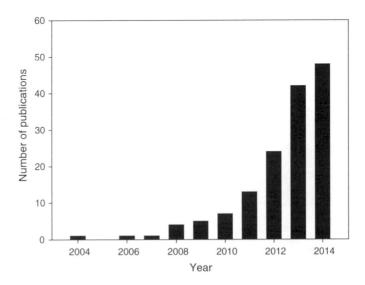

Fig. 7.3 Number of publications on microplastics over time (2004–2014). Modified from GES-AMP (2014), courtesy of S. Gall, Plymouth University

SEI   15kV   WD11mm   SS40   x35   500μm
Plym EMC

Fig. 7.4 Scanning electron microscope image of microbeads isolated from cosmetics (Photo: A. Bakir and R.C. Thompson, Plymouth University)

1996). It is estimated that in the US alone around 100 tons of microplastics might enter the oceans annually (Gouin et al. 2011). In addition to the direct release of primary microplastics, larger items of plastic debris will progressively become brittle under the action of ultraviolet light and heat and then fragment with physical action from wind and waves (Andrady 2015). Hence large items of debris are likely to represent a considerable source of microplastics (Andrady 2003, 2011). In addition to fragmentation in the environment, some items also fragment in use resulting in particles of microplastic being released to the environment as a consequence of everyday usage or cleaning. This has been demonstrated for the release of fibres from garments as a consequence of washing (Browne et al. 2011). It is now evident that, as a collective consequence of these diverse inputs, microplastics are widespread in natural habitats and in the organisms living there, including benthic invertebrates, commercially important lobsters, numerous species of fish, sea birds and marine mammals (Murray and Cowie 2011; Possatto et al. 2011; van Franeker et al. 2011; Foekema et al. 2013; Lusher et al. 2013; Rebolledo et al. 2013).

Our understanding about microplastics has advanced considerably over the last decade, but is still in its infancy and our knowledge of the relative importance of various sources, spatial trends in distribution and abundance, temporal trends, or effects on biota are still quite limited (Law and Thompson 2014). Initial work describing microplastics indicated a small increase in the abundance of this debris over time and that in laboratory conditions a range of invertebrates would ingest the material (Thompson et al. 2004). Subsequent work has described the range of habitats (Law et al. 2010; Browne et al. 2011; Van Cauwenberghe et al. 2013) and organisms (Graham and Thompson 2009; Murray and Cowie 2011; van Franeker et al.

2011; Lusher et al. 2013) that are contaminated by microplastic in the environment. These early studies have been pioneering in nature providing proof of concept, but are difficult to use as a base line because of the inevitable lack of consistency in methods. In parallel, there have been laboratory studies which have exposed organisms to microplastics in order to determine the potential for this debris to result in harm to the creatures that encounter it in the natural environment (Browne et al. 2008, 2013; Rochman et al. 2013; Wright et al. 2013a). The main route of concern is currently as a consequence of ingestion, which could lead to physical (Wright 2014) and toxicological effects on biota (Teuten et al. 2007; Browne et al. 2008). Plastics are known to sorb persistent organic pollutants (Mato et al. 2001; Ogata et al. 2009; Teuten et al. 2009) and metals (Holmes et al. 2012) from seawater and organic pollutants can become orders of magnitude more concentrated on the surface of the plastic than in the surrounding water (Mato et al. 2001; Ogata et al. 2009; Teuten et al. 2009). There is evidence from laboratory studies that these chemicals can be transferred from plastics to organisms upon ingestion (Teuten et al. 2009) and that this can result in harm (Browne et al. 2013; Rochman et al. 2013; Wright et al. 2013a). The potential for transfer varies according to the specific combination of plastic and contaminant with some polymers such as polyethylene having considerable potential for transport (Bakir et al. 2012). Subsequent desorption will also vary according to physiological conditions upon ingestion with the presence of gut surfactants and increased temperature leading to increased desorption (Teuten et al. 2007; Bakir et al. 2012). However, modeling studies suggest that when compared to the transport of persistent organic pollutants (POPs) by other pathways such as respiration and food that plastics are not likely to be a major vector in the transport of POPs from seawater to organisms (Gouin et al. 2011; Koelmans et al. 2013). A second toxicological issue is that some plastics contain chemical additives that are potentially harmful (Rochman 2015). These additives can be present in concentrations much greater than is likely to result from sorption of POPs and there is concern that additives might be released to organisms upon ingestion (Oehlmann et al. 2009; Thompson et al. 2009; Rochman and Browne 2013). There is evidence that such chemicals can be present, for example as leachates from landfill sites, in aquatic habitats at concentrations that are sufficient to cause harm (Oehlmann et al. 2009). There is also evidence that chemical additives can transfer from plastics to sea birds (Tanaka et al. 2013). However, it is not clear whether ingestion of plastics themselves could result in sufficient transfer of additive chemicals to cause harm. This would require experiments with plastics for which the composition of chemical constituents is known.

7.2 Definitions of Microplastics

When reported in 2004 the term microplastics was used to describe fragments of plastic around 20 μm in diameter. These were reported in intertidal and shallow subtidal sediments and in surface waters in northwestern Europe (Thompson et al. 2004; Fig. 7.1). Subsequent research showed that similar sized particles were present in shallow waters around Singapore (Ng and Obbard 2006). However, while these early reports referred to truly microscopic particles they did not give a specific definition of microplastic. In 2008, the National Oceanographic and Atmospheric Agency (NOAA) of the US hosted the first International Microplastics Workshop in Washington and as part of this meeting formulated a broader working definition to include all particles less than 5 mm in diameter (Arthur et al. 2009). Particles of this size (i.e. <5 mm) have been very widely reported including publications that considerably pre-dated the use of the term "microplastics" (Carpenter et al. 1972; Colton et al. 1974). There is still some debate over the most appropriate upper size bound to use in a formal definition of microplastics, with perhaps a more intuitive boundary following the SI classification of <1 mm. The European Union have followed the US and adopted a 5-mm upper bound for categorization of microplastics within the Marine Strategy Framework Directive (MSFD, Galgani et al. 2010). There is a similar lack of clarity when considering the lower size bound for a definition of microplastics. Operationally, this, by default, has been assumed to be the mesh size of the particular net or sieve used to separate the microplastic from the bulk medium of sediment or water column (see review by Hidalgo-Ruz et al. 2012). However, as a necessity of construction, collection devices with meshes in the sub-millimetre size range have a high ratio of net/sieve material compared to apertures and as a consequence they will trap particles much smaller than the size of the apertures/mesh size. Hence, it is not sensible to define the minimum size captured on the basis of the mesh used to collect the sample. Within the EU MSFD a pragmatic approach has been taken based on that used by researchers sampling benthic infauna and sediments with sieves (e.g. Wentworth graduated sieves), where the organisms '*retained*' by a particular sieve are reported. In summary, there is no universally agreed definition of microplastic size, but most workers consider microplastic to be particles of plastic <5 mm in size. There is little consensus on the lower size bound.

While defining parameters is essential for consistent monitoring, in the wider context of marine debris and concerns about the potential harmful effects of microplastic it may actually be unwise to specify the size definitions precisely at the present time. Differently sized particles are likely to have differing effects. For example, smaller particles could have consequences that are fundamentally different to larger particles, since the particles themselves can accumulate in tissues and/ or may cause disruption of physiological processes (Browne et al. 2008; Wright et al. 2013c). From a monitoring science, rather than a curiosity-driven perspective, a logical rationale for sampling is to consider abundance in relation to any associated impacts. Since our understanding of the potential impacts of microplastics is currently in its infancy it could, for the time being, be unwise to set a formal limit to lower size boundary and, until there is better understanding about which types/sizes of microplastics are of concern a sensible strategy could be to collect from the bulk medium any particles <5 mm and then quantify microplastics according to size categories.

7.3 Spatial and Temporal Patterns in the Abundance of Microplastics

Our understanding about the distribution and the factors affecting the distribution of microplastics in the oceans is limited and much of the sampling to date has been opportunistic utilizing existing research programs (research cruises, educational programs, routine plankton monitoring) to collect material. There has also been some targeted microplastic sampling and attempts to make formal comparisons in the abundance of microplastics between locations (Browne et al. 2010, 2011). Existing data indicate that microplastics are widely distributed in surface waters, in shallow waters (Browne et al. 2011; Hidalgo-Ruz et al. 2012), in deep-sea sediments (Van Cauwenberghe et al. 2013) and in the digestive tract of a range of organisms living within these habitats (Lusher 2015). With the exception of heavily contaminated areas such as shipbreaking yards (Reddy et al. 2006), the abundance of microplastics would appear to be relatively low in surface waters and sediments (see Lusher 2015). By volume it is apparent, however, that sediments are more contaminated than surface waters.

However, because of their ubiquity, the total quantity of microplastics in the environment is considerable and in some locations represents the most numerous type of debris present (Browne et al. 2010). This ubiquity is also demonstrated by encounters when considered by marine species, of which around 10 % are with microplastics (Secretariat of the Convention on Biological Diversity and Scientific and Technical Advisory Panel GEF 2012). In terms of spatial patterns in abundance, at a global scale Browne et al. (2011) detected a weak relationship between the abundance of microplastics and human population density. Extensive sampling by Law et al. (2010) demonstrated the role of large-scale physical factors leading to increased abundance in the North Atlantic gyre far from the nearest land. She matched abundance data form the ocean surface with model predictions based on physical factors indicating that, at large scales, factors driving the abundance of debris can be used to make predictions about relative abundance (Law et al. 2010; Fig. 7.5). Formal comparisons also demonstrated patterns at smaller spatial scales with locations previously used for the dumping of sewage sludge having greater quantities of microplastic than control areas (Browne et al. 2011). In addition, intertidal sediments on shores that were downwind in relation to prevailing wind direction can have greater quantities of microplastic than those on shorelines that were up-wind (Browne et al. 2010). Targeted sampling has also indicated extremely high microplastic abundance near to a plastic processing plant in Sweden (Norén 2008). However, while the role of some potential sources including sewage and industrial spillage have been demonstrated together with the influence of physical factors leading to accumulation of debris in particular locations, our collective understanding of the relative importance of these factors in influencing spatial patterns of distribution or in making predictions about such is limited.

Only a handful of studies have considered temporal patterns in the abundance of microplastics. Thompson et al. (2004) in the northeast Atlantic and Goldstein

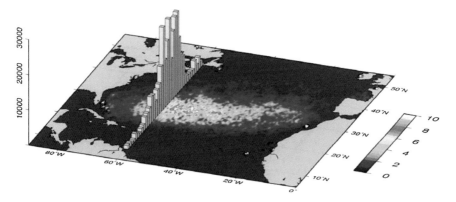

Fig. 7.5 Average plastic concentration as a function of latitude (bars, units of pieces km^{-2}), and concentration, C (*color shading*), of initially homogeneous (C = 1) surface tracer after 10-year model integration. Averages and standard errors were computed in one-degree latitude bins. The highest plastic concentrations were observed in subtropical latitudes (22–38°N) where model tracer concentration is also a maximum. Reproduced from Law et al. (2010) with permission

et al. (2012) in the North Pacific both report on an increase in abundance over time. While examination of a very extensive data set by Law et al. (2010) revealed no clear temporal trend in abundance over two decades of sampling in the North Atlantic, Thompson et al. (2004) used samples collected by the continuous plankton recorder to examine temporal changes in surface waters to the north of Scotland and showed a significant increase in the abundance of microplastics when comparing between the 1960s and 1970s with the 1980s and 1990s. Goldstein et al. (2012) compared abundance in heavily contaminated areas of the Pacific and also recorded and increase in abundance over time. However, sampling methodology differed between sampling dates making it difficult to clearly identify the underlying trends in microplastic abundance (Goldstein et al. 2012). It is clear that the abundance of microplastic is likely to vary considerably in space and in time, but we have little understanding of the associated scales of variation, neither do we have a clear understanding about the relative importance of, or interactions among, the various factors affecting distribution or about which, if any, types of microplastic might be hazardous. Such uncertainty considerably limits our ability to implement monitoring programs necessary to assess changes in abundance over time and in relation to regulatory measures.

7.4 Anticipated Future Trends

Global production of plastic has increased from around 5 million tons per year during the 1950s to over 280 million tons today (Thompson et al. 2009; PlasticsEurope 2011). However, the majority of this is used to make single-use items, which are disposed of within a year of production (Thompson et al. 2009).

Hence, considerable quantities of end-of-life plastics are accumulating in land fills and in the natural environment. The quantity of end-of-life plastic in the marine environment is substantial but as yet there are few reliable estimates of the total amount, or the relative proportions of different types of debris such as microplastic. Recent studies have attempted to assess global distributions (Cózar et al. 2014; Eriksen et al. 2014), the logical next step could be to estimate total production, current tonnage in use and accumulated disposal via recognized waste management in order to establish via a mass balance the amount of plastic that is missing and potentially in the environment (Jambeck et al. 2015). It is apparent that end-of-life plastic items are abundant and widely distributed in the oceans and that these items are progressively fragmenting into small pieces which are now abundant in the environment (Fig. 7.6). In some locations, it is evident that microplastics are numerically, as opposed to by mass, the most abundant type of solid debris present (Browne et al. 2010). However, despite the deterioration of plastic items into plastic fragments, conventional plastics will not readily biodegrade and it is considered that all of the plastics that have ever been produced are still present

Fig. 7.6 Accumulation of plastic debris on a shoreline in Europe. Small fragments of plastic including microplastics pieces <5 mm are often overlooked during routine beach monitoring, but are now the most abundant items on many shorelines (Photo: R.C. Thompson)

on the planet (unless they have bene incinerated) (Thompson et al. 2005). Hence, even if we were to cease using plastic items, which is not something I would advocate, the quantity of microplastic will continue to increase as a consequence of fragmentation of existing larger items (Thompson et al. 2009; STAP 2011).

From a personal perspective my interest in what we now describe as microplastics started in the in the mid-1990s. I was well aware that over the previous decades we had shifted to a very disposable society with considerable generation of waste. It was apparent that waste items including plastics were entering the oceans on a daily basis. These plastic items were resistant to degradation and I became curious as to where all the end-of-life single-use plastic items were accumulating in the natural environment. At that time, as is still largely the case, there was a distinct lack of data indicating any increasing temporal trends in the abundance of plastic debris and I considered that a substantial proportion may be accumulating as fragments, which were being missed by routine litter surveys (Fig. 7.6). These observations inspired the research leading to my paper in 2004 entitled '*Lost at sea where is all the plastic?*'. In this paper I suggested that one reason we were not seeing a temporal trend was because the smaller fragments that were forming from larger items were not being recorded in routine monitoring. Ten years on it seems likely that accumulation of microplastics represents an important sink where the fragments of larger items reside in a size range that has seldom been monitored. However, while widely distributed in the marine environment the densities of microplastic recorded in the habitats studied to date are relatively low and indicate that if microplastics are indeed the ultimate end-product of our disposable society then some of the major sinks of this material are yet to be discovered. Many consider the deep sea likely to be a major sink and there is growing evidence that substantial quantities of macroplastic are accumulating there (Galgani et al. 1996). An initial survey suggested abundance in the deep sea may be lower than in shallow water habitats (Van Cauwenberghe et al. 2013), however using different approaches to record fibres there is recent evidence that the deep sea could be a substantial sink for microplastics (Woodall et al. 2014). Clearly more investigation is required to confirm the relative importance of the deep sea as a sink for microplastics, to understand their long-term fate in the deep sea and the extent of any subsequent deterioration or biodegradation over extended timescales (Zettler et al. 2013).

7.5 Conclusions

It is evident that microplastic pieces now contaminate marine habitats worldwide. This debris is ingested by a wide range of organisms and for some species a major proportion of the population contains plastic fragments. There are concerns about the physical and toxicological harm that ingesting this debris might cause and laboratory experiments have demonstrated harmful effects. However, the relative importance of plastics as a vector for chemical transport or their importance as an agent causing physical harm to organisms in the natural environment are much less clear (Koelmans 2015).

Our understanding of potential future trends in the abundance of microplastic debris is limited. While it seems inevitable that the quantities of microplastic will increase in the environment as a consequence of further direct introductions of primary microplastic and fragmentation of larger items the likely trajectories and potential sinks or hot spots of accumulation are not clear. In conclusion, 10 years after the term microplastic widely entered the published literature and after a considerable body of research, there remain more questions than answers about the accumulation and consequences of microplastic contamination in the environment (Law and Thompson 2014). Ultimately, however, there is broad recognition that plastic debris does not belong in the ocean. It is also clear that the numerous societal benefits that are derived from every-day-use of plastics can be achieved without the need for emissions of plastic waste to the environment. Since 8 % of the global oil production is currently used to make plastic items it seems clear that we urgently need to change the way we produce, use and dispose of plastic items. There is also a growing realization that the solution to two major environmental problems, our non-sustainable use of fossil carbon and accumulation of debris lie in utilizing end-of-life plastics as a raw material for new production. Such principles are central to the philosophy of developing a more circular economy and some believe that rethinking our use of plastic materials in line with this philosophy has considerable potential to bring much greater resource efficiency (European Commission 2012).

References

Andrady, A. L. (Ed.). (2003). *Plastics in the environment. Plastics in the Environment.* New Jersey: Wiley.

Andrady, A. L. (2011). Microplastics in the marine environment. *Marine Pollution Bulletin, 62,* 1596–1605.

Andrady, A. L. (2015). Persistence of plastic litter in the oceans. In M. Bergmann., L. Gutow, M. Klages (Eds.), *Marine anthropogenic litter* (pp. 57–72). Berlin: Springer.

Arthur, C., Baker, J., & Bamford, H. (2009). *Proceedings of the international research workshop on the occurrence, effects and fate of microplastic marine debris.* Sept 9–11, 2008, NOAA Technical Memorandum NOS-OR&R30.

Bakir, A., Rowland, S. J., & Thompson, R. C. (2012). Competitive sorption of persistent organic pollutants onto microplastics in the marine environment. *Marine Pollution Bulletin, 64,* 2782–2789.

Barnes, D. K. A., Galgani, F., Thompson, R. C., & Barlaz, M. (2009). Accumulation and fragmentation of plastic debris in global environments. *Philosophical Transactions of the Royal Society B, 364,* 1985–1998.

Blight, L. K., & Burger, A. E. (1997). Occurrence of plastic particles in seabirds from the eastern North Pacific. *Marine Pollution Bulletin, 34,* 323–325.

Bourne, W. R. P., & Imber, M. J. (1982). Plastic pellets collected by a prion on Gough Island, Central South-Atlantic ocean. *Marine Pollution Bulletin, 13*, 20–21.

Browne, M. A., Galloway, T. S., & Thompson, R. C. (2007). Microplastic—An emerging contaminant of potential concern. *Integrated Environmental Assessment and Management, 3*, 559–566.

Browne, M. A., Dissanayake, A., Galloway, T. S., Lowe, D. M., & Thompson, R. C. (2008). Ingested microscopic plastic translocates to the circulatory system of the mussel, *Mytilus edulis* (L.). *Environmental Science and Technology, 42*, 5026–5031.

Browne, M. A., Galloway, T. S., & Thompson, R. C. (2010). Spatial patterns of plastic debris along estuarine shorelines. *Environmental Science and Technology, 44*, 3404–3409.

Browne, M. A., Crump, P., Niven, S. J., Teuten, E., Tonkin, A., Galloway, T. S., et al. (2011). Accumulation of microplastic on shorelines worldwide: Sources and sinks. *Environmental Science and Technology, 45*, 9175–9179.

Browne, M. A., Niven, S. J., Galloway, T. S., Rowland, S. J., & Thompson, R. C. (2013). Microplastic moves pollutants and additives to worms, reducing functions linked to health and biodiversity. *Current Biology, 23*, 2388–2392.

Carpenter, E. J., Anderson, S. J., Harvey, G. R., Miklas, H. P., & Bradford, B. P. (1972). Polystyrene spherules in coastal waters. *Science, 178*, 749–750.

Cole, M., Lindeque, P., Halsband, C., & Galloway, T. S. (2011). Microplastics as contaminants in the marine environment: A review. *Marine Pollution Bulletin, 62*, 2588–2597.

Collignon, A., Hecq, J. H., Galgani, F., Voisin, P., Collard, F., & Goffart, A. (2012). Neustonic microplastic and zooplankton in the North Western Mediterranean Sea. *Marine Pollution Bulletin, 64*, 861–864.

Colton, J. B., Knapp, F. D., & Burns, B. R. (1974). Plastic particles in surface waters of the Northwestern Atlantic. *Science, 185*, 491–497.

Cózar, A., Echevarria, F., Gonzalez-Gordillo, J. I., Irigoien, X., Ubeda, B., Hernandez-Leon, S., et al. (2014). Plastic debris in the open ocean. *Proceedings of the National Academy of Sciences of the United States of America, 111*, 10239–10244.

Eriksen, M., Mason, S., Wilson, S., Box, C., Zellers, A., Edwards, W., et al. (2013). Microplastic pollution in the surface waters of the Laurentian Great Lakes. *Marine Pollution Bulletin, 77*, 177–182.

Eriksen, M., Lebreton, L. C. M., Carson, H. S., Thiel, M., Moore, C. J., Borerro, J. C., et al. (2014). Plastic pollution in the world's oceans: more than 5 trillion plastic pieces weighing over 250,000 tons afloat at sea. *PLoS ONE, 9*, 111913.

European Commission. (2012). Manifesto for a resource-efficient Europe, p. 2. Brussels.

Foekema, E. M., De Gruijter, C., Mergia, M. T., van Franeker, J. A., Murk, A. J., & Koelmans, A. A. (2013). Plastic in North sea fish. *Environmental Science and Technology, 47*, 8818–8824.

Galgani, F., Fleet, D., Van Franeker, J., Katsanevakis, S., Maes, T., Mouat, J., et al. (2010). Marine strategy framework directive, task group 10 report: Marine litter. In N. Zampoukas (Ed.), *JRC scientific and technical reports*. Ispra: European Comission Joint Research Centre.

Galgani, F., Souplet, A., & Cadiou, Y. (1996). Accumulation of debris on the deep sea floor off the French Mediterranean coast. *Marine Ecology Progress Series, 142*, 225–234.

Goldstein, M., Rosenberg, M., & Cheng, L. (2012). Increased oceanic microplastic debris enhances oviposition in an endemic pelagic insect. *Biology Letters, 11*, doi:10.1098/rsbl.2012.0298.

Gouin, T., Roche, N., Lohmann, R., & Hodges, G. (2011). A thermodynamic approach for assessing the environmental exposure of chemicals absorbed to microplastic. *Environmental Science and Technology, 45*, 1466–1472.

Graham, E. R., & Thompson, J. T. (2009). Deposit- and suspension-feeding sea cucumbers (Echinodermata) ingest plastic fragments. *Journal of Experimental Marine Biology and Ecology, 368*, 22–29.

Gregory, M. R. (1996). Plastic 'scrubbers' in hand cleansers: A further (and minor) source for marine pollution identified. *Marine Pollution Bulletin, 32*, 867–871.

Harper, P. C., & Fowler, J. A. (1987). Plastic pellets in New Zealand storm-killed prions (*Pachyptila* spp.), 1958–1998. *Notornis, 34*, 65–70.

Hays, H., & Cormons, G. (1974). Plastic particles found in tern pellets, on coastal beaches and at factory sites. *Marine Pollution Bulletin, 5*, 44–46.

Hidalgo-Ruz, V., Gutow, L., Thompson, R. C., & Thiel, M. (2012). Microplastics in the marine environment: A review of the methods used for identification and quantification. *Environmental Science and Technology, 46*, 3060–3075.

Holmes, L. A., Turner, A., & Thompson, R. C. (2012). Adsorption of trace metals to plastic resin pellets in the marine environment. *Environmental Pollution, 160*, 42–48.

Imhof, H. K., Ivleva, N. P., Schmid, J., Niessner, R., & Laforsch, C. (2013). Contamination of beach sediments of a subalpine lake with microplastic particles. *Current Biology, 23*, R867–R868.

Ivar do Sul, J. A., & Costa, M. F. (2014). The present and future of microplastic pollution in the marine environment. *Environmental Pollution, 185*, 352–364.

Ivar do Sul, J. A., Costa, M. F., Barletta, M., & Cysneiros, F. J. A. (2013). Pelagic microplastics around an archipelago of the equatorial Atlantic. *Marine Pollution Bulletin, 75*, 305–309.

Jambeck J. R., Geyer, R., Wilcox, C., Siegler, T. R, Perryman, M., Andrady, A., et al. (2015). Plastic waste inputs from land into the ocean. *Science, 347*, 6223, 768–771.

Koelmans, A. A. (2015). Modeling the role of microplastics in bioaccumulation of organic chemicals to marine aquatic organisms. Critical review. In M. Bergmann., L. Gutow., & M. Klages (Eds.), *Marine anthropogenic litter* (pp. 313–328). Berlin: Springer.

Koelmans, A. A., Besseling, E., Wegner, A., & Foekema, E. M. (2013). Plastic as a carrier of POPs to aquatic organisms: A model analysis. *Environmental Science and Technology, 47*, 7812–7820.

Law, K. L., & Thompson, R. C. (2014). Microplastics in the seas. *Science, 345*, 144–145.

Law, K. L., Morét-Ferguson, S., Maximenko, N. A., Proskurowski, G., Peacock, E. E., Hafner, J., et al. (2010). Plastic accumulation in the North Atlantic subtropical gyre. *Science, 329*, 1185–1188.

Lusher, A. (2015). Microplastics in the marine environment: distribution, interactions and effects. In M. Bergmann., L. Gutow., & M. Klages (Eds.), *Marine anthropogenic litter* (pp. 245–308). Berlin: Springer.

Lusher, A., McHugh, M., & Thompson, R. C. (2013). Occurrence of microplastics in the gastrointestinal tract of pelagic and demersal fish from the English Channel. *Marine Pollution Bulletin, 67*, 94–99.

Mato, Y., Isobe, T., Takada, H., Kanehiro, H., Ohtake, C., & Kaminuma, T. (2001). Plastic resin pellets as a transport medium for toxic chemicals in the marine environment. *Environmental Science and Technology, 35*, 318–324.

Murray, F., & Cowie, P. R. (2011). Plastic contamination in the decapod crustacean *Nephrops norvegicus* (Linnaeus, 1758). *Marine Pollution Bulletin, 62*, 1207–1217.

Ng, K. L., & Obbard, J. P. (2006). Prevalence of microplastics in Singapore's coastal marine environment. *Marine Pollution Bulletin, 52*, 761–767.

Norén, F. (2008). *Small plastic particles in coastal Swedish waters* (p. 11). Lysekil: KIMO Sweden.

Obbard, R. W., Sadri, S., Wong, Y. Q., Khitun, A. A., Baker, I. & Thompson, R. C. (2014). Global warming releases microplastic legacy frozen in Arctic Sea ice. Earth's Future, 2, 315–320.

Oehlmann, J., Schulte-Oehlmann, U., Kloas, W., Jagnytsch, O., Lutz, I., Kusk, K. O., et al. (2009). A critical analysis of the biological impacts of plasticizers on wildlife. *Philosophical Transactions of the Royal Society B, 364*, 2047–2062.

Ogata, Y., Takada, H., Master, M. K., Hirai, H., Iwasa, S., Endo, S., et al. (2009). International pellet watch: Global monitoring of persistent organic pollutants (POPs) in coastal waters. 1. Initial phase data on PCBs, DDTs, and HCHs. *Marine Pollution Bulletin, 58*, 1437–1466.

PlasticsEurope. (2011). *Plastics the facts 2011. An analysis of European plastics production, demand and recovery for 2010, PlasticsEurope*, p. 32.

Possatto, F. E., Barletta, M., Costa, M. F., do Sul, J. A. I., & Dantas, D. V. (2011). Plastic debris ingestion by marine catfish: An unexpected fisheries impact. *Marine Pollution Bulletin, 62*, 1098–1102.

Rebolledo, E. L. B., Van Franeker, J. A., Jansen, O. E., & Brasseur, S. M. J. M. (2013). Plastic ingestion by harbour seals (*Phoca vitulina*) in The Netherlands. *Marine Pollution Bulletin, 67*, 200–202.

Reddy, M. S., Basha, S., Adimurthy, S., & Ramachandraiah, G. (2006). Description of the small plastics fragments in marine sediments along the Alang-Sosiya ship-breaking yard, India. *Estuarine, Coastal and Shelf Science, 68*, 656–660.

Rochman, C. M. (2015). The complex mixture, fate and toxicity of chemicals associated with plastic debris in the marine environment. In M. Bergmann., L. Gutow., & M. Klages (Eds.), *Marine anthropogenic litter* (pp. 117–140). Berlin: Springer.

Rochman, C. M., & Browne, M. A. (2013). Classify plastic waste as hazardous. *Nature, 494*, 169–171.

Rochman, C. M., Hoh, E., Kurobe, T., & Teh, S. J. (2013). Ingested plastic transfers hazardous chemicals to fish and induces hepatic stress. *Nature Scientific Reports, 3*, 3263.

Sadri, S. S., & Thompson, R. C. (2014). On the quantity and composition of floating plastic debris entering and leaving the Tamar Estuary, Southwest England. *Marine Pollution Bulletin, 81*, 55–60.

Santos, I. R., Friedrich, A., & do Sul, J. A. I. (2009). Marine debris contamination along undeveloped tropical beaches from northeast Brazil. *Environmental Monitoring and Assessment, 148*, 455–462.

Secretariat of the Convention on Biological Diversity and Scientific and Technical Advisory Panel GEF. (2012). *Impacts of marine debris on biodiversity: Current status and potential solutions* (Vol. 67, p. 61). Montreal.

Shiber, J. G. (1987). Plastic pellets and tar on spain mediterranean beaches. *Marine Pollution Bulletin, 18*, 84–86.

STAP. (2011). *Marine debris as a global environmental problem: Introducing a solutions based framework focused on plastic. In a STAP information document* (p. 40). Washington, DC: Global Environment Facility.

Tanaka, K., Takada, H., Yamashita, R., Mizukawa, K., Fukuwaka, M., & Watanuki, Y. (2013). Accumulation of plastic-derived chemicals in tissues of seabirds ingesting marine plastics. *Marine Pollution Bulletin, 69*, 219–222.

Teuten, E. L., Rowland, S. J., Galloway, T. S., & Thompson, R. C. (2007). Potential for plastics to transport hydrophobic contaminants. *Environmental Science and Technology, 41*, 7759–7764.

Teuten, E. L., Saquing, J. M., Knappe, D. R. U., Barlaz, M. A., Jonsson, S., Björn, A., et al. (2009). Transport and release of chemicals from plastics to the environment and to wildlife. *Philosophical Transactions of the Royal Society B, 364*, 2027–2045.

Thompson, R. C., Olsen, Y., Mitchell, R. P., Davis, A., Rowland, S. J., John, A. W. G., et al. (2004). Lost at sea: Where is all the plastic? *Science, 304*, 838.

Thompson, R. C., Moore, C., Andrady, A., Gregory, M., Takada, H., & Weisberg, S. (2005). New directions in plastic debris. *Science, 310*, 1117.

Thompson, R. C., Moore, C., vom Saal, F. S., & Swan, S. H. (2009). Plastics, the environment and human health: Current consensus and future trends. *Philosophical Transactions of the Royal Society B, 364*, 2153–2166.

Van Cauwenberghe, L., Vanreusel, A., Mees, J., & Janssen, C. R. (2013). Microplastic pollution in deep-sea sediments. *Environmental Pollution (Barking, Essex: 1987), 182*, 495–499.

van Franeker, J. A., Blaize, C., Danielsen, J., Fairclough, K., Gollan, J., Guse, N., et al. (2011). Monitoring plastic ingestion by the northern fulmar *Fulmarus glacialis* in the North Sea. *Environmental Pollution, 159*, 2609–2615.

Woodall, L. C., Sanchez-Vidal, A., Canals, M., Paterson, G. L. J., Coppock, R., Sleight, V., et al. (2014). The deep sea is a major sink for microplastic debris. *Royal Society Open Science 1*, 140317.

Wright, S. L., Rowe, D., Thompson, R. C., & Galloway, T. S. (2013a). Microplastic ingestion decreases energy reserves in marine worms. *Current Biology, 23*, 1031–1033.

Wright, S. L., Thompson, R. C., & Galloway, T. S. (2013b). The physical impacts of microplastics on marine organisms: A review. *Environmental Pollution, 178,* 483–492.

Wright, S. L., Rowe, D. Thompson, R. C. & Galloway T. S. (2014). Microplastic ingestion decreases energy reserves in marine worms. *Current Biology, 23,* R1031–R1033.

Zarfl, C., Fleet, D., Fries, E., Galgani, F., Gerdts, G., Hanke, G., et al. (2011). Microplastics in oceans. *Marine Pollution Bulletin, 62,* 1589–1591.

Zettler, E. R., Mincer, T. J., & Amaral-Zettler, L. A. (2013). Life in the "Plastisphere": Microbial communities on plastic marine debris. *Environmental Science and Technology, 47,* 7137–7146.

Zitko, V., & Hanlon, M. (1991). Another source of pollution by plastics—Skin cleaners with plastic scrubbers. *Marine Pollution Bulletin, 22,* 41–42.

Chapter 8
Methodology Used for the Detection and Identification of Microplastics—A Critical Appraisal

Martin G.J. Löder and Gunnar Gerdts

Abstract Microplastics in aquatic ecosystems and especially in the marine environment represent a pollution of increasing scientific and societal concern, thus, recently a substantial number of studies on microplastics were published. Although first steps towards a standardization of methodologies used for the detection and identification of microplastics in environmental samples are made, the comparability of data on microplastics is currently hampered by a huge variety of different methodologies, which result in the generation of data of extremely different quality and resolution. This chapter reviews the methodology presently used for assessing the concentration of microplastics in the marine environment with a focus on the most convenient techniques and approaches. After an overview of non-selective sampling approaches, sample processing and treatment in the laboratory, the reader is introduced to the currently applied techniques for the identification and quantification of microplastics. The subsequent case study on microplastics in sediment samples from the North Sea measured with focal plane array (FPA)-based micro-Fourier transform infrared (micro-FTIR) spectroscopy shows that only 1.4 % of the particles visually resembling microplastics were of synthetic polymer origin. This finding emphasizes the importance of verifying the synthetic polymer origin of potential microplastics. Thus, a burning issue concerning current microplastic research is the generation of standards that allow for the assessment of reliable data on concentrations of microscopic plastic particles and the involved polymers with analytical laboratory techniques such as micro-FTIR or micro-Raman spectroscopy.

Keywords Detection of microplastics · Focal plane array-based micro-FTIR imaging · Identification of microplastics · Microplastic · Microplastic analysis · Micro-FTIR spectroscopy

M.G.J. Löder
Animal Ecology I, University of Bayreuth, Universitätsstraße 30, 95440 Bayreuth, Germany

M.G.J. Löder (✉) · G. Gerdts
Biologische Anstalt Helgoland, Alfred Wegener Institute Helmholtz Centre
for Polar and Marine Research, Marine Station, PO Box 180, 27483 Helgoland, Germany
e-mail: Martin.Loeder@uni-bayreuth.de

M. Bergmann et al. (eds.), *Marine Anthropogenic Litter*,
DOI 10.1007/978-3-319-16510-3_8

201

8.1 Introduction

Since the middle of the 20th century, the increasing global production of plastics is accompanied by an accumulation of plastic litter in the marine environment (Barnes et al. 2009; Thompson et al. 2004). Being dispersed by currents and winds, persistent plastics, whether deliberately dumped or accidentally lost, are rarely degraded but become fragmented over time (Thompson 2015). Together with micro-sized primary plastic litter from consumer products, these degraded secondary micro-fragments lead to an increasing amount of small plastic particles, so called "microplastics" (i.e. particles <5 mm) in the oceans (Andrady 2011). Microplastics are further divided according to their size into "large microplastics" (1–5 mm) and "small microplastics" (20 µm–1 mm) (Hanke et al. 2013).

The distribution of microplastics in the marine environment is strongly dependent on their density. The density of a virgin-polymer particle is often altered during the manufacturing process (e.g. density increase due to addition of inorganic fillers, density decrease due to foaming of the polymer) as well as through ageing or biofouling processes (Harrison et al. 2011; Morét-Ferguson et al. 2010; Gregory 1983). Since most synthetic polymers have a lower density than seawater, microplastic particles mostly float at the sea surface (0.022–8,654 items m^{-3}) but occur to a lower extent suspended in the water column (0.014–12.51 items m^{-3}). Sediments seem to represent a sink for microplastics (18,000–125,000 items m^{-3} in subtidal sediments) while beaches, as intermediate environments, can accumulate floating, neutrally buoyant as well as sinking plastics (185–80,000 items m^{-3}) (Hidalgo-Ruz et al. 2012).

The massive accumulation of microplastics in the oceans has been recognized by scientists and authorities worldwide, and previous studies have demonstrated the ubiquitous presence of microplastics in the marine environment (Browne et al. 2010; Hidalgo-Ruz and Thiel 2013; Ng and Obbard 2006; Claessens et al. 2011; Van Cauwenberghe et al. 2013; Vianello et al. 2013). Thus, with the Marine Strategy Framework Directive (MSFD-indicator 10.1.3) the EU prescribes a mandatory monitoring of microplastics (Zarfl et al. 2011), and the EU Technical Subgroup on Marine Litter (TSG-ML) proposed a standardized monitoring strategy for microplastics in the EU (Hanke et al. 2013).

Researchers worldwide report on the uptake of microplastics by various marine organisms (Cole et al. 2013; Ugolini et al. 2013; Foekema et al. 2013; Murray and Cowie 2011; Browne et al. 2008). Ingestion of microplastics may lead to "… potentially fatal injuries such as blockages throughout the digestive system or abrasions from sharp objects…" (Wright et al. 2013), which, in contrast to macroplastics, mainly affect microorganisms, smaller invertebrates or larvae. In addition to these physical effects on single organisms, the ecological implications can be even more severe as microplastics can release toxic additives upon degradation and accumulate persistent organic pollutants (POPs) (Bakir et al. 2012; Engler 2012; Rios et al. 2007; Teuten et al. 2009; Rochman et al. 2013). Because of their minute size, microplastics harbor the risk of entering marine food webs at low trophic levels and propagating toxic substances up the food web (Besseling et al. 2013; Mato et al. 2001). However, this is discussed controversially in the literature and several studies suggest that this issue is of minor importance from a risk assessment perspective (compare, e.g. Gouin et al. 2011; Koelmans et al. 2013, Koelmans 2015). Nevertheless, microplastics harbor the risk of transporting POPs to human food (Engler 2012). Because of their long residence time at sea plastics can travel long distances (Ebbesmeyer and Ingraham 1994) and thus function as vectors for dispersal of toxins and/or pathogenic microorganisms (Harrison et al. 2011; Zettler et al. 2013). However, although the potential risks associated with marine microplastics have recently been acknowledged the manifold impacts of microplastics on the ecosystems of the oceans have not been investigated in detail and are thus only poorly understood.

Current microplastic research suffers from insufficient reliable data on concentrations of microplastics in the marine environment and on the composition of involved polymers because standard operation protocols (SOP) for microplastic sampling and detection are not available (Hidalgo-Ruz et al. 2012; Claessens et al. 2013; Imhof et al. 2012; Nuelle et al. 2014). Although first steps towards a standardization have been made, e.g. in the European Union by TSG-ML (Hanke et al. 2013), the comparability of data on microplastics is still hampered by a huge variety of different methods that lead to the generation of data of extremely different quality and resolution.

In this chapter, we will critically review the methodology presently used for assessing the concentration of microplastics in the marine environment. We will focus on the most convenient techniques and approaches recently applied for the identification of microplastics. After an overview of non-selective sampling approaches and sample processing in the laboratory, we will introduce the reader to currently applied detection techniques for microplastics. Finally, we present a case study to emphasize the importance of verifying the synthetic polymer origin of potential microplastics by e.g. micro-Fourier transform infrared (micro-FTIR) spectroscopy. In an outlook, we will address important gaps in knowledge concerning the detection of microplastics and how these could be potentially filled.

8.2 Sampling for Microplastics

Today, synthetic polymers are omnipresent and daily life without plastics is inconceivable. As a consequence, even microplastic sampling, preparation and analysis procedures themselves are affected by the ubiquity of synthetic polymers in the environment. Hence, a multitude of contamination sources from sampling equipment through clothes or airborne particles can compromise the analysis of microplastics in the environment. This can lead to a great overestimation of concentrations of microplastics in samples. Because of their ability to hover in air, especially fibres have a high contamination potential and can cause problems during microplastic analysis (Hidalgo-Ruz et al. 2012; Nuelle et al. 2014; Norén

2007; Norén and Naustvoll 2010). Thus, a special focus should be laid on the prevention of contamination (Hidalgo-Ruz et al. 2012). Potential sources of contamination should be avoided by replacing plastic devices or laboratory ware by non-plastic material and the strict use of control samples is highly recommended. Analysis of control samples facilitates the identification of the source in case a contamination has occurred.

8.2.1 Water Samples

Because of their relatively low concentrations in the environment sampling of microplastic particles generally requires large sample volumes. Thus, samples from the open water are usually taken with plankton nets of different mesh sizes. The sea surface is sampled for floating microplastics by manta trawls (Eriksen et al. 2013a, b; Doyle et al. 2011) or neuston nets (Morét-Ferguson et al. 2010; Carpenter and Smith 1972; Colton et al. 1974). While neuston catamarans (Fig. 8.1a) can be operated even in higher waves, a manta trawl (Fig. 8.1b) is best used in calm waters to prevent hopping on waves and damage to the device. The volume filtered by a net is usually recorded by a flowmeter mounted at the net opening, enabling the normalization to the filtered water volume and thus a calculation of concentrations of microplastics (items/grams) per unit water volume. Relating concentrations to sampled area is also possible by multiplying trawl distance by the horizontal width of the net opening. The water column can be sampled for suspended microplastics by trawling with different plankton nets, e.g. CalCOFI (California Cooperative Oceanic Fisheries Investigations) or Bongo nets (Doyle et al. 2011). Trawling speed depends on weather conditions and currents, but usually lies between 1 and 5 knots. Trawling time depends on seston concentrations and lies between a few minutes up to several hours (Boerger et al. 2010). The plankton sample is concentrated in the cod-end of the net and after recovery the net has to be carefully rinsed from the exterior to assure that all plankton and debris are washed into the cod-end (Doyle et al. 2011). It is important to assure that no residual sample is left in the net, which would lead to a carryover of microplastics to the next sample. The content of the codend is finally transferred to a sample container and fixed with plastic friendly fixatives (e.g. formalin) or stored frozen. If the particles are directly sorted they should be dried and kept in the dark until further analysis (Hidalgo-Ruz et al. 2012).

The size of the particles retained and also the filterable volume is a direct consequence of the mesh size used. The mesh sizes used for sampling in previous studies varied between 50 and 3000 µm (Hidalgo-Ruz et al. 2012). Another factor influencing the filtered volume is the net size, i.e. the area, which acts as filter. Depending on the seston concentration in the water, a few thousand litres to several hundred cubic metres can be filtered until a net becomes clogged. Seasons with red tides or plankton and jellyfish blooms are generally unfavorable for sampling large volumes of water. Nets are usually 3–4.5 m long and a mesh size of around 300 µm is most commonly used. These nets do not sample microplastic particles <300 µm quantitatively but allow for sampling of larger volumes of water. In order to avoid the risk of

Fig. 8.1 A neuston catamaran (**a**) and a manta trawl (**b**) during trawling

clogging nets at small mesh sizes, only few studies used mesh sizes <300 μm. The non-standardized use of different nets and mesh sizes seriously impedes the comparability of data sets on pelagic microplastic concentrations.

Besides common net sampling, other techniques are occasionally used for assessing microplastic concentrations in the water column: bulk sampling with subsequent filtration (Ng and Obbard 2006; Dubaish and Liebezeit 2013), screening Continuous Plankton Recorder (CPR) samples (Thompson et al. 2004) or using direct in situ filtration (Norén and Naustvoll 2010). A highly promising technique, currently under development, is the use of direct fractionated pressure filtering of large (>1 m^3) volumes of water through a filter cascade (developed by -4H-JENA engineering GmbH). This approach theoretically allows for the simultaneous sampling of different size fractions of microplastics down to <10 μm and thus enables a more comprehensive resolution of the size spectrum of microplastics.

8.2.2 Sediment Samples

Microplastics in sediments or beaches are currently more frequently analyzed than microplastics in the water column (Hidalgo-Ruz et al. 2012). Sampling approaches depend on the sampling location i.e. sampling sediments directly on beaches or sampling subtidal sediments from a ship.

8.2.2.1 Beaches

Sampling beaches for microplastics is relatively easy and requires nothing more than a non-plastic sampling tool (tablespoon, trowel or small shovel), a frame or a corer to specify the sampling area and a container (if possible non-plastic) to store the sample. The quantity of samples reported in the literature varies between less than 500 g to up to 10 kg (Hidalgo-Ruz et al. 2012). While sampling on a beach poses no problem per se, the positioning of the sample location on the beach is still a matter of scientific debate as the distribution of microplastic particles is as dynamic as the beach itself (Hidalgo-Ruz et al. 2012). The high-tide line where flotsam accumulates is sampled mostly (Browne et al. 2010). Commonly applied sampling strategies include random sampling at several locations on the beach, on transects perpendicular or parallel to the water or in single squares. Often, several samples are pooled for an integrated estimate of the microplastic contamination of a beach. Every single sampling location for the pooled sample is then defined as described above. Another point of concern is the sampling depth. Sampling the top five centimetres is a common approach (as also suggested by the TSG-ML), but sampling to a depth of 0.3 m is also reported in the literature (Claessens et al. 2011). If corers are used for sampling, different depth layers can be sampled so that microplastic concentrations can be related to sediment depth and eventually to the age of the corresponding sediment layer. The units of microplastic abundance reported depend on the sampling approach. Thus, abundance is normalized to sampling area, sediment weight or volume. Sampling sediments for microplastics at beaches might appear trivial. However, currently no standard protocol exists for sampling microplastics with respect to location, sampling technique and sample quantity, and thus the comparability of the data produced is limited. Accordingly, there is an urgent need for the development of standardized sampling approaches. Because of the patchy distribution of microplastics at beaches a standardized, spatially integrating sampling design appears reasonable and would facilitate the generation of comparable data. A first step towards the standardization of sampling microplastics at beaches in the EU has been made by the TSG-ML (Hanke et al. 2013). They recommend to monitor microplastics at sandy beaches at the strandline with a minimum of five replicate samples separated by at least five metres and to distinguish two size categories: large microplastics (1–5 mm) and small microplastics (20 µm–1 mm). Small microplastics should be sampled from the top five centimetres with a metal spoon by combining several scoops at arm length in an arc-shaped area at the strand line to collect ca. 250 g of sediment; large microplastics should be sampled from the top five centimetres and several kilograms of sediment sample can be reduced by sieving over a 1-mm sieve directly at the beach.

8.2.2.2 Subtidal Sediments

Subtidal sediments can be sampled from vessels with grabs, e.g. Van Veen or Ekman grab or corers of different design, e.g. a multiple corer. Grabs tend to disturb the sediment and are suited for surface (e.g. top five centimetres) or bulk sampling, whereas undisturbed core samples enable the simultaneous sampling of surface and depth layers but yield smaller sample volumes. The size of the instrument applied as well as the time needed for its retrieval depends strongly on the water depth at the sampling location. The use of corers enables sampling to a water depth of more than 5,000 m (Van Cauwenberghe et al. 2013). Sediment samples are usually stored frozen or dried and kept in the dark until further analysis.

8.2.3 Biota

The ingestion of microplastics by various marine vertebrates and invertebrates under laboratory conditions and in the field has been reported in the literature (Lusher 2015). As sampling strategies are variable and depend strongly on the organism targeted we give only a short overview of possible sample organisms for the ingestion of microplastics with a focus on field sampling. For a detailed review please refer to Wright et al. 2013 and Ivar do Sul and Costa 2013.

Laboratory studies on the ingestion of microplastics by marine biota frequently use microscopic plastic beads of known polymer origin, which can be easily recognized and counted under the microscope in gut contents and excretions or—in the case of transparent planktonic organisms—in the organism itself (Cole et al. 2013). In this context, the use of fluorescent particles facilitates the recovery and enumeration of the particles.

Investigations on the ingestion of microplastics by vertebrates in the field require a substantially higher effort and thus studies in this area of research are rare (Wright et al. 2013). The target for sampling is the content of the digestive tract or the excretions of an organism. Larger organisms that are directly sampled for microplastics are mainly fish, which are usually sampled by nets or traps.

Stranded carcasses (e.g. birds, seals, cetaceans) can be collected and examined for ingested microplastics. This has been done for stranded northern fulmars (*Fulmarus glacialis*) in the North Sea for more than three decades (Wright et al. 2013; Kühn et al. 2015). After dissection, the gut content or the entire digestive tract has to be conserved or frozen for later analysis. Another possibility used for the sampling of mammals and birds is to "indirectly" sample organisms for the ingestion of microplastics by collecting their boluses, casts or faeces. Bond and Lavers (2013) used emetics to monitor plastic ingestion in storm-petrel chicks (*Oceanodroma leucorhoa*) enabling the authors to study plastic ingestion non-lethally. Smaller invertebrate organisms such as worms, mussels and snails can be directly collected in the field (Besseling et al. 2013; Claessens et al. 2013) with nets or traps and are best frozen as a whole until analyzed. Biological samples can be conserved by using plastic-friendly fixatives (e.g. formalin) or best be frozen or dried and kept dark until analysis.

8.3 Laboratory Preparation of Samples

8.3.1 Extraction of Microplastics

The densities of common consumer-plastic polymers range between 0.8 (silicone) and 1.4 g cm^{-3} (e.g. polyethylene terephthalate (PET), polyvinyl chloride (PVC)) while expanded plastic foams have only a fraction of the densities of the original polymer (e.g. expanded polystyrene (EPS) <0.05 g cm^{-3}). Microplastic particles can thus be separated from matrices with higher densities, such as sediments (2.65 g cm^{-3}), by flotation with saturated salt solutions of high density. The dried sediment sample is mixed with the concentrated salt solution and agitated (e.g. by stirring, shaking, aeration) for a certain amount of time. Plastic particles float to the surface or stay in suspension while heavy particles such as sand grains settle quickly. Subsequently, microplastics are recovered by removing the supernatant. Depending on the solution used, different fractions of the range of consumer polymers are targeted—the higher the density of the solution the more polymer types can be extracted. Often a saturated NaCl solution is used for the extraction of microplastics (Thompson et al. 2004; Browne et al. 2010; Ng and Obbard 2006; Claessens et al. 2011; Browne et al. 2011). Although being an inexpensive and environment-friendly approach, not all common polymers are extracted (e.g. PVC, PET, polycarbonate (PC), polyurethane (PUR)) because of the relatively low density of the solution (~1.2 g cm^{-3}). Other solutions used include sodium polytungstate solution (1.4 g cm^{-3}) (Corcoran et al. 2009), zinc chloride solution (1.5–1.7 g cm^{-3}) (Imhof et al. 2012; Liebezeit and Dubaish 2012) or sodium iodine solution (1.8 g cm^{-3}) (Nuelle et al. 2014). These high-density solutions are suitable for the extraction of most of the common user plastics. For financial/environmental reasons the use of zinc chloride and the recycling of the saturated solution by pressure filtration is highly recommended.

There is great variability in the extraction techniques applied. The approaches range from simply stirring the sediment sample in a saturated salt solution (classical setup) (Thompson et al. 2004; Claessens et al. 2011) to the use of an elutriation/fluidisation with subsequent flotation (Claessens et al. 2013; Nuelle et al. 2014) or the extraction with a novel instrument, the "Microplastic Sediment Separator" (MPSS) (Imhof et al. 2012). The extraction efficiencies vary between the techniques used but also depend on the particle shape, size and the polymer origin of the model particles used during recovery experiments. The classic extraction setup reaches recoveries of 80–100 % (Fries et al. 2013) but recovers small microplastics insufficiently (mean recovery rate 40 %, mean particle size 40–309 µm) (Imhof et al. 2012), whereas new approaches achieve high recovery rates of 68–99 % (Nuelle et al. 2014), 96–100 % (Imhof et al. 2012) and 98–100 % (Claessens et al. 2013). Small particles (<500 μm) are more difficult to extract from sediments. Therefore, time-consuming repeated extraction steps are recommended to maximize recovery (Claessens et al. 2013; Nuelle et al. 2014; Browne et al. 2011). Only the MPSS showed a recovery rate of 96 % for small microplastics in a single extraction step (Imhof et al. 2012).

8.3.2 Size Fractionation

Irrespective of the technique used for later identification of microplastics the fractionation of samples (water, sediment, biota) into (at least) two size classes, e.g. >500 μm and <500 μm, is reasonable (Hidalgo-Ruz et al. 2012). For EU monitoring purposes, a separation into fractions of 1–5 mm and 20 μm–1 mm was recently suggested (Hanke et al. 2013). Water samples can be fractionated easily by sieving. If large amounts of biological matrix (e.g. gut contents, tissue, large plankton) clog the sieve a purification step prior to sieving can be helpful. Microplastics from sediment samples are easily size-fractionated after extraction. If the sediment sample matrix consists mainly of smaller grains (<500 μm) it can be sieved after drying (or wet) to reduce the volume for later extraction. In this case, the sample must be handled with care during sieving to avoid the mechanical generation of additional microplastic particles from larger, brittle plastic material. A 500 μm sieve, ideally made of steel, can be used for size separation. The use of a sieve cascade of different mesh sizes allows for size separation and quantification of different size classes of microplastics (Moore et al. 2002; McDermid and McMullen 2004).

Microplastic particles >500 μm can be sorted out manually under a stereomicroscope using forceps and subsequently analyzed (visually, spectroscopically, other techniques). The effort involved in the manual sorting of particles increases for the fraction <500 μm owing to difficulties in handling small particles. Furthermore, an increasing amount of background matrix particles of different organic or inorganic origin may impede a proper separation. Therefore, this fraction should be purified and concentrated on filters for further analysis by, e.g. spectroscopy. The suggested size separation (>500 μm; <500 μm) is accounted for by the techniques that can be used for later identification. Additionally, the standardized application of size fractionation enables an inter-comparison between different studies, at least for the larger fraction, even if the smaller fraction is not of interest for the study (Hidalgo-Ruz et al. 2012).

8.3.3 Sample Purification

The purification of microplastic samples is obligatory, especially, for instrumental analyses (FTIR/Raman spectroscopy, pyrolysis-GC/MS). Biofilms and other organic and inorganic adherents have to be removed from the microplastic particles to avoid artifacts that impede a proper identification. Furthermore, the purification step is necessary to minimize the non-plastic filter residue on filters on which the microplastic fraction <500 µm is concentrated. The most gentle way to clean plastic samples is stirring and rinsing with freshwater (McDermid and McMullen 2004). The use of ultrasonic cleaning (Cooper and Corcoran 2010) should be carefully considered because aged and brittle plastic material might break during treatment resulting in the artificial generation of secondary microplastics (Löder and Gerdts, personal observation). A treatment with 30 % hydrogen peroxide of the dried sediment sample (Liebezeit and Dubaish 2012), the sample filter (Imhof et al. 2012; Nuelle et al. 2014) or the microplastic particles themselves removes large amounts of natural organic debris. Andrady (2011) suggests the use of mineral acids to disintegrate organic impurities in samples. For the digestion of the soft tissue of biotic samples Claessens et al. (2013) used either acid, base and oxidizer (hydrogen peroxide) or a specific mixture thereof. Hot acid digestion with *HNO3* resulted in the best purification results (Claessens et al. 2013). However, several plastic polymers (e.g. polyamide, polyoxymethylene, polycarbonate) react to strong acidic or alkaline solutions (Claessens et al. 2013; Liebezeit and Dubaish 2012), which limits the applicability of these reagents. More promising is the use of a sequential enzymatic digestion as a plastic-friendly purification step. A first attempt of enzymatic purification of samples has been made by Cole et al. (2014) who used only a single enzymatic step (proteinase-K). Sample purification with different technical enzymes (lipase, amylase, proteinase, chitinase, cellulase) prior to micro-FTIR spectroscopy has successfully been applied by our group. This approach reduces the biological matrix of plankton and sediment samples (including chitin, which is present especially in marine samples) as well as the matrix of biological tissue samples to a minimum and thus proved to be a very valuable technique to minimize matrix artifacts during FTIR measurements (Löder and Gerdts, unpublished data).

8.4 Identification of Microplastics

8.4.1 Visual Identification

According to Hidalgo-Ruz et al. (2012) visual sorting to separate potential microplastics from other organic or inorganic material in the sample residues is an obligatory step for the identification of microplastics. If large microplastics are the target of a study this can be done by visual inspection (Morét-Ferguson et al. 2010) whereas smaller microplastic particles should generally be sorted out under a dissection microscope (Doyle et al. 2011). Sorting of aqueous samples can be facilitated by the use of sorting chambers (e.g. Bogorov counting chamber). Generally, if no more accurate methods (e.g. FTIR or Raman spectroscopy) are used to verify synthetic polymer origin of potential microplastic particles the visual identification should not be applied to particles <500 μm as the probability of a misidentification is very high. Hidalgo-Ruz et al. (2012) thus suggest an even higher size limit of 1 mm for visual identification. According to Norén (2007), selection of particles according to standardized criteria in connection with a strict and conservative examination reduces the possibility of misidentification. He suggests the following criteria: (1) no structures of organic origin should be visible in the plastic particle or fibre, (2) fibres should be equally thick and have a three-dimensional bending to exclude a biological origin, (3) particles should be clear and homogeneously colored, (4) transparent or whitish particles must be examined under high magnification and with the help of fluorescence microscopy to exclude a biological origin (Norén 2007). General aspects that are used to describe visually sorted microplastics are source, type, shape, degradation stage, and color of the particles (Hidalgo-Ruz et al. 2012).

It is strongly recommended to subsequently analyze sorted particles by techniques that facilitate a proper identification of plastics (Hidalgo-Ruz et al. 2012; Dekiff et al. 2014) because the quality of the data produced by visual sorting depends strongly on (1) the counting person, (2) the quality and magnification of the microscope and (3) the sample matrix (e.g. plankton, sediment, gut content). Another fundamental drawback of visual sorting is the size limitation, i.e. particles below a certain size cannot be discriminated visually from other material or be sorted because they are unmanageable because of their minuteness. Furthermore, visual sorting is extremely time-consuming. In summary, even an experienced person cannot discriminate all potential microplastic particles unambiguously from sand grains, chitin fragments, diatom frustule fragments, etc. Thus the error rate of visual sorting reported in the literature ranges from 20 % (Eriksen et al. 2013a) to 70 % (Hidalgo-Ruz et al. 2012) and increases with decreasing particle size.

8.4.2 Identification of Microplastics by Their Chemical Composition

The repetitive fingerprint-like molecular composition of plastic polymers allows for a clear assignment of a sample to a certain polymer origin. In the following we will give a short overview of methods applied for polymer identification with a focus on the nowadays frequently used FTIR and Raman analyses of microplastics.

8.4.2.1 Density Separation with Subsequent C:H:N Analysis

Morét-Ferguson et al. (2010) used the specific densities of particles to identify the polymer origin of visually sorted microplastics. For this purpose, the sample was placed in distilled water and, depending on the density of the sample, either ethanol or concentrated solutions of calcium or strontium chloride were added until the sample was neutrally buoyant. The density of the particle was indirectly assessed by weighing a certain volume of the solution. This facilitated the determination of the density with high precision. Different groups of polymers possess a characteristic elemental composition, which was used to identify the plastic origin of a particle by a subsequent C:H:N analysis. By comparison with the densities and C:H:N ratios of virgin-polymer samples the particle could be assessed as either plastic or not and assigned to a group of potential polymers (Morét-Ferguson et al. 2010). This approach represents an approximation to the identification of microplastic particles by narrowing the search for the potential polymer type but not a rigorous chemical analysis. Further drawbacks are the relatively high time effort, which hampers a high sample throughput and that this technique is not applicable to smaller particles.

8.4.2.2 Pyrolysis-GC/MS

Pyrolysis-gaschromatography (GC) in combination with mass spectrometry (MS) can be used to assess the chemical composition of potential microplastic particles by analyzing their thermal degradation products (Fries et al. 2013). The pyrolysis of plastic polymers results in characteristic pyrograms, which facilitate an identification of the polymer type. This analytical approach is already used after extraction and visual sorting of microplastics from sediments. The polymer origin of particles is then identified by comparing their characteristic combustion products with reference pyrograms of known virgin-polymer samples (Nuelle et al. 2014; Fries et al. 2013). If a thermal desorption step precedes the final pyrolysis organic plastic additives can be analyzed simultaneously during pyrolysis-GC/MS runs (Fries et al. 2013). Although the pyrolysis-GC/MS approach allows for a relatively good assignment of potential microplastics to polymer type it has the disadvantage that particles have to be manually placed into the pyrolysis tube. Since only particles of a certain minimum size can be manipulated manually this results in a lower size limitation of particles that can be analyzed. Furthermore, the technique allows only for the analysis of one particle per run and is thus not suitable for processing large sample quantities, which are collected during sampling campaigns or routine monitoring programs. However, currently promising pyrolysis-GC/MS approaches for the qualitative/quantitative analysis of microplastics on whole environmental sample filters are being developed (Scholz-Böttcher, personal communication).

8.4.2.3 Raman Spectroscopy

Raman spectroscopy is a straightforward technique that has been successfully used to identify microplastic particles in different environmental samples with high reliability (Van Cauwenberghe et al. 2013; Cole et al. 2013; Murray and Cowie 2011; Imhof et al. 2012, 2013). During the analysis with Raman spectroscopy the sample is irradiated with a monochromatic laser source. The laser depends on the system used: available laser wavelengths usually range between 500 and 800 nm. The interaction of the laser light with the molecules and atoms of the sample (vibrational, rotational, and other low-frequency interactions) results in differences in the frequency of the backscattered light when compared to the irradiating laser frequency. This so-called Raman shift can be detected and leads to substance-specific Raman spectra. Since plastic polymers possess characteristic Raman spectra the technique can be applied to identify plastic polymers within minutes by comparison with reference spectra. Raman spectroscopy is a "surface technique", thus large, visually sorted microplastic particles can be analyzed and the technique can also be coupled with microscopy. Accordingly, micro-Raman spectroscopy allows for the identification of a broad range of size classes down to very small plastic particles of sizes below 1 μm (Cole et al. 2013). If Raman microscopy is combined with Raman spectral imaging it is possible to generate spatial chemical images based on the Raman spectra of a sample. Micro-Raman imaging theoretically allows for the spectral analysis of whole membrane filters at a spatial resolution below 1 μm. This would facilitate the detection of even the smallest microplastic particles in environmental samples, but the applicability for microplastic research has yet to be demonstrated. Raman spectroscopy can also be coupled with confocal laser-scanning microscopy to locate polymer particles within biological tissues with subcellular precision (Cole et al. 2013). One drawback of Raman spectroscopy is that fluorescent samples excited by the laser (e.g. residues of biological origin from samples) cannot be measured as they prevent the generation of interpretable Raman spectra. Generally, lower laser wave lengths, which transfer a high energy result in high signal intensity but also in a high fluorescence. The fluorescence can be minimized by using lasers with higher wave lengths (>1,000 nm). However, the lower energy of the laser results in a lower signal of the polymer sample. More research is necessary to find the optimum laser wave length for a compromise between suppressed fluorescence and low signal intensity for assessments of microplastics in environmental samples. Generally, a purification step of samples to prevent fluorescence is thus recommended prior to measurements for a clear identification of the polymer type of microplastic particles with Raman spectroscopy.

8.4.2.4 IR Spectroscopy

Similar to Raman spectroscopy, infrared (IR) or Fourier-transform infrared (FTIR) spectroscopy offers the possibility of accurate identification of plastic polymer particles according to their characteristic IR spectra (Thompson et al. 2004; Ng and Obbard 2006; Vianello et al. 2013; Harrison et al. 2012; Frias et al. 2010; Reddy et al. 2006). FTIR and Raman spectroscopy are complementary techniques. Molecular vibrations, which are Raman inactive are IR active and vice versa and can thus provide complementary information on microplastic samples. IR spectroscopy takes advantage of the fact that infrared radiation excites molecular vibrations when interacting with a sample. The excitable vibrations depend on the composition and molecular structure of a substance and are wave-length specific. The energy of the IR radiation that excites a specific vibration will—depending on the wave length—be absorbed to a certain amount, which enables the measurement of characteristic IR spectra. Plastic polymers possess highly specific IR spectra with distinct band patterns making IR spectroscopy an optimal technique for the identification of microplastics (Hidalgo-Ruz et al. 2012). FTIR spectroscopy can provide further information on physico-chemical weathering of sampled plastic particles by detecting the intensity of oxidation (Corcoran et al. 2009).

As for Raman spectroscopy the comparison with reference spectra is necessary for polymer identification. Large particles can be easily analyzed by an FTIR surface technique—"attenuated total reflectance" (ATR) FTIR spectroscopy—at high accuracy in less than one minute. A step forward with respect to the characterization of small-sized particles is the application of FTIR microscopy. In this context, the use of two measuring modes is feasible: reflectance and transmittance. The reflectance mode bears the disadvantage that measurements of irregularly-shaped microplastics may result in non-interpretable spectra due to refractive error (Harrison et al. 2012). The transmittance mode needs IR transparent filters (e.g. aluminium oxide) and is, owing to total absorption patterns, limited by a certain thickness of the microplastics sample. However, the additional use of micro-ATR objectives in combination with microscopy can circumvent this as IR spectra are collected at the surface of a particle enabling the direct measurement on the sample filter without the need for manual handling of particles. Thus, an approach combining transmittance measurements with micro-ATR measurement of particles that show total absorption could be a promising solution for the measurement of particles <500 μm collected on filters. Although micro-FTIR mapping, i.e. the sequential measurement of IR spectra at spatially separated, user-defined points on the sample surface, has been successfully applied for microplastics identification (Levin and Bhargava 2005) this technique is still extremely time-consuming when targeting the whole sample filter surface at a high spatial resolution because it uses only a single detector element (Vianello et al. 2013; Harrison et al. 2012). Harrison et al. (2012) concluded that a highly promising FTIR extension, focal plane array (FPA)-based FTIR imaging (Levin and Bhargava 2005), allows for detailed and unbiased high throughput analysis of total microplastics on a sample filter. This technique enables the simultaneous recording of several thousand spectra within an area with a single measurement and thus the generation of chemical images (see the below case study for a successful application of this technique by Löder and Gerdts). By combining FPA fields, whole sample filters can be analyzed via FTIR imaging. It should be noted, that the lateral resolution of micro-FTIR spectroscopy is diffraction limited (e.g. 10 μm at 1000 cm^{-1}) and, in contrast to Raman spectroscopy, samples must be dried prior to measurement via IR spectroscopy as water strongly absorbs IR radiation. Because of their high IR absorption the IR measurement of black particles is difficult. As for Raman measurements, samples should be purified for proper identification of the polymer type of microplastic particles by IR spectroscopy.

8.5 Case Study

The aim of the case study was to evaluate the applicability of FPA-based micro-FTIR imaging for the measurement of microplastic particles in environmental samples. We tested this technique on particles, which were purified and pre-extracted from sediment samples.

8.5.1 Materials and Methods

All steps preceding the micro-FTIR analysis (sampling, preparation, counting etc.) were done by the Lower Saxony Water Management, Coastal Defense and Nature Conservation Agency (NLWKN) or its contractors, which provided the samples.

8.5.1.1 Sediment Sampling

During a large-scale microplastics baseline assessment of the NLWKN, North Sea sediments were sampled at 101 stations along the German coast of Lower Saxony and on the East Frisian Islands between November 2011 and March 2012. Seventy-three stations were sampled on the islands of Baltrum, Juist, Kachelotplate, Mellum, Minsener Oog and Spiekeroog. On the islands, samples were taken in transects with five sub-samples at each station from the low-water line towards the vegetation zone of the first dune. Sixteen eulittoral samples were taken in the back-barrier tidal flat of Norderney and Spiekeroog and the tidal-flats "Hoher Weg" and "Wurster Watt". Twelve sublittoral samples were taken on north-south transects at depths of 5, 10 and 20 m off the islands Baltrum, Juist and Spiekeroog and at one station in a tidal inlet in the back-barrier tidal flat of each island.

Beach and eulittoral samples were obtained at the onshore stations by sampling of five replicate areas of 10×10 cm within 1 m^2 down to a depth of 1 cm using a metal frame of the abovementioned dimensions. Sublittoral samples at the offshore stations were obtained by a van Veen grab and the recovered material was also sampled on an area of 10×10 cm and down to 1 cm depth. Five grab samples were taken at each sublittoral station. No plastic equipment was used during sampling and all samples were stored in aluminium foil until further processing.

8.5.1.2 Extraction of Microplastics

After drying at 60 °C the samples were screened over a 500 µm metal sieve as suggested by Hidalgo-Ruz et al. (2012) to obtain two size fractions of microplastics.

The fraction >500 μm was stored and not considered further in this study. The material passing through the sieve was homogenized and analyzed.

For the extraction of microplastics 10 g of each sediment sample were treated with 50 ml 30 % hydrogen peroxide overnight to remove natural organic material. After a second drying step, microplastic particles were extracted via density separation in zinc chloride solution (1.5 g cm^{-3}) in a 100-ml glass beaker. After stirring the sample was treated in an ultrasonic bath for 15 min and the beaker kept covered overnight for the sedimentation of sand particles. Potential microplastic particles that accumulated at the surface of the zinc chloride solution were sampled with a syringe and finally filtered onto gray, pre-washed cellulose nitrate filters with a pore size of 1.2 μm and a grid of 3.1 mm. The filters were dried for further analysis.

8.5.1.3 Visual Quantification of Microplastics

The pre-washed filters were weighed to obtain the amount of potential microplastics per 10 g sediment sample. Particles were counted under the bright-field microscope at 20–80-fold magnifications differentiating visually between granular/spherical microplastics, fragments and fibrous microplastics. After analysis, the material on the filters was gently transferred into small glass vials for storage prior to FTIR analyses.

8.5.1.4 FPA-based Micro-FTIR Spectroscopy

Large numbers of particles, especially granular material, were detected when counting the material on the filters. Since a solely optical classification does not enable a reliable identification of particles as microplastics, the NLWKN selected ten samples (beach, eulittoral and sublittoral samples, for details see Table 8.1) for FTIR analysis to determine the polymer origin and composition of the microplastic particles in the samples. Each complete microplastic sample or a subsample thereof as large as possible was carefully transferred to a circular calcium fluoride crystal sample carrier (13 mm diameter, 1 mm thickness) for FTIR analysis. The objective of the FTIR analyses was to identify the polymer origin of at least 20 particles of each of the aforementioned classification types (spherical-granular, fragments and fibres) distinguished during previous counting. Prior to the FTIR measurements, a microscopic overview picture (40-fold magnification; bright-field) of each sample was taken to identify sample regions with conspicuous particles (Fig. 8.2). Because of the general scarcity of fragments and fibres and the high abundance of granular particles, areas with those rare particle types present were selected. Thus the measurements were not random. Black particles showed hardly interpretable IR spectra because of their high absorption of IR radiation and were, thus, not given priority during the measurements.

FTIR spectra of particles were recorded using a Bruker HYPERION 3000 FTIR microscope equipped with a liquid nitrogen cooled 64 × 64 detector elements focal

Table 8.1 Sampling locations of the sediment samples

Station	Location	Zone	Sampling date	Latitude (decimal degree)	Longitude (decimal degree)	Water depth (m)
MP_BAL_W20	Baltrum	Sublittoral	01/03/2012	53.81237	7.39095	21.8
MP_BAL_W05	Baltrum	Sublittoral	01/03/2012	53.75822	7.38965	5.3
BAL_01	Baltrum	Beach, high-water mark	05/02/2012	53.73173	7.42765	–
KP_02	Kachelot-plate	Beach, high-water mark	02/11/2011	53.64873	6.82507	–
MEL_11	Mellum	Beach, dune	07/06/2012	53.71392	8.15507	–
MEL_09	Mellum	Beach, low-water mark	07/06/2012	53.71282	8.15812	–
Nney_EU_01	Norderney	Eulittoral, mud	15/02/2012	53.70352	7.23642	–
Nney_EU_02	Norderney	Eulittoral, sand	15/02/2012	53.70098	7.23493	–
HoWe_EU_02	Hoher Weg	Eulittoral, mud	06/02/2012	53.59462	8.22840	–
WuKu_EU_04	Wurster Watt	Eulittoral, sand	06/02/2012	53.79787	8.52068	–

Fig. 8.2 Overview of a sample prepared for measurement by FPA-based micro-FTIR spectroscopy (sample WuKu_EU_04, scale bar: 1000 µm)

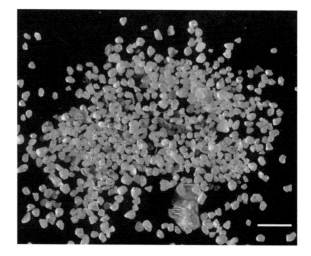

plane array detector (FPA) with a 15 × IR objective. The microscope was coupled with a TENSOR 27 spectrometer. IR spectra were recorded in transmission mode as average spectra by calculating the arithmetic mean of 32 scans in the range 3850–900 cm^{-1} with a resolution of 4 cm^{-1} and a 6 mm aperture.

Depending on the amount of regions with conspicuous particles, between 20 and 65 FPA fields were measured per sample after manual focusing of the microscope. A single FPA field covered an area of 170 × 170 μm with a lateral pixel resolution of 2.7 μm (Bruker Optics) and ca. 40 s were needed for a measurement. Background measurements were performed on the blank sample carrier prior to each sample measurement. The measurements were processed with the software "OPUS 7.0" (Bruker Optics). The polymer origin of particles was identified by comparison with a self-generated polymer library in OPUS 7.0. For creating this library, we measured (ATR-FTIR) pre-production plastic pellets, powders and films of the most commonly used consumer-plastic polymers provided by different plastic polymer manufacturers. Currently, the library consists of 128 plastic polymer records and several other marine abiotic and biotic materials (e.g. cellulose, quartz, chitin, silicate, keratin) and will be expanded in the future.

8.5.2 Results

8.5.2.1 FPA-based Micro-FTIR Analysis of Pre-extracted Particles in Sediment Samples

Between 29 and 64 particles per sample were analyzed by FTIR microscopy resulting in a total number of 404 particles analyzed from the ten samples provided by the NLWKN. Only 16 particles (12 fragments and 4 fibres), i.e. 4 % of the total number of investigated particles, differed in their shape from the abundant granular material.

8.5.2.2 Granular Particles

With 388 counts the majority (96 %) of the particles could clearly be assigned to the category of granular particles. Most of the granular particles were transparent to whitish-opaque and between 100 and 300 μm in size (Fig. 8.2). Of these, 320 particles showed a spectrum that strongly differed from common polymer IR spectra. By comparison with the IR spectrum of laboratory quartz (p.a. grade, Merck) we were able to show, that all these particles were in fact quartz sand and not microplastics (Fig. 8.3). 68 further granular particles that were very similar in shape and appearance to the quartz particles also displayed non-polymer spectra. These spectra were similar to the quartz spectra (Fig. 8.4) and were thus certainly also of mineral origin. However, because of the lack of reference spectra a definite identification was not possible in those cases.

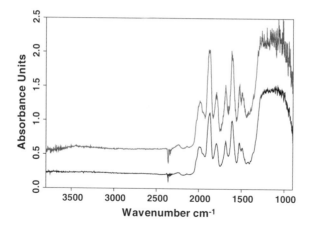

Fig. 8.3 Comparison of the spectrum of laboratory quartz (p.a. grade Merck) (*blue*) and a spectrum obtained from the measurement of a typical granular particle (*red*) by FPA-based micro-FTIR spectroscopy

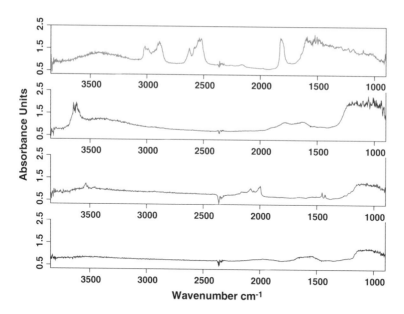

Fig. 8.4 Four examples of IR spectra of granular particles that were very similar in shape and appearance to quartz particles but displayed different non-polymer spectra when analyzed by FPA-based micro-FTIR spectroscopy

8.5.2.3 Fragments and Fibres

Within the 16 particles that differed from the granular material, we found four black malleable fragments with a size between 260 and 390 μm, which exhibited high absorbance and showed non-interpretable transmission IR spectra (Fig. 8.5). According to their bitumen-like, malleable properties and coloration, those particles are most likely oil residues from ship spillages or road wear. In total, six particles were of biological origin. Four particles of fibrous appearance (length: 290–900 μm, width 16–100 μm) showed organic IR spectra similar to cellulose and were most likely of plant origin. Two particles (200–800 μm) exhibited chitin-like IR spectra and were thus assigned to particles of animal origin.

Within the differentially shaped particles, only six fragments (i.e. 1.4 %) of the analyzed particles, displayed a plastic polymer spectrum. Three fragment-shaped particles in the size between 440–1200 μm were polystyrene and another three fragments between 180–450 μm were polyethylene (Fig. 8.6). With the help of chemical imaging the localization of characteristic spectra within the measured area could be visualized accurately and the microplastic particles could be identified (Fig. 8.7).

8.5.3 Summary

As a consequence of the applied sample extraction and purification techniques, it was assumed that the provided samples should have comprised mainly microplastic fragments and other organic material of low density. However, plastic fragments were exceptionally rare as they made up only 1.4 % of all analyzed particles. Since we manually screened the bright-field microscopic images for areas with heterogeneous particle appearance and measured at these locations, it can be assumed, that in total, the abundance of microplastics in the samples was

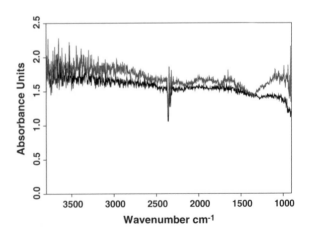

Fig. 8.5 IR spectra of bitumen-like black particles that showed high absorbance when measured by FPA-based micro-FTIR spectroscopy

Fig. 8.6 Examples of the IR spectra of the six microplastic particles found in the sediment samples, which were measured by FPA-based micro-FTIR spectroscopy: upper panel polystyrene, lower panel polyethylene (FPA-measured spectra in *red*, ATR reference spectra in *blue*)

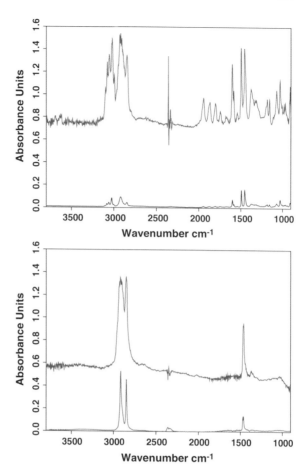

even lower. Surprisingly, the majority of particles could be clearly identified as quartz particles, although these should have been excluded via the density separation step. While it remains to be clarified why the samples contained such large amounts of sand grains, the case study proves the absolute necessity of spectroscopic (e.g. FTIR) measurements for microplastics analysis. In this context FPA-based micro-FTIR spectroscopy proved to be a very promising technique for verifying polymer origin of microplastic particles and thus should be mandatorily included in future monitoring programs on microplastics after SOPs for the purification and measurement approaches have been established.

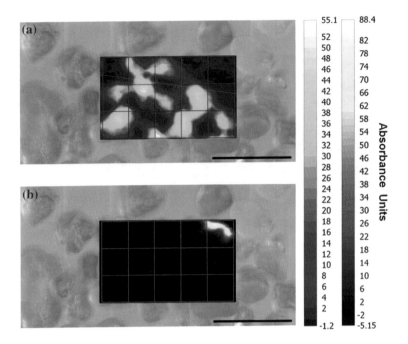

Fig. 8.7 Chemical imaging of an integration of typical spectral bands of **a** quartz (1931–1832 cm^{-1}), **b** polyethylene (2981–2780 cm^{-1}) in the sample Wu_Ku_Eu 4 analyzed by FPA-based micro-FTIR spectroscopy. The color scale represents the intensity of the chosen spectral bands according to peak area (*bluish color scale* intensity of the peak 1,931–1,832 cm^{-1}; *reddish color scale* intensity of the peak 2,981–2,780 cm^{-1}) and clearly shows quartz particles in *light blue* (**a**) and a polyethylene particle in *orange* (**b**). Scalebars: 500 µm

8.6 Conclusions

Currently, the reliability and comparability of data on marine microplastic concentrations is hampered by the huge variety of different methodologies applied, which lead to the generation of data of extremely different quality (Hidalgo-Ruz et al. 2012). Thus, one basic aim must be the standardization of methodologies for identification and quantification of microplastics in the marine environment and the subsequent formulation of standard operating procedures (SOPs). This involves the whole cycle of the assessment of microplastics in environmental samples from sampling procedures to sample purification and identification of microplastic particles.

More research is needed in the field of representative sampling designs, especially when it comes to sampling sediments on beaches. Although sampling in itself is no challenge, choosing the appropriate sampling locations and number of replicates to representatively describe the plastic contamination of a highly dynamic environment such as a beach is indeed challenging. The suggestions by the TSG-ML (Hanke et al. 2013) are first steps towards a standardization of beach sampling within the microplastics monitoring programs of the member states of the EU.

Depending on the mesh size used, plankton or neuston nets allow for adequate sampling of a certain size class of microplastics from the water column or from surface waters. However, since nets can easily clog, especially the biologically important fraction of particles <100 µm is often severely underrepresented. Sequential size fractionated filtering of large volumes of water is a very promising technique that could provide a remedy for this problem although the applicability has yet to be proven in the field. Techniques for sampling microplastics in biological material are as variable as the biota themselves and thus a general standardization cannot be achieved. For the monitoring of microplastics in single species, such as the northern fulmar, long-term observations could lead to the establishment of standardized sampling procedures.

Sample preparation should involve a size fractionation of particles larger and smaller than 500 µm as suggested by Hidalgo-Ruz et al. (2012) (alternatively 1000 µm, TSG-ML, Hanke et al. 2013) to enable comparability of data from different studies including studies that focus on particles <500 µm. Currently, no SOP exists for extraction solutions and techniques used to extract microplastic particles from samples. The use of extraction solutions of lower densities (e.g. NaCl) results in underestimation of microplastics of higher densities. Zinc chloride is a relatively cheap and recyclable solution that can be used to produce high-density solutions, which extract the whole spectrum of common user plastics. Furthermore, we suggest using separation techniques with high recovery rates in the whole size spectrum of microplastic particles at low temporal and personal effort such as the Microplastic Sediment Separator (Imhof et al. 2012).

Identification by spectroscopy is essential, especially for particles <500 µm. The quality of the spectroscopic results relies strongly on effective sample purification. Purifying steps with agents that negatively affect certain plastic polymers (e.g. strong acidic or alkaline solutions) should be avoided to conserve the whole spectrum of plastic contamination within an environmental sample. Gentle approaches such as sequential enzymatic digestion have been successfully applied by the authors and provide efficient purification for subsequent spectroscopic analysis. The contamination of microplastic samples during the whole assessment procedure from sampling to analysis is a highly urgent topic that needs to be addressed by standardised contamination prevention techniques. In this context, the avoidance of plastic material as far as possible as well as the use of control samples should be implemented as key elements of microplastics investigations.

An example of recent research by the authors showed that the commonly applied visual inspection of samples alone is insufficient to identify microplastic particles in environmental samples. This is especially the case for particles <500 µm, which have to be concentrated on filters for analysis. In this context, a spectroscopic verification of microplastic particles by micro-Raman or micro-FTIR spectroscopy is essential and has also been used successfully for analyzing marine samples. Spectroscopic techniques create an added value as they also provide information on the polymer composition. Although FTIR spectroscopy has been used more frequently in microplastics research (Hidalgo-Ruz et al. 2012) Raman spectroscopy is of equal value for analyzing microplastic samples.

By combining spectroscopic techniques with fast area-resolved measurements, with e.g. FPA detectors, and chemical imaging, it is possible to scan whole sample filters in a fraction of time compared to chemical mapping with single detectors (Harrison et al. 2012). However, SOPs for the use of this highly promising novel techniques are not accessible at present but will soon be published (Löder et al. 2015). The standardized use of spectroscopy in microplastic research finally enables the generation of valid data on concentrations, particle size distribution, involved polymers and distribution among different marine habitats and in marine biota. This is essential to guarantee comparability of different data sets. Furthermore, the data generated will serve as basis for the design of realistic laboratory experiments on the impact of microplastics on marine biota. These steps are crucial for a reliable evaluation and the assessment of potential impacts and risks of microplastics in the world's oceans.

Acknowledgments   The authors would like to thank the German Federal Ministry of Education and Research (BMBF) and the Alfred-Wegener-Institut Helmholtz-Zentrum für Polar- und Meeresforschung (AWI) for funding the project MICROPLAST. Furthermore, we would like to thank the NLWKN for preparation and providing the microplastics samples for the case study and Dr. Sonja Oberbeckmann for providing the manta trawl picture. This is publication number 37791 of the Alfred-Wegener-Institut Helmholtz-Zentrum für Polar- und Meeresforschung.

References

Andrady, A. L. (2011). Microplastics in the marine environment. *Marine Pollution Bulletin, 62*(8), 1596–1605.

Bakir, A., Rowland, S. J., & Thompson, R. C. (2012). Competitive sorption of persistent organic pollutants onto microplastics in the marine environment. *Marine Pollution Bulletin, 64*(12), 2782–2789.

Barnes, D. K., Galgani, F., Thompson, R. C., & Barlaz, M. (2009). Accumulation and fragmentation of plastic debris in global environments. *Philosophical Transactions of the Royal Society of London. Series B, 364*(1526), 1985–1998.

Besseling, E., Wegner, A., Foekema, E. M., van den Heuvel-Greve, M. J., & Koelmans, A. A. (2013). Effects of microplastic on fitness and PCB bioaccumulation by the lugworm *Arenicola marina* (L.). *Environmental Science and Technology, 47*(1), 593–600.

Boerger, C. M., Lattin, G. L., Moore, S. L., & Moore, C. J. (2010). Plastic ingestion by planktivorous fishes in the North Pacific Central Gyre. *Marine Pollution Bulletin, 60*(12), 2275–2278.

Bond, A. L., & Lavers, J. L. (2013). Effectiveness of emetics to study plastic ingestion by Leach's storm-petrels (*Oceanodroma leucorhoa*). *Marine Pollution Bulletin, 70*(1–2), 171–175.

Browne, M. A., Dissanayake, A., Galloway, T. S., Lowe, D. M., & Thompson, R. C. (2008). Ingested microscopic plastic translocates to the circulatory system of the mussel, *Mytilus edulis* (L.). *Environmental Science and Technology, 42*(13), 5026–5031.

Browne, M. A., Galloway, T. S., & Thompson, R. C. (2010). Spatial patterns of plastic debris along estuarine shorelines. *Environmental Science and Technology, 44*(9), 3404–3409.

Browne, M. A., Crump, P., Niven, S. J., Teuten, E., Tonkin, A., Galloway, T., et al. (2011). Accumulation of microplastic on shorelines worldwide: Sources and sinks. *Environmental Science and Technology, 45*(21), 9175–9179.

Carpenter, E. J., & Smith, K. L, Jr. (1972). Plastics on the Sargasso Sea surface. *Science, 175*(4027), 1240–1241.

Claessens, M., De Meester, S., Van Landuyt, L., De Clerck, K., & Janssen, C. R. (2011). Occurrence and distribution of microplastics in marine sediments along the Belgian coast. *Marine Pollution Bulletin, 62*(10), 2199–2204.

Claessens, M., Van Cauwenberghe, L., Vandegehuchte, M. B., & Janssen, C. R. (2013). New techniques for the detection of microplastics in sediments and field collected organisms. *Marine Pollution Bulletin, 70*(1–2), 227–233.

Cole, M., Lindeque, P., Fileman, E., Halsband, C., Goodhead, R., Moger, J., et al. (2013). Microplastic ingestion by zooplankton. *Environmental Science and Technology, 47*(12), 6646–6655.

Cole, M., Webb, H., Lindeque, P. K., Fileman, E. S., Halsband, C., & Galloway, T. S. (2014). Isolation of microplastics in biota-rich seawater samples and marine organisms. *Scientific Reports, 4*, 4528.

Colton, J. B., Knapp, F. D., & Burns, B. R. (1974). Plastic particles in surface waters of the northwestern Atlantic. *Science, 185*(4150), 491–497.

Cooper, D. A., & Corcoran, P. L. (2010). Effects of mechanical and chemical processes on the degradation of plastic beach debris on the island of Kauai, Hawaii. *Marine Pollution Bulletin, 60*(5), 650–654.

Corcoran, P. L., Biesinger, M. C., & Grifi, M. (2009). Plastics and beaches: A degrading relationship. *Marine Pollution Bulletin, 58*(1), 80–84.

Dekiff, J. H., Remy, D., Klasmeier, J., & Fries, E. (2014). Occurrence and spatial distribution of microplastics in sediments from Norderney. *Environmental Pollution, 186*, 248–256.

Doyle, M. J., Watson, W., Bowlin, N. M., & Sheavly, S. B. (2011). Plastic particles in coastal pelagic ecosystems of the Northeast Pacific Ocean. *Marine Environmental Research, 71*(1), 41–52.

Dubaish, F., & Liebezeit, G. (2013). Suspended microplastics and black carbon particles in the Jade system, Southern North Sea. *Water, Air, and Soil pollution, 224*(2), 1–8.

Ebbesmeyer, C. C., & Ingraham, W. J. (1994). Pacific toy spill fuels ocean current pathways research. *EOS, Transactions American Geophysical Union, 75*(37), 425–430.

Engler, R. E. (2012). The complex interaction between marine debris and toxic chemicals in the ocean. *Environmental Science and Technology, 46*(22), 12302–12315.

Eriksen, M., Mason, S., Wilson, S., Box, C., Zellers, A., Edwards, W., et al. (2013a). Microplastic pollution in the surface waters of the Laurentian Great Lakes. *Marine Pollution Bulletin, 77*(1–2), 177–182.

Eriksen, M., Maximenko, N., Thiel, M., Cummins, A., Lattin, G., Wilson, S., et al. (2013b). Plastic pollution in the South Pacific subtropical gyre. *Marine Pollution Bulletin, 68*(1–2), 71–76.

Foekema, E. M., De Gruijter, C., Mergia, M. T., van Franeker, J. A., Murk, A. J., & Koelmans, A. A. (2013). Plastic in North Sea fish. *Environmental Science and Technology, 47*(15), 8818–8824.

Frias, J., Sobral, P., & Ferreira, A. M. (2010). Organic pollutants in microplastics from two beaches of the Portuguese coast. *Marine Pollution Bulletin, 60*(11), 1988–1992.

Fries, E., Dekiff, J. H., Willmeyer, J., Nuelle, M. T., Ebert, M., & Remy, D. (2013). Identification of polymer types and additives in marine microplastic particles using pyrolysis-GC/MS and scanning electron microscopy. *Environmental Science-Processes & Impacts, 15*(10), 1949–1956.

Gouin, T., Roche, N., Lohmann, R., & Hodges, G. (2011). A thermodynamic approach for assessing the environmental exposure of chemicals absorbed to microplastic. *Environmental Science and Technology, 45*(4), 1466–1472.

Gregory, M. R. (1983). Virgin plastic granules on some beaches of Eastern Canada and Bermuda. *Marine Environmental Research, 10*(2), 73–92.

Hanke, G., Galgani, F., Werner, S., Oosterbaan, L., Nilsson, P., Fleet, D., et al. (2013). MSFD GES technical subgroup on marine litter. Guidance on monitoring of marine litter in European Seas. Luxembourg: Joint Research Centre–Institute for Environment and Sustainability, Publications Office of the European Union.

Harrison, J. P., Sapp, M., Schratzberger, M., & Osborn, A. M. (2011). Interactions between microorganisms and marine microplastics: A call for research. *Marine Technology Society Journal, 45*(2), 12–20.

Harrison, J. P., Ojeda, J. J., & Romero-Gonzalez, M. E. (2012). The applicability of reflectance micro-Fourier-transform infrared spectroscopy for the detection of synthetic microplastics in marine sediments. *Science of the Total Environment, 416*, 455–463.

Hidalgo-Ruz, V., & Thiel, M. (2013). Distribution and abundance of small plastic debris on beaches in the SE Pacific (Chile): A study supported by a citizen science project. *Marine Environmental Research, 87–88*, 12–18.

Hidalgo-Ruz, V., Gutow, L., Thompson, R. C., & Thiel, M. (2012). Microplastics in the marine environment: A review of the methods used for identification and quantification. *Environmental Science and Technology, 46*(6), 3060–3075.

Imhof, H. K., Schmid, J., Niessner, R., Ivleva, N. P., & Laforsch, C. (2012). A novel, highly efficient method for the separation and quantification of plastic particles in sediments of aquatic environments. *Limnology and Oceanography-Methods, 10*, 524–537.

Imhof, H. K., Ivleva, N. P., Schmid, J., Niessner, R., & Laforsch, C. (2013). Contamination of beach sediments of a subalpine lake with microplastic particles. *Current Biology, 23*(19), R867–R868.

Ivar do Sul, J. A., & Costa, M. F. (2013). The present and future of microplastic pollution in the marine environment. *Environmental Pollution, 185*, 352–364.

Koelmans, A. A. (2015). Modeling the role of microplastics in bioaccumulation of organic chemicals to marine aquatic organisms. Critical review. In M. Bergmann, L. Gutow & M. Klages (Eds.), *Marine anthropogenic litter* (pp. 313–328). Springer, Berlin.

Koelmans, A. A., Besseling, E., Wegner, A., & Foekema, E. M. (2013). Plastic as a carrier of POPs to aquatic organisms: a model analysis. *Environmental Science and Technology, 47*(14), 7812–7820.

Kühn, S., Bravo Rebolledo, E. L., & van Franeker, J. A. (2015). Deleterious effects of litter on marine life. In M. Bergmann, L. Gutow & M. Klages (Eds.), *Marine anthropogenic litter* (pp. 75–116). Berlin: Springer.

Levin, I. W., & Bhargava, R. (2005). Fourier transform infrared vibrational spectroscopic imaging: Integrating microscopy and molecular recognition. *Annual Review of Physical Chemistry, 56*(1), 429–474.

Liebezeit, G., & Dubaish, F. (2012). Microplastics in beaches of the East Frisian Islands Spiekeroog and Kachelotplate. *Bulletin of Environmental Contamination and Toxicology, 89*(1), 213–217.

Löder, M.G.J., Kuczera, M., Mintenig, S., Lorenz, C., & Gerdts, G. (2015). FPA-based micro-FTIR imaging for the analysis of microplastics in environmental samples. *Environmental Chemistry*, in press.

Lusher, A. (2015). Microplastics in the marine environment: distribution, interactions and effects. In M. Bergmann, L. Gutow & M. Klages (Eds.), *Marine anthropogenic litter* (pp. 245–308). Springer, Berlin.

Mato, Y., Isobe, T., Takada, H., Kanehiro, H., Ohtake, C., & Kaminuma, T. (2001). Plastic resin pellets as a transport medium for toxic chemicals in the marine environment. *Environmental Science and Technology, 35*(2), 318–324.

McDermid, K. J., & McMullen, T. L. (2004). Quantitative analysis of small-plastic debris on beaches in the Hawaiian archipelago. *Marine Pollution Bulletin, 48*(7–8), 790–794.

Moore, C. J., Moore, S. L., Weisberg, S. B., Lattin, G. L., & Zellers, A. F. (2002). A comparison of neustonic plastic and zooplankton abundance in Southern California's coastal waters. *Marine Pollution Bulletin, 44*(10), 1035–1038.

Morét-Ferguson, S., Law, K. L., Proskurowski, G., Murphy, E. K., Peacock, E. E., & Reddy, C. M. (2010). The size, mass, and composition of plastic debris in the Western North Atlantic Ocean. *Marine Pollution Bulletin, 60*(10), 1873–1878.

Murray, F., & Cowie, P. R. (2011). Plastic contamination in the decapod crustacean *Nephrops norvegicus* (Linnaeus, 1758). *Marine Pollution Bulletin, 62*(6), 1207–1217.

Ng, K. L., & Obbard, J. P. (2006). Prevalence of microplastics in Singapore's coastal marine environment. *Marine Pollution Bulletin, 52*(7), 761–767.

Norén, F. (2007). Small plastic particles in coastal Swedish waters. Lysekil, Sweden: KIMO Sweden, N-Research.

Norén, F., & Naustvoll, L.-J. (2010). Survey of microscopic anthropogenic particles in Skagerrak. *Klima- og forurensningsdirektoratet TA, 2779–2011*, 1–20.

Nuelle, M.-T., Dekiff, J. H., Remy, D., & Fries, E. (2014). A new analytical approach for monitoring microplastics in marine sediments. *Environmental Pollution, 184*, 161–169.

Reddy, M. S., Basha, S., Adimurthy, S., & Ramachandraiah, G. (2006). Description of the small plastics fragments in marine sediments along the Alang-Sosiya ship-breaking yard, India. *Estuarine Coastal and Shelf Science, 68*(3–4), 656–660.

Rios, L. M., Moore, C., & Jones, P. R. (2007). Persistent organic pollutants carried by synthetic polymers in the ocean environment. *Marine Pollution Bulletin, 54*(8), 1230–1237.

Rochman, C. M., Hoh, E., Kurobe, T., & Teh, S. J. (2013). Ingested plastic transfers hazardous chemicals to fish and induces hepatic stress. *Scientific Reports, 3*, 3263.

Teuten, E. L., Saquing, J. M., Knappe, D. R., Barlaz, M. A., Jonsson, S., Bjorn, A., et al. (2009). Transport and release of chemicals from plastics to the environment and to wildlife. *Philosophical Transactions of the Royal Society of London B, 364*(1526), 2027–2045.

Thompson, R. C. (2015). Microplastics in the marine environment: Sources, consequences and solutions. In M. Bergmann, L. Gutow & M. Klages (Eds.), *Marine anthropogenic llitter* (pp. 185–200). Springer, Berlin.

Thompson, R. C., Olsen, Y., Mitchell, R. P., Davis, A., Rowland, S. J., John, A. W. G., et al. (2004). Lost at sea: Where is all the plastic? *Science, 304*(5672), 838.

Ugolini, A., Ungherese, G., Ciofini, M., Lapucci, A., & Camaiti, M. (2013). Microplastic debris in sandhoppers. *Estuarine, Coastal and Shelf Science, 129*, 19–22.

Van Cauwenberghe, L., Vanreusel, A., Mees, J., & Janssen, C. R. (2013). Microplastic pollution in deep-sea sediments. *Environmental Pollution, 182*, 495–499.

Vianello, A., Boldrin, A., Guerriero, P., Moschino, V., Rella, R., Sturaro, A., et al. (2013). Microplastic particles in sediments of Lagoon of Venice, Italy: First observations on occurrence, spatial patterns and identification. *Estuarine, Coastal and Shelf Science, 130*, 54–61.

Wright, S. L., Thompson, R. C., & Galloway, T. S. (2013). The physical impacts of microplastics on marine organisms: A review. *Environmental Pollution, 178*, 483–492.

Zarfl, C., Fleet, D., Fries, E., Galgani, F., Gerdts, G., Hanke, G., et al. (2011). Microplastics in oceans. *Marine Pollution Bulletin, 62*(8), 1589–1591.

Zettler, E. R., Mincer, T. J., & Amaral-Zettler, L. A. (2013). Life in the 'Plastisphere': Microbial communities on plastic marine debris. *Environmental Science and Technology, 47*(13), 7137–7146.

Chapter 9
Sources and Pathways of Microplastics to Habitats

Mark A. Browne

Abstract Identifying and eliminating the sources of microplastic to habitats is crucial to reducing the social, environmental and economic impacts of this form of debris. Although eliminating sources of pollution is a fundamental component of environmental policy in the U.S.A. and Europe, the sources of microplastic and their pathways into habitats remain poorly understood compared to other persistent, bioaccumulative and/or toxic substances (i.e. priority pollutants; EPA in U.S. Environmental Protection Agency 2010–2014 Pollution Prevention (P2) Program Strategic Plan. Washington, USA, pp. 1–34, 2010; EU in Official J Eur Union L334:17–119, 2010). This chapter reviews our understanding of sources and pathways of microplastic, appraises terminology, and outlines future directions for meaningfully integrating research, managerial actions and policy to understand and reduce the infiltration of microplastic to habitats.

Keywords Hypothesis · Micrometre · Emissions · Sewage · Storm-water · Textile · Exfoliants

M.A. Browne (✉)
National Center for Ecological Analysis and Synthesis, University of California, Santa Barbara, 735 State Street, Suite 300, Santa Barbara, CA 93101-3351, USA
e-mail: markanthonybrowne@gmail.com

M.A. Browne
Evolution and Ecology Research Centre, School of Biological, Earth and Environmental Sciences, University of New South Wales, Sydney, NSW 2052, Australia

9.1 Defining Sources and Pathways of Microplastic

Since the first review of microplastic (Browne et al. 2007), a number of terms have been used to describe and categorize sources of microplastic. Some authors have used the terms "primary" and "secondary" to distinguish between sources of microplastic, in which they borrow terminology from atmospheric sciences (Arthur et al. 2009; Cole et al. 2011). In these cases, "primary sources" are those in which microplastic is intentionally produced through extrusion or grinding, either as precursors to other products (e.g. plastic pellets; Costa et al. 2010) or for direct use (e.g. abrasives in cleaning products or roto-milling), whilst "secondary sources" of microplastics are those formed in the environment from the fragmentation of larger plastic material into ever-smaller pieces (Arthur et al. 2009; Cole et al. 2011). Using similar ideas, Andrady (2011) described runoff as a "direct source" (sewage or storm water) whilst fragmentation of existing plastic debris was described as "indirect source" of microplastic to the environment.

Although using adjectives to categorize sources may be helpful, these terms introduce jargon without clearly identifying the actual sources and conflate sources with the pathway by which microplastic enter habitats, which may in turn confuse scientists, public, industry and government. In this chapter, I argue that these problems can be overcome if we choose alternative terms for *sources* that identify the place, person, company, or product where the microplastic originates and use separate terms to describe the *pathways* of microplastic from its source to a habitat.

Based upon our current understanding there are four types of sources (i) larger plastic litter, (ii) cleaning products (Zitko and Hanlon 1991; Gregory 1996; Derraik 2002); (iii) medicines; and (iv) textiles. For the latter, I have chosen to use a global case-study to illustrate how one can gain a more meaningful and scientific understanding of the sources and pathways of microplastic through developing better programs of research and monitoring that integrate advances in forensics (e.g. vibrational spectroscopy to identify the shape and type of microplastic), logic and experimental design (i.e. making observations, developing explanatory models, testing explicit hypotheses about composition and spatial patterns; Underwood 1997) and statistics. Throughout the chapter, I have chosen to define microplastic as micrometre-sized particles of plastic because this is consistent with previous work on this topic (Browne et al. 2007, 2008, 2010, 2011, 2013; Costa et al. 2009; Claessens et al. 2011; Rochman et al. 2013; Van Cauwenberghe 2013) and the globally ratified use of the prefix "micro" for measures of length under the International System of Units. Other authors have chosen to ignore the International System of Units definition of "micro" and have instead chosen to use <5 mm to define microplastic, a philosophical discussion about which people should use is beyond the scope of this review.

9.2 Larger Plastic Litter

Sources. Larger plastic debris originates from maritime activities including shipping, fishing (e.g. Merrell Jr 1980, Ramirez-Llodra et al. 2013; Galgani et al. 2015), recreation and offshore industries (e.g. oil, gas). These sources are, however, likely to be much smaller than terrestrial sources. Whatever the source, larger plastic litter (including millimetre-sized pre-production pellets) is likely to be an important source of microplastic. Irregularly shaped fragments are abundant in intertidal and oceanic habitats with the size-frequency of plastic debris skewed towards smaller debris. This suggests that microplastic can originate from the fragmentation of larger objects that causes ever smaller pieces of plastic to be present in the environment (Browne et al. 2007, 2010; Morét-Ferguson et al. 2010; Collignon et al. 2012; Fig. 9.1a–c).

Pathways. Fragmentation is the pathway, by which plastic debris breaks into smaller pieces, which is distinct from the more subtle processes of degradation that reduces the molecular mass of plastic debris (Andrady 2011, 2015). These processes occur through the action of light (photolysis), heat and oxygen (thermal-oxidation), water (hydrolysis), organisms (see review by Andrady (2011) for more details about weathering experiments with pieces of plastic >1 mm) and physical abrasion by particles of sediment. Laboratory and field experiments are required to determine the relative importance of these processes in generating the sizes and shapes of micrometre-sized plastic that we find in habitats. One such laboratory experiment by Davidson (2012) showed that each time a single isopod (*Sphaeroma quoianum*) burrows into a floating dock made of expanded polystyrene they can produce and release between 4900 and 6300 micrometre-sized fragments of polystyrene. There is also a modest literature on plastic degradation in marine habitats (Andrady and Pegram 1989; Andrady 2003; Gregory and Andrady 2003; Corcoran et al. 2009; Cooper and Corcoran 2010; O'Brine and Thompson 2010). But the rigour of these

Fig. 9.1 Fragments (a–b) and sizes (c) of plastic debris found in the Tamar Estuary. Reprinted adapted with permission from Browne et al. (2010). © 2010 American Chemical Society

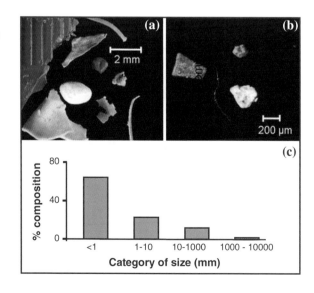

studies varies with some lacking many of the ecological developments in designing, implementing and analysing manipulative field experiments (Underwood 1997). So far, no one has experimentally deployed micrometre-sized plastic (or millimetre-sized pellets) in habitats to determine the rates at which it fragments into smaller pieces. Thus, the role of pellets as major sources of microplastic is unclear.

9.3 Cleaning Products

Sources. Another source of microplastic is from industrial and domestic cleaning products that use microplastic as an abrasive scrubber (Browne et al. 2007). For instance, surfaces of buildings, machinery and boats can be cleaned and prepared (e.g. smoothed, roughened, shaped) using 'media blasting', where small plastics (e.g. polystyrene, acrylic, polyester, poly-allyl-diglycol-carbonate, urea-, melamine- and phenol-formaldehyde; 0.25–1.7 mm; DOD 1992) and other types of granules (e.g. sand) are propelled onto a surface using a centrifugal wheel or pressurized fluid/gas (Wolbach and McDonald 1987; Abbott 1992; Gregory 1996; Neulicht and Shular 1997; Anonymous 1998). Although 'media blasting' has been suspected of being a source of microplastic to habitats there has been no scientific work to (i) characterize the number of industries using this technique, (ii) the size, shape and amount of microplastic used in the process of cleaning and (iii) the quantity of particles emitted into, or found within, the environment through this source.

More work has been done for microplastics used as physical abrasives in domestic products. Fendall and Sewell (2009) qualitatively showed that the size and shape of microplastic in such products varies (Fig. 9.2). By examining four different facial cleansers with labels that indicated they contained particles of polyethylene, they found that the size of the particles ranged from 4.1 to 1240 µm in diameter, and consisted of uniform spheres, ellipses, rods, fibres and granules (Fig. 9.2). For granules this presents a problem because it will be very difficult to differentiate whether they come from cleaning products or from the fragmentation of larger

Fig. 9.2 Microplastic (polyethylene) fragments found in facial cleansers (Photo: M. Sewell, University of Auckland)

articles of plastic debris. Using vibrational spectrometry Zitko and Hanlon (1991) found that 47 % of the mass of the contents of a single bottle of skin cleanser was made up of irregular fragments of polystyrene (100–200 μm). A separate study that used vibrational spectrometry showed replicate formulations of hand cleansers between 0.2 and 4 % of their mass made up of polyethylene, whilst for facial cleansers it was 2–3 % (Gregory 1996), though it is important to note that this study did not report particle numbers <63 μm in size, which may account for the smaller amounts recorded. Gouin et al. (2011) estimated the emission of microplastic from cleaning products in the U.S. by combining estimates of sales figures and assuming proportions of polyethylene were 10 % by volume. From this, the authors calculated that each year the U.S. could be emitting 263 t of micrometre-sized fragments of polyethylene from domestic cleaning products. Given that the type of polymer (e.g. polyethylene, polystyrene) and proportions of microplastic can vary from 0.2 to 47 %, it seems that more work is needed to test individual products and different batches so that we can provide precise, accurate and ground-truthed estimates of microplastic emissions from cleaning products.

Pathways. Microplastics used in cleaning products are thought to transfer to habitats through sewage and storm water (Fig. 9.3). The quantities of microplastic, however, in water or sediment from habitats, sewage or storm water are unknown because they are interspersed with large concentrations of organic matter, and because it is difficult to distinguish uniform spheres, ellipses and granules with a biofilm from natural particles. Some of these problems may be overcome with the application of chemical techniques to remove organic matter (Claessens et al. 2013) and vibrational spectroscopes that can map microplastic in environmental samples (Harrison et al. 2012).

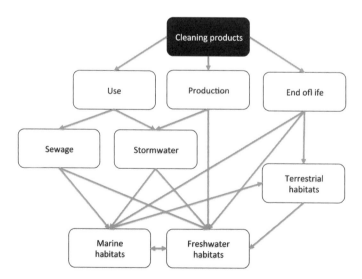

Fig. 9.3 Sources and pathways of microplastic from cleaning products into habitats. *Gray arrows* indicate hypothesized pathways. There are no black arrows because there is currently no research showing evidence of these pathways

9.4 Medicines

Sources. Ingestible and inhalable medicines containing microplastic are used to deliver drugs to the organs of humans and farmed animals (terrestrial and aquatic) because the microplastic can translocate from their lungs or guts into their circulatory system (Thanoo et al. 1993; Dalmon et al. 1995; Curley et al. 1996; Hussain et al. 2001; Matsusaki et al. 2001; Wen et al. 2003; Kockisch et al. 2003; Corbanie et al. 2006). There has, however, been no work to synthesize information about the different types of polymers and sizes of microplastic used in medicines because this information does not seem to be readily available.

Pathways. Like microplastics in cleaning products, microplastics from medicines are likely to transfer to habitats through sewage and storm water, or more directly through treating diseased animals in aquaculture and farming. There is, however, no quantitative work evaluating how much plastic is taken up by animals and excreted compared to those retained in tissues. As such, the quantities of medical microplastic in water or sediment sampled from habitats, sewage or storm water are unknown. Research is needed to synthesize a complete inventory of these polymers so that samples from humans, sewage, storm water, wildlife and habitats can be tested for the presence of these polymers. Some of the polymers used are thought to be biodegradable (Matsusaki et al. 2001), whist others can be composed of more durable polymers such as polycarbonate and polystyrene (Thanoo et al. 1993; Dunn et al. 1994). The rates and mechanisms of degradation inside human tissues may not be the same as in wildlife or habitats. Moreover, just because a polymer degrades, does not necessarily mean that the resulting metabolites are not toxic themselves: So, research is needed to determine how safe these particles are in humans, wildlife and habitats.

9.5 Textiles

Many people have attempted to examine the sources of microplastic but a lack of a hypothesis-driven framework has meant that sources and pathways are poorly understood. More useful understanding about the sources and pathways of microplastic to habitats comes from work done on fibres that originate from textiles and clothing (Browne et al. 2011). The following case study is provided to illustrate how one can understand better the sources and pathways of microplastic (Browne et al. 2011). The work was done in four phases by examining microplastic in (i) sediment from sandy shores worldwide; (ii) sediment from replicated sub-tidal areas where sewage sludge had, and had not, been discharged; (iii) effluent from replicate treatment plants; and finally; (iv) effluent from manipulative experiments involving washing machines.

(i) *A global program to sample sediment from sandy shores.* Between 2004 and 2007 samples of sediment were collected from sandy beaches in Australia, Oman, United Arab Emirates, Chile, Philippines, Azores, USA, South Africa,

Mozambique and the U.K. During collection (and in all work), cotton clothing was worn rather than synthetic items to prevent samples being contaminated by plastic fibres. Samples were collected by working down-wind to the
particular part of the highest strandline deposited by the previous tide. Using
established techniques, sediment was sampled to a depth of 1 cm and microplastic and sediment was quantified using established techniques (Browne
et al. 2010, 2011). Two explanatory models for pathways of microplastic in
habitats were put forward to explain spatial patterns of microplastic.

If spatial patterns of microplastic result from the transportation of natural particulates by currents of water (*Model 1*), we expected shores that accumulate smaller-
sized particles of sediment would accumulate more microplastic (*Hypothesis 1*).
Alternatively, spatial patterns may be influenced by sources of microplastic
(*Model 2*). Over the last 50 years the global population density of humans had
increased by 250 % from 19 to 48 individuals km^{-2} (UN 2008), and during this
time the abundance of microplastic had increased in pelagic habitats (Thompson
et al. 2004). Previous observations had suggested that there was a greater abundance of larger items of debris along shorelines adjacent to densely populated
areas (Barnes 2005). This led to the prediction that there would be more microplastic along shorelines adjacent to densely populated areas (*Hypothesis 2*). The
work showed that microplastic contaminated all 18 shores examined (Fig. 9.4)
with more microplastic in sediments collected from densely populated areas (linear regression, $F_{1,16} = 8.36$, $P < 0.05$, n = 18, $r^2 = 0.34$; Fig. 9.5), but there was
no relationship with the quantity of smaller-sized particles of sediment. Thus,
there was evidence to support *Model 2* but not *Model 1*.

To examine the pathway of this microplastic onto shorelines, forensic analysis was used to gather crucial observations about the shapes and types of polymers that made up the microplastic. This showed that the microplastic was mostly
made up of synthetic fibres that consisted of polyester (56 %), acrylic (23 %),

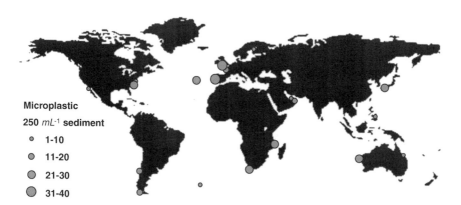

Fig. 9.4 Spatial extent of microplastic in sediments from 18 sandy shores. The size of filled-cir-
cles represents number of microplastic particles found. Reprinted adapted with permission from
Browne et al. (2011). © 2011 American Chemical Society

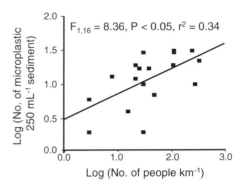

Fig. 9.5 Relationship between population-density and number of microplastic particles in sediment from sandy beaches. Reprinted adapted with permission from Browne et al. (2011). © 2011 American Chemical Society

polypropylene (7 %), polyethylene (6 %) and polyamide (3 %). Previous observations had shown that coastal habitats receive millions of tonnes of sewage each year (CEFAS 1997) and that sewage can contain microplastic fibres (Habib et al. 1996; Zubris and Richards 2005) because although larger debris is removed during the treatment of sewage, filters are not specifically designed to retain microplastic. *Model 2* was therefore refined to include sewage as the pathway of synthetic fibres to marine habitats causing greater quantities of microplastic fibres in areas adjacent to densely populated areas (*Model 2.1*). The next step was then to test this model to determine whether there was evidence to support the model that the discharge of sewage was an important pathway of microplastic fibres into marine habitats.

(ii) *Comparison of sub-tidal areas where sewage sludge had been discharged with reference areas.* Previous observations from coastal habitats of the U.K. suggested that each year treatment plants discharge >11 km^3 of sewage effluent into coastal habitats (CEFAS 1997) and for nearly 30 years, a quarter of U.K. sewage sludge was dumped at 13 designated sub-tidal disposal-sites around the coast, until this stopped in 1998 through implementation of The Urban Waste Water Treatment Regulations 1994 (Fig. 9.6; CEFAS 1997; British Government 1994). Using replicate disposal-sites and reference-sites allowed us to test *Hypothesis 2.1* that sediments from disposal-sites would contain larger quantities of fibres in their sediments and that the shape and types of polymers that make up the microplastic would resemble those found on shores. For this, van Veen grabs deployed from boats collected replicate samples of sediment from two reference-sites and two disposal-sites in the English Channel and the North Sea (U.K.). Despite sewage not being added for more than a decade, disposal-sites still contained >250 % more fibres than reference sites (Fig. 9.7; ANOVA, $F_{1,16} = 4.50$, $P < 0.05$). Again the types of fibres were dominated by polyester (78 %) and acrylic (22 %).

During discussions with the sewage treatment authorities they explained that filters are not specifically designed to retain microplastic, which suggests that discharges of sewage effluent could also be a pathway of fibres from treatment plants

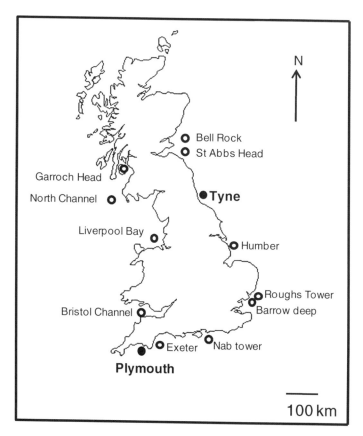

Fig. 9.6 Locations of U.K. sewage-sludge disposal sites (1970–1998) (CEFAS 1997). Plymouth (English Channel) and Tyne (North Sea) disposal sites presented as *filled black circles*, whereas the other 11 sites are with *open circles*

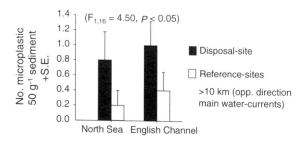

Fig. 9.7 Abundance of microplastic in sediments from disposal sites for sewage and reference sites at two locations in the U.K. (Tyne and Plymouth). Values expressed as mean ± S.E. Reprinted adapted with permission from Browne et al. (2011). © 2011 American Chemical Society

Fig. 9.8 Abundance of
microplastic in effluent
discharged from two separate
tertiary-level treatment
plants (West Hornsby and
Hornsby Heights, NSW,
Australia). Values expressed
as mean ± S.E

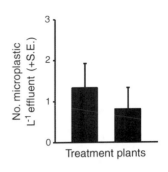

Fig. 9.9 Number of
polyester fibres discharged
into wastewater from using
washing machines with
blankets, fleeces, and shirts
(all polyester). Reprinted
adapted with permission from
Browne et al. (2011). © 2011
American Chemical Society

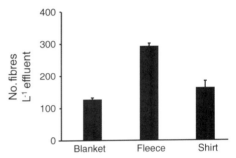

to habitats (*Model 2.2*). To test this model we examined the hypothesis that discharges of sewage effluent would contain similar proportions of polyester and acrylic fibres. As expected, polyester (67 %) and acrylic (17 %) fibres polyamide (16 %) dominated. These proportions of polyester and acrylic fibres resembled those contaminating intertidal and subtidal habitats (Fig. 9.8) suggesting that these microplastic fibres were mainly derived from sewage *via* washing-clothes rather than fragmentation or cleaning products. In recent years, the clothing industry has used textiles that contain >170 % more synthetics than natural fibres (e.g. cotton, wool, silk) and because proportions of fibres found in marine habitats and sewage resembled those used for textiles (78 % polyester, 9 % polyamide, 7 % polypropylene, 5 % acrylic; Oerlikon 2009) we counted the number of fibres discharged into wastewater from using clothes and garments.

(iii) *Experiments with washing machines.* Here, experimental work counted the number of fibres discharged into waste water from domestic washing machines used to launder clothing. To estimate the number of fibres entering wastewater from washing clothes and garments, three replicate washing machines were used with and without cloth (polyester blankets, fleeces, shirts). Effluent was filtered and microplastic counted. The experiments showed all garments released >100 fibres per litre of effluent, with >180 % more from fleeces (>1900 fibres per wash; Fig. 9.9), demonstrating that a large proportion of microplastic fibres found in marine habitats may be derived from sewage as a consequence of washing of clothes.

9.6 Outlook and Conclusion

In the future, contamination by microplastic is likely to continue to increase. Populations of humans are predicted to double in the next 40 years (UN 2008) and further concentrate in large coastal cities that will discharge larger volumes of sewage into marine habitats. The last case study provides a useful approach for identifying and quantifying sources and pathways of microplastic that should be extended to other sources, including medical and cleaning products, by screening sewage, storm water, habitats, wildlife and humans for the types of microplastic found in these products.

In parallel to this, I believe work is needed to reduce and eliminate sources and pathways of microplastic through (i) establishing and controlling inventories of materials; (ii) modifying the process of production by redesigning products so that they contain less hazardous substances; and (iii) using novel equipment and technology. This section now explores some of current opportunities for the public, scientists, engineers, industry and government to reduce sources and pathways of microplastic.

(i) *Establishing and controlling inventories that detail the use and emissions of microplastics in products.* Inventories are frequently used by European (EA 2012) and U.S. government agencies (EPA 2010, 2012) to control emissions of pollutants. An open-access online inventory is urgently required for textiles, medicines and cleaning products containing microplastic so that we have accurate information about emissions of microplastic during their production, use and disposal. This should include information about the use and emissions of microplastic, in terms of dimensions of size (i.e. minimum, maximum, median, mode and mean) shape, numbers, mass, types of polymers and sales figures. Because industry has, on occasion, been unwilling to provide this information when requested (Rosner 2008), this will probably require a change in policy and specific funding for representative sampling so that measures are accurate and precise (Figs. 9.10 and 9.11).

(ii) *Modifying the process of production and redesigning products so that they contain less hazardous substances.* Currently there are no published data on the effectiveness of modifying the process of production of products to reduce emissions of microplastic, since microplastic is not currently considered hazardous by policy-makers (Rochman et al. 2013). In response to advocacy from scientists and activists, several companies who make domestic cleaning products (e.g. Unilever, Johnson & Johnson) have agreed to replace microplastic with non-plastic particles. It is, however, unclear what alternatives they will use (Alumina; pumice; seeds of strawberries, blueberries, cranberry, evening primrose, grapes, kiwi or raspberry; stones of apricots, avocados, olives or peaches; peel of oranges or mandarin; castor or jojoba beads, shells of cocoa, coconuts, almonds or walnuts; coir; corn cob; salt; sugar; luffa; rice; macadamia nuts) or whether they will be more or less toxic to humans and wildlife, so scientific research is needed to find the most cost-effective alternative. Similar research is needed within the textile and clothing industry so that they produce cost-effective clothing that sheds fewer and less toxic fibres.

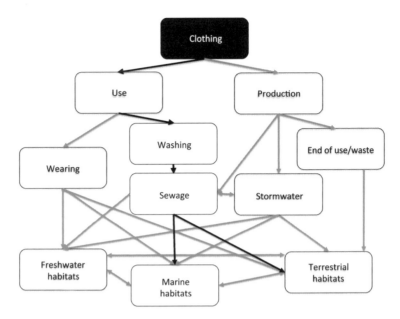

Fig. 9.10 Sources and pathways of fibres from textiles into habitats. *Gray arrows* indicate hypothesized pathways, *black arrows* indicate research that has been showing evidence of these pathways

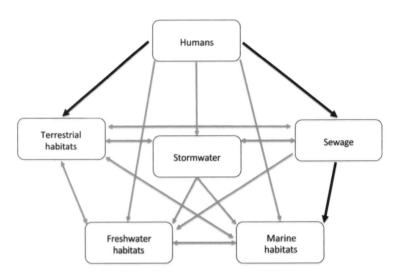

Fig. 9.11 Overview of sources and pathways of microplastic to habitats. *Gray arrows* indicate hypothesized pathways, *black arrows* indicate research that has been showing evidence of these pathways

These types of research are important to ensure decisions by policy-makers are based on robust scientific information as opposed to untested perceptions of the hazards of synthetic materials over those made from natural products.

(iii) *Novel equipment and technology to reduce pathways of microplastic.* Filters for washing machines are a promising prospect for reducing emissions of fibres to sewage (www.environmentalenhancements.com); however, their effectiveness in reducing emissions has not been tested as yet. Work is also needed to determine how effective different types of sewage treatment are at removing the different sizes, shapes and types of polymers that represent the microplastic found in sewage. However, unless the microplastic can be isolated from the sludge or effluent there are still likely to be problems because sewage is added to soil as a fertilizer. To identify the place, company or product where the microplastic originates requires government, industry and scientists to work together and share information. For this to happen there needs to be policies that (i) provide funded frameworks to measure (and if necessary manage) sources and pathways of microplastic into the environment; (ii) balance the needs of industries, society and the environment. In the U.S., 16 persistent, bioaccumulative and toxic chemicals (i.e. aldrin, benzoperylene, chlordane, heptachlor, hexachlorobenzene, isodrin, lead, mercury, methoxychlor, octachlorostyrene, pendimethalin, pentachlorobenzene, polychlorinated biphenyl, tetrabromobisphenol A, toxaphene and trifluralin) are controlled in this way using the Emergency Planning and Community Right-to-Know Act 1986. The Community Right-to-Know provisions help to increase public's knowledge and access to information on chemicals at individual facilities, their uses, and releases into the environment. Through this facilities are obliged to share information with government, scientists and public to improve chemical safety and protect public health and the environment. Similar requirements are required in Europe under article 5 of the Directive 2008/105/EC where member states are obliged to establish an inventory of emissions, discharges and losses of priority pollutants listed in part A of Annex I (EU 2012). For microplastic, several other solutions could also help. For instance, polymers could be designed with unique chemical fingerprints (that identify particular sites of production or use), which remain even after the polymer has been physically or chemically degraded during its time in the environment. Alternatively, prior to a product being licensed for sale on the market place, information on the composition of polymers (and additives) in commercial applications could be made available to environmental scientists so that the environmental sources and pathways of materials can be quantified and managed if needed. Whatever developments take place, hypothesis-driven frameworks are required to identify and falsify sources and pathways.

In conclusion, if we are to use terms to describe the sources of microplastic they should aim to identify the origin (e.g. larger plastic litter, cleaning products, medicines, textiles, etc.) and separate terms should be used for the pathways (e.g. storm water, sewage). As researchers use more integrated hypothesis-driven frameworks and the chemical methods and inventories improve, it may be possible to be more specific (e.g. the place, person or company).

Acknowledgments I am grateful to Mary Sewell (University of Auckland, New Zealand) for providing pictures of microplastic found in cleaning products. I thank the reviewers for their constructive comments.

References

Abbott, K. E. (1992). Plastic media blasting—state of the technology (reprinted from society of automotive engineers). *Materials Performance, 31*, 38–39.

Andrady, A. L., & Pegram, J. E. (1989). Outdoor weathering of selected polymeric materials under marine exposure conditions. *Polymer Degradation and Stability, 26*, 333.

Andrady, A. L. (2003). *Plastics and the environment*. In A.L. Andrady (Ed.). Hoboken, NJ: John Wiley and Sons.

Andrady, A. L. (2011). Microplastics in the marine environment. *Marine Pollution Bulletin, 62*, 1596–1605.

Andrady, A. L. (2015). Persistence of plastic litter in the oceans. In M. Bergmann, L. Gutow & M. Klages (Eds.), *Marine anthropogenic litter* (pp. 57–72), Berlin: Springer.

Anonymous. (1998). Plastic media blasting of composite honeycomb sandwich structures. *Aerospace Engineering, 18*, 41–44.

Arthur, C., Baker, J., & Bamford, H. (2009). In *Proceedings of the International Research Workshop on the Occurrence, Effects and Fate of Microplastic Marine Debris*, 9–11 September 2008, NOAA Technical, Memorandum NOS-OR&R30.

Browne, M. A., Galloway, T.S., & Thompson, R.C. (2007). Microplastic—an emerging contaminant of potential concern? *Integrated Environmental Assessment and Management, 3*, 559–561.

Barnes, D. K. A. (2005). Remote Islands reveal rapid rise of Southern Hemisphere, sea debris. *Scientific World Journal, 5*, 915–921.

Browne, M. A., Dissanayake, A., Galloway, T. S., Lowe, D. M., & Thompson, R. C. (2008). Ingested microscopic plastic translocates to the circulatory system of the mussel, *Mytilus edulis. Environmental Science and Technology, 42*, 5026–5031.

Browne, M. A., Galloway, T. S., & Thompson, R. C. (2010). Spatial patterns of plastic debris along estuarine shorelines. *Environmental Science and Technology, 44*, 3404–3409.

Browne, M. A., Crump, P., Niven, S. J., Teuten, E. L., Tonkin, A., & Galloway, T., et al. (2011). Accumulation of microplastic on shorelines worldwide: Sources and sinks. *Environmental Science and Technology, 45*, 9175–9179.

Browne, M. A., Niven, S. J., Galloway, T. S., Rowland, S. J., & Thompson, R. C. (2013). Ingested microplastic moves pollutants and additives into marine worms compromising functions linked to health and biodiversity. *Current Biology, 23*, 2388–2392.

CEFAS. (1997). Centre for Environment, Fisheries and Science. *Monitoring and surveillance of non-radioactive contaminants in the aquatic environment and activities regulating the disposal of wastes at sea*. Lowestoft: CEFAS.

Claessens, M., Meester, S. D., Landuyt, L. V., Clerck, K. D., & Janssen, C. R. (2011). Occurrence and distribution of microplastics in marine sediments along the Belgian coast. *Marine Pollution Bulletin, 62*, 2199–2204.

Claessens, M., Van Cauwenberghe, L., Vandegehuchte, M. B., & Janssen, C. R. (2013). New techniques for the detection of microplastics in sediments and field collected organisms. *Marine Pollution Bulletin, 70*, 227–233.

Cole, M., Lindeque, P., Halsband, C., & Galloway, T. S. (2011). Microplastics as contaminants in the marine environment: A review. *Marine Pollution Bulletin, 62*(12), 2588–2597.

Collignon, A., Hecq, J. H., Galgani, F., Voisin, P., Collard, F., & Goffart, A. (2012). Neustonic microplastic and zooplankton in the North Western Mediterranean Sea. *Marine Pollution Bulletin, 64*, 861–864.

Corbanie, E. A., Matthijs, M. G. R., van Eck, J. H. H., Remon, J. P., Landman, W. J. M., & Vervaet, C. (2006). Deposition of differently sized airborne microspheres in the respiratory tract of chickens. *Avian Pathology, 35*, 475–485.

Corcoran, P. L., Biesinger, M. C., & Grifi, M. (2009). Plastics and beaches: A degrading relationship. *Marine Pollution Bulletin, 58*, 80–84.

Cooper, D. A., & Corcoran, P. L. (2010). Effects of mechanical and chemical processes on the degradation of plastic beach debris on the island of Kauai, Hawaii. *Marine Pollution Bulletin, 60*, 650–654.

Costa, M., Ivar do Sul, J., Silva-Cavalcanti, J., Araújo, M., Spengler, Â., & Tourinho, P. (2010). On the importance of size of plastic fragments and pellets on the strandline: A snapshot of a Brazilian beach. *Environmental Monitoring and Assessment, 168*, 299–304.

Curley, J., Castillo, J., Hotz, J., Uezono, M., Hernandez, S., Lim, J. O., et al. (1996). Prolonged regional nerve blockade—injectable biodegradable bupivacaine/polyester microspheres. *Anesthesiology, 84*, 1401–1410.

Dalmon, R. A., McQueen Leifson, R., & Bøgwald, J. (1995). Microspheres as antigen carriers: Studies on intestinal absorption and tissue localization of polystyrene microspheres in Atlantic salmon, *Salmo salar* L. *Journal of Fish Diseases, 18*, 87–91.

Davidson, T. M. (2012). Boring crustaceans damage polystyrene floats under docks polluting marine waters with microplastic. *Marine Pollution Bulletin, 64*, 1821–1828.

Derraik, J. G. B. (2002). The pollution of the marine environment by plastic debris: A review. *Marine Pollution Bulletin, 44*, 842–852.

DOD. (1992). MIL-P-85891A, Military specification, plastics media, for removal of organic coatings. http://www.everyspec.com/MIL-SPECS/MIL-SPECS-MIL-P/MIL-P-85891A_8399/

Dunn, S. E., Brindley, A., Davis, S. S., Davies, M. C., & Illum, L. (1994). Polystyrene-poly (ethylene glycol) (PS-PEG2000) particles as model systems for site specific drug delivery. 2. The effect of PEG surface density on the in vitro cell interaction and in vivo biodistribution. *Pharmaceutical Research, 11*, 1016–1022.

EA [UK Environment Agency]. (2012). Pollution inventory reporting—guidance for operators of waste transfer stations Environmental Permitting (England and Wales) Regulations 2010 Regulation 60(1) Version 5 December. Bristol, U.K.: Environment Agency. https://www.gov.uk/government/uploads/system/uploads/attachment_data/file/296990/LIT_7759_f0b081.pdf

EPA [US Environmental Protection Agency]. (2010). U.S. Environmental Protection Agency 2010–2014 Pollution Prevention (P2) Program Strategic Plan (pp. 1–34). Washington, USA.

EPA. (2012). The Emergency Planning and Community Right-to-Know Act 1986. Washington, USA. http://www2.epa.gov/sites/production/files/2013-08/documents/epcra_fact_sheet.pdf

EU [European Union]. (2010). Part 1: List of chemicals subject to export notification procedure Directive 2010/75/EU of the European Parliament and of the council. *Official Journal of the European Union, L334*, 17–119.

EU. (2012). Part 1: List of chemicals subject to export notification procedure. http://edexim.jrc.ec.europa.eu/list_annex_i_chemicals.php?id_part=1

Fendall, L. S., & Sewell, M. A. (2009). Contributing to marine pollution by washing your face: Microplastics in facial cleansers. *Marine Pollution Bulletin, 58*, 1225–1228.

Galgani, F., Hanke, G., & Maes, T. (2015). Global distribution, composition and abundance of marine litter. In M. Bergmann, L. Gutow & M. Klages (Eds.), *Marine anthropogenic litter* (pp. 29–56). Berlin: Springer.

Gouin, T., Roche, N., Lohmann, R., & Hodges, G. A. (2011). Thermodynamic approach for assessing the environmental exposure of chemicals absorbed to microplastic. *Environmental Science and Technology, 45*, 1466–1472.

Gregory, M. R. (1996). Plastic 'scrubbers' in hand cleansers: a further (and minor) source for marine pollution identified. *Marine Pollution Bulletin, 32*, 867–871.

Gregory, M. R., & Andrady, A. L. (2003). Plastics in the marine environment. In A. L. Andrady (Ed.), *Plastics and the Environment*. Hoboken, NJ: John Wiley and Sons.

Habib, B., Locke, D. C., & Cannone, L. J. (1996). Synthetic fibers as indicators of municipal sewage sludge, sludge products and sewage treatment plant effluents. *Water, Air, Soil Pollution, 103*, 1–8.

Harrison, J. P., Ojeda, J. J., & Romero-González, M. E. (2012). The applicability of reflectance micro-Fourier-transform infrared spectroscopy for the detection of synthetic microplastics in marine sediments. *Science of the Total Environment, 416*, 455–463.

Hussain, N., Jaitley, V., & Florence, A. T. (2001). Recent advances in the understanding of uptake of microparticulates across the gastrointestinal lymphatics. *Advanced Drug Delivery Reviews, 50*, 107–142.

Kockisch, S., Rees, G. D., Young, S. A., Tsibouklis, J., & Smart, J. D. (2003). Polymeric microspheres for drug delivery to the oral cavity: An in-vitro evaluation of mucoadhesive potential. *Journal of Pharmaceutical Sciences, 92*, 1614–1623.

Merrell, T. R. Jr. (1980). Accumulation of plastic litter on beaches of Amchitka Island, Alaska. *Marine Environmental Research, 3*, 171–184.

Matsusaki, M., Kishida, A., Stainton, N., Ansell, C. W. G., & Akashi, M. (2001). Synthesis and characterisation of novel biodegradable polymers composed of hydroxycinnamic acid and D, L-lactic acid. *Journal of Applied Polymer Science, 82*, 2357–2364.

Morét-Ferguson, S., Law, K. L., Proskurowski, G., Murphy, E. K., Peacock, E. E., & Reddy, C. M. (2010). The size, mass, and composition of plastic debris in the western North Atlantic Ocean. *Marine Pollution Bulletin, 60*, 1873–1878.

Neulicht, R., & Shular, J. (1997). Emission factor documentation for AP-42. Section 13.2.6. Abrasive blasting final report (pp. 1–38). Washington, USA: EPA. http://www.epa.gov/ttnchie1/ap42/ch13/bgdocs/b13s02-6.pdf

O'Brine, T., & Thompson, R. C. (2010). Degradation of plastic carrier bags in the marine environment. *Marine Pollution Bulletin, 60*, 2279–2283.

Oerlikon. (2009). The Fiber Year 2008/09: A world-survey on textile and nonwovens industry. Switzerland: Oerlikon.

Ramirez-Llodra, E., Company, J. B., Sard, F., De Mol, B., Coll, M., & Sardà, F. (2013). Effects of natural and anthropogenic processes in the distribution of marine litter in the deep Mediterranean Sea. *Progress in Oceanography, 118*, 273–287.

Rochman, C. M., Hoh, E., Kurobe, T., & Teh, S. J. (2013). Ingested plastic transfers hazardous chemicals to fish and induces hepatic stress. *Scientific Reports, 3*, 3263.

Rosner, H. (2008). Scrubbing out sea life. http://www.slate.com/articles/business/moneybox/2008/06/scrubbing_out_sea_life.html

Thanoo, B. C., Sunny, M. C., & Jayakrishnan, A. (1993). Oral sustained-release drug delivery systems using polycarbonate microspheres capable of floating on the gastric fluid. *Journal of Pharmacy and Pharmacology, 45*, 21–24.

Thompson, R. C., Olsen, Y., Mitchell, R. P., Davis, A., Rowland, S. J., John, A. W. G., et al. (2004). Lost at sea: Where is all the plastic? *Science, 304*(5672), 838.

Underwood, A. J. (1997). Experiments in ecology—Their logical design and interpretation using analysis of variance. Cambridge: Cambridge University Press.

UN [United Nation]. (2008). World Population Prospects: The 2008 Revision Population Database, New York, 2008. www.esa.un.org/unpp. Accessed 1 Oct 2008.

Van Cauwenberghe, L., Vanreusel, A., Mees, J., & Janssen, C. R. (2013). Microplastic pollution in deep-sea sediments. *Environmental Pollution, 182*, 495–499.

Wen, J., Kim, G. J. A., & Leong, K. W. (2003). Poly(D, Lactide-co-ethyl ethylene phosphate)s as new drug carriers. *Journal of Controlled Release, 92*, 39–48.

Wolbach, C. D., & McDonald, C. (1987). Reduction of total toxic organic discharges and VOC emissions from paint stripping operations using plastic media blasting. *Journal of Hazardous Materials, 17*, 109–113.

Zitko, V., & Hanlon, M. (1991). Another source of pollution by plastics—skin cleaners with plastic scrubbers. *Marine Pollution Bulletin, 22*, 41–42.

Zubris, K. A. V., & Richards, B. K. (2005). Synthetic fibers as an indicator of land application of sludge. *Environmental Pollution, 138*, 201–211.

Chapter 10
Microplastics in the Marine Environment: Distribution, Interactions and Effects

Amy Lusher

Abstract Microplastics are an emerging marine pollutant. It is important to understand their distribution in the marine environment and their implications on marine habitats and marine biota. Microplastics have been found in almost every marine habitat around the world, with plastic composition and environmental conditions significantly affecting their distribution. Marine biota interact with microplastics including birds, fish, turtles, mammals and invertebrates. The biological repercussions depend on to the size of microplastics encountered, with smaller sizes having greater effects on organisms at the cellular level. In the micrometre range plastics are readily ingested and egested, whereas nanometre-sized plastics can pass through cell membranes. Despite concerns raised by ingestion, the effects of microplastic ingestion in natural populations and the implications for food webs are not understood. Without knowledge of retention and egestion rates of field populations, it is difficult to deduce ecological consequences. There is evidence to suggest that microplastics enter food chains and there is trophic transfer between predators and prey. What is clear is that further research on a variety of marine organisms is required to understand the environmental implications of microplastics in more detail and to establish effects in natural populations.

Keywords Distribution · Ingestion · Trophic transfer · Habitat alterations · Biomagnification · Bioaccumulation

A. Lusher (✉)
Marine and Freshwater Research Centre, Galway-Mayo Institute of Technology, Dublin Road, Galway, Ireland
e-mail: amy.lusher@research.gmit.ie

© The Author(s) 2015
M. Bergmann et al. (eds.), *Marine Anthropogenic Litter*,
DOI 10.1007/978-3-319-16510-3_10

10.1 Introduction

With the increasing reliance on plastics as an everyday item, and rapid increase in their production and subsequent disposal, the environmental implications of plastics are a growing concern. The benefits of plastics, including their durability and resistance to degradation, inversely result in negative environmental impacts. As user-plastics are primarily "*single use*" items they are generally disposed of within one year of production, and whilst some plastic waste is recycled, the majority ends up in land-fill. Concerns arise when plastics enter the marine environment through indiscriminate disposal and it has been estimated that up to 10 % of plastic debris produced will enter the sea (Thompson 2006). Interactions between litter and the marine environment are complex. The impacts of larger plastic debris are discussed by Kühn et al. (2015) and consequences include aesthetic, social and economic issues (Newman et al. 2015), and numerous environmental impacts on marine biota (Derraik 2002; Barnes et al. 2009). However, with an ever increasing reliance on plastic products, and as plastic production, use and disposal continue, microplastics are of increasing concern (Sutherland et al. 2010). Microplastics enter the sea from a variety of sources (Browne 2015) and distributed by oceans currents; these ubiquitous contaminants are widespread (Cózar et al. 2014). The amount of microplastics in the sea will continue to rise, leading to gradual but significant accumulation in coastal and marine environments (Andrady and Neal 2009).

Increasing evidence of microplastics in the sea has led to a need to understand its environmental impacts as a form of marine pollution. A recent review of marine debris research found only 10 % of publications to focus on microplastics, the majority of which were from the last decade (CBD 2012). Even though plastic is the primary constituent of marine debris, microplastics are considered under-researched due to difficulties in assessing their distribution and abundance (Doyle et al. 2011). It has only been in recent years that international, national and regional efforts were made to quantify microplastics in the sea. The Marine Strategy Framework Directive (MSFD, 2008/56/EC) has highlighted concerns for environmental implications of marine litter and one of the key attributes of the MSFD is to determine the ecological harm caused by microplastics and their associated chemicals (Zarfl et al. 2011).

Microplastics were first described as microscopic particles in the region of 20 μm diameter (Thompson et al. 2004). For the purpose of this study, microplastic refers to items <5 mm in size using the criteria developed by US National Oceanic and Atmospheric Administration (NOAA) (Arthur et al. 2009). The small size of microplastics makes them available for interaction with marine biota in different trophic levels. By inhabiting different marine habitats, a range of organisms are vulnerable to exposure (Wright et al. 2013a). At the millimetre and micrometre scale, sorption of microplastics is dominated by bulk portioning, with effects including blockages when fibres or fragments form aggregates.

Whereas at smaller size ranges, specifically the nanometre scale, there is a potential for microplastics to cause harm to organisms (Galloway 2015; Koelmans et al. 2015). Additionally, the consequences of exposure to chemicals associated with plastics are being investigated (Rochman 2015). A widely cited hypothesis explores how the large surface area to volume ratio of microplastics leaves them prone to adsorbing waterborne organic pollutants and the potential for toxic plasticisers to leach from polymer matrices into organisms tissues (Teuten et al. 2007). It was further hypothesized that if subsequently ingested, microplastics may act as a route for toxin introduction to the food chain (Teuten et al. 2009). Whether microplastics act as vectors depends on the gradient between microplastics and biota lipids (Koelmans 2015).

It is important to understand the transport and distribution of microplastics before understanding their fate, including the physical and chemical effects they could have on marine organisms. The objectives of this chapter are to assess the environmental impact of microplastic in the sea by: (1) summarising the distribution of marine microplastics, including the use of models to understand the distribution; (2) determine the interaction of microplastics with marine organisms.

10.2 The Global Distribution of Microplastics in the Sea

From strandlines on beaches to the deep seafloor and throughout the water column, microplastic research is dominated by studies monitoring microplastic distribution and abundance in the marine environment (Ivar du Sol and Costa 2014). A recent estimate suggested there could be between 7000 and 35,000 tons of plastic floating in the open ocean (Cózar et al. 2014). Another study estimated that more than five trillion pieces of plastic and >250,000 t are currently floating in the oceans (Eriksen et al. 2014). Once in the sea microplastics are transported around the globe by ocean currents where they persist and accumulate. Microplastics are suspended in the water column (e.g. Lattin et al. 2004), surface waters (e.g. Cózar et al. 2014), coastal waters (e.g. Ng and Obbard 2006), estuaries (e.g. Browne et al. 2010), rivers (Sadri and Thompson 2014), beaches (e.g. Browne et al. 2011) and deep-sea sediments (Van Cauwenberghe et al. 2013b; Woodall et al. 2014; Fischer et al. 2015). Suspended in the water column, microplastics can become trapped by ocean currents and accumulate in central ocean regions (e.g. Law et al. 2010). Ocean gyres and convergent zones are noteworthy areas of debris accumulation, as the rotational pattern of currents cause high concentrations of plastics to be captured and moved towards the centre of the region (Karl 1999). As gyres are present in all of the world's oceans, microplastic accumulation can occur at a global scale and has been documented during the past four decades. Distribution is further influenced by wind mixing, affecting the vertical movement of plastics

(Kukulka et al. 2012). Physical characteristics of plastic polymers, including their density, can influence their distribution in the water column and benthic habitats (Murray and Cowie 2011). Buoyant plastics float at the surface, whereas more dense microplastics or those fouled by biota sink to the sea floor. It has recently been estimated that 50 % of the plastics from municipal waste have a higher density than seawater such that it will readily sink to the seafloor (Engler 2012). It is currently not economically feasible nor is it desirable to remove microplastics from the ocean.

A number of concerns have been raised regarding the assessment of microplastic distribution. There are multiple pathways for the introduction of microplastics into the marine environment which do not have accurate timescales for the rate of degradation (Ryan et al. 2009). Quantification is complicated by the size of the oceans in relation to the size of plastics being assessed (Cole et al. 2011), which are further confounded by ocean currents and seasonal patterns introducing spatial and temporal variability (Doyle et al. 2011). As a result, there are various techniques applied to the sampling of microplastics in the marine environment (Löder and Gerdts 2015). Results of studies have been reported in different dimensions, e.g. the number of microplastics in a known water volume (particles m^{-3}) or area measurements (particles km^{-2}). This discrepancy presents a problem when comparing between studies, as it is not possible to compare results directly. For the purpose of this review, which aims to carry out a critical assessment of the global knowledge of microplastic distribution, a conversion was made to enable comparisons between the different dimensions of measurement. It is reasonable to assume that surface samples are collected in the top 0.20 m of water and therefore by making a simple calculation to add a third dimension (firstly converting particle km^{-2} to m^{-2}, then multiplying by 0.20 m to convert to a volume measurement, m^{-3}) we are able to compare different sampling methods in a variety of geographical locations. However, because of current directions in relation to boats, and approximate vessel speeds, it is difficult to calculate the amount of water passing through a net. As nets can ride out of the water, the exact volume of water passing through is unknown: the calculations have to be considered, at best, estimations.

It is important to understand the distribution of microplastics in the sea to grasp their potential impacts. This section will present a number of studies documenting microplastics in geographical regions including the Pacific, Atlantic, European Seas and the Mediterranean Sea, Indian Ocean and polar regions. It will introduce modelling strategies that have been utilised to understand microplastic distribution and accumulation around the globe.

10.2.1 Microplastics in the Pacific Ocean

Numerous studies on microplastics have been undertaken in the Pacific Ocean, the world's largest water basin (Table 10.1). One area which has received considerable attention is the North Pacific Central Gyre (NPCG) located off the west coast of

Table 10.1 Mean abundance (±SD, unless stated otherwise) of microplastic debris in the surface waters of the Pacific Ocean

Location	Equipment used	Amount (±SD)	Particles (m^{-3})	Source
North Pacific				
Bering Sea	Ring net	[a]80 (±190) km^{-2}	0.000016	Day and Shaw (1987)
Bering Sea	Ring/neuston net	1.0 (± 4.2) km^{-2}	0.0000002	Day et al. (1990)
Bering Sea	Sameota sampler/ manta net	Range: 0.004–0.19 m^{-3}	0.004–0.19	Doyle et al. (2011)
Subarctic N.P.	Ring net	[a]3,370 (±2,380) km^{-2}	0.00067	Day and Shaw (1987)
Subarctic N.P.	Ring/neuston net	61.4 (±225.5) km^{-2}	0.000012	Day et al. (1990)
Eastern North Pacific				
Vancouver Island, Canada	Underway sampling	279 (±178) m^{-3}	279	Desforges et al. (2014)
Eastern North Pacific	Plankton net	Estimated 21,290 t afloat	/	Law et al. (2014)
N.P. transitional water	Ring/neuston net	291.6 (±714.4) km^{-2}	0.00012	Day et al. (1990)
N.P. central gyre	Manta net	334,271 km^{-2}	*2.23	Moore et al. (2001)
N.P. central gyre	Manta net	85,184 km^{-2}	0.017	Carson et al. (2013)
N.P. subtropical gyre 1999–2010	Plankton net/manta net/neuston net	Median: 0.116 m^{-3}	0.12	Goldstein et al. (2012)
South Californian current system	Manta net	Median: 0.011–0.033 m^{-3}	0.011–0.033	Gilfillan et al. (2009)
Santa Monica Bay, California, USA	Manta net	3.92 m^{-3}	3.92	Lattin et al. (2004)
Santa Monica Bay, California, USA	Manta net	7.25 m^{-3}	7.25	Moore et al. (2002)
N.P. subtropical gyre	Manta net	Median: 0.02–0.45 m^{-2}	0.0042–0.089	Goldstein et al. (2013)
South Equatorial current	Neuston net	137 km^{-2}	0.000027	Spear et al. (1995)
Equatorial counter current		24 km^{-2}	0.0000048	
Western North Pacific				
Subtropical N.P.	Ring net	[a]96,100 (±780,000) km^{-2}	0.019	Day and Shaw (1987)
Subtropical N.P.	Ring/neuston net	535.1 (±726.1) km^{-2}	0.00011	Day et al. (1990)
Near-shore waters, Japan	Ring/neuston net	128.2 (±172.2) km^{-2}	0.000026	Day et al. (1990)
Kuroshio current system	Neuston net	174,000 (±467,000) km^{-2}	0.034	Yamashita and Tanimura (2007)

(continued)

Table 10.1 (continued)

Location	Equipment used	Amount (±SD)	Particles (m^{-3})	Source
Yangtze estuary system, East China Sea	Neuston net	4,137.3 (±8.2 × 10^4) m^{-3}	4137.3	Zhao et al. (2014)
Geoje Island, South Korea	Bulk sampling, hand-net, manta net	16,000 (±14 × 10^3) m^{-3}	16,000	Song et al. (2014)
South Pacific				
South Pacific subtropical gyre	Manta net	26,898 (±60,818) km^{-2}	0.0054	Eriksen et al. (2013)
Australian coast	Neston net	[b]4,256.3 (±757.8) km^{-2}	0.00085	Reisser et al. (2013)
	Manta net			

If particles in m^{-3} were not reported, the values have been converted as follows: (1) km^{-2} to m^{-2}: by division by 1,000,000 followed by multiplication by 0.2 m; (2) m^{-2} to m^{-3} carried out by 0.2 multiplication
[a]Mean ±95 % confidence intervals
[b]Mean ± standard error

California, USA. The gyre contains possibly the most well publicised area of plastic accumulation, known as the *"Great Pacific Garbage Patch"* (Kaiser 2010). Microplastic concentrations in the NPCG have increased by two orders of magnitude in the last four decades (Goldstein et al. 2012). In comparison, microplastic abundance in the North Pacific subtropical gyre (NPSG) is widespread and spatially variable, but values are two orders of magnitude lower than in the NPCG (Goldstein et al. 2013). Microplastic studies in the south Pacific are limited to the subtropical gyre where an increasing trend of microplastics was found towards the centre of the gyre (5.38 particles m^{-3} [1] Eriksen et al. 2013). In a similar way to macroplastic debris, oceanographic features strongly affect the distribution of microplastics in open oceans and areas of upwelling create oceanographic convergence zones for marine debris.

Coastal ecosystems of the Pacific appear to be impacted by microplastics in areas of nutrient upwelling (Doyle et al. 2011) and influenced by local weather systems (Moore et al. 2002; Lattin et al. 2004). Microplastic load increased further inshore, reflecting the inputs from terrestrial runoff and particles re-suspended from sediments following storms (Lattin et al. 2004). Microplastics are in turn transported by ocean currents from populated coastal areas (Reisser et al. 2013). This is also reflected in offshore subsurface waters which had 4–27 times less plastics than coastal sites in the northeast Pacific (Desforges et al. 2014).

Pre-production plastic resin pellets and fragments wash up on coastlines worldwide and have been recovered from several Pacific beaches (Table 10.2). Plastic pellets, typically 3–5 mm in size, are made predominantly from the polymers polyethylene and polypropylene (Endo et al. 2005; Ogata et al. 2009). The average

---

[1] Calculated from km^{-2}.

Table 10.2 Mean microplastic abundance (±SD, unless otherwise stated) in sediments from the Pacific

Location	Types	Amount (±SD)	Source
North Pacific			
Pacific beaches	Fragments 10 mm	/	Hirai et al. (2011)
9 beaches, Hawaiian islands	Fragments 1–15 mm	[a]37.8 kg^{-1}	[a]McDermid and McMullen (2004)
	Pellets 1–15 mm	[a]4.9 kg^{-1}	
Hawaiian islands	Pellets and fragments	/	Rios et al. (2007)
Kauai, Hawaiian islands	Fragments and pellets 0.8–6.5 mm	/	Corcoran et al. (2009)
Kauai, Hawaiian islands	Fragments <1 cm	/	Cooper and Corcoran (2010)
Kamillo Beach, Hawaii	Pellets and fragments	Total: 248	Carson et al. (2011)
Northeast Pacific			
Los Angeles, California, USA	Pellets and fragments	/	Rios et al. (2007)
San Diego, California, USA	Pellets and fragments <5 mm	/	Van et al. (2012)
Beaches, western USA	Pellets	/	Ogata et al. (2009)
Guadalupe Island, Mexico	Pellets and fragments	/	Rios et al. (2007)
Northwest Pacific			
Coastal beaches, Russia	Fragments and pellets	[b]29 m^{-2}	Kusui and Noda (2003)
Tokyo, Japan	Pellets	>1,000 m^{-2}	Kuriyama et al. (2002)
Coastal beaches, Japan	Pellets	/	Mato et al. (2001)
Coastal beaches, Japan	Pellets	[b]0.52 m^{-2}	Kusui and Noda (2003)
	Fragments	[b]1.1 m^{-2}	
Coastal beaches, Japan	Pellets <5 mm	>100 per beach	Endo et al. (2005)
Korean Strait			
Heugnam Beach, South Korea	PS spheres	874 (±377) m^{-2}	Heo et al. (2013)
	Fragments	25 (±10) m^{-2}	
	Pellets	41 (±19) m^{-2}	
South China Sea			
Ming Chau Island, Vietnam	Pellets	/	Ogata et al. (2009)
Hong Kong, China	Pellets	/	Ogata et al. (2009)
South Pacific			
Coastal beaches, New Zealand	Pellets <5 mm	>1,000 m^{-1}	Gregory (1978)
Coastal beaches, Chile	Fragments and pellets 1–10 mm	30 m^{-2}	Hidalgo-Riz and Thiel (2013)
Easter Island, Chile	Fragments and pellets 1–10 mm	805 m^{-2}	Hidalgo-Riz and Thiel (2013)

[a]Calculated from total plastic collected from an overall total of 440 L of beach sediment
[b]Calculated from total plastics found over total survey area

abundance of plastic fragments on beaches in the southeast Pacific was greater in isolated areas (Easter Island: >800 items m^{-2}) than on beaches from continental Chile (30 items m^{-2}) (Hidalgo-Ruz and Thiel 2013). This trend has been seen in the Hawaiian archipelago, where the remotest beaches on Midway Atoll and Moloka'I contained the highest quantity of plastic particles (McDermid and McMullen 2004; Corcoran et al. 2009; Cooper and Corcoran 2010).

10.2.2 Microplastics in the Atlantic Ocean

Research on microplastic distribution in the Atlantic is less extensive than in the Pacific (Table 10.3), but includes a number of long-term studies. A time-series conducted in the north Atlantic and Caribbean Sea identified microplastics in 62 % of the trawls conducted with densities reaching 580,000 particles km^{-2} (Law et al. 2010). Distinct patterns emerged with the highest concentration (83 % of plastics) in subtropical latitudes, 22°N and 88°N, of the north Atlantic gyre marking the presence of a large-scale convergence zone (Law et al. 2010; Morét-Ferguson et al. 2010) similar to the south Pacific (Eriksen et al. 2013). Converging surface currents driven by winds are assumed to be the driving force of this accumulation. To assess long-term trends in abundance, a time-series data set of continuous plankton recorder (CPR) samples from north Atlantic shipping routes were re-examined and microplastics were identified from the 1960s with a significant increase over time (Thompson et al. 2004). Regular sampling schemes have begun to monitor the spatial and temporal trends of microplastics in the northeast Atlantic and found microplastics to be widespread and abundant (Lusher et al. 2014).

Microplastics accumulate in the coastal pelagic zones of the Atlantic (Table 10.3). Water samples from the Portuguese coast identified microplastics in 61 % of the samples with higher concentrations found in Costa Vicentina and Lisbon (0.036 and 0.033 particles m^{-3}, respectively) than in the Algarve and Aveiro (0.014 and 0.002 particles m^{-3}, respectively). These results are probably related to the proximity to urban areas and river runoff (Frias et al. 2014), which is similar to the trend seen in the Pacific. Following a MARMAP cruise in the south Atlantic, microplastic beads were present in 14.6–34.2 % of tows conducted (van Dolah et al. 1980). Pelagic subsurface plankton samples from a geographically isolated archipelago, Saint Peter and Saint Paul, were not free of microplastic fragments. Modelling studies suggested that oceanographic mechanisms promote the topographic trapping of zooplankton and therefore microplastics might be retained by small-scale circulation patterns (Ivar do Sul et al. 2013). Additionally, research in the Firth of Clyde (U.K.) indicated that intense environmental sampling regimes are necessary to encompass the small-scale and temporal variation in coastal microplastic abundance (Welden, *pers. comm.*).

Microplastic granules and pellets have been identified on Atlantic beaches since the 1980s (Table 10.4). It was hypothesised that pre-production pellets are

Table 10.3 Mean abundance (±SD, unless stated otherwise) of microplastic debris in the surface waters of the Atlantic Ocean

Location	Equipment used	Amount (± SD)	Particles (m^{-3})	Source
North Atlantic				
North Atlantic gyre (29–31°N)	Plankton net	20,328 (±2,324) km^{-2}	0.0041	Law et al. (2010)
North Atlantic	Continuous plankton recorder (CPR)	1960–1980: 0.01 m^{-3}	0.01	Thompson et al. (2004)
		1980–2000: 0.04 m^{-3}	0.04	
Northwest Atlantic				
Northwest Atlantic	Neuston net	[a]490 km^{-2}	0.00098	Wilber (1987)
Block Island Sound, USA	Plankton net	Range: 14–543 m^{-3}	14–543	Austin and Stoops-Glass (1977)
Gulf of Maine	Plankton net	1534 (±200) km^{-2}	0.00031	Law et al. (2010)
New England, USA	Plankton net	Mean ranges: 0.00–2.58 m^{-3}	0.00–2.58	Carpenter et al. (1972)
Continental shelf, west coast USA	Neuston net	2,773 km^{-2}	0.00056	Colton et al. (1974)
Western Sargasso Sea	Neuston net	3,537 km^{-2}	0.00071	Carpenter and Smith (1972)
Caribbean Sea				
Caribbean	Neuston net	60.6–180 km^{-2}	0.000012–0.000036	Colton et al. (1974)
Caribbean	Plankton net	1,414 (±112) km^{-2}	0.00028	Law et al. (2010)
Northeast Atlantic				
Offshore, Ireland	Underway sampling	2.46 m^{-3}	2.46	Lusher et al. (2014)
English Channel, U.K.	Plankton net	0.27 m^{-3}	0.27	Cole et al. (2014a)
Bristol Channel, U.K.	Lowestoft plankton sampler	Range: 0–100 m^{-3}	0–>100	Morris and Hamilton (1974)
Severn Estuary, U.K.				Kartar et al. (1973, 1976)
Portuguese coast	Neuston net/ CPR	0.02–0.036 m^{-3}	0.02–0.036	Frias et al. (2014)
Equatorial Atlantic				
St. Peter and St. Paul Archipelago, Brazil	Plankton net	0.01 m^{-3}	0.01	Ivar do Sul et al. (2013)

(continued)

Table 10.3 (continued)

Location	Equipment used	Amount ($\pm$ SD)	Particles (m^{-3})	Source
South Atlantic				
South Atlantic Bight	Neuston net	Mean weight: 0.03–0.08 mg m^{-2}		van Dolah et al. (1980)
Cape Basin, South Atlantic	Neuston sledge	1,874.3 km^{-2}	0.00037	Morris (1980)
Cape Province, South Africa	Neuston net	3,640 km^{-2}	0.00073	Ryan (1988)
Fernando de Noronha, Abrolhos and Trindade, Brazil	Manta net	0.03 m^{-3}	0.03	Ivar do Sul et al. (2014)
Gioana estuary, Brazil	Conical plankton net	26.04–100 m^{-3}	0.26	Lima et al. (2014)

If particles in m^{-3} were not reported, the values have been converted as follows: (1) km^{-2} to m^{-2}: by division by 1,000 000, followed by multiplication by 0.2 m; (2) m^{-2} to m^{-3} carried out multiplication by 0.2
[a]This value is for pellets only, although fragments >5 mm were also reported

Table 10.4 Mean microplastic abundance ($\pm$SD, unless stated otherwise) in sediments from the Atlantic

Location	Types	Amount	Source
North Atlantic			
Nova Scotia, Canada	Pellets	Max: <10 m^{-1}	Gregory (1983)
Nova Scotia, Canada	Fibres	200–800 fibres kg^{-1}	Mathalon and Hill (2014)
Beaches, eastern USA	Pellets		Ogata et al. (2009)
Factory beaches, New York, USA	Spheres		Hays and Cormons (1974)
*Maine, USA	Pellets and fragments	105 kg^{-1}	Graham and Thompson (2009)
*Florida, USA	Pellets and fragments	214 kg^{-1}	Graham and Thompson (2009)
Florida Keys, USA	Pellets and fragments	100–1,000 m^{-2}	Wilber (1987)
Cape Cod, USA	Pellets and fragments	100–1,000 m^{-2}	Wilber (1987)
North Carolina, USA	Fragments <5 cm	60 % of debris in size class	Viehman et al. (2011)
Bermuda	Pellets	>5,000 m^{-1}	Gregory (1983)
Bermuda	Pellets and fragments	2,000–10,000 m^{-2}	Wilber (1987)
Bahamas	Pellets and fragments	Windward: 500–1,000 m^{-2}	Wilber (1987)
		Leeward: 200–500 m^{-2}	
Lesser Antilles	Pellets and fragments	Windward: 100–5,000 m^{-2}	Wilber (1987)
		Leeward: 50–100 m^{-2}	

(continued)

Table 10.4 (continued)

Location	Types	Amount	Source
Le Havre, France	Pellets		Endo et al. (2013)
Costa Nova, Portugal	Pellets		Ogata et al. (2009)
Lisbon, Portugal	Fibres and pellets		Frias et al. (2010)
Portuguese coast	Pellets and fragments	185.1 m^{-2}	Martins and Sobral (2011)
Portuguese coast	Pellets 3–6 mm	1,289 m^{-2}	Antunes et al. (2013)
*Porcupine abyssal plain	Fragments	[a]40 item m^{-2}	Van Cauwenberghe et al. (2013b)
Canary Islands, Spain	Pellets and fragments <5 mm	<1 g kg^{-1}– >40 g kg^{-1}	Baztan et al. (2014)
English Channel			
Estuarine sediment, U.K.	Fragments and fibres	Maximum: 31 kg^{-1}	Thompson et al. (2004)
*Subtidal sediments, U.K.	Fragments and fibres	Maximum: 86 kg^{-1}	Thompson et al. (2004)
Plymouth, U.K.	Pellets		Ogata et al. (2009)
South Devon, U.K.	Pellets	~100	Ashton et al. (2010)
Tamar estuary, U.K.	Fragments <1 mm	65 % of total debris	Browne et al. (2010)
Southwest England, U.K.	Pellets	~100 at each location	Holmes et al. (2012)
South Atlantic			
Fernando de Noronha, Brazil	Pellets 23 %	[b]3.5 kg^{-1}	Ivar do Sul et al. (2009)
	Fragments 65 %	[b]9.63 kg^{-1}	
	Nylon monofilament 5 %	[b]0.73 kg^{-1}	
Recife, Brazil	Fragments 96.7 %	[c]300,000 m^{-3}	Costa et al. (2010)
	Pellets 3.3 %		
Northeast Brazil	Fragments 1–10 mm	59 items m^{-3}	Costa et al. (2011)
*Southern Atlantic	Fragments	[a]40 items m^{-2}	Van Cauwenberghe et al. (2013b)
Santos Bay, Brazil	Pellets	0–2,500 m^{-3}	Turra et al. (2014)

All sediments are beach sediments unless annotated with *, which refers to benthic or subtidal sediment. d.w. is dry weight of sediment. When originally reported in l, values were converted to kg
[a]Estimated from 1 item 25 cm^{-2}
bCalculated from total weight of sand (13,708 g)
cCalculated from 0.3 items cm^{-3}

transported by trans-oceanic currents before being washed ashore in areas such as the mid-Atlantic Archipelago, Fernando de Noronha (Ivar do Sul et al. 2009). Fragments make up a considerable proportion of marine debris on saltmarsh beaches in North Carolina (Viehman et al. 2011), the Canary Islands (Baztan et al. 2014) and beaches and intertidal plains in Brazil (Costa et al. 2010, 2011). Whereas, fibres were primarily identified in sediment samples from an intertidal ecosystem in Nova Scotia, Canada (Mathalon and Hill 2014).

Table 10.5 Mean microplastic abundance in surface waters of the Mediterranean and European seas

Location	Equipment used	Amount	Particles (m^{-3})	Source
West coast, Sweden	Manta net (80 µm)	Range: 150–2,400 m^{-3}	150–2400	Norén (2007)
	Manta net (450 µm)	Range: 0.01–0.14 m^{-3}	0.01–0.14	
Skagerrak, Sweden	Submersible in situ pump	Maximum: 102,000 m^{-3}	102,000	Norén and Naustvoll (2011)
Northwest Mediterranean	Manta net	1.33 m^{-2}	0.27	Collignon et al. (2012)
Bay of Calvi, Corsica, France	wp2 net	0.062 m^{-2}	0.012	Collignon et al. (2014)
Gulf of Oristano, Sardinia, Italy	Manta net	0.15 m^{-3}	0.15	de Lucia et al. (2014)
North Sea, Finland	Manta net	Range: 0–0.74 m^{-3}	0–0.74	Magnusson (2014)

If particles in m^{-3} were not reported, the values have been converted as follows: (1) km^{-2} to m^{-2}: by division by 1,000,000 followed by multiplication by 0.2 m; (2) m^{-2} to m^{-3} carried out multiplication by 0.2

10.2.3 Microplastics in European Seas and the Mediterranean Sea

Marine litter including microplastic is a serious concern in the Mediterranean, with plastics accounting for 70–80 % of litter identified (Fossi et al. 2014). This enclosed water basin is not free of microplastic contamination (Table 10.5). Levels of microplastics in surface waters of the northwest Mediterranean were similar to those reported for the NPCG, (0.27 particles m^{-3} [2] Collignon et al. 2012), and areas far away from point sources of pollution have high microplastic abundance (0.15 particles m^{-3}; de Lucia et al. 2014). Interestingly, fewer particles were recorded from surface waters from coastal Corsica (0.012 particles m^{-3} [3]; Collignon et al. 2014). Microplastic distribution is strongly influenced by wind stress, which may redistribute particles in the upper layers of the water column and preclude sampling by surface tows (Collignon et al. 2012). Oceanographic influences may affect the distribution of microplastics in the Mediterranean. Further research will help to clarify if the new hypothesis by de Lucia et al. (2014) holds, which suggests that upwelling dilutes the amount of plastic in the surface waters.

Microplastics, including beads and pellets, have been widely reported for sedimentary habitats and beaches in European Seas and the Mediterranean Sea (Table 10.6). Microplastics have been extracted from sediments from Norderney, in the North Sea (Dekiff et al. 2014; Fries et al. 2013) and samples taken at the East Frisian

---

[2]Calculated from 1.334 particles m^{-2}.

[3]Calculated from 0.062 particles m^{-2}.

Table 10.6 Mean microplastic abundance (±SD, unless stated otherwise) in sediments from the Mediterranean and European seas

Location	Types	Amount	Source
North Sea			
Harbor sediment, Sweden	Fragments	[a]20 and 50 kg^{-1}	Norén (2007)
Industrial harbor sediment, Sweden	Pellets	[a]3320 kg^{-1}	Norén (2007)
Industrial coastal sediment, Sweden	Pellets	[a]340 kg^{-1}	Norén (2007)
Spiekeroog, Germany	Fibres and granules	[b]3,800 kg^{-1} d.w.	Liebezeit and Dubaish (2012)
Jade System, Germany	Fibres	88 (±82) kg^{-1}	Dubaish and Liebezeit (2013)
	Granules	64 (±194) kg^{-1}	
Norderney, Germany	Fragments	/	Fries et al. (2013)
Norderney, Germany	Fragments	1.3, 1.7, 2.3 kg^{-1} d.w.	Dekiff et al. (2014)
Zandervoord, Netherlands	Pellets	/	Ogata et al. (2009)
*Harbor, Belgium	Fibres, granules, films, spheres	116.7 (±92.1) kg^{-1} d.w.	Claessens et al. (2011)
*Continental shelf, Belgium	Fibres, granules, films	97.2 (±18.6) kg^{-1} d.w.	Claessens et al. (2011)
Beach, Belgium	Fibres, granules, films	92.8 (±37.2) kg^{-1} d.w.	Claessens et al. (2011)
Beach, Belgium	Pellets and fragments	17 (±11) kg^{-1}	Van Cauwenberghe et al. (2013a)
Forth estuary, U.K.	Pellets	/	Ogata et al. (2009)
Mediterranean Sea			
8 beaches, Malta	Pellets	0.7–167 m^{-2}	Turner and Holmes (2011)
Sicily, Italy	Pellets	/	Ogata et al. (2009)
Venice lagoon, Italy	Fragments and fibres	672–2,175 kg^{-1} d.w.	Vianello et al. (2013)
*Nile deep sea fan, Mediterranean	Fragments	[c]40 items m^{-2}	Van Cauwenberghe et al. (2013b)
Lesvos, Greece	Pellets	/	Karapangioti and Klontza (2007)
Kato Achaia, Greece	Pellets	/	Ogata et al. (2009)
Beaches, Greece	Pellets	/	Karapanagioti et al. (2011)
Kea Island, Greece	Pellets	10, 43, 218, 575 m^{-2}	Kaberi et al. (2013)
Tripoli-Tyre, Lebanon	Pellets and fragments	/	Shiber (1979)
Costa del Sol, Spain	Pellets	/	Shiber (1982)
18 beaches, western Spain	Pellets	/	Shiber (1987)
Izmir, Turkey	Pellets	/	Ogata et al. (2009)

All sediments are beach sediments unless annotated with *, which refers to benthic or subtidal sediment. d.w. is dry weight of sediment. When originally reported in l, values were converted to kg
[a]Calculated from 100 ml sediment
[b]Calculated from 10 g sediment
[c]Estimated from 1 item 25 cm^{-2}

Islands, where tidal flats were more contaminated than sandy beaches (Liebezeit and Dubaish 2012). Areas of low hydrodynamics appear to have high microplastic abundance, such as the Venice lagoon (Vianello et al. 2013). Reduced water movement could also be attributed to the difference between concentrations of microplastics in Belgium: higher concentrations of microplastics were identified in sediments from Belgium harbors (Claessens et al. 2011) than in beach samples (Van Cauwenberghe et al. 2013a). Lastly, microplastics were recorded in deep offshore sediments (Van Cauwenberghe et al. 2013b; Fischer et al. 2015), which shows that microplastics sink to the deep seafloor. In fact, the deep seafloor may be considered a major sink for microplastic debris (Woodall et al. 2014) and explain the current mismatch between estimated global inputs of plastic debris to the oceans (Jambeck et al. 2015) and field data (Cózar et al. 2014; Eriksen et al. 2014), which refer largely to floating litter.

10.2.4 Microplastics in the Indian Ocean and Marginal Seas

To date there are few large-scale reports on microplastics from the Indian Ocean. Reddy et al. (2006) reported microplastic fragments from a ship-breaking yard in the Arabian Sea, and microplastics accounted for 20 % of the plastics recorded on sandy beaches in Mumbai (Jayasiri et al. 2013). Pellets were also recorded on

Table 10.7 Mean microplastic abundance (±SD, unless stated otherwise) in sediments from the Indian Ocean and marginal seas

Location	Types	Amount	Source
Arabian Sea			
Ship-breaking yard, Alang-Sosiya, India	Fragments	81 mg kg^{-1}	Reddy et al. (2006)
Mumbai, Chennai and Sunderbans, India	Pellets	/	Ogata et al. (2009)
Mumbai, India	Fragments	41.85 % of total plastics	Jayasiri et al. (2013)
East Asian Marginal Seas			
Coastline, Singapore	Fragments	/	Ng and Obbard (2006)
Coastline, Singapore	Fibres, grains, fragments	36.8 ± 23.6 kg^{-1}	Mohamed Nor and Obbard (2014)
Selangor, Malaysia	Pellets	<18 m^{-2}	Ismail et al. (2009)
Lang Kawi, Penang and Borneo, Malaysia	Pellets	/	Ogata et al. (2009)
Rayong, Thailand	Pellets	/	Ogata et al. (2009)
Jakarta Bay, Indonesia	Pellets	/	Ogata et al. (2009)
Southern Indian Ocean	Pellets	/	Ogata et al. (2009)
Mozambique	Pellets	/	Ogata et al. (2009)
Gulf of Oman	Pellets	>50–200 m^{-2}	Khordagui and Abu-Hilal (1994)
Arabian Gulf	Pellets	>50–80,000 m^{-2}	

All sediments are beach sediments

Malaysian beaches (Ismail et al. 2009). Most of the studies shown in Table 10.7 are part of the *"International Pellet Watch"* (Takada 2006; Ogata et al. 2009). Shoreline surveys conducted in surface waters and sediments on Singapore's coasts identified microplastics >2 μm (Ng and Obbard 2006). This highlights an area that requires further investigation to obtain a wider picture of microplastic distribution around the globe.

10.2.5 Microplastics in Polar Regions

Prior to 2014, there had been no direct studies of microplastics in either the Arctic or Antarctica; the plastic flux into the Arctic Ocean has been calculated to range between 62,000 and 105,000 tons per year, with variation due to spatial heterogeneity, temporal variability and different sampling methods (Zarfl and Matthies 2010). With the estimated value four to six orders of magnitude below the atmospheric transport and ocean current fluxes, the study concluded that plastic transport levels to the Arctic are negligible and that plastics are not a likely vector for organic pollutants to the Arctic. However, Obbard et al. (2014) published results from ice cores collected from remote locations in the Arctic Ocean. The levels of microplastics observed (range: 38–234 particles m^{-3}) were two orders of magnitude greater than previously reported in the Pacific gyre (Goldstein et al. 2012). Macroplastics have been identified floating in surface waters of Antarctica. However, trawls for microplastics did not catch any particles (Barnes et al. 2010). Dietary studies of birds from the Canadian Arctic have reported ingested plastics (Mallory et al. 2006; Provencher et al. 2009, 2010), and macroplastics were observed on the deep Arctic seafloor (Bergmann and Klages 2012). This indirect evidence suggests that microplastics have already entered polar regions. A modelling study even suggests the presence or formation of a sixth garbage patch in the Barents Sea (van Sebille et al. 2012).

10.2.6 Modelling the Distribution of Microplastics

Studies have highlighted the interaction of oceanographic and environmental variables on the distribution of microplastics (e.g. Eriksen et al. 2013). As polymer densities affect the distribution of plastics in the water column, it is important to understand how microplastics are transported at the surface and at depths. Knowledge of point-source pollution, including riverine input and sewage drainage into marine and coastal environments, can be useful in understanding the extent to which certain ecosystems are affected. Furthermore, knowledge of plastic accumulation on beaches will benefit the study of microplastics. For example, a study of plastic litter washed onto beaches developed a particle tracking model, which indicated that, if levels of plastic outflow remain constant over the coming decade, plastic litter quantity on beaches would continue to increase, and in some cases (3 % of all east Asian beaches) could see a 250-fold increase in plastic litter (Kako et al. 2014). If not removed, these larger items of plastic litter will break down into microplastics over time.

The fate of plastics in the marine environment is affected by poorly understood geophysical processes, including ocean mixing of the sea-surface boundary layer, re-suspension from sediments, and sinking rates plastics denser than seawater. Modelling approaches are required to further understand, and accurately estimate the global distribution, residence time, convergence zones, and ecological consequences of microplastics (Ballent et al. 2013). Models predicting the breakdown, fragmentation, and subsequent mixing and re-suspension of microplastics in sediments and seawater could provide an estimation of microplastic accumulation over short and long time scales; as well as an estimation of the dispersal patterns of microplastics in the marine environment. Generalized linear models have indicated that oceanographic mechanisms may promote topographic trapping of zooplankton and microplastics, which may be retained by small-scale circulation patterns in the Equatorial Atlantic, suggesting there is an outward gradient of microplastics moving offshore (Ivar do Sul et al. 2013). The recovery of plastic from surface seawater is dependent on wind speeds: stronger winds resulted in the capture of fewer plastics because wind-induced mixing of the surface layer vertically distributes plastics (Kulkula et al. 2012). Furthermore, by integrating the effect of vertical wind mixing on the concentrations of plastics in Australian waters, researchers estimated depth-integrated plastic concentrations, with high concentrations expected at low wind speeds. Thus, with the inverse relationship between wind force and plastic concentration, net tow concentrations of microplastics increased by a factor of 2.8 (Reisser et al. 2013).

Ballent et al. (2013) used the MOHID modelling system to predict the dispersal of non-buoyant pellets in Portugal using their density, settling velocity and re-suspension characteristics. Researchers simulated the transport of microplastic pellets over time using oceanographic processes, scales and systems. Model predictions suggest that the bottom topography restricts pellet movement at the head of the Nazaré Canyon with a potential area of accumulation of plastics pellets on the seafloor, implying long-term exposure of benthic ecosystems to microplastics. Tidal forces, as well as large-scale oceanographic circulation patterns are likely to transport microplastics up and down the Nazaré Canyon, which may be greatly increased during mass transport of waters linked to storms (Ballent et al. 2013) or deep-water cascading events (Durrieu de Madron et al. 2013).

With residence times from decades to centuries predicted for microplastics in the benthic environment (Ballent et al. 2013), future studies should assess the degradation of microplastics on the seafloor to be able to estimate residence times in those potential sink environments. Coupled with observations of microplastics in surface waters, the total oceanic plastic concentrations might be underestimated because of limited but growing knowledge of the geophysical and oceanographic processes in the surface waters. Furthermore, as microplastics degrade towards a nanometre scale, transport properties may be affected, and as a result, long-term transport models will need to be corrected. Modelling should be adapted to bring in ecological consequences of microplastics in benthic environments and the water column.

Research should focus on critical areas such as biodiversity hotspots and socio-economic hotspots that could affect vulnerable marine biota and coastal communities.

10.2.7 Summary

Microplastics have been documented in almost every habitat of the open oceans and enclosed seas, including beaches, surface waters, water column and the deep seafloor. Although most water bodies have been investigated, there is a lack of published work from polar regions and the Indian Ocean. Further research is required to accurately estimate the amount of different types of microplastics in benthic environments around the globe. Distribution of microplastics depends on environmental conditions including ocean currents, horizontal and vertical mixing, wind mixing and biofilm formation, as well as the properties of individual plastic polymers. A number of modelling approaches have been considered in the recent literature, which highlighted the effect of wind on the distribution of microplastics in the ocean. Oceanographic modelling of floating debris has shown accumulation in ocean gyres, and the distribution of microplastics within the water column appears to be dependent on the composition, density and shape of plastic polymers affecting their buoyancy. Further modelling studies may help to identify and predict regions with ecological communities and fisheries more vulnerable to the potential consequences of plastic contamination. The distribution of microplastic plays a significant role in terms of which organisms and habitats are affected. Widespread accumulation and distribution of microplastics raises concerns regarding the interaction and potential effects on marine organisms.

10.3 Interactions of Microplastics with Marine Organisms

Recently, Wright et al. (2013a) discussed the biological factors, which could enhance microplastic bioavailability to marine organisms: the varying density of microplastics allows them to occupy different areas of the water column and benthic sediments. As microplastics interact with plankton and sediment particles, both suspension and deposit feeders may be at risk of accidentally or selectively ingesting marine debris. However, the relative impacts are likely to vary across the size spectrum of microplastic in relation to the organisms affected, which is dependent on the size of the microplastic particles encountered. Microplastics in the upper end of the size spectrum (1–5 mm) may compromise feeding and digestion. For example, Codina-García et al. (2013) isolated such pellets and fragments from the stomachs of seabirds. Particles <20 µm are actively ingested by small invertebrates (e.g. Thompson et al. 2004) but they are also egested (e.g. Lee et al. 2013). Studies have shown that nanoparticles can translocate (e.g. Wegner et al. 2012) and model simulations have indicated that nano-sized polystyrene (PS) particles may permeate into the lipid membranes of organisms, altering the membrane structure, membrane protein activity, and therefore cellular function (Rossi et al. 2013). The following section deals with incidences of ingestion, trophic transfer and provision of new habitat by the presence of microplastics in the marine environment. Although the sections contain examples, comprehensive lists of microplastics ingestion are included in the corresponding tables.

10.3.1 Ingestion

Ingestion is the most likely interaction between marine organisms and microplastics. Microplastics' small size gives them the potential to be ingested by a wide range of biota in benthic and pelagic ecosystems. In some cases, organisms feeding mechanisms do not allow for discrimination between prey and anthropogenic items (Moore et al. 2001). Secondly, organisms might feed directly on microplastics, mistaking them for prey or selectively feed on microplastics in place of food (Moore 2008). If there is a predominance of microplastic particles associated with planktonic prey items, organisms could be unable to differentiate or prevent ingestion. A number of studies have reported microplastics from the stomachs and intestines of marine organisms, including fish and invertebrates. Watts et al. (2014) showed that shore crabs (*Carcinus maenas*) will not only ingest microplastics along with food (evidence in the foregut) but also draw plastics into the gill cavity because of their ventilation mechanism: this highlights that it is important to consider all sorts of routes of exposure to microplastics. If organisms ingest microplastics they could have adverse effects on individuals by disrupting feeding and digestion (GESAMP 2010). Laboratory (Table 10.8) and field (Table 10.9) studies highlighted that microplastics are mistaken for food by a wide variety of animals including birds, fish, turtles, mammals and invertebrates. Despite concerns raised regarding microplastic ingestion, few studies specifically examined the occurrence of microplastic in natural, in situ, populations as it is methodologically challenging to assess microplastic ingestion in the field (Browne et al. 2008).

10.3.1.1 Planktonic Invertebrates

Microplastics can enter the very base of the marine food web via absorption. Such was observed when charged nano-polystyrene beads were absorbed into the cellulose of a marine alga (*Scenedesmus* spp.), which inhibited photosynthesis and caused oxidative stress (Bhattacharya et al. 2010). Microplastics can also affect the function and health of marine zooplankton (Cole et al. 2013; Lee et al. 2013). Decreased feeding was observed following ingestion of polystyrene beads by zooplankton (Cole et al. 2013). Furthermore, adult females and nauplius larvae of the copepod (*Tigriopus japonicus*) survived acute exposure, but increased

Table 10.8 Laboratory studies exposing organisms to microplastics

Organism	Size of ingested material	Exposure concentration	Effect	Source
Phylum Chlorophyta				
Scenedesmus spp.	20 nm	1.6–40 mg mL^{-1}	Absorption, ROS increased, photosynthesis affected	Bhattacharya et al. (2010)
Phylum Haptophyta				
Isochrysis galbana	2 μm PS	9×10^4 mL^{-1}	Microspheres attached to algae, no negative effect observed	Long et al. (2014)
Phylum Dinophyta				
Heterocapsa triquetra	2 μm PS	9×10^4 mL^{-1}	Microspheres attached to algae, no negative effect observed	Long et al. (2014)
Phylum Cryptophyta				
Rhodomonas salina	2 μm PS	9×10^4 mL^{-1}	Microspheres attached to algae, no negative effect observed	Long et al. (2014)
Phylum Ochrophyta				
Chaetoceros neogracilis	2 μm PS	9×10^4 mL^{-1}	Microspheres attached to algae, no negative effect observed	Long et al. (2014)
Phylum Ciliophora				
Strombidium sulcatum	0.41–10 μm	5–10 % ambient bacteria concentration	Ingestion	Christaki et al. (1998)
Tintinnopsis lobiancoi	10 μm PS	$1,000$, $2,000$, $10,000$ mL^{-1}	Ingestion	Setälä et al. (2014)
Phylum Rotifera				
Synchaeta spp.	10 μm PS	$2,000$ mL^{-1}	Ingestion	Setälä et al. (2014)
Phylum Annelida				
Class Polychaete				
Lugworm (*Arenicola marina*)	20–2000 μm	1.5 g L^{-1}	Ingestion	Thompson et al. (2004)

(continued)

Table 10.8 (continued)

Organism	Size of ingested material	Exposure concentration	Effect	Source
Arenicola marina	130 μm UPVC	0–5 % by weight	Ingestion, reduced feeding, increased phagocytic activity, reduced available energy reserves, lower lipid reserves	Wright et al. (2013b)
Arenicola marina	230 μm PVC	1500 g of sediment mixture	Ingestion, oxidative stress	Browne et al. (2013)
Arenicola marina	400–1300 μm PS	0, 1, 10, 100 g L^{-1}	Ingestion, reduced feeding, weight loss	Besseling et al. (2013)
Fan worm (*Galeolaria caespitosa*)	3–10 μm	5 microspheres μL^{-1}	Ingestion	Bolton and Havenhand (1998)
Galeolaria caespitosa	3 and 10 μm PS	635, 2,240, 3,000 beads mL^{-1}	Ingestion, size selection, egestion	Cole et al. (2013)
Mud worms (*Marenzelleria* spp.)	10 μm PS	2,000 mL^{-1}	Ingestion	Setälä et al. (2014)
Phylum Mollusca				
Class Bivalvia				
Blue mussel (*Mytilus edulis*)	30 nm PS	0, 0.1, 0.2, and 0.3 g L^{-1}	Ingestion, pseudofaeces, reduced filtering	Wegner et al. (2012)
Mytilus edulis	θ80 μm HDPE	2.5 g L^{-1}	Ingestion, retention in digestive tract, transferred to lymph system, immune response	von Moos et al. (2012) Köhler (2010)
Mytilus edulis	0.5 μm PS	50 μL per 400 ml seawater	Ingestion, trophic transfer → *Carcinus maenas*	Farrell and Nelson (2013)
Mytilus edulis	3, 9.6 μm	0.51 g L^{-1}	Ingestion, retention in digestive tract, transferred to lymph system	Browne et al. (2008)
Mytilus edulis	10 μm PS	2 × 10^4 mL^{-1} 1,000 mL^{-1}	Ingestion, egestion	Ward and Tagart (1989) Ward and Kach (2009)
Mytilus edulis	10, 30 μm PS	3.10 × 10^5 mL^{-1} 8.65 × 10^4 mL^{-1}	Ingestion	Claessens et al. (2013)

(continued)

Table 10.8 (continued)

Organism	Size of ingested material	Exposure concentration	Effect	Source
Bay mussel (*Mytilus trossulus*)	10 μm PS	/	Ingestion	Ward et al. (2003)
Atlantic Sea scallop (*Placopecten magellanicus*)	15, 10, 16, 18, 20 μm PS	1.05 mL^{-1}	Ingestion, retention, egestion	Brilliant and MacDonald (2000, 2002)
Eastern oyster (*Crassostrea virginica*)	10 μm PS	1,000 mL^{-1}	Ingestion, egestion	Ward and Kach (2009)
Pacific oyster (*Crassostrea gigas*)	2, 6 μm PS	1,800 mL^{-1} for the 2 μm size; 200 mL^{-1} for the 6 μm size	Increased filtration and assimilation, reduced gamete quality (sperm mobility, oocyte number and size, fecundation yield), slower larval rearing for larvae from MP exposed parents	Sussarellu et al. (2014)
Phylum Echinodermata				
Class Holothuridea				
Giant Californian sea cucumber (*Apostichopus californicus*)	10, 20 μm PS	2.4 μL^{-1}	Ingestion, retention	Hart (1991)
Stripped sea cucumber (*Thyonella gemmata*)	0.25–15 mm PVC shavings, nylon line, resin pellets	10 g PVC shavings, 60 g resin pellets	Selective ingestion	Graham and Thompson (2009)
Grey sea cucumber (*Holothuria (Halodeima) grisea*)				
Florida sea cucumber (*Holothuria floridana*)		2 g nylon line added to 600 mL of silica sand		
Orange footed sea cucumber (*Cucumaria frondosa*)				
Class Echinoidea				

(continued)

Table 10.8 (continued)

Organism	Size of ingested material	Exposure concentration	Effect	Source
Collector urchin (*Tripneustes gratilla*)	32–35 μm PE	1, 10, 100, 300 mL^{-1}	Ingestion, egestion	Kaposi et al. (2014)
Eccentric sand dollar (*Dendraster excentricus*)	10, 20 μm PS	2.4 μL^{-1}	Ingestion, retention	Hart (1991)
Sea urchin (*Strongylocentrotus* sp.)	10, 20 μm PS	2.4 μL^{-1}	Ingestion, retention	Hart (1991)
Class Ophiuroidea				
Crevice brittlestar (*Ophiopholis aculeata*)	10, 20 μm PS	2.4 μL^{-1}	Ingestion, retention	Hart (1991)
Class Asteriodea				
Leather star (*Dermasterias imbricata*)	10, 20 μm PS	2.4 μL^{-1}	Ingestion, retention	Hart (1991)
Phylum Arthropoda				
Subphylum Crustacea				
Class Maxillopoda				
Barnacle (*Semibalanus balanoides*)	20–2,000 μm	1 g L^{-1}	Ingestion	Thompson et al. (2004)
Subclass Copepoda				
Tigriopus japonicus	0.05 μm PS	9.1 × 10^{11} mL^{-1}	Ingestion, egestion, mortality, decreased fecundity	Lee et al. (2013)
	0.5 μm PS	9.1 × 10^{8} mL^{-1}		
	6 μm PS	5.25 × 10^{5} mL^{-1}		
Acartia (*Acanthacartia*) *tonsa*	10–70 μm	3,000–4,000 beads mL^{-1}	Ingestion, size selection	Wilson (1973)
Acartia spp.	10 μm PS	2,000 mL^{-1}	Ingestion	Setälä et al. (2014)
Eurytemora affinis	10 μm PS	1,000, 2,000, 10,000 mL^{-1}	Ingestion, egestion	Setälä et al. (2014)

(continued)

Table 10.8 (continued)

Organism	Size of ingested material	Exposure concentration	Effect	Source
Limnocalanus macrurus	10 µm PS	1,000, 2,000, 10,000 mL^{-1}	Ingestion	Setälä et al. (2014)
Temora longicornis	20 µm PS	100 mL^{-1}	Ingestion 10.7 ± 2.5 beads per individual	Cole et al. (2014a)
Calanus helgolandicus	20 µm PS	75 mL^{-1}	Egestion, ingestion	Cole et al. (2014b)
Class Malacostraca				
Orchestia gammarellus	20–2000 µm	1 g per individual (n = 150)	Ingestion	Thompson et al. (2004)
Talitrus saltator	10–45 µm PE	10 % weight food (0.06–0.09 g dry fish food)	Ingestion, egestion after 2 h	Ugolini et al. (2013)
Allorchestes compressa	11–700 µm	0.1 g	Ingestion, egestion within 36 h	Chua et al. (2014)
Neomysis integer	10 µm PS	2,000 spheres mL^{-1}	Ingestion	Setälä et al. (2014)
Mysis relicta	10 µm PS	2,000 spheres mL^{-1}	Ingestion, egestion	Setälä et al. (2014)
Shore crab (*Carcinus maenas*)	8–10 µm PS	4.0×10^4 L^{-1} ventilation	Ingestion through gills and gut, retention and excretion, no biological effects measured	Watts et al. (2014)
		1.0×10^6 g^{-1} feeding		
Norway lobster (*Nephrops norvegicus*)	5 mm PP fibres	10 fibres per 1 cm^3 fish	Ingestion	Murray and Cowie (2011)
Nephrops norvegicus	500–600 µm PE loaded with 10 µg of PCBs	150 mg microplastics in gelatin food	Ingestion, 100 % egestion. Increase of PCB level in the tissues. Same increase for positive control. No direct effect of microplastics	Devriese et al. (2014)
Class Branchipoda				
Bosmina coregoni	10 µm PS	2,000, 10,000 spheres mL^{-1}	Ingestion	Setälä et al. (2014)

(continued)

Table 10.8 (continued)

Organism	Size of ingested material	Exposure concentration	Effect	Source
Phylum Chordata				
Common goby (*Pomatoschistus microps*)	1–5 μm PE	18.4, 184 μg L^{-1}	Ingestion, modulation bioavailability or biotransformation of pyrene, decreased energy, inhibited AChE activity	Oliveira et al. (2013)
Atlantic cod (*Gadus morhua*)	2, 5 mm	/	Ingestion, egestion, 5 mm held for prolonged periods, emptying of plastics improved by food consumption additional meals	Dos Santos and Jobling (1992)
Japanese medaka (*Oryzias latipes*)	3 mm LDPE	Ground up as 10 % of diet	Liver toxicity, pathology, hepatic stress	Rochman et al. (2013)
Oryzias latipes	PE pellets	Two months chronic exposure	Altered gene expression, decreased choriogenin regulation in males and decreased vitellogenin and choriogenin in females	Rochman et al. (2014)
Seabass larvae (*Dicentrarchus labrax*)	10–45 μm PE	0–105 g^{-1} incorporated with food	Ingestion, no significant increase in growth, effect on survival of larvae. Possible gastric obstruction	Mazurais et al. (2014)

For comparison the size of ingested material increases within species

Table 10.9 Evidence of microplastic ingestion by field studies organisms

Species	Number studied	Percentage with plastic (%)	Mean number of particles per individual (±SD)	Type and size ingested (mm)	Location	Source
Phylum Mollusca						
Humbolt squid (*Dosidicus gigas*)	30	26.7	Max: 11	Nurdles: 3–5 mm	British Columbia, Canada	Braid et al. (2012)
Blue mussel (*Mytilus edulis*)	45	/	3.7 per 10 g mussel	Fibres 300–1,000 µm	Belgium, The Netherlands	De Witte et al. (2014)
Mytilus edulis	36	/	0.36 (±0.07) g^{-1}	5–25 µm	North Sea, Germany	Van Cauwenberghe and Janssen (2014)
Pacific oyster (*Crassostrea gigas*)	11	/	0.47 (±0.16) g^{-1}	5–25 µm	Atlantic Ocean	Van Cauwenberghe and Janssen (2014)
Phylum Crustacea						
Goosneck barnacle (*Lepas spp.*)	385	33.5	1–30	1.41	North Pacific	Goldstein and Goodwin (2013)
Norway lobster (*Nephrops norvegicus*)	120	83	/	/	Clyde, U.K.	Murray and Cowie (2011)
Brown shrimp (*Crangon crangon*)	110	/	11.5 fibres per 10 g shrimp	95 % fibres, 5 % films 300–1000 µm	Belgium	Devriese et al. (2014)
Phylum Chaetognatha						
Arrow worm (*Parasagitta elegans*)	1	100	/	0.1–3 mm PS	New England, USA	Carpenter et al. (1972)
Phylum Chordata						
Class Mammalia						
Harbor seal (*Phoca vitulina*)	100 stomachs, 107 intestines	S:11.2 / I: 1	Max: 8 items / Max: 7 items	>0.1	The Netherlands	Bravo Rebolledo et al. (2013)

(continued)

Table 10.9 (continued)

Species	Number studied	Percentage with plastic (%)	Mean number of particles per individual (±SD)	Type and size ingested (mm)	Location	Source
Fur seal (*Arctocephalus* spp.)	145 scat	100	1–4 per scat	4.1	Macquarie Island, Australia	Eriksen and Burton (2003)
Class Reptilia						
Green turtle (*Chelonia mydas*)	24	/	Total: 11 pellets	<5 mm	Rio Grande do Sul, Brazil	Tourinho et al. 2010
Class Actinoptergii						
Order Atheriniformes						
Atlantic silversides (*Menidia menidia*)	9	33	/	0.1–3 mm PS	New England, USA	Carpenter et al. (1972)
Order Aulopiformes						
Longnosed lancetfish (*Alepisaurus ferox*)	144	24	2.7 (±2.0)	68.3 (±91.1)	North Pacific	Choy and Drazen (2013)
Order Beloniformes						
Cololabis saira	52	*35	3.2 (±3.05)	1–2.79	North Pacific	Boerger et al. (2010)
Order Clupeiformes						
Atlantic herring (*Clupea harengus*)	2	100	1	0.1–3 mm PS	New England, USA	Carpenter et al. (1972)
Clupea harengus	566	2	1–4	0.5–3	North Sea	Foekema et al. (2013)
Anchovy (*Stolephorus commersonnii*)	16	37.5	/	1.14–2.5	Alappuzha, India	Kripa et al. (2014)
Order Gadiformes						
Saithe (*Pollachius virens*)	1	100	1	0.1–3 mm PS	New England, USA	Carpenter et al. (1972)
Five-bearded rockling (*Ciliata mustela*)	113	0–10	/	1 mm PS	Severn Estuary, U.K.	Kartar (1976)

(continued)

Table 10.9 (continued)

Species	Number studied	Percentage with plastic (%)	Mean number of particles per individual (±SD)	Type and size ingested (mm)	Location	Source
Whiting (*Merlangius merlangus*)	105	6	1–3	1.7 (±1.5)	North Sea	Foekema et al. (2013)
Merlangius merlangus	50	32	1.75 (±1.4)	2.2 (±2.3)	English Channel	Lusher et al. (2013)
Haddock (*Melanogrammus aeglefinus*)	97	6	1.0	0.7 (±0.3)	North Sea	Foekema et al. (2013)
Cod (*Gadus morhua*)	80	13	1–2	1.2 (±1.2)	North Sea	Foekema et al. (2013)
Blue whiting (*Micromesistius poutassou*)	27	51.9	2.07 (±0.9)	2.0 (±2.4)	English Channel	Lusher et al. (2013)
Poor cod (*Trisopterus minutus*)	50	40	1.95 (±1.2)	2.2 (±2.2)	English Channel	Lusher et al. (2013)
Order Lampriformes						
Lampris sp. (*big eye*)	115	29	2.3 (±1.6)	49.1 (±71.1)	North Pacific	Choy and Drazen (2013)
Lampris sp. (*small eye*)	24	5	5.8 (±3.9)	48.8 (±34.5)	North Pacific	Choy and Drazen (2013)
Order Myctophiformes						
Hygophum reinhardtii	45	*35	1.3 (±0.71)	1–2.79	North Pacific	Boerger et al. (2010)
Loweina interrupta	28	*35	1.0	1–2.79	North Pacific	Boerger et al. (2010)
Myctophum aurolaternatum	460	*35	6.0 (±8.99)	1–2.79	North Pacific	Boerger et al. (2010)
Symbolophorus californiensis	78	*35	7.2 (±8.39)	1–2.79	North Pacific	Boerger et al. (2010)
Anderson's lanternfish (*Diaphus anderseni*)	13	15.4	1	/	North Pacific	Davison and Asch (2011)
Lanternfish (*Diaphus fulgens*)	7	28.6	1	/	North Pacific	Davison and Asch (2011)

(continued)

Table 10.9 (continued)

Species	Number studied	Percentage with plastic (%)	Mean number of particles per individual (±SD)	Type and size ingested (mm)	Location	Source
Boluin's lanternfish (*Diaphus phillipsi*)	1	100	1	Longest dimension 0.5	North Pacific	Davison and Asch (2011)
Coco's lanternfish (*Lobianchia gemellarii*)	3	33.3	1	/	North Pacific	Davison and Asch (2011)
Pearly lanternfish (*Myctophum nitidulum*)	25	16	1.5	Longest dimension 5.46	North Pacific	Davison and Asch (2011)
Order Perciformes						
White perch (*Morone americana*)	12	33	/	0.1–3 mm PS	New England, USA	Carpenter et al. (1972)
Bergall (*Tautogolabrus adspersus*)	6	<83	/	0.1–3 mm PS	New England, USA	Carpenter et al. (1972)
Goby (*Pomatoschistus minutus*)	200	0–25	/	1 mm PS	Severn estuary, U.K.	Kartar et al. (1976)
Stellifer brasiliensis	330	9.2	0.33–0.83	<1	Goiana estuary, Brazil	Dantas et al. (2012)
Stellifer stellifer	239	6.9	0.33–0.83	<1	Goiana estuary, Brazil	Dantas et al. (2012)
Eugerres brasilianus	240	16.3	1–5	1–5	Goiana estuary, Brazil	Ramos et al. (2012)
Eucinostomus melanopterus	141	9.2	1–5	1–5	Goiana estuary, Brazil	Ramos et al. (2012)
Diapterus rhombeus	45	11.1	1–5	1–5	Goiana estuary, Brazil	Ramos et al. (2012)
Horse mackerel (*Trachurus trachurus*)	100	1	1.0	1.52	North Sea	Foekema et al. (2013)

(continued)

Table 10.9 (continued)

Species	Number studied	Percentage with plastic (%)	Mean number of particles per individual (±SD)	Type and size ingested (mm)	Location	Source
Trachurus trachurus	56	28.6	1.5 (±0.7)	2.2 (±2.2)	English Channel	Lusher et al. (2013)
Yellowtail amberjack (*Seriola lalandi*)	19	10.5	1	0.5–10	North Pacific	Gassel et al. (2013)
Dragonet (*Callionymus lyra*)	50	38	1.79 (±0.9)	2.2 (±2.2)	English Channel	Lusher et al. (2013)
Red band fish (*Cepola macrophthalma*)	62	32.3	2.15 (±2.0)	2.0 (±1.9)	English Channel	Lusher et al. (2013)
Order Pleuronectiformes						
Winter flounder (*Pseudopleuronectes americanus*)	95	2.1	/	0.1–3 mm PS	New England, USA	Carpenter et al. (1972)
Flounder (*Platichthys flesus*)	/	/	/	1 mm PS	Severn estuary, U.K.	Kartar et al. (1973)
Platichthys flesus	1090	0–20.7	/	1 mm PS	Severn estuary, U.K.	Kartar et al. (1976)
Solenette (*Buglossidium luteum*)	50	26	1.23 (±0.4)	1.9 (±1.8)	English Channel	Lusher et al. (2013)
Thickback sole (*Microchirus variegatus*)	51	23.5	1.58 (±0.8)	2.2 (±2.2)	English Channel	Lusher et al. (2013)
Order Scorpaeniformes						
Grubby (*Myoxocephalus aenaeus*)	47	4.2	/	0.1–3 mm PS	New England, USA	Carpenter et al. (1972)
Striped searobin (*Prionotus evolans*)	1	100	1	0.1–3 mm PS	New England, USA	Carpenter et al. (1972)
Sea snail (*Liparis liparis liparis*)	220	0–25	/	1 mm PS	Severn estuary, U.K.	Kartar et al. (1976)

(continued)

Table 10.9 (continued)

Species	Number studied	Percentage with plastic (%)	Mean number of particles per individual (±SD)	Type and size ingested (mm)	Location	Source
Red gurnard (*Chelidonichthys cuculus*)	66	51.5	1.94 (±1.3)	2.1 (±2.1)	English Channel	Lusher et al. (2013)
Order Siluriformes						
Madamago sea catfish (*Cathorops spixii*)	60	18.3	0.47	1–4	Goiana estuary, Brazil	Possatto et al. (2011)
Catfish (*Cathorops* spp.)	60	33.3	0.55	1–4	Goiana estuary, Brazil	Possatto et al. (2011)
Pemecoe catfish (*Sciades herzbergii*)	62	17.7	0.25	1–4	Goiana estuary, Brazil	Possatto et al. (2011)
Order Stomiiformes						
Astronesthes indopacificus	7	*35	1.0	1–2.79	North Pacific	Boerger et al. (2010)
Hatchetfish (*Sternoptyx diaphana*)	4	25	1	Longest dimension 1.58 mm	North Pacific	Davison and Asch (2011)
Highlight hatchetfish (*Sternoptyx pseudobscura*)	6	16.7	1	Longest dimension 4.75 mm	North Pacific	Davison and Asch (2011)
Pacific black dragon (*Idiacanthus antrostomus*)	4	25	1	Longest dimension 0.5 mm	North Pacific	Davison and Asch (2011)
Order Zeiformes						
John Dory (*Zeus faber*)	46	47.6	2.65 (±2.5)	2.2 (±2.2)	English Channel	Lusher et al. (2013)

If mean not available range is reported. Standard deviation is reported where possible. *Represents percentage ingestion by total number of individuals, not separated by species

mortality rates were observed following a two-generation chronic toxicity test (12.5 μg mL^{-1}) (Lee et al. 2013). Although a third of gooseneck barnacle (*Lepas* spp.) stomachs examined contained microplastics, no adverse effect was reported for these filter feeders (Goldstein and Goodwin 2013). Interestingly, the stomachs of mass stranded Humboldt squids (*Dosidicus gigas*) contained plastic pellets (Braid et al. 2012). This large predatory cephalopod usually feeds at depth between 200 and 700 m. The route of uptake is unclear; the squid may have fed directly on sunken pellets, or on organisms with pellets in their digestive system.

10.3.1.2 Benthic Invertebrates

A number of benthic invertebrates have been studied under laboratory conditions to investigate the consequences of microplastic ingestion (Table 10.8). Laboratory feeding and retention trials have focused on direct exposure of invertebrates to microplastic particles (as summarised by Cole et al. 2011; Wright et al. 2013a). Exposure studies demonstrated that benthic invertebrates including lugworms (*Arenicola marina*), amphipods (*Orchestia gammarellus*) and blue mussels (*Mytilus edulis*) feed directly on microplastics (Thompson et al. 2004; Wegner et al. 2012), and deposit-feeding sea cucumbers even selectively ingested microplastic particles (Graham and Thompson 2009).

Although microplastic uptake was recorded for a number of species, organisms appear to reject microplastics before digestion and excrete microplastics after digestion. Pseudofaeces production is a form of rejection before digestion but requires additional energetic cost. Furthermore, prolonged pseudofaeces production could lead to starvation (Wegner et al. 2012). On the other hand, polychaete worms, sea cucumbers and sea urchins are able to excrete unwanted materials through their intestinal tract without suffering obvious harm (Thompson et al. 2004; Graham and Thompson 2009; Kaposi et al. 2014). Adverse effects of microplastic ingestion were reported for lugworms: weight loss was positively correlated with concentration of spiked sediments (40–1300 μm polystyrene) (Besseling et al. 2013). Similarly, Wright et al. (2013b) recorded significantly reduced feeding activity and significantly decreased energy reserves in lugworm exposed to 5 % un-plasticised polyvinyl chloride (U-PVC). Supressed feeding reduced energy assimilation, compromising fitness. At the chronic exposure level, either fewer particles were ingested overall or a lack of protein coating on the U-PVC may have weakened particle adhesion to the worm's feeding apparatus.

Several studies have raised concern for microplastic retention and transference between organisms' tissues. For example, microplastics were retained in the digestive tract of mussels, and transferred to the haemolymph system after three days (Browne et al. 2008). However, negative effects on individuals were not detected. Von Moos et al. (2012) tracked particles of high density polyethylene (HDPE) into the lysosomal system of mussels after three hours of exposure; particles were taken up by the gills and transferred to the digestive tract and lysosomal system, again triggering an inflammatory immune response. It should be noted, however, that while these studies succeeded in determining the pathways of microplastics in organisms the exposure concentrations used to achieve this goal exceeded those expected in the field, such that the results have to be treated with care.

Studies of microplastic ingestion by benthic invertebrates in the field are less common than laboratory studies. Murray and Cowie (2011) identified fibres of monofilament plastics that could be sourced to fibres of trawls and fragments of plastic bags in the intestines of the commercially valuable Norway lobster (*Nephrops norvegicus*). These results indicated that normal digestive processes do not eliminate some of the filaments as they cannot pass through the gastric mill system. Norway lobsters have various feeding modes, including scavenging and predation, and are not adapted to cut flexible filamentous materials (Murray and Cowie 2011). The identification of microplastics in organisms that are caught for commercial purposes and subsequently consumed whole (including guts) highlights the potential human health implications. For example, field-caught brown shrimps (*Crangon crangon*) (Pott 2014) and farmed and store-brought bivalves (De Witte et al. 2014; Van Cauwenberghe and Janssen 2014) had microplastics in their digestive system.

Invertebrates could be used as indicator species for environmental contamination. Species such as *Nephrops* are able to integrate seasonal variation in microplastic abundance, providing an accurate measure of environmental contamination (Welden, *pers. comm.*). Additional studies are required to understand the flux of microplastic within benthic sediments and the interaction between different species of benthic infauna feeding in/or manipulating the sediment, such as bivalves and worms. Benthic infauna could ingest and/or excrete microplastics, the individuals or their faecal pellets may in turn be ingested by secondary consumers, thus affecting higher trophic levels.

10.3.1.3 Fish

Some of the earliest studies noting ingestion of microplastics by wild-caught fish include coastal species from the USA (Carpenter et al. 1972) and the U.K. (Kartar et al. 1973, 1976). More recent studies from the NPCG reported microplastic (fibres, fragments and films) ingestion by mesopelagic fish (Boerger et al. 2010; Davison and Asch 2011; Choy and Drazen 2013). Estuarine environments and their inhabitants are also prone to plastic contamination, which is hardly surprising given the riverine input (e.g. Morritt et al. 2014). Estuarine fish affected include catfish, Ariidae, (23 % of individuals examined) and estuarine drums, Scianenidae, (7.9 % of individuals examined), which spend their entire life cycle in estuaries (Possatto et al. 2011; Dantas et al. 2012). Similarly, 13.4 % of bottom-feeding fish (Gerreidae) from a tropical estuary in northeast Brazil contained microplastics in their stomachs (Ramos et al. 2012). The authors suggested that ingestion occurred during suction feeding on biofilms.

Lusher et al. (2013) reported microplastic polymers from 10 fish species from the English Channel. Of the 504 fish examined, 37 % had ingested a variety of microplastics, the most common being polyamide and the semi-synthetic material rayon. Similarly, Boerger et al. (2010) recorded microplastics in 35 % planktivorous fish examined from the NPCG (94 % of which were plastic fragments). Fish from the northern North Sea ingested microplastics at significantly lower levels (1.2 %) compared to those from the southern North Sea (5.4 %) (Foekema et al. 2013). All the studies cited suggest direct ingestion as the prime route of exposure, either targeted as food or mistaken for prey items. No adverse effects of ingestion were reported. Consequently, studies are required to follow the route of microplastic ingestion in fish, to assess if microplastics are egested in faecal pellets as seen in invertebrates. Dos Santos and Jobling (1992) showed that microplastic beads (2 mm) were excreted quickly following ingestion, whereas larger beads (5 mm) were held for prolonged periods of time. This implies that larger items of plastic might pose a greater risk following ingestion whereas smaller microplastics are likely to be excreted along with natural faeces.

10.3.1.4 Sea Birds

Numerous studies have dealt with the ingestion of marine debris by sea birds (see Kühn et al. 2015). Microplastics and small plastic items have been isolated from birds targeted deliberately for dietary studies, dead cadavers, regurgitated samples and faeces (Table 10.10). Nearly 50 species of Procellariiformes (fulmars, petrels, shearwaters, albatrosses), known to feed opportunistically at the sea surface had microplastics in their stomachs. Ingested microplastics appeared to comprise primarily of pellets and user-fragments (Ryan 1987; Robards et al. 1995) although there was a decrease in the proportion of pellets ingested by birds from the south Atlantic between the 1980s and 2006 (Ryan 2008). This trend is also true for short-tailed shearwater (*Puffinus tenuirostris*) from the North Sea (Vlietstra and Parga 2002). In this case however, the mass of industrial plastics (pellets) have decreased by half and the mass of plastic fragments has tripled (van Franeker et al. 2011). It is possible that the shift in the type of plastic consumed may be explained by fragmentation of larger user-plastics into smaller microplastics, the accumulation of user-plastic over time and a decreased disposal of industrial plastics (Thompson et al. 2004), or simply by a stronger awareness of the presence of microplastics.

Seabirds appear to be able to remove microplastics from their digestive tracts as regurgitation has been observed in the boluses of glaucous-winged gulls (*Larus glaucescens*) (Lindborg et al. 2012). However, this suggests that parents expose their offspring to plastics during feeding. Juveniles of northern fulmars (*Fulmarus glacialis*) had more plastic in their intestines than adults (Kühn and van Franeker 2012), with higher quantities in areas of higher fishing and shipping traffic (van Franeker et al. 2011). Still, as the majority of birds examined did not die as a direct result of microplastic uptake, it can be concluded that microplastic ingestion does not affect seabirds as severely as macroplastic ingestion. To date, there have been no studies demonstrating nanometre-sized microplastics in sea birds. This could be because it is extremely difficult to control laboratory conditions in terms of contamination.

10.3.1.5 Marine Mammals

Only one study on microplastic ingestion by marine mammals has been published to date. Bravo Rebolledo et al. (2013) recorded microplastics in stomachs (11 %, n = 100) and intestines (1 %, n = 107) of harbour seals (*Phoca vitulina*). Direct microplastic ingestion by other species of marine mammals has not been observed. However, larger plastics items were identified in the stomachs of numerous ceta-ceans (46 % of all species; Baulch and Perry 2014, see also Kühn et al. 2015). The frequency of microplastic uptake by marine mammals is hitherto unknown, but could occur through filter feeding, inhalation at the water-air interface, or via trophic transfer from prey items. As baleen whales (Mysticetes) strain water between baleen plates, to trap planktonic organisms and small fish (Nemoto 1970), they may incidentally trap microplastics. Thus, their feeding mode may ren-der baleen whales more susceptible to direct microplastic ingestion than toothed (Odotocetes) or beaked whales (Ziphiids) which are active predators of squid and fish (Pauly et al. 1998). It is also likely that marine mammals are exposed to microplastic via trophic transfer from prey species. For example, microplastics were recorded from the scats of fur seals (*Arctocephalus* spp.) believed to origi-nate from lantern fish (*Electrona subaspera*) (Eriksson and Burton 2003).

Cetaceans were suggested as sentinels for microplastic pollution (Fossi et al. 2012a; Galgani et al. 2014). However, it is notoriously difficult to extract and sub-sequently assess microplastics from cetacean stomachs, the often large size and decomposition rate of stomachs make sampling almost impossible. Furthermore, strandings are infrequent and unpredictable. Although adaption of sampling meth-ods for smaller organisms such as fish and birds have the potential to be imple-mented, further work is necessary. The assessment of phthalate concentrations in the blubber of stranded fin whales (*Balaenoptera physalus*) (Fossi et al. 2012b, 2014) could serve as an indicator for the uptake of microplastics, but this raises other concerns as it is not possible to distinguish the origin of the phthalates. Exposure routes could be via micro- or macroplastics or simply from direct uptake of chemicals from the surrounding seawater into the blubber. Further work is essential to assess if microplastics significantly affect marine mammals.

10.3.1.6 Sea Turtles

Although all species of marine turtle ingest macroplastics (Derraik 2002; Schuyler et al. 2014; Kühn et al. 2015), only one study reported plastic pellets in the stom-achs of the herbivorous green turtles (*Chelonia mydas*) (Tourinho et al. 2010).

Table 10.10 Evidence of microplastic ingestion by seabirds mean (±SD unless * = SE)

Species	Number studied	Percentage with plastic (%)	Mean number of particles per individual	Type and mean size ingested (mm)	Location	Source
Order Procellariiformes						
Family Procellariidae						
Kerguelen petrel (*Aphrodroma brevirostris*)	26	3.8	1	Pellet	North Island, New Zealand	Reid (1981)
Aphrodroma brevirostris	13	8	0.2	Pellets max. mass: 0.0083 g	Gough Island, U.K. South Atlantic	Furness (1985b)
Aphrodroma brevirostris	63	22.2	/	20 % pellet	Breeding grounds, Southern Ocean	Ryan (1987)
Aphrodroma brevirostris	28	7	/	Fragments and pellets 3–6 mm	Antarctica	Ainley et al. (1990)
Cory's shearwater (*Calonectris diomedea*)	7	42.8	/	Pellets 46 %	Breeding grounds, Southern Ocean	Ryan (1987)
Calonectris diomedea	147	24.5	Stomach = 2 Gizzard = 3.1	Beads 63.7 %	North Carolina, USA	Moser and Lee (1992)
Calonectris diomedea	5	100	/	<10	Rio Grande do Sul, Brazil	Colabuono et al. (2009)
Calonectris diomedea	85	83	8 (±7.9)	3.9 (±3.5)	Canary Islands, Spain	Rodríguez et al. (2012)
Calonectris diomedea	49	96	14.6 (±24.0)	2.5 (±6.0[a])	Catalan coast, Mediterranean	Codina-García et al. (2013)
Cape petrel (*Daption capense*)	18	83.3	/	Pellets 48 %	Breeding grounds, Southern Ocean	Ryan (1987)
Daption capense	30	33	1.0	5.0	Ardery Island, Antarctica	van Franeker and Bell (1988)
Daption capense	105	14	/	Fragments and pellets 3–6 mm	Antarctica	Ainley et al. (1990)
Northern fulmar (*Fulmarus glacialis*)	3	100	7.6	Pellets 1–4 mm	California, USA	Baltz and Morejohn (1976)

(continued)

Table 10.10 (continued)

Species	Number studied	Percentage with plastic (%)	Mean number of particles per individual	Type and mean size ingested (mm)	Location	Source
Fulmarus glacialis	79	92	11.9	Pellets 50 %	The Netherlands, Arctic	van Franeker (1985)
Fulmarus glacialis	8	50	3.9	Pellets	St. Kilda, U.K.	Furness (1985a)
Fulmarus glacialis	13	92.3	10.6	Pellets	Foula. U.K.	Furness (1985a)
Fulmarus glacialis	1	100	1	Pellet, 4 mm	Oregon, USA	Bayer and Olson (1988)
Fulmarus glacialis	44	86.4	Stomach = 3 Gizzard = 14	Beads 91.9 %	North Carolina, USA	Moser and Lee (1992)
Fulmarus glacialis	19	84.2	Max: 26	Pellets 36 %	Alaska, USA	Robards et al. (1995)
Fulmarus glacialis	3	100	7.7	Pellets 48 %	Offshore, eastern North Pacific	Blight and Burger (1997)
Fulmarus glacialis	15	36	3.6 (±2.7)	7 (±4.0)	Davis Strait, Canadian Arctic	Mallory et al. (2006)
Fulmarus glacialis	1295	95	14.6 (±2.0*)– 33.2 (±3.3*)	>1.0	North Sea	van Franeker et al. (2011)
Fulmarus glacialis	67	92.5	36.8 (±9.8*)	>0.5	Eastern North Pacific	Avery-Gomm et al. (2012)
Fulmarus glacialis	58	79	6.0 (±0.9*)	>1.0	Westfjords, Iceland	Kühn and van Franeker (2012)
Fulmarus glacialis	176	93	26.6 (±37.5)	Fragments and pellets	Nova Scotia, Canada	Bond et al. (2014)
Antarctic fulmar (*Fulmarus glacialoides*)	84	2	/	Fragments and pellets 2–6 mm	Antarctica	Ainley et al. (1990)
Fulmarus glacialoides	9	79	/	<10	Rio Grande do Sul, Brazil	Colabuono et al. (2009)
Blue petrel (*Halobaena caerulea*)	27	100	/	Pellets	New Zealand	Reid (1981)
Halobaena caerulea	74	85.1	/	Pellets 69 %	Southern Ocean	Ryan (1987)

(continued)

Table 10.10 (continued)

Species	Number studied	Percentage with plastic (%)	Mean number of particles per individual	Type and mean size ingested (mm)	Location	Source
Halobaena caerulea	62	56	/	Fragments and pellets 3–6 mm	Antarctica	Ainley et al. (1990)
Prions *Pachyptila* spp.	/	/	/	Pellets	Gough Island, U.K. South Atlantic	Bourne and Imber (1982)
Salvin's prion (*Pachyptila salvini*)	663	20	/	Pellets 2.5–3.5 mm	Wellington, New Zealand	Harper and Fowler (1987)
Pachyptila salvini	31	51.6	/	Pellets 49 %	Breeding grounds, Southern Ocean	Ryan (1987)
Thin-billed prion (*Pachyptila belcheri*)	152	6.6	/	Pellets 2.5–3.5 mm	Wellington, New Zealand	Harper and Fowler (1987)
Pachyptila belcheri	32	68.7	/	Pellets 38 %	Breeding grounds, Southern Ocean	Ryan (1987)
Broad-billed prion (*Pachyptila vittata*)	31	39	0.6	Pellets max mass: 0.066	Gough Island, U.K. South Atlantic	Furness (1985b)
Pachyptila vittata	310	16.5	/	Pellets 2.5–3.5 mm	Wellington, New Zealand	Harper and Fowler (1987)
Pachyptila vittata	137	20.4	/	56 % pellet	Breeding grounds, Southern Ocean	Ryan (1987)
Pachyptila vittata	69	10	/	Fragments and pellets 3–6 mm	Antarctica	Ainley et al. (1990)
Pachyptila vittata	149	/	1987–1989 [b]1.73 ± 3.58	Pellets 43.6 %	Breeding grounds, Southern Ocean	Ryan (2008)
Pachyptila vittata	86	/	1999 [b]2.93 ± 3.80	Pellets 37.3 %	Breeding grounds, Southern Ocean	Ryan (2008)
Pachyptila vittata	95	/	2004 [b]2.66 ± 5.34	Pellets 15.4 %	Breeding grounds, Southern Ocean	Ryan (2008)

(continued)

Table 10.10 (continued)

Species	Number studied	Percentage with plastic (%)	Mean number of particles per individual	Type and mean size ingested (mm)	Location	Source
Antarctic prion (*Pachyptila desolata*)	35	14.3	/	Pellets 2.5–3.5 mm	Wellington. New Zealand	Harper and Fowler (1987)
Pachyptila desolata	88	47.7	/	Pellets 53 %	Breeding grounds, Southern Ocean	Ryan (1987)
Pachyptila desolata	2	100	1.0	6–8.1 mm	Heard Island, Australia	Auman et al. (2004)
Fairy prion (*Pachyptila turtur*)	105	96.2	/	Pellets 2.5–3.5 mm	Wellington, New Zealand	Harper and Fowler (1987)
Snow petrel (*Pagodroma nivea*)	363	1	/	Fragments and pellets 3–6 mm	Antarctica	Ainley et al. (1990)
White-chinned petrel (*Procellaria aequinoctialis*)	193	/	1983–1985 [b]1.66 (±3.04)	Pellets 38.2 %	Breeding grounds, Southern Ocean	Ryan (1987, 2008)
Procellaria aequinoctialis	526	/	2005–2006 [b]1.39 (±3.25)	16.2 % pellets	Breeding grounds, Southern Ocean	Ryan (2008)
Procellaria aequinoctialis	41	/	/	<10	Rio Grande do Sul, Brazil	Colabuono et al. (2009)
Procellaria aequinoctialis	34	44	/	<10	Rio Grande do Sul, Brazil	Colabuono et al. (2010)
Spectacled petrel (*Procellaria conspicillata*)	3	33	/	<10	Rio Grande do Sul, Brazil	Colabuono et al. (2010)
Procellaria conspicillata	9	/	/	<10	Rio Grande do Sul, Brazil	Colabuono et al. (2009)
Tahiti petrel (*Pseudobulweria rostrata*)	121	<1	1	Fragments	Tropical, North Pacific	Spear et al. (1995)
Atlantic petrel (*Pterodroma incerta*)	13	8	0.1	Pellets max mass: 0.0053 g	Gough Island, U.K. South Atlantic	Furness (1985b)

(continued)

Table 10.10 (continued)

Species	Number studied	Percentage with plastic (%)	Mean number of particles per individual	Type and mean size ingested (mm)	Location	Source
Pterodroma incerta	20	5	/	Pellets	Breeding grounds, Southern Ocean	Ryan (1987)
Great-winged petrel (*Pterodroma macroptera*)	13	7.6	/	Pellets	Breeding grounds, Southern Ocean	Ryan (1987)
Soft-plumaged petrel (*Pterodroma mollis*)	29	20.6	/	Pellets 22 %	Breeding grounds, Southern Ocean	Ryan (1987)
Pterodroma mollis	18	6	0.1	Pellets max. mass: 0.014 g	Gough Island, U.K. South Atlantic	Furness (1985b)
Juan Fernández petrel (*Pterodroma externa*)	183	<1	1	Pellets 3–5 mm	Offshore, North Pacific	Spear et al. (1995)
White-necked petrel (*Pterodroma cervicalis*)	12	8.3	5	Fragments 3–4 mm	Offshore, North Pacific	Spear et al. (1995)
Pycroft's petrel (*Pterodroma pycrofti*)	5	40	2.5 (±0.7)	Fragments Pellets 3–5 mm	Offshore, North Pacific	Spear et al. (1995)
White-winged petrel (*Pterodroma leucoptera*)	110	11.8	2.2 (±3.0)	Fragments 2–5 mm	Offshore, North Pacific	Spear et al. (1995)
Collared petrel (*Pterodroma brevipes*)	3	66.7	1	Pellets 2–5 mm	Offshore, North Pacific	Spear et al. (1995)
Black-winged petrel (*Pterodroma nigripenni*)	66	4.5	3.0 (±3.5)	Fragments 3–5 mm	Offshore, North Pacific	Spear et al. (1995)
Stejneger's petrel (*Pterodroma longirostris*)	46	73.9	6.8 (± 8.6)	Fragments and pellets 2–5 mm	Offshore, North Pacific	Spear et al. (1995)
Audubon's shearwater (*Puffinus lherminieri*)	119	5	Stomach = 1 Gizzard = 4.4	Beads 50 %	North Carolina, USA	Moser and Lee (1992)
Little shearwater (*Puffinus assimilis*)	13	8	0.8	Pellets max. mass: 0.12 g	Gough Island, U.K. South Atlantic	Furness (1985b)

(continued)

Table 10.10 (continued)

Species	Number studied	Percentage with plastic (%)	Mean number of particles per individual	Type and mean size ingested (mm)	Location	Source
Buller's shearwater (*Puffinus bulleri*)	3	100	8.5 (±8.6)	Fragments and pellets 2–8 mm	Tropical, North Pacific	Spear et al. (1995)
Pink-footed shearwater (*Puffinus creatopus*)	5	20	2.2	Pellets 1–4 mm	California, USA	Baltz and Morejohn (1976)
Great shearwater (*Puffinus gravis*)	24	100	/	Beads	Briar Island, Nova Scotia	Brown et al. (1981)
Puffinus gravis	13	85	12.2	Pellets max. mass: 1.13 g	Gough Island, U.K. South Atlantic	Furness (1985b)
Puffinus gravis	55	63.6	Stomach = 1 Gizzard = 13.2	Beads 91.2 %	North Carolina, USA	Moser and Lee (1992)
Puffinus gravis	50	66	1983–1985 [b]16.5 (±19.0)	Pellets 64.3 %	Breeding grounds, Southern Ocean	Ryan (1987, 2008)
Puffinus gravis	53	/	2005–2006 [b]11.8 (±18.9)	Pellets 11.3 %	Breeding grounds, Southern Ocean	Ryan (2008)
Puffinus gravis	19	89	/	<10 mm	Rio Grande do Sul, Brazil	Colabuono et al. (2009)
Puffinus gravis	6	100	/	Pellets < 3.2–5.3 mm	Rio Grande do Sul, Brazil	Colabuono et al. (2009)
Puffinus gravis	84	88	11.8 (±16.9)	Fragments and pellets	Nova Scotia, Canada	Bond et al. (2014)
Sooty shearwater (*Puffinus griseus*)	21	43	5.05	Pellets 1–4 mm	California, USA	Baltz and Morejohn (1976)
Puffinus griseus	5	100	/	Beads	Briar Island, Nova Scotia, Canada	Brown et al. (1981)
Puffinus griseus	36	58.3	11.4 (±12.2)	Fragments and pellets 3–20 mm	Tropical, North Pacific	Spear et al. (1995)
Puffinus griseus	218	88.5	/	Pellets 25.4 %	Offshore, North Pacific	Ogi (1990)

(continued)

Table 10.10 (continued)

Species	Number studied	Percentage with plastic (%)	Mean number of particles per individual	Type and mean size ingested (mm)	Location	Source
Puffinus griseus	20	75	3.4	Pellets 38 %	Offshore eastern North Pacific	Blight and Burger (1997)
Puffinus griseus	50	72	2.48 (±2.7)	Fragments and pellets	Nova Scotia, Canada	Bond et al. (2014)
Balearic shearwater (*Puffinus mauretanicus*)	46	70	2.5 (±2.9)	3.5 (±10.5[a])	Catalan coast, Mediterranean	Codina-García et al. (2013)
Christmas shearwater (*Puffinus nativitatis*)	5	40	1	Pellet 3–5 mm / Fragment 4 mm	Tropical, North Pacific	Spear et al. (1995)
Wedge-tailed shearwater (*Puffinus pacificus*) dark phase	23	4	2.5 (±2.1)	Fragments	Tropical, North Pacific	Spear et al. (1995)
	62	24.2	3.5 (±2.7)	Fragments and pellets		
Puffinus pacificus	20	60	Max: 11	Pellets 2–4 mm	Hawaii, USA	Fry et al. (1987)
Manx shearwater (*Puffinus puffinus*)	10	30	0.4	Pellets	Rhum, U.K.	Furness (1985a)
Puffinus puffinus	25	60	/	<10 mm	Rio Grande do Sul, Brazil	Colabuono et al. (2009)
Puffinus puffinus	6	17	/	Fragments	Rio Grande do Sul, Brazil	Colabuono et al. (2009)
Short-tailed shearwater (*Puffinus tenuirostris*)	6	100	19.8	Pellets 1–4 mm	California, USA	Baltz and Morejohn (1976)
Puffinus tenuirostris	324	81.8	/	Pellets 67.2 %	Offshore, North Pacific	Ogi (1990)
Puffinus tenuirostris	330	83.9	5.8 (±0.4*)	Pellets 2–5 mm	Bering Sea, North Pacific	Vlietstra and Parga (2002)
Puffinus tenuirostris	5	80	/	Fragments and pellets	Alaska, USA	Robards et al. 1995
Puffinus tenuirostris	99	100	15.1 (±13.2)	>2 mm	Offshore, North Pacific	Yamashita et al. (2011)
Puffinus tenuirostris	129	67	Adults: 4.5 / Juvenile: 7.1	Fragments 0.97–80.8 mm	North Stradbroke Island, Australia	Acampora et al. (2013)

(continued)

Table 10.10 (continued)

Species	Number studied	Percentage with plastic (%)	Mean number of particles per individual	Type and mean size ingested (mm)	Location	Source
Puffinus tenuirostris	12	100	27	>2 mm	Offshore, North Pacific	Tanaka et al. (2013)
Yelkouan shearwater (*Puffinus yelkouan*)	31	71	4.9 (±7.3)	4.0 (±13.0[a])	Catalan coast, Mediterranean	Codina-García et al. (2013)
Antarctic petrel (*Thalassoica antarctica*)	184	<1	/	Fragments and pellets 3–6 mm	Antarctica	Ainley et al. (1990
Family Hydrobatidae						
White-bellied storm petrel (*Fregetta grallaria*)	13	38	1.2	Pellets max. mass: 0.042 g	Gough Island, U.K. South Atlantic	Furness (1985b)
Fregetta grallaria	296	<1	1	Fragment	Offshore, North Pacific	Spear et al. (1995)
Fregetta grallaria	318	/	1987–89 [b]0.63 ± 1.13	Pellets 33.3 %	Breeding grounds, Southern Ocean	Ryan (2008)
Fregetta grallaria	137	/	1999 [b]0.63 ± 1.37	Pellets 20.9 %	Breeding grounds, Southern Ocean	Ryan (2008)
Fregetta grallaria	95	/	2004 [b]0.72 ± 1.87	Pellets 16.2 %	Breeding grounds, Southern Ocean	Ryan (2008)
Grey-backed storm petrel (*Garrodia nereis*)	11	27	0.3	Pellets max. mass: 0.010 g	Gough Island, U.K. South Atlantic	Furness (1985b)
Garrodia nereis	12	8.3	/	Pellets	Breeding grounds, Southern Ocean	Ryan (1987)
Fork-tailed storm petrel (*Oceanodroma furcata*)	/	/	/	<5 mm	Aleutian Islands, USA	Ohlendorf et al. (1978)
Oceanodroma furcata	21	85.7	Max.: 12	Pellets 22 %	Alaska, USA	Robards et al. (1995)
Oceanodroma furcata	7	100	20.1	Pellets 16 %	Offshore, eastern North Pacific	Blight and Burger (1997)
Leach's storm petrel (*Oceanodroma leucorhoa*)	15	40	1.66 (±1.2)	2–5 mm	Newfoundland, Canada	Rothstein (1973)

(continued)

Table 10.10 (continued)

Species	Number studied	Percentage with plastic (%)	Mean number of particles per individual	Type and mean size ingested (mm)	Location	Source
Oceanodroma leucorhoa	17	58.8	2.9	Pellets	St. Kilda, U.K.	Furness (1985a)
Oceanodroma leucorhoa	354	19.8	3.5 (±2.6)	Fragments and pellets 2–5 mm	Offshore, North Pacific	Spear et al. (1995)
Oceanodroma leucorhoa	64	48.4	Max.: 13	Monofilament line, fragments, pellets	Alaska, USA	Robards et al. (1995)
Wilson's storm petrel (*Oceanites oceanicus*)	20	75	4.4	2.9 mm	Ardery Island, Antarctica	van Franeker and Bell (1988)
Oceanites oceanicus	91	19	/	Fragments and pellets 3–6 mm	Antarctica	Ainley et al. (1990)
Oceanites oceanicus	133	38.3	Stomach = 1.4; Gizzard = 5.4	26 % beads	North Carolina, USA	Moser and Lee (1992)
White-faced storm petrel (*Pelagodroma marina*)	19	84	11.7	Pellets max. mass: 0.34 g	Gough Island, U.K. South Atlantic	Furness (1985b)
Pelagodroma marina	15	73.3	13.2 ± 9.5	Pellets 2–5 mm	Offshore, North Pacific	Spear et al. (1995)
Pelagodroma marina	24	20.8	/	Pellets 41 %	Southern Hemisphere	Ryan (1987)
Pelagodroma marina	253		1987–89 [b]3.98 ± 5.45	Pellets 69.6 %	Breeding grounds, Southern Ocean	Ryan (2008)
Pelagodroma marina	86	/	1999 [b]4.06 ± 5.93	Pellets 37.5 %	Breeding grounds, Southern Ocean	Ryan (2008)
Pelagodroma marina	5	/	2004 [b]2.52 ± 4.43	Pellets 13.5 %	Breeding grounds, Southern Ocean	Ryan (2008)
Family Diomedeidae						
Sooty albatross (*Phoebetria fusca*)	73	42.7	/	Pellets 34 %	Breeding grounds, Southern Ocean	Ryan (1987)
Laysan albatross (*Phoebastria immutabilis*)	/	52	/	Pellets 2–5 mm	Hawaiian Islands, USA	Sileo et al. (1990)

(continued)

Table 10.10 (continued)

Species	Number studied	Percentage with plastic (%)	Mean number of particles per individual	Type and mean size ingested (mm)	Location	Source
Black-footed albatross (*Phoebastria nigripes*)	/	12	/	Pellets 2–5 mm	Hawaiian Islands, USA	Sileo et al. (1990)
Phoebastria nigripes	3	100	5.3	Pellets 50 %	Offshore, eastern North Pacific	Blight and Burger (1997)
Black-browed albatross (*Thalassarche melanophris*)	2	100	3	Pellets 50 %	Rio Grande do Sul, Brazil	Tourinho et al. (2010)
Order Charadriiformes						
Family Laridae						
Audouin's gull (*Larus audouinii*)	15	13	49.3 (±77.7)	2.5 (±5.0*)	Catalan coast, Mediterranean	Codina-García et al. (2013)
Glaucous-winged gull (*Larus glaucescens*)	589 boluses	12.2	/	<10 mm	Protection Island, USA	Lindborg et al. (2012)
Heermann's Gull (*Larus heermanni*)	15	7	1	Pellets 1–4 mm	California, USA	Baltz and Morejohn (1976)
Mediterranean gull (*Larus melanocephalus*)	4	25	3.7 (±7.5)	3.0 (±5.0*)	Catalan coast, Mediterranean	Codina-García et al. (2013)
Yellow-legged gull (*Larus michahellis*)	12	33	0.9 (±1.5)	2.0 (±8.0*)	Catalan coast, Mediterranean	Codina-García et al. (2013)
Red-legged kittiwake (*Rissa brevirostris*)	15	26.7	/	Pellets 5 % Mean size: 5.87 mm	Alaska, USA	Robards et al. (1995)
Black-legged kittiwake (*Rissa tridactyla*)	8	8	4.0	Pellets 1–4 mm	California, USA	Baltz and Morejohn (1976)
Rissa tridactyla	256	7.8	Max.: 15	Pellets 5 %	Alaska, USA	Robards et al. (1995)
Rissa tridactyla	4	50	1.2 (±1.9)	3.0 (±5.0*)	Catalan coast, Mediterranean	Codina-García et al. (2013)

(continued)

Table 10.10 (continued)

Species	Number studied	Percentage with plastic (%)	Mean number of particles per individual	Type and mean size ingested (mm)	Location	Source
Family Alcidae						
Parakeet auklet (*Aethia psittacula*)	/	/	/	<5 mm	Aleutians Islands, USA	Ohlendorf et al. (1978)
Aethia psittacula	208	93.8	17.1	Pellets > 80 % 4.08 mm	Alaska, USA	Robards et al. (1995)
Tufted puffin (*Fratercula cirrhata*)	489	24.5	Max.: 51	Pellets 90 % 4.10 mm	Alaska, USA	Robards et al. (1995)
Fratercula cirrhata	9	89	3.3	Pellets 43 %	Offshore, North Pacific	Blight and Burger (1997)
Horned puffin (*Fratercula corniculata*)	/	/	/	<5 mm	Aleutian Islands, USA	Ohlendorf et al. (1978)
Fratercula corniculata	120	36.7	Max.:14	Pellets 40 % 5.03 mm	Alaska, USA	Robards et al. (1995)
Fratercula corniculata	2	50	1.5	Pellets	Offshore, North Pacific	Blight and Burger (1997)
Common murre (*Uria aalge*)	1	100	2011–2012 1	6.6 (±2.2)	Newfoundland, Canada	Bond et al. (2013)
Thick-billed murre (*Uria lomvia*)	186	11	0.2 (±0.8)	4.5 (±3.8)	Canadian Arctic	Provencher et al. (2010)
Uria lomvia	3	100	2011–2012 1	6.6 (±2.2)	Newfoundland, Canada	Bond et al. (2013)
Uria lomvia	1249	7.7	1985–1986 0.14 (±0.7*)	10.1 (±7.4)	Newfoundland, Canada	Bond et al. (2013)
Family Stercorariidae						
Brown skua (*Stercorarius antarcticus*)	494	22.7	/	Pellets 67 %	Breeding grounds, Southern Ocean	Ryan (1987)
Tristan skua (*Stercorarius hamiltoni*)	11	9	0.3 Max.: 3	Pellets Max. mass: 0.064 g	Gough Island, U.K. South Atlantic	Furness (1985b)

(continued)

Table 10.10 (continued)

Species	Number studied	Percentage with plastic (%)	Mean number of particles per individual	Type and mean size ingested (mm)	Location	Source
Long-tailed skua (*Stercorarius longicaudus*)	2	50	5	Fragments and pellets	Offshore, eastern North Pacific	Spear et al. (1995)
Arctic skua (*Stercorarius parasiticus*)	2	50	/	Pellets 50 %	Breeding grounds, Southern Ocean	Ryan (1987)
Family Scolopacidae						
Grey phalarope (*Phalaropus fulicarius*)	20	100	Max.: 36	Beads 1.7–4.4 mm	California, USA	Bond (1971)
Phalaropus fulicarius	7	85.7	5.7	Pellets	California, USA	Connors and Smith (1982)
Phalaropus fulicarius	2	50	/	Pellets	Breeding grounds, Southern Ocean	Ryan (1987)
Phalaropus fulicarius	55	69.1	Stomach = 1 Gizzard = 6.7	Beads 16.7 %	North Carolina, USA	Moser and Lee (1992)
Red-necked phalarope (*Phalaropus lobatus*)	36	19.4	Stomach = 0 Gizzard = 3.7	Beads 16.7 %	North Carolina, USA	Moser and Lee (1992)
Family Sternidae						
Sooty tern (*Onychoprion fuscatus*)	64	1.6	2	Pellets 4 mm	Offshore, eastern North Pacific	Spear et al. (1995)
White tern (*Gygis alba*)	8	12.5	5	Fragments 3–4 mm	Offshore, eastern North Pacific	Spear et al. (1995)
Order Suliformes						
Family Phalacrocoracidae						
Macquarie Shag (*Phalacrocorax (atriceps) purpurascens*)	64 boluses	7.8	1 per bolus	Polystyrene spheres	Macquarie Island, Australia	Slip et al. (1990)

[a]Median (±0.8) 95 % confidence intervals. Plastics found in total of 28 % birds

[b]This is total mean abundance of plastics, including pellets and user fragments; sizes of pellets are assumed to be 2–5 mm, according to recent literature

It is highly likely that other species of sea turtle also ingest microplastics incidentally or directly, depending on their feeding habits (Schuyler et al. 2014). Neonatal and oceanic post-hatchlings are generalist feeders (Bjorndal 1997), targeting plankton from surface waters and microplastic uptake may occur. Trophic transfer from prey items could be a pathway to larger individuals; loggerhead (*Caretta caretta*) and Kemp's Ridley (*Lepidochelys kempii*) turtles are carnivores, feeding on crustaceans and bivalves (Bjorndal 1997), which ingest microplastics (e.g. Browne et al. 2008). Flatbacks (*Natator depressa*) are also carnivores but feed on soft bodied invertebrates (Bjorndal 1997), including sea cucumbers, which again, ingest microplastics (Graham and Thompson 2009). Leatherbacks (*Dermochelys coriacea*) feed on gelatinous organisms (Bjorndal 1997) and are thus more likely to ingest macroplastics because of their size and similarity to prey items. If microplastics are ingested they could affect sea turtle growth and development if they are not egested. Additional work is required to understand whether turtles actively ingest microplastics, and if so, the extent of the harm caused.

10.3.2 Trophic Transfer

Absorption and ingestion of microplastics by organisms from the primary trophic level, e.g. phytoplankton and zooplankton, could be a pathway into the food chain (Bhattacharya et al. 2010). Many species of zooplankton undergo a diurnal migration. Migrating zooplankton could be considered a vector of microplastic contamination to greater depths of the water column and its inhabitants, either through predation or the production of faecal pellets sinking to the seafloor (Wright et al. 2013a). Only a few studies deal with the potential for microplastics to be transferred between trophic levels following ingestion. Field observation highlighted the presence of microplastics in the scat of fur seals (*Arctocephalus* spp.) and Eriksson and Burton (2003) suggested that microplastics had initially been ingested by the fur seals' prey, the plankton feeding Mycophiids. In feeding experiments, Farrell and Nelson (2013) identified microplastic in the gut and haemolymph of the shore crab (*Carcinus maenas*), which had previously been ingested by blue mussels (*Mytilus edulis*). There was large variability in the number of microspheres in tissues samples, and the results have to be treated with caution as the number of individuals was low and the exposure levels used exceeded those from the field. Similarly, *Nephrops*-fed fish, which had been seeded with microplastic strands of polypropylene rope were found to ingest but not to excrete the strands (Murray and Cowie 2011), again implying potential trophic transfer. As mentioned above, microplastics were also detected in cod, whiting, haddock, bivalves and brown shrimp, which are consumed by humans and raises concerns about trophic transfer to humans and human exposure (see Galloway 2015). Further studies are required to increase our understanding of trophic transfer.

10.3.3 Microplastic Effect on Habitats

Surfaces of buoyant microplastics provide habitats for rafting organisms. For example, pelagic insects (*Halobates micans* and *H. sericeus*) utilize microplastic pellets for oviposition (Goldstein et al. 2012; Majer et al. 2012). Indeed, Goldstein et al. (2012) attributed an overall increase in *H. sericeus* and egg densities in the NPCG to high concentrations of microplastics. Likewise, plastics serve as a floating habitat for bacterial colonisation (Lobelle and Cunliffe 2011). Microorganisms including Bacillus bacteria (mean: 1664 ± 247 individuals mm^{-2}) and pennate diatoms (mean: 1097 ± 154 individuals mm^{-2}) were identified on plastic items from the North Pacific gyre (Carson et al. 2013). These studies suggest that microplastics affect the distribution and dispersal of marine organisms and may represent vectors to alien invasion. Plastics colonised by pathogenic viruses or bacteria may spread the potential for disease, but there is currently no evidence to support this hypothesis.

Microplastic buried in sediments could have fundamental impacts on marine biota as they increase the permeability of sediment and decrease thermal diffusivity (Carson et al. 2011). This may affect temperature-dependent processes. For example, altered temperatures during incubation can bias the sex ratios of sea turtle eggs. At 30 °C, equal numbers of males and female embryos develop, whereas at temperatures <28 °C all embryos become male (Yntema and Mrosovsky 1982). With microplastics in sediments it will take longer to reach maximum temperatures because of its increased permeability. Therefore, eggs may require a longer incubation period, with more male hatchlings because of the insulating effect. Microplastic concentrations as low as 1.5 can decrease maximum temperatures by 0.75 °C (Carson et al. 2011), which has important implications for sexual bias in sea turtles including loggerhead turtles (*Caretta caretta*) and hawksbill turtles (*Eretmochelys imbricata*) (Yntema and Mrosovsky 1982; Mrosovsky et al. 1992). Changes in the sediment temperatures could also affect infaunal organisms as it may affect enzymatic and other physiological processes, feeding and growth rates, locomotory speeds, reproduction and ultimately population dynamics. However, this remains speculative until further researched.

10.3.4 Summary

Microplastic ingestion has been documented for a range of marine vertebrates and invertebrates (Fig. 10.1). Interactions were recorded primarily during controlled laboratory studies, but results from field sampling of wild populations also indicate microplastic ingestion. In the case of some invertebrates, adverse physiological and biological effects were reported. The biological repercussions depend on to the size of microplastics with smaller sizes having greater effects on organisms at the cellular level. In the micrometre range, plastics are readily ingested and egested whereas

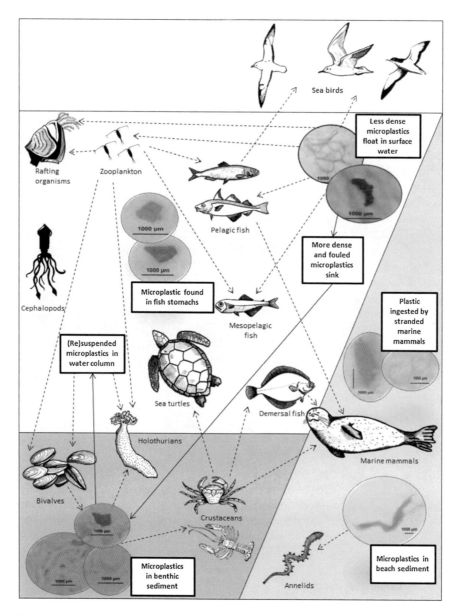

Fig. 10.1 Microplastic interactions in the marine environment including environmental links (*solid arrows*) and biological links (*broken arrows*), which highlights potential trophic transfer (Photos of microplastics: A. Lusher)

nanometre-sized plastics can pass through cell membranes. Acute exposure experiments demonstrated significant biological effects including weight loss, reduced feeding activity, increased phagocytic activity and transference to the lysosomal (storage) system. Larger microplastics (2–5 mm) may take longer to pass from the stomachs of organisms and could be retained in the digestive system, potentially increasing the exposure time to adsorbed toxins (see Rochman 2015).

It is important to determine the ecological effects of microplastic ingestion. Studies are required to assess the contamination of more species of fish, marine mammals and sea turtles, as well as consequences of microplastic uptake and retention. Further research is necessary to determine the limits of microplastic translocation between tissues, and assess the differences between multiple polymer types and shapes. It is likely that additional species of invertebrate ingest microplastics in wild populations, as fibres and fragments found in the field are actively selected in experiments. Although some organisms appear to be able to differentiate between microplastics and prey, and microplastic excretion has been recorded. Without knowledge of retention and egestion rates of field populations, it is difficult to deduce ecological consequences. There is some evidence to suggest that microplastics enter the food chain and transfer of microplastics between trophic levels implies bioaccumulation and biomagnification. Despite concerns raised by ingestion in the marine environment, the effects of microplastic ingestion in natural populations and the implications for food webs are not understood. Such knowledge is crucial in order to be able to develop and implement effective management strategies (Thompson et al. 2009). Additional studies are required to understand the flux of microplastic from benthic sediments to the infauna. Lastly, microplastics provide open ocean habitats for colonisation by invertebrates, bacteria and viruses. As a result, these organisms can be transported over large distances by ocean currents and/or through the water column (Kiessling et al. 2015).

10.4 Conclusion

Microplastics have been found in almost every marine habitat around the world, and plastic density along with ocean currents appears to have a significant effect on their distribution. Modelling studies suggest that floating debris accumulates in ocean gyres but this is dependent on the composition and shape of individual polymers. The widespread distribution and accumulation of microplastics raises concerns regarding the interaction and potential effects of microplastics on marine organisms. As microplastics interact with plankton and sediments, both suspension and deposit feeders may accidentally or selectively ingest microplastics. Despite concerns regarding ingestion, only a limited number of studies examined microplastic ingestion in the field. Knowledge of the retention rates of microplastics would enable estimations of the impacts of microplastic uptake. If rejection occurs before digestion, microplastics might pose less of a threat to organisms than initially assumed. However, there could be energetic costs associated with the production of pseudofaeces. Laboratory studies can be used to determine the end point of microplastic ingestion, and would benefit from using multiple types of microplastics to simulate field conditions. Unfortunately, it is difficult to establish a direct link between microplastics and adverse effects on marine biota experimentally. Furthermore, due to the difficult nature of field studies, it will be harder to understand effects on natural populations.

As microplastic research is still in its infancy, there are many more unanswered questions, the answers to which are required to build on current knowledge to develop a clearer picture of the impact of microplastics in the sea.

Acknowledgments   The author would like to thank Emily Lorch, Julia Hemprich, Chelsea Rochman, Rick Officer and Ian O'Connor for their useful comments on an earlier draft. Bart Koelmans and an anonymous reviewer improved an earlier version of the manuscript. Marta Bolgan for the illustrations in Fig. 10.1. The author was awarded an Irish Research Council postgraduate scholarship [Project ID: GOIPG/2013/284] and a Galway-Mayo Institute of Technology 40th Anniversary studentship to conduct her Ph.D. research.

References

Acampora, H., Schuyler, Q. A., Townsend, K. A., & Denise, B. (2013). Comparing plastic inges-tion in juvenile and adult stranded short-tailed shearwaters (*Puffinus tenuirostris*) in eastern Australia. *Marine Pollution Bulletin, 78*, 63–68.

Ainley, D. G., Fraser, W. R. & Spear, L. B. (1990). The incidence of plastic in the diets of Antarctic Seabirds. In *Proceedings of the Second International Conference on Marine Debris* (pp. 682–691). Honolulu, Hawaii, 2–7 April 1989.

Andrady, A. L., & Neal, M. A. (2009). Applications and societal benefits of plastics. *Philosophical Transactions of the Royal Society of London B, 364*(1526), 1977–1984.

Antunes, J. C., Frias, J. G. L., Micaelo, A. C., & Sobral, P. (2013). Resin pellets from beaches of the Portuguese coast and adsorbed persistent organic pollutants. *Estuarine, Coastal and Shelf Science, 130*, 62–69.

Arthur, C., Baker, J. & Bamford, H. (2009). In *Proceedings of the International Research Workshop on the Occurrence, Effects and Fate of Microplastic Marine Debris, NOAA Technical Memorandum NOS-OR & R-30.NOAA* (p. 530). Silver Spring, September 9–11, 2008.

Ashton, K., Holmes, L., & Turner, A. (2010). Association of metals with plastic production pel-lets in the marine environment. *Marine Pollution Bulletin, 60*(11), 2050–2055.

Auman, H. J., Woehler, E. J., Riddle, M. J., & Burton, H. (2004). First evidence of ingestion of plastic debris by seabirds at sub-Antarctic Heard Island. *Marine Ornithology, 32*(1), 105–106.

Austin, H. M., & Stoops-Glass, P. M. (1977). The distribution of polystyrene spheres and nibs in Block Island Sound during 1972–1973. *Chesapeake Science, 18*, 89–91.

Avery-Gomm, S., O'Hara, P. D., Kleine, L., Bowes, V., Wilson, L. K., & Barry, K. L. (2012). Northern fulmars as biological monitors of trends of plastic pollution in the eastern North Pacific. *Marine Pollution Bulletin, 64*(9), 1776–1781.

Ballent, A., Pando, S., Purser, A., Juliano, M. F., & Thomsen, L. (2013). Modelled transport of benthic marine microplastic pollution in the Nazaré Canyon. *Biogeosciences, 10*(12), 7957–7970.

Baltz, D. M., & Morejohn, G. V. (1976). Evidence from seabirds of plastic particle pollution off central California. *Western Birds, 7*(3), 111–112.

Barnes, D. K. A., Galgani, F., Thompson, R. C., & Barlaz, M. (2009). Accumulation and fragmentation of plastic debris in global environments. *Philosophical Transactions of the Royal Society of London B, 364*(1526), 1985–1998.

Barnes, D. K. A., Walters, A., & Gonçalves, L. (2010). Macroplastics at sea around Antarctica. *Marine Environmental Research, 70*(2), 250–252.

Bayer, R. D., & Olson, R. E. (1988). Plastic particles in 3 Oregon fulmars. *Oregon Birds, 14*, 155–156.

Baulch, S., & Perry, C. (2014). Evaluating the impacts of marine debris on cetaceans. *Marine Pollution Bulletin, 80*(1), 210–221.

Baztan, J., Carrasco, A., Chouinard, O., Cleaud, M., Gabaldon, J. E., Huck, T., et al. (2014). Protected areas in the Atlantic facing the hazards of micro-plastic pollution: First diagnosis of three islands in the Canary Current. *Marine Pollution Bulletin, 80*(1–2), 302–311.

Bergmann, M., & Klages, M. (2012). Increase of litter at the Arctic deep-sea observatory HAUSGARTEN. *Marine Pollution Bulletin, 64*(12), 2734–2741.

Besseling, E., Wegner, A., Foekema, E. M., van den Heuvel-Greve, M. J., & Koelmans, A. A. (2013). Effects of microplastic on fitness and PCB bioaccumulation by the Lugworm *Arenicola marina* (L.). *Environmental Science & Technology, 47*, 593–600.

Bhattacharya, P., Turner, J. P., & Ke, P.-C. (2010). Physical adsorption of charged plastic nanoparticles affects algal photosynthesis. *The Journal of Physical Chemistry C, 114*(39), 16556–16561.

Bjorndal, K. A. (1997). Foraging ecology and nutrition of sea turtles. In P. L. Lutz & J. A. Musick (Eds.), *The biology of sea turtles* (pp. 199–231). Boca Raton, Florida: CRC Press.

Blight, L. K., & Burger, A. E. (1997). Occurrence of plastic particles in sea-birds from the eastern North Pacific. *Marine Pollution Bulletin, 34*(5), 323–325.

Boerger, C. M., Lattin, G. L., Moore, S. L., & Moore, C. J. (2010). Plastic ingestion by planktivorous fishes in the North Pacific Central Gyre. *Marine Pollution Bulletin, 60*(12), 2275–2278.

Bolton, T. F., & Havenhand, J. N. (1998). Physiological versus viscosity-induced effects of an acute reduction in water temperature on microsphere ingestion by trochophore larvae of the serpulid polychaete *Galeolaria caespitosa*. *Journal of Plankton Research, 20*(11), 2153–2164.

Bond, S. I. (1971). Red phalarope mortality in Southern California. *Western Birds, 2*(3), 97.

Bond, A. L., Provencher, J. F., Elliot, R. D., Ryan, P. C., Rowe, S., Jones, I. L., et al. (2013). Ingestion of plastic marine debris by common and thick-billed Murres in the northwestern Atlantic from 1985 to 2012. *Marine Pollution Bulletin, 77*(1), 192–195.

Bond, A. L., Provencher, J. F., Daoust, P. Y. & Lucas, Z. N. (2014). Plastic ingestion by fulmars and shearwaters at Sable Island, Nova Scotia, Canada. *Marine Pollution Bulletin, 87*(1), 68–75.

Bourne, W. R. P., & Imber, M. J. (1982). Plastic pellets collected by a prion on Gough Island, central South Atlantic Ocean. *Marine Pollution Bulletin, 13*(1), 20–21.

Braid, H. E., Deeds, J., DeGrasse, S. L., Wilson, J. J., Osborne, J., & Hanner, R. H. (2012). Preying on commercial fisheries and accumulating paralytic shellfish toxins: a dietary analysis of invasive *Dosidicus gigas* (Cephalopoda Ommastrephidae) stranded in Pacific Canada. *Marine Biology, 159*(1), 25–31.

Bravo Rebolledo, E. L., van Franeker, J. A., Jansen, O. E., & Brasseur, S. M. (2013). Plastic ingestion by harbour seals (*Phoca vitulina*) in The Netherlands. *Marine Pollution Bulletin, 67*(1), 200–202.

Brilliant, M. G. S., & MacDonald, B. A. (2000). Postingestive selection in the sea scallop *Placopecten magellanicus* (Gmelin): The role of particle size and density. *Journal of Experimental Marine Biology and Ecology, 253*, 211–227.

Brilliant, M. G. S., & MacDonald, B. A. (2002). Postingestive selection in the sea scallop (*Placopecten magellanicus*) on the basis of chemical properties of particles. *Marine Biology, 141*, 457–465.

Brown, R. G. B., Barker, S. P., Gaskin, D. E., & Sandeman, M. R. (1981). The foods of great and sooty shearwaters *Puffinus gravis* and *P. griseus* in eastern Canadian waters. *Ibis, 123*(1), 19–30.

Browne, M. A. (2015). Sources and pathways of microplastic to habitats. In M. Bergmann, L. Gutow, & M. Klages (Eds.), *Marine anthropogenic litter* (pp. 229–244). Springer, Berlin.

Browne, M. A., Dissanayake, A., Galloway, T. S., Lowe, D. M., & Thompson, R. C. (2008). Ingested microscopic plastic translocates to the circulatory system of the mussel, *Mytilus edulis* (L.). *Environmental Science & Technology, 42*(13), 5026–5031.

Browne, M. A., Galloway, T. S., & Thompson, R. C. (2010). Spatial patterns of plastic debris along estuarine shorelines. *Environmental Science & Technology, 44*(9), 3404–3409.

Browne, M. A., Crump, P., Niven, S. J., Teuten, E., Tonkin, A., Galloway, T., et al. (2011). Accumulation of microplastic on shorelines worldwide: Sources and sinks. *Environmental Science & Technology, 45*(21), 9175–9179.

Browne, M. A., Niven, S. J., Galloway, T. S., Rowland, S. J., & Thompson, R. C. (2013). Microplastic moves pollutants and additives to worms, reducing functions linked to health and biodiversity. *Current Biology, 23*(23), 2388–2392.

Carpenter, E. J., & Smith, K. L. (1972). Plastics on the Sargasso Sea surface. *Science, 175*, 1240–1241.

Carpenter, E. J., Anderson, S. J., Harvey, G. R., Miklas, H. P., & Peck, B. B. (1972). Polystyrene spherules in coastal waters. *Science, 178*, 749–750.

Carson, H. S., Colbert, S. L., Kaylor, M. J., & McDermid, K. J. (2011). Small plastic debris changes water movement and heat transfer through beach sediments. *Marine Pollution Bulletin, 62*(8), 1708–1713.

Carson, H. S., Nerheim, M. S., Carroll, K. A., & Eriksen, M. (2013). The plastic-associated microorganisms of the North Pacific Gyre. *Marine Pollution Bulletin, 75*(1), 126–132.

CBD (2012). Secretariat of the Convention on Biological Diversity and the Scientific and Technical Advisory Panel-GEF. Impacts of marine debris on biodiversity: Current status and potential solutions, Montreal, Technical Series No. 67, 61 pp.

Chua, E., Shimeta, J., Nugegoda, D., Morrison, P. D., & Clarke, B. O. (2014). Assimilation of Polybrominated Diphenyl Ethers from microplastics by the marine amphipod, *Allorchestes compressa*. *Environmental Science & Technology, 48*(14), 8127–8134.

Choy, C. A., & Drazen, J. C. (2013). Plastic for dinner? Observations of frequent debris ingestion by pelagic predatory fishes from the central North Pacific. *Marine Ecology Progress Series, 485*, 155–163.

Christaki, U., Dolan, J. R., Pelegri, S., & Rassoulzadegan, F. (1998). Consumption of picoplankton-size particles by marine ciliates: effects of physiological state of the ciliate and particle quality. *Limnology and Oceanography, 43*(3), 458–464.

Claessens, M., De Meester, S., Van Landuyt, L., De Clerck, K., & Janssen, C. R. (2011). Occurrence and distribution of microplastics in marine sediments along the Belgian coast. *Marine Pollution Bulletin, 62*(10), 2199–2204.

Claessens, M., Van Cauwenberghe, L., Vandegehuchte, M. B., & Janssen, C. R. (2013). New techniques for the detection of microplastics in sediments and field collected organisms. *Marine Pollution Bulletin, 70*, 227–233.

Codina-García, M., Militão, T., Moreno, J., & González-Solís, J. (2013). Plastic debris in Mediterranean seabirds. *Marine Pollution Bulletin, 77*(1), 220–226.

Colabuono, F. I., Barquete, V., Domingues, B. S., & Montone, R. C. (2009). Plastic ingestion by Procellariiformes in southern Brazil. *Marine Pollution Bulletin, 58*(1), 93–96.

Colabuono, F. I., Taniguchi, S., & Montone, R. C. (2010). Polychlorinated biphenyls and organochlorine pesticides in plastics ingested by seabirds. *Marine Pollution Bulletin, 60*(4), 630–634.

Cole, M., Lindeque, P., Halsband, C., & Galloway, T. S. (2011). Microplastics as contaminants in the marine environment: A review. *Marine Pollution Bulletin, 62*(12), 2588–2597.

Cole, M., Lindeque, P. K., Fileman, E. S., Halsband, C., Goodhead, R., Moger, J., et al. (2013). Microplastic ingestion by zooplankton. *Environmental Science & Technology, 47*(12), 6646–6655.

Cole, M., Webb, H., Lindeque, P. K., Fileman, E. S., Halsband, C. & Galloway, T. S. (2014a). Isolation of microplastics in biota-rich seawater samples and marine organisms. *Scientific Reports, 4*(4528).

Cole, M., Lindeque, P. K., Fileman, E. S., Halsband, C. & Galloway, T. S. (2014b). Impact of microplastics on feeding, function and decundity in the copepod *Calanus helgolandicus*. Platform presentation, International workshop on fate and impact of microplastics in marine ecosystems (MICRO2014). Plouzane (France), 13–15 January 2014.

Collignon, A., Hecq, J.-H., Glagani, F., Voisin, P., Collard, F., & Goffart, A. (2012). Neustonic microplastic and zooplankton in the North Western Mediterranean Sea. *Marine Pollution Bulletin, 64*(4), 861–864.

Collignon, A., Hecq, J. H., Galgani, F., Collard, F., & Goffart, A. (2014). Annual variation in neustonic micro-and meso-plastic particles and zooplankton in the Bay of Calvi (Mediterranean–Corsica). *Marine Pollution Bulletin, 79*(1–2), 293–298.

Colton, J. B., Burns, B. R., & Knapp, F. D. (1974). Plastic particles in surface waters of the northwestern Atlantic. *Science, 185*(4150), 491–497.

Connors, P. G., & Smith, K. G. (1982). Oceanic plastic particle pollution: suspected effect on fat deposition in red phalaropes. *Marine Pollution Bulletin, 13*(1), 18–20.

Cooper, D. A., & Corcoran, P. L. (2010). Effects of mechanical and chemical processes on the degradation of plastic beach debris on the island of Kauai, Hawaii. *Marine Pollution Bulletin, 60*(5), 650–654.

Corcoran, P. L., Biesinger, M. C., & Grifi, M. (2009). Plastics and beaches: A degrading relationship. *Marine Pollution Bulletin, 58*(1), 80–84.

Costa, M. F., Ivar, J. A., Christina, M., Ângela, B. A., Paula, S., Ivar do Sul, J. A., et al. (2010). On the importance of size of plastic fragments and pellets on the strandline: a snapshot of a Brazilian beach. *Environmental Monitoring and Assessment, 168*(1–4), 299–304.

Costa, M. F., Silva-Cavalcanti, J. S., Barbosa, C. C., Portugal, J. L. & Barletta, M. (2011). Plastics buried in the inter-tidal plain of a tropical estuarine ecosystem. *Journal of Coastal Research SI, 64*, 339–343.

Cózar, A., Echevarría, F., González-Gordillo, J. I., Irigoien, X., Úbeda, B., Hernández-León, S., et al. (2014). Plastic debris in the open ocean. *Proceedings of the National Academy of Sciences of the United States of America, 111*(28), 10239–10244.

Dantas, D. V., Barletta, M., & Da Costa, M. F. (2012). The seasonal and spatial patterns of ingestion of polyfilament nylon fragments by estuarine drums (Sciaenidae). *Environmental Science and Pollution Research International, 19*(2), 600–606.

Davison, P., & Asch, R. (2011). Plastic ingestion by mesopelagic fishes in the North Pacific Subtropical Gyre. *Marine Ecology Progress Series, 432*, 173–180.

Day, R. H., & Shaw, D. G. (1987). Patterns in the abundance of pelagic plastic and tar in the North Pacific Ocean, 1976–1985. *Marine Pollution Bulletin, 18*(6), 311–316.

Day, R. H., Shaw, D. G. & Ignell, S. E. (1990). The quantitative distribution and characteristics of neuston plastic in the North Pacific Ocean, 1985–88. In R. S. Shomura & M. L. Godfrey (Eds.) *Proceedings of the Second International Conference on Marine Debris* (pp. 2–7). Honolulu, Hawaii. U.S Dep. Commerce., NOAA Technical. Memorandum. NMFS, NOAA-TM-SWFSC-154, 2–7 April 1989.

Dekiff, J. H., Remy, D., Klasmeier, J., & Fries, E. (2014). Occurrence and spatial distribution of microplastics in sediments from Norderney. *Environmental Pollution, 186*, 248–256.

de Lucia, G., Caliani, I., Marra, S., Camedda, A., Coppa, S., Alcaro, L., et al. (2014). Amount and distribution of neustonic micro-plastic off the Western Sardinian coast (Central-Western Mediterranean Sea). *Marine Environmental Research, 100*, 10–16.

De Witte, B., Devriese, L., Bekaert, K., Hoffman, S., Vandermeersch, G., Cooreman, K., et al. (2014). Quality assessment of the blue mussel (*Mytilus edulis*): Comparison between commercial and wild types. *Marine Pollution Bulletin, 85*(1), 146–155.

Derraik, J. G. B. (2002). The pollution of the marine environment by plastic debris: A review. *Marine Pollution Bulletin, 44*(9), 842–852.

Desforges, J. P. W., Galbraith, M., Dangerfield, N., & Ross, P. S. (2014). Widespread distribution of microplastics in subsurface seawater in the NE Pacific Ocean. *Marine Pollution Bulletin, 79*(1–2), 94–99.

Devriese, L., Vandendriessche, S., Theetaert, H., Vandermeersch, G., Hostens, K. & Robbens, J. (2014). Occurrence of synthetic fibres in brown shrimp on the Belgian part of the North Sea. Platform presentation, International workshop on fate and impact of microplastics in marine ecosystems (MICRO2014). Plouzane (France), 13–15 January 2014.

Dos Santos, J., & Jobling, M. (1992). A model to describe gastric evacuation in cod (*Gadus morhua* L.) fed natural prey. *ICES Journal of Marine Science: Journal du Conseil, 49*(2), 145–154.

Doyle, M. J., Watson, W., Bowlin, N. M., & Sheavly, S. B. (2011). Plastic particles in coastal pelagic ecosystems of the Northeast Pacific ocean. *Marine Environmental Research, 71*(1), 41–52.

Dubaish, F., & Liebezeit, G. (2013). Suspended microplastics and black carbon particles in the Jade System, Southern North Sea. *Water, Air, and Soil pollution, 224*(2), 1–8.

Durrieu de Madron, X. D., Houpert, L., Puig, P., Sanchez-Vidal, A., Testor, P., Bosse, A., et al. (2013). Interaction of dense shelf water cascading and open-sea convection in the Northwestern Mediterranean during winter 2012. *Geophysical Research Letters, 40*(7), 1379–1385.

Endo, S., Takizawa, R., Okuda, K., Takada, H., Chiba, K., Kanehiro, H., et al. (2005). Concentration of polychlorinated biphenyls (PCBs) in beached resin pellets: Variability among individual particles and regional differences. *Marine Pollution Bulletin, 50*(10), 1103–1114.

Endo, S., Yuyama, M., & Takada, H. (2013). Desorption kinetics of hydrophobic organic contaminants from marine plastic pellets. *Marine Pollution Bulletin, 74*(1), 125–131.

Engler, R. E. (2012). The complex interaction between marine debris and toxic chemicals in the ocean. *Environmental Science & Technology, 46*(22), 12302–12315.

Eriksen, M., Lebreton, L. C. M., Carson, H. S., Thiel, M., Moore, C. J., Borerro, J. C., et al. (2014). Plastic pollution in the world's oceans: More than 5 Trillion plastic pieces weighing over 250,000 tons afloat at sea. *PLoS ONE, 9*, e111913.

Eriksen, M., Maximenko, N., & Thiel, M. (2013). Plastic pollution in the South Pacific subtropical gyre. *Marine Pollution Bulletin, 68*(1–2), 71–76.

Eriksson, C., & Burton, H. (2003). Origins and biological accumulation of small plastic particles in fur seals from Macquarie Island. *AMBIO: A Journal of the Human Environment, 32*, 380–384.

Farrell, P., & Nelson, K. (2013). Trophic level transfer of microplastic: *Mytilus edulis* (L.) to *Carcinus maenas* (L.). *Environmental Pollution, 177*, 1–3.

Fischer, V., Elsner, N. O., Brenke, N., Schwabe, E., & Brandt, A. (2015). Plastic pollution of the Kuril-Kamchatka-trench area (NW Pacific). *Deep-Sea Research II, 111*, 399–405.

Foekema, E. M., De Gruijter, C., Mergia, M. T., van Franeker, J. A., Murk, T. J., & Koelmans, A. A. (2013). Plastic in north sea fish. *Environmental Science & Technology, 47*(15), 8818–8824.

Fossi, M. C., Casini, S., Caliani, I., Panti, C., Marsili, L., Viarengo, A., et al. (2012a). The role of large marine vertebrates in the assessment of the quality of pelagic marine ecosystems. *Marine Environmental Research, 77*, 156–158.

Fossi, M. C., Panti, C., Guerranti, C., Coppola, D., Giannetti, M., Marsili, L., et al. (2012b). Are baleen whales exposed to the threat of microplastics? A case study of the mediterranean fin whale (*Balaenoptera physalus*). *Marine Pollution Bulletin, 64*(11), 2374–2379.

Fossi, M. C., Coppola, D., Baini, M., Giannetti, M., Guerranti, C., Marsili, L., et al. (2014). Large filter feeding marine organisms as indicators of microplastic in the pelagic environment: The case studies of the Mediterranean basking shark (*Cetorhinus maximus*) and fin whale (*Balaenoptera physalus*). *Marine Environmental Research, 100*, 17–24.

Frias, J. P. G. L., Sobral, P., & Ferreira, A. M. (2010). Organic pollutants in microplastics from two beaches of the Portuguese coast. *Marine Pollution Bulletin, 60*(11), 1988–1992.

Frias, J. P. G. L., Otero, V., & Sobral, P. (2014). Evidence of microplastics in samples of zooplankton from Portuguese coastal waters. *Marine Environmental Research, 95*, 89–95.

Fries, E., Dekiff, J. H., Willmeyer, J., Nuelle, M. T., Ebert, M., & Remy, D. (2013). Identification of polymer types and additives in marine microplastic particles using pyrolysis-GC/MS and scanning electron microscopy. *Environmental Science: Processes & Impacts, 15*(10), 1949–1956.

Furness, R. W. (1985a). Plastic particle pollution: accumulation by Procellariiform seabirds at Scottish colonies. *Marine Pollution Bulletin, 16*(3), 103–106.

Furness, R. W. (1985b). Ingestion of plastic particles by seabirds at Gough Island, South Atlantic Ocean. *Environmental Pollution Series A, Ecological and Biological, 38*(3), 261–272.

Fry, D. M., Fefer, S. I., & Sileo, L. (1987). Ingestion of plastic debris by Laysan albatrosses and wedge-tailed shearwaters in the Hawaiian Islands. *Marine Pollution Bulletin, 18*(6), 339–343.

Galgani, F., Claro, F., Depledge, M. & Fossi, C. (2014). Monitoring the impact of litter in large vertebrates in the Mediterranean Sea within the European Marine Strategy Framework Directive (MSFD): Constraints, specificities and recommendations. *Marine Environmental Research, 100*, 3–9.

Galloway, T. S. (2015). Micro- and nano-plastics and human health. In M. Bergmann, L. Gutow, M. Klages (Eds.), *Marine anthropogenic litter*, (pp. 347–370). Springer, Berlin.

Gassel, M., Harwani, S., Park, J. S., & Jahn, A. (2013). Detection of nonylphenol and persistent organic pollutants in fish from the North Pacific Central Gyre. *Marine Pollution Bulletin, 73*(1), 231–242.

GESAMP (2010). Proceedings of the GESAMP international workshop on plastic particles as a vector in transporting persistent, bio-accumulating and toxic substances in the oceans, 28–30th June 2010, UNESCO-IOC Paris. In T. Bowmer & P. J. Kershaw (Eds.), GESAMP Reports and Studies, 68 pp.

Gilfillan, L. R., Ohman, M. D., Doyle, M. J., & Watson, W. (2009). Occurrence of plastic micro-debris in the Southern California Current system. *California Cooperative Oceanic and Fisheries Investigations, 50*, 123–133.

Goldstein, M. C., Rosenberg, M., & Cheng, L. (2012). Increased oceanic microplastic debris enhances oviposition in an endemic pelagic insect. *Biology Letters, 8*(5), 817–820.

Goldstein, M. C. & Goodwin, D. S. (2013). Gooseneck barnacles (*Lepas* spp.) ingest microplastic debris in the North Pacific Subtropical Gyre. *PeerJ, 1*(e841).

Goldstein, M. C., Titmus, A. J., & Ford, M. (2013). Scales of spatial heterogeneity of plastic marine debris in the Northeast Pacific Ocean. *PLoS ONE, 8*(11), e80020. doi:10.1371/journal.pone.0080020.

Graham, E. R., & Thompson, J. T. (2009). Deposit- and suspension-feeding sea cucumbers (Echinodermata) ingest plastic fragments. *Journal of Experimental Marine Biology and Ecology, 368*(1), 22–29.

Gregory, M. R. (1978). Accumulation and distribution of virgin plastic granules on beaches. *New Zealand Journal of Marine and Freshwater Research, 12*, 399–414.

Gregory, M. R. (1983). Virgin plastic granules on some beaches of Eastern Canada and Bermuda. *Marine Environmental Research, 10*(2), 73–92.

Harper, P. C., & Fowler, J. A. (1987). Plastic pellets in New Zealand storm-killed prions (*Pachyptila* spp.) 1958–1977. *Notornis, 34*(1), 65–70.

Hart, M. W. (1991). Particle capture and the method of suspension feeding by echinoderm larvae. *Biology Bulletin, 180*(1), 12–27.

Hays, H., & Cormons, G. (1974). Plastic particles found in tern pellets, on coastal beaches and at factory sites. *Marine Pollution Bulletin, 5*(3), 44–46.

Heo, N. W., Hong, S. H., Han, G. M., Hong, S., Lee, J., Song, Y. K., et al. (2013). Distribution of small plastic debris in cross-section and high strandline on Heungnam beach, South Korea. *Ocean Science Journal, 48*(2), 225–233.

Hidalgo-Ruz, V., & Thiel, M. (2013). Distribution and abundance of small plastic debris on beaches in the SE Pacific (Chile): A study supported by a citizen science project. *Marine Environmental Research, 87–88*, 12–18.

Hirai, H., Takada, H., Ogata, Y., Yamashita, R., Mizukawa, K., Saha, M., et al. (2011). Organic micropollutants in marine plastics debris from the open ocean and remote and urban beaches. *Marine Pollution Bulletin, 62*(8), 1683–1692.

Holmes, L. A., Turner, A., & Thompson, R. C. (2012). Adsorption of trace metals to plastic resin pellets in the marine environment. *Environmental Pollution, 160*(1), 42–48.

Ismail, A., Adilah, N. M. B., & Nurulhudha, M. J. (2009). Plastic pellets along Kuala Selangor-Sepang coastline. *Malaysian Applied Biology Journal, 38*, 85–88.

Ivar Do Sul, J. A., Spengler, A. & Costa, M. F. (2009). Here, there and everywhere. Small plastic fragments and pellets on beaches of Fernando de Noronha (Equatorial Western Atlantic). *Marine Pollution Bulletin, 58*(8), 1236–1238.I.

Ivar do Sul, J. A., Costa, M. F., Barletta, M. & Cysneiros, F. J. A. (2013). Pelagic microplastics around an archipelago of the Equatorial Atlantic. *Marine Pollution Bulletin, 75*(1), 305–309.

Ivar do Sul, J. A. & Costa, M. F. (2014). The present and future of microplastic pollution in the marine environment. *Environmental Pollution, 185*, 352–364.

Ivar do Sul, J. A., Costa, M. F. & Fillmann, G. (2014). Microplastics in the pelagic environment around oceanic islands of the Western Tropical Atlantic Ocean. *Water, Air & Soil Pollution, 225*(7), 1–13.

Jambeck, J. R., Geyer, R., Wilcox, C., Siegler, T. R., Perryman, M., Andrady, A., et al. (2015). Plastic waste inputs from land into the ocean. *Science, 347*, 768–771.

Jayasiri, H. B., Purushothaman, C. S., & Vennila, A. (2013). Quantitative analysis of plastic debris on recreational beaches in Mumbai, India. *Marine Pollution Bulletin, 77*(1), 107–112.

Kaberi, H., Tsangaris, C., Zeri, C., Mousdis, G., Papadopoulos, A., & Streftaris, N. (2013). Microplastics along the shoreline of a Greek island (Kea island, Aegean Sea): Types and densities in relation to beach orientation, characteristics and proximity to sources. In *Proceedings of the 4th International Conference on Environmental Management, Engineering, Planning and Economics (CEMEPE) and SECOTOX Conference*. Mykonos island, Greece, June 24–28, ISBN:978-960-6865-68-8.

Kaiser, J. (2010). The dirt on ocean garbage patches. *Science, 328*(5985), 1506.

Kako, S. I., Isobe, A., Kataoka, T., & Hinata, H. (2014). A decadal prediction of the quantity of plastic marine debris littered on beaches of the East Asian marginal seas. *Marine Pollution Bulletin, 81*(1), 174–184.

Kaposi, K. L., Mos, B., Kelaher, B., & Dworjanyn, S. A. (2014). Ingestion of microplastic has limited impact on a marine larva. *Environmental Science & Technology, 48*(3), 1638–1645.

Karapanagioti, H. K., & Klontza, I. (2007). Investigating the properties of resin pellets found in the coastal areas of Lesvos Island. *Global Nest. The International Journal, 9*(1), 71–76.

Karapanagioti, H. K., Endo, S., Ogata, Y., & Takada, H. (2011). Diffuse pollution by persistent organic pollutants as measured in plastic pellets sampled from various beaches in Greece. *Marine Pollution Bulletin, 62*(2), 312–317.

Karl, D. M. (1999). A sea of change: Biogeochemical variability in the North Pacific Subtropical Gyre. *Ecosystems, 2*, 181–214.

Kartar, S., Milne, R. A., & Sainsbury, M. (1973). Polystyrene waste in the Severn Estuary. *Marine Pollution Bulletin, 4*(9), 144.

Kartar, S., Abou-Seedo, F., & Sainsbury, M. (1976). Polystyrene Spherules in the Severn Estuary—A progress report. *Marine Pollution Bulletin, 7*(3), 52.

Khordagui, H. K., & Abu-Hilal, A. H. (1994). Industrial plastic on the southern beaches of the Arabian Gulf and the western beaches of the Gulf of Oman. *Environmental Pollution, 84*(3), 325–327.

Kiessling, T., Gutow, L., & Thiel, M. (2015). Marine litter as a habitat and dispersal vector. In M. Bergmann, L. Gutow, & M. Klages (Eds.), *Marine anthropogenic litter* (pp. 141–181). Springer, Berlin.

Koelmans, A. A. (2015). Modeling the role of microplastics in bioaccumulation of organic chemicals to marine aquatic organisms. Critical review. In M. Bergmann, L. Gutow, M. Klages (Eds.), *Marine anthropogenic litter* (pp. 313–328). Springer, Berlin.

Koelmans, A. A., Besseling, E., & Shim, W. J. (2015) Nanoplastics in the aquatic environment. In M. Bergmann, L. Gutow, & M. Klages (Eds.), *Marine anthropogenic litter* (pp. 329–344). Springer, Berlin.

Köhler, A. (2010). Cellular fate of organic compounds in marine invertebrates. *Comparative Biochemistry and Physiology—Part A: Molecular & Integrative Physiology, 157*(Supplement), 8–11.

Kühn, S., Bravo Rebolledo, E. L., & van Franeker, J. A. (2015). Deleterious effects of litter on marine life. In M. Bergmann, L. Gutow, & M. Klages (Eds.), *Marine anthropogenic litter* (pp. 75–116). Springer, Berlin.

Kühn, S., & van Franeker, J. A. (2012). Plastic ingestion by the northern fulmar (*Fulmarus glacialis*) in Iceland. *Marine Pollution Bulletin, 64*(6), 1252–1254.

Kukulka, T., Proskurowski, G., Morét-Ferguson, S., Meyer, D. W., & Law, K. L. (2012). The effect of wind mixing on the vertical distribution of buoyant plastic debris. *Geophysical Research Letters, 39*(7), L07601.

Kripa, V., Nair, P. G., Dhanya, A. M., Pravitha, V. P., Abhilash, K. S., Mohammed, A. A., et al. (2014). Microplastics in the gut of anchovies caught from the mud bank area of Alappuzha, Kerala. *Marine Fisheries Information Service; Technical and Extension Series, 219*, 27–28.

Kuriyama, Y., Konishi, K., Kanehiro, H., Otake, C., Kaminuma, T., Mato, Y., et al. (2002). Plastic pellets in the marine environment of Tokyo Bay and Sagami Bay. *Bulletin of the Japanese Society of Scientific Fisheries, 68*(2), 164–171.

Kusui, T., & Noda, M. (2003). International survey on the distribution of stranded and buried litter on beaches along the Sea of Japan. *Marine Pollution Bulletin, 47*(1), 175–179.

Lattin, G. L., Moore, C. J., Zellers, A. F., Moore, S. L., & Weisberg, S. B. (2004). A comparison of neustonic plastic and zooplankton at different depths near the southern California shore. *Marine Pollution Bulletin, 49*(4), 291–294.

Law, K. L., Morét-Ferguson, S., Maximenko, N. A., Proskurowski, G., Peacock, E. E., Hafner, J., et al. (2010). Plastic accumulation in the North Atlantic subtropical gyre. *Science, 329*(5996), 1185–1188.

Law, K. L., Morét-Ferguson, S., Goodwin, D. S., Zettler, E. R., DeForce, E., Kukulka, T., et al. (2014). Distribution of surface plastic debris in the eastern Pacific Ocean from an 11-year dataset. *Environmental Science & Technology, 48*(9), 44732–44738.

Lee, K.-W., Shim, W. J., Kwon, O. Y., & Kang, J.-H. (2013). Size-dependent effects of micro polystyrene particles in the marine copepod *Tigriopus japonicus. Environmental Science & Technology, 47*(19), 11278–11283.

Liebezeit, G., & Dubaish, F. (2012). Microplastics in beaches of the East Frisian Islands Spiekeroog and Kachelotplate. *Bulletin of Environmental Contamination & Toxicology, 89*(1), 213–217.

Lima, A. R. A., Costa, M. F., & Barletta, M. (2014). Distribution patterns of microplastics within the plankton of a tropical estuary. *Environmental Research, 132*, 146–155.

Lindborg, V. A., Ledbetter, J. F., Walat, J. M., & Moffett, C. (2012). Plastic consumption and diet of Glaucous-winged gulls (*Larus glaucescens*). *Marine Pollution Bulletin, 64*(11), 2351–2356.

Lobelle, D., & Cunliffe, M. (2011). Early microbial biofilm formation on marine plastic debris. *Marine Pollution Bulletin, 62*(1), 197–200.

Löder, M. G. J., & Gerdts, G. (2015). Methodology used for the detection and identification of microplastics—A critical appraisal. In M. Bergmann, L. Gutow, & M. Klages (Eds.), *Marine anthropogenic litter* (pp. 201–227). Springer, Berlin.

Long, M., Hégaret, H., Lambert, C., Le Goic, N., Huvetm, A., Robbens, J., et al. (2014). Can phytoplankton species impact microplastic behaviour within water column? Platform presentation, International workshop on fate and impact of microplastics in marine ecosystems (MICRO2014) 13–15 January 2014. Plouzane (France).

Lusher, A. L., Burke, A., O'Connor, I., & Officer, R. (2014). Microplastic pollution in the Northeast Atlantic Ocean: validated and opportunistic sampling. *Marine Pollution Bulletin, 88*(1–2), 325–333.

Lusher, A. L., McHugh, M., & Thompson, R. C. (2013). Occurrence of microplastics in the gastrointestinal tract of pelagic and demersal fish from the English Channel. *Marine Pollution Bulletin, 67*(1–2), 94–99.

Magnusson, K. (2014). Microlitter and other microscopic anthropogenic particles in the sea area off Rauma and Turku, Finland. Swedish Environmental Institute Report U4645, 17 pp.

Majer, A. P., Vedolin, M. C., & Turra, A. (2012). Plastic pellets as oviposition site and means of dispersal for the ocean-skater insect *Halobates*. *Marine Pollution Bulletin, 64*(6), 1143–1147.

Mallory, M. L., Roberston, G. J., & Moenting, A. (2006). Marine plastic debris in northern fulmars from Davis Strait, Nunavut, Canada. *Marine Pollution Bulletin, 52*(7), 813–815.

Martins, J., & Sobral, P. (2011). Plastic marine debris on the Portuguese coastline: A matter of size? *Marine Pollution Bulletin, 62*(12), 2649–2653.

Mathalon, A., & Hill, P. (2014). Microplastic fibers in the intertidal ecosystem surrounding Halifax Harbor, Nova Scotia. *Marine Pollution Bulletin, 81*(1), 69–79.

Mato, Y., Isobe, T., Takada, H., Kanehiro, H., Ohtake, C., & Kaminuma, T. (2001). Plastic resin pellets as a transport medium for toxic chemicals in the marine environment. *Environmental Science & Technology, 35*(2), 318–324.

Mazurais, D., Huvet, A., Madec, L., Quazuguel, P., Severe, A., Desbruyeres, E., et al. (2014). Impact of polyethylene microbeads ingestion on seabass larvae development Platform presentation, International workshop on fate and impact of microplastics in marine ecosystems (MICRO2014), 13–15 January 2014. Plouzane (France).

McDermid, K. J., & McMullen, T. L. (2004). Quantitative analysis of small-plastic debris on beaches in the Hawaiian Archipelago. *Marine Pollution Bulletin, 48*(7), 790–794.

Mohamed Nor, H., & Obbard, J. P. (2014). Microplastics in Singapore's coastal mangrove ecosystems. *Marine Pollution Bulletin, 79*, 278–283.

Moore, C. J. (2008). Synthetic polymers in the marine environment: A rapidly increasing, long-term threat. *Environmental Research, 108*(2), 131–139.

Moore, C. J., Moore, S. L., Leecaster, M. K., & Weisberg, S. B. (2001). A comparison of plastic and plankton in the north Pacific central gyre. *Marine Pollution Bulletin, 42*(12), 1297–1300.

Moore, C. J., Moore, S. L., Weisberg, S. B., Lattin, G. L., & Zellers, A. F. (2002). A comparison of neustonic plastic and zooplankton abundance in southern California's coastal waters. *Marine Pollution Bulletin, 44*(10), 1035–1038.

Morét-Ferguson, S., Law, K. L., Proskurowski, G., Murphy, E. K., Peacock, E. E., & Reddy, C. M. (2010). The size, mass, and composition of plastic debris in the western North Atlantic Ocean. *Marine Pollution Bulletin, 60*(10), 1873–1878.

Morris, A., & Hamilton, E. (1974). Polystyrene spherules in the Bristol Channel. *Marine Pollution Bulletin, 5*(2), 26–27.

Morris, R. J. (1980). Plastic debris in the surface waters of the South Atlantic. *Marine Pollution Bulletin, 11*(6), 164–166.

Morritt, D., Stefanoudis, P. V., Pearce, D., Crimmen, O. A., & Clark, P. F. (2014). Plastic in the Thames: A river runs through it. *Marine Pollution Bulletin, 78*(1), 196–200.

Moser, M. L., & Lee, D. S. (1992). A fourteen-year survey of plastic ingestion by western North Atlantic seabirds. *Colonial Waterbirds, 15*(1), 83–94.

Mrosovsky, N., Bass, A., Corliss, L. A., Richardson, J. I., & Richardson, T. H. (1992). Pivotal and beach temperatures for hawksbill turtles nesting in Antigua. *Canadian Journal of Zoology, 70*, 1920–1925.

Murray, F., & Cowie, P. R. (2011). Plastic contamination in the decapod crustacean *Nephrops norvegicus* (Linnaeus, 1758). *Marine Pollution Bulletin, 62*(6), 1207–1217.

Nemoto, T. (1970). Feeding pattern of baleen whales in the ocean. In J. H. Steele (Ed.), *Marine food chains* (pp. 241–252). Edinburgh: Oliver and Boyd.

Newman, S., Watkins, E., Farmer, A., Ten Brink, P., & Schweitzer, J.-P. (2015). The economics of marine litter. In M. Bergmann, L. Gutow, & M. Klages (Eds.), *Marine anthropogenic litter* (pp. 371–398). Springer, Berlin.

Ng, K. L., & Obbard, J. P. (2006). Prevalence of microplastics in Singapore's coastal marine environment. *Marine Pollution Bulletin, 52*(7), 761–767.

Norén, F. (2007). Small Plastic Particles in Coastal Swedish Waters. N-Research report, commissioned by KIMO, Sweden. 11 pp.

Norén, F., & Naustvoll, L.-J. (2011). *Survey of microscopic anthropogenic particles in Skagerrak* (p. 21). Flødevigen, Norway: Institute of Marine Research.

Obbard, R. W., Sadri, S., Wong, Y. Q., Khitun, A. A., Baker, I., & Thompson, R. C. (2014). Global warming releases microplastic legacy frozen in Arctic Sea ice. *Earth's Future, 2*(6), 315–320.

Ogata, Y., Takada, H., Mizukawa, K., Hirai, H., Iwasa, S., Endo, S., et al. (2009). International Pellet Watch: global monitoring of persistent organic pollutants (POPs) in coastal waters. 1. Initial phase data on PCBs, DDTs, and HCHs. *Marine Pollution Bulletin, 58*(10), 1437–1446.

Ogi, H. (1990). Ingestion of plastic particles by sooty and short-tailed shearwaters in the North Pacific. In *Proceedings of the Second International Conference on Marine Debris* (pp. 635–652). Honolulu, Hawaii, 2–7 April 1989.

Ohlendorf, H. M., Risebrough, R. W., & Vermeer, K. (1978). Exposure of marine birds to environmental pollutants [Oil, organochlorines, heavy metals, toxicity]. Wildlife Research Report (USA). no. 9.

Oliveira, M., Ribeiro, A., Hylland, K., & Guilhermino, L. (2013). Single and combined effects of microplastics and pyrene on juveniles (0+ group) of the common goby *Pomatoschistus microps* (Teleostei, Gobiidae). *Ecological Indicators, 34*, 641–647.

Pauly, D., Trites, A. W., Capuli, E., & Christensen, V. (1998). Diet composition and trophic levels of marine mammals. *ICES Journal of Marine Science, 55*(3), 467–481.

Possatto, F. E., Barletta, M., Costa, M. F., Ivar do Sul, J. A. & Dantas, D. V. (2011). Plastic debris ingestion by marine catfish: An unexpected fisheries impact. *Marine Pollution Bulletin, 62*(5), 1098–1102.

Pott, A. (2014). A new method for the detection of microplastics in the North Sea brown shrimp (*Crangon crangon*) by Fourier Transform Infrared Spectroscopy (FTIR). M.Sc. thesis, RWTH Aachen University/Alfred Wegener Institute Helmholtz Centre for Polar and Marine Research, 61 pp.

Provencher, J. F., Gaston, A. J., & Mallory, M. L. (2009). Evidence for increased ingestion of plastics by northern fulmars (*Fulmarus glacialis*) in the Canadian Arctic. *Marine Pollution Bulletin, 58*(7), 1092–1095.

Provencher, J. F., Gaston, A. J., Mallory, M. L., O'hara, P. D., & Gilchrist, H. G. (2010). Ingested plastic in a diving seabird, the thick-billed murre (*Uria lomvia*), in the eastern Canadian Arctic. *Marine Pollution Bulletin, 60*(9), 1406–1411.

Ramos, J. A. A., Barletta, M., & Costa, Monica F. (2012). Ingestion of nylon threads by Gerreidae while using a tropical estuary as foraging grounds. *Aquatic Biology, 17*, 29–34.

Reddy, M. S., Basha, S., Adimurthy, S., & Ramachandraiah, G. (2006). Description of the small plastics fragments in marine sediments along the Alang-Sosiya ship-breaking yard, India. *Estuarine, Coastal and Shelf Science, 68*(3), 656–660.

Reid, S. (1981). Wreck of kerguelen and blue petrels. *Notornis, 28*, 239–240.

Reisser, J., Shaw, J., Wilcox, C., Hardesty, B. D., Proietti, M., Thums, M., et al. (2013). Marine plastic pollution in waters around Australia: Characteristics, concentrations, and pathways. *PLoS ONE, 8*(11), e80466.

Rios, L. M., Moore, C., & Jones, P. R. (2007). Persistent organic pollutants carried by synthetic polymers in the ocean environment. *Marine Pollution Bulletin, 54*(8), 1230–1237.

Robards, M. D., Piatt, J. F., & Wohl, K. D. (1995). Increasing frequency of plastic particles ingested by seabirds in the subarctic North Pacific. *Marine Pollution Bulletin, 30*(2), 151–157.

Rochman, C. M. (2015). The complex mixture, fate and toxicity of chemicals associated with plastic debris in the marine environment. In M. Bergmann, L. Gutow, & M. Klages (Eds.), *Marine anthropogenic litter* (pp. 117–140). Springer, Berlin.

Rochman, C. M., Hoh, E., Kurobe, T., & Teh, S. J. (2013). Ingested plastic transfers hazardous chemicals to fish and induces hepatic stress. *Scientific Reports, 3*(3263).

Rochman, C. M., Kurobe, T., Flores, I., & Teh, S. J. (2014). Early warning signs of endocrine disruption from the ingestion of plastic debris in the adult Japanese medaka (*Oryzias latipes*). *Science of the Total Environment, 493*, 656–661.

Rodríguez, A., Rodríguez, B., & Nazaret Carrasco, M. (2012). High prevalence of parental delivery of plastic debris in Cory's shearwaters (*Calonectris diomedea*). *Marine Pollution Bulletin, 64*(10), 2219–2223.

Rossi, G., Barnoud, J., & Monticelli, L. (2013). Polystyrene nanoparticles perturb lipid membranes. *The Journal of Physical Chemistry Letters, 5*(1), 241–246.

Rothstein, S. I. (1973). Plastic particle pollution of the surface of the Atlantic Ocean: Evidence from a seabird. *Condor, 75*(344), 5.

Ryan, P. G. (1987). The incidence and characteristics of plastic particles ingested by seabirds. *Marine Environmental Research, 23*(3), 175–206.

Ryan, P. G. (1988). The characteristics and distribution of plastic particles at the sea-surface off the southwestern Cape Province, South Africa. *Marine Environmental Research, 25*(4), 249–273.

Ryan, P. G. (2008). Seabirds indicate changes in the composition of plastic litter in the Atlantic and south-western Indian Oceans. *Marine Pollution Bulletin, 56*(8), 1406–1409.

Ryan, P. G., Moore, C. J., van Franeker, J. A., & Moloney, C. L. (2009). Monitoring the abundance of plastic debris in the marine environment. *Philosophical Transactions of the Royal Society of London B, 364*(1526), 1999–2012.

Sadri, S. S., & Thompson, R. C. (2014). On the quantity and composition of floating plastic debris entering and leaving the Tamar Estuary, Southwest England. *Marine Pollution Bulletin, 81*(1), 55–60.

Schuyler, Q., Hardesty, B. D., Wilcox, C., & Townsend, K. (2014). Global analysis of anthropogenic debris ingestion by sea turtles. *Conservation Biology, 28*(1), 129–139.

Setälä, O., Fleming-Lehtinen, V., & Lehtiniemi, M. (2014). Ingestion and transfer of microplastics in the planktonic food web. *Environmental Pollution, 185*, 77–83.

Shiber, J. G. (1979). Plastic pellets on the coast of Lebanon. *Marine Pollution Bulletin, 10*(1), 28–30.

Shiber, J. G. (1982). Plastic pellets on Spain's 'Costa del Sol' beaches. *Marine Pollution Bulletin, 13*(12), 409–412.

Shiber, J. G. (1987). Plastic pellets and tar on Spain's Mediterranean beaches. *Marine Pollution Bulletin, 18*(2), 84–88.

Sileo, L., Sievert, P. R., Samuel, M. D., & Fefer, S. I. (1990). Prevalence and characteristics of plastic ingested by Hawaiian seabirds. In *Proceedings of the Second International Conference on Marine Debris* (pp. 2–7). Honolulu, Hawaii, 2–7 April 1989.

Slip, D. J., Green, K., & Woehler, E. J. (1990). Ingestion of anthropogenic articles by seabirds at Macquarie Island. *Marine Ornithology, 18*(1), 74–77.

Song, Y. K., Hong, S. H., Kang, J. H., Kwon, O. Y., Jang, M., Han, G. M., et al. (2014). Large accumulation of micro-sized synthetic polymer particles in the sea surface microlayer. *Environmental Science & Technology, 48*(16), 9014–9021.

Spear, L. B., Ainley, D. G., & Ribic, C. A. (1995). Incidence of plastic in seabirds from the tropical pacific, 1984–1991: Relation with distribution of species, sex, age, season, year and body weight. *Marine Environmental Research, 40*(2), 123–146.

Sussarellu, R., Soudant, P., Lambert, C., Fabioux, C., Corporeau, C., Laot, C., et al. (2014). Microplastics: effects on oyster physiology and reproduction. Platform presentation, International workshop on fate and impact of microplastics in marine ecosystems (MICRO2014), 13–15 January 2014. Plouzane (France).

Sutherland, W. J., Clout, M., Côté, I. M., Daszak, P., Depledge, M. H., Fellman, I., et al. (2010). A horizon scan of global conservation issues for 2010. *Trends in Ecology & Evolution*, 25, 1–7.

Takada, H. (2006). Call for pellets! International pellet watch global monitoring of POPs using beached plastic resin pellets. *Marine Pollution Bulletin, 52*(12), 1547–1548.

Tanaka, K., Takada, H., Yamashita, R., Mizukawa, K., Fukuwaka, M.-A., & Watanuki, Y. (2013). Accumulation of plastic-derived chemicals in tissues of seabirds ingesting marine plastics. *Marine Pollution Bulletin, 69*(1–2), 219–222.

Teuten, E. L., Rowland, S. J., Galloway, T. S., & Thompson, R. C. (2007). Potential for plastics to transport hydrophobic contaminants. *Environmental Science & Technology, 41*(22), 7759–7764.

Teuten, E. L., Saquing, J. M., Knappe, D. R. U., Barlaz, M. A., Jonsson, S., Björn, A., et al. (2009). Transport and release of chemicals from plastics to the environment and to wildlife. *Philological Transactions of the Royal Society London B, 364*(1526), 2027–2045.

Thompson, R. C. (2006). Plastic debris in the marine environment: Consequences and solutions. In J. C. Krause, H. von Nordheim, & S. Bräger (Eds.), Marine Nature Conservation in Europe (pp. 107–115). Stralsund, Germany: Federal Agency for Nature Conservation.

Thompson, R. C., Olsen, Y., Mitchell, R. P., Davis, A., Rowland, S. J., John, A. W. G., et al. (2004). Lost at sea: where is all the plastic? *Science, 304*(5672), 838.

Thompson, R. C., Moore, C. J., Saal, F. S., & Swan, S. H. (2009). Plastics, the environment and human health: current consensus and future trends. *Philosophical Transactions of the Royal Society of London B, 364*, 2153–2166.

Tourinho, P. S., Ivar do Sul, J. A. & Fillmann, G. (2010). Is marine debris ingestion still a problem for the coastal marine biota of southern Brazil? *Marine Pollution Bulletin, 60*(3), 396–401.

Turner, A., & Holmes, L. (2011). Occurrence, distribution and characteristics of beached plastic production pellets on the island of Malta (central Mediterranean). *Marine Pollution Bulletin, 62*(2), 377–381.

Turra, A., Manzano, A. B., Dias, R. J. S., Mahiques, M. M., Barbosa, L., Balthazar-Silva, D., & Moreira, F. T. (2014). Three-dimensional distribution of plastic pellets in sandy beaches: shifting paradigms. *Scientific Reports, 4*(4435).

Ugolini, A., Ungherese, G., Ciofini, M., Lapucci, A., & Camaiti, M. (2013). Microplastic debris in sandhoppers. *Estuarine, Coastal and Shelf Science, 129*, 19–22.

Van, A., Rochman, C. M., Flores, E. M., Hill, K. L., Varges, E., Vargas, S. A., et al. (2012). Persistent organic pollutants in plastic marine debris found on beaches in San Diego, California. *Chemosphere, 86*(3), 258–263.

van Cauwenberghe, L., Claessens, M., Vandegehuchte, M. B., Mees, J., & Janssen, C. R. (2013a). Assessment of marine debris on the Belgian continental shelf. *Marine Pollution Bulletin, 73*(1), 161–169.

van Cauwenberghe, L., Vanreusel, A., Mees, J., & Janssen, C. R. (2013b). Microplastic pollution in deep-sea sediments. *Environmental Pollution, 182*, 495–499.

van Cauwenberghe, L., & Janssen, C. R. (2014). Microplastics in bivalves cultured for human consumption. *Environmental Pollution, 193*, 65–70.

van Dolah, R. F., Burrell, V. G., & West, S. B. (1980). The distribution of pelagic tars and plastics in the south Atlantic bight. *Marine Pollution Bulletin, 11*(12), 352–356.

van Franeker, J. A. (1985). Plastic ingestion in the North Atlantic fulmar. *Marine Pollution Bulletin, 16*(9), 367–369.

van Franeker, J. A., & Bell, P. J. (1988). Plastic ingestion by petrels breeding in Antarctica. *Marine Pollution Bulletin, 19*(12), 672–674.

van Franeker, J. A., Blaize, C., Danielsen, J., Fairclough, K., Gollan, J., Guse, N., et al. (2011). Monitoring plastic ingestion by the northern fulmar *Fulmarus glacialis* in the North Sea. *Environmental Pollution, 159*(10), 2609–2615.

van Sebille, E., England, M. H., & Froyland, G. (2012). Origin, dynamics and evolution of ocean garbage patches from observed surface drifters. *Environmental Research Letters, 7*, 044040.

Vianello, A., Boldrin, A., Guerriero, P., Moschino, V., Rella, R., Sturaro, A., et al. (2013). Microplastic particles in sediments of Lagoon of Venice, Italy: First observations on occurrence, spatial patterns and identification. *Estuarine, Coastal and Shelf Science, 130*, 54–61.

Viehman, S., Vander, J. L., Schellinger, J., & North, C. (2011). Characterization of marine debris in North Carolina salt marshes. *Marine Pollution Bulletin, 62*(12), 2771–2779.

Vlietstra, L. S., & Parga, J. A. (2002). Long-term changes in the type, but not amount, of ingested plastic particles in short-tailed shearwaters in the Southeastern Bering Sea. *Marine Pollution Bulletin, 44*(9), 945–955.

von Moos, N., Burkhardt-Holm, P., & Köhler, A. (2012). Uptake and effects of microplastics on cells and tissue of the blue mussel *Mytilus edulis* L. after an experimental exposure. *Environmental Science & Technology, 46*(20), 11327–11335.

Ward, J. E., & Targett, N. M. (1989). Influence of marine microalgal metabolites on the feeding behavior of the blue mussel *Mytilus edulis*. *Marine Biology, 101*, 313–321.

Ward, J. E., Levinton, J. S., & Shumway, S. E. (2003). Influence of diet on pre-ingestive particle processing in bivalves: I: Transport velocities on the ctenidium. *Journal of Experimental Marine Biology and Ecology, 293*(2), 129–149.

Ward, J. E., & Kach, D. J. (2009). Marine aggregates facilitate ingestion of nanoparticles by suspension-feeding bivalves. *Marine Environmental Research, 68*(3), 137–142.

Watts, A., Lewis, C., Goodhead, R. M., Beckett, D. J., Moger, J., Tyler, C., et al. (2014). Uptake and retention of microplastics by the shore crab *Carcinus maenas*. *Environmental Science & Technology, 48*(15), 8823–8830.

Wegner, A., Besseling, E., Foekema, E. M., Kamermans, P., & Koelmans, A. A. (2012). Effects of nanopolystyrene on the feeding behaviour of the blue mussel (*Mytilus edulis* L.). *Environmental Toxicology and Chemistry, 31*, 2490–2497.

Wilber, R. J. (1987). Plastic in the North Atlantic. *Oceanus, 30*(3), 61–68.

Wilson, D. S. (1973). Food size selection among copepods. *Ecology, 54*(4), 909–914.

Woodall, L. C., Sanchez-Vidal, A., Canals, M., Paterson, G. L. J., Coppock, R., Sleight, V., et al. (2014). The deep sea is a major sink for microplastic debris. *Royal Society Open Science, 1*, 140317.

Wright, S. L., Thompson, R. C., & Galloway, T. S. (2013a). The physical impacts of microplastics on marine organisms: A review. *Environmental Pollution, 178*, 483–492.

Wright, S. L., Rowe, D., Thompson, R. C., & Galloway, T. S. (2013b). Microplastic ingestion decreases energy reserves in marine worms. *Current Biology, 23*(23), R1031–R1033.

Yamashita, R., & Tanimura, A. (2007). Floating plastic in the Kuroshio Current area, western North Pacific Ocean. *Marine Pollution Bulletin, 54*(4), 485–488.

Yamashita, R., Takada, H., Fukuwaka, M.-A., & Watanuki, Y. (2011). Physical and chemical effects of ingested plastic debris on short-tailed shearwaters, *Puffinus tenuirostris*, in the North Pacific Ocean. *Marine Pollution Bulletin, 62*, 2845–2849.

Yntema, C., & Mrosovsky, N. (1982). Critical periods and pivotal temperatures for sexual differentiation in loggerhead sea turtles. *Canadian Journal of Zoology, 60*, 1012–1016.

Zarfl, C., & Matthies, M. (2010). Are marine plastic particles transport vectors for organic pollutants to the Arctic? *Marine Pollution Bulletin, 60*(10), 1810–1814.

Zarfl, C., Fleet, D., Fries, E., Galgani, F., Gerdts, G., Hanke, G., et al. (2011). Microplastics in oceans. *Marine Pollution Bulletin, 62*, 1589–1591.

Chapter 11
Modeling the Role of Microplastics in Bioaccumulation of Organic Chemicals to Marine Aquatic Organisms. A Critical Review

Albert A. Koelmans

Abstract It has been shown that ingestion of microplastics may increase bioaccumulation of organic chemicals by aquatic organisms. This paper critically reviews the literature on the effects of plastic ingestion on the bioaccumulation of organic chemicals, emphasizing quantitative approaches and mechanistic models. It appears that the role of microplastics can be understood from chemical partitioning to microplastics and subsequent bioaccumulation by biota, with microplastic as a component of the organisms' diet. Microplastic ingestion may either clean or contaminate the organism, depending on the chemical fugacity gradient between ingested plastic and organism tissue. To date, most laboratory studies used clean test organisms exposed to contaminated microplastic, thus favouring chemical transfer to the organism. Observed effects on bioaccumulation were either insignificant or less than a factor of two to three. In the field, where contaminants are present already, gradients can be expected to be smaller or even opposite, leading to cleaning by plastic. Furthermore, the directions of the gradients may be opposite for the different chemicals present in the chemical mixtures in microplastics and in the environment. This implies a continuous trade-off between slightly increased contamination and cleaning upon ingestion of microplastic, a trade-off that probably attenuates the overall hazard of microplastic ingestion. Simulation models have shown to be helpful in mechanistically analysing these observations and scenarios, and are discussed in detail. Still, the literature on parameterising such models is limited and further experimental work is required to better constrain the parameters in these models for the wide range of organisms and chemicals acting in the aquatic environment. Gaps in knowledge and recommendations for further research are provided.

A.A. Koelmans (✉)
Aquatic Ecology and Water Quality Management Group, Department of Environmental Sciences, Wageningen University, P.O. Box 47, 6700 AA Wageningen, The Netherlands
e-mail: bart.koelmans@wur.nl

A.A. Koelmans
IMARES—Institute for Marine Resources and Ecosystem Studies, Wageningen UR, P.O. Box 68, 1970 AB IJmuiden, The Netherlands

© The Author(s) 2015
M. Bergmann et al. (eds.), *Marine Anthropogenic Litter*,
DOI 10.1007/978-3-319-16510-3_11

Keywords Additives · Bioaccumulation · Chemical transfer · Microplastic · Persistent organic pollutants

11.1 Introduction

Pollution with plastic debris and microplastic fragments has been recognized as a major problem in fresh water and marine systems (Derraik 2002; Andrady 2011; Koelmans et al. 2014a). Negative effects may relate to entanglement in plastic wires or nets, or to ingestion, which has been reported for benthic invertebrates, birds, fish, mammals and turtles. Extensive overviews of the deleterious effects of litter on marine life are provided by Kühn et al. (2015) and by Lusher (2015). Furthermore, it is generally assumed that microplastic may act as a vector for transport of chemicals associated with the plastic particles, such as persistent organic pollutants (POPs) or additives, residual monomers or oligomers of the component molecules of the plastics (hereafter referred to as 'additives') (Gouin et al. 2011; Teuten et al. 2007, 2009; Hammer et al. 2012; Browne et al. 2013; Rochman 2015; Lusher 2015). Hydrophobic chemicals including polychlorobiphenyls (PCB), polycyclic aromatic hydrocarbons (PAH) or polybrominated diethyl ethers (PBDEs), are known to concentrate in polymers such as polyvinylchloride (PVC), polyethylene (PE), polystyrene (PS) or polyoxymethylene (POM), which is the basis of using the latter materials in passive sampling devices (e.g. Hale et al. 2010). Microplastic particles present in seas and oceans have been found to contain considerable quantities of these chemicals (e.g. Ogata et al. 2009; Hirai et al. 2011). Concentrations of additives such as nonylphenol (NP), bisphenol A (BPA), PBDEs and phthalates also have been reported to be high in marine plastics, rendering them a potential source to the environment and marine biota. The question whether microplastic-mediated chemical transfer poses a serious actual hazard, however, depends on several other factors. First, for transport of the chemicals from plastic to an organism, a gradient that drives the chemical from plastic to the organism is required (Gouin et al. 2011; Koelmans et al. 2013a, b). If, however, a reverse gradient existed, ingestion would lead to cleaning of the organism and ingestion would in this sense be beneficial. Second, the chemical uptake through ingestion of plastic should be substantial compared to other exposure pathways, i.e. by food ingestion or uptake from ambient water. Because POPs as well as additives are ubiquitous in many environments, a dominant role of plastic ingestion is not self-evident (Koelmans et al. 2014b). Third, the chemical hazard of microplastic ingestion should relate to all the chemicals in the plastic-organism system, that is, the chemical mixture transferred to or from the organism by ingestion and chemicals should not be considered in isolation. A plastic additive may leach from a heavily contaminated plastic particle, but clean the organism from its body burden of legacy POPs at the same time. This means that there may be a trade-off between positive and negative effects of microplastic ingestion.

To date, a few controlled experimental studies have been published confirming transfer of chemicals from microplastic to marine organisms. Besseling et al. (2013) mimicked natural conditions by exposing relatively clean worms to mixtures of a natural marine sediment and PS microplastic, which were pre-equilibrated with PCBs, thus providing realistic exposure conditions. The presence of microplastic caused a small (factor of three) increase in bioaccumulation. However, bioaccumulation decreased again at higher concentrations. The authors argued that PS may not have caused PCB transfer but that the increased bioaccumulation probably had a biological cause, such as a change in lipid content or feeding rates. Browne et al. (2013) did not use natural sediment but exposed clean lug worms (*Arenicola marina*) to sand with 5 % of PVC microplastic that was presorbed with high concentrations of nonylphenol, phenanthrene, triclosan and/or PBDE-47. Because by using clean worms, a gradient from the PVC to the organism was created, chemical transfer from the particles to the worms occurred, but uptake from sand was larger than that from the PVC microplastic. Rochman et al. (2013b) exposed fish (Japanese medaka; *Oryzias latipes*) to contaminated food, to contaminated food mixed with 10 % virgin low density PE (LDPE) and to contaminated food mixed with LDPE that was pre-equilibrated in seawater. They observed an increase in body burdens up to a factor of 2.4 after two months, which was statistically significant for chrysene, PCB28 and most PBDEs. Chua et al. (2014) observed that adding PBDE-spiked microplastics to seawater with amphipods (*Allorchestes compressa*) in closed vials resulted in PBDE uptake by the amphipods, which was however only statistically significant compared to the controls when spiked concentrations were ten times higher than environmentally relevant concentrations. Addition of clean plastic to the same closed systems yet pre-contaminated with PBDEs resulted in a decreased uptake.

Considering the complex processes involved, modelling approaches have been proven useful for the interpretation of experimental data as well as for prognostic assessments of the possible hazards caused by plastic ingestion. Model-based scenario studies have helped to define in which cases plastic ingestion may be relevant, dependent on plastic type, chemical properties and species traits. The aim of this chapter is to present and critically discuss the model approaches used to quantify the effect of plastic on bioaccumulation of POPs and additives. This includes a mathematical description of the processes at play, a review of the model-based inferences described in the literature, and an outlook to future work and recommendations.

11.2 Models to Assess the Importance of Microplastic Ingestion

In the literature several processes have been identified as important to address when modelling effects of microplastic on the bioaccumulation of chemicals. These studies typically consider biota lipids as the target tissue for chemical accumulation. First of all, plastic has been reported to act as an additional sorbent for POPs and additives (Andrady 2011). Upon addition of clean plastic in any closed system, chemicals will bind to the plastic thus lowering the chemical concentration in other media or compartments present, such as water, sediment and biota (Teuten et al. 2007, 2009; Gouin et al. 2011; Koelmans et al. 2013a; Chua et al. 2014). This mechanism of repartitioning thus causes a decrease in exposure of aquatic organisms to chemicals in water, sediment organic matter and food. Conversely, if the plastic carries high enough concentrations of chemicals to act as a source, these chemicals will be released and redistribute among the various media present, possibly increasing the chemical concentrations in the other compartments, including biota (Hammer et al. 2012; Koelmans et al. 2014b). Consequently, whether plastic acts as a source or a sink depends on the gradient between the chemical concentration in the plastic and the ambient water. Second, plastic items may slowly disintegrate and degrade under the influence of turbulence and UV radiation or by microbial activity (Andrady 2011, 2015). This means that the chemical mass held by the plastic being degraded will be released, even if no *a priori* gradient between the chemical concentration in plastic and in ambient water exists. In turn, smaller and/or weathered plastic items may have different sorption properties compared to their pristine original state (Teuten et al. 2009; Rochman et al. 2013a). Leaching and weathering for instance may change the polymers' structure and overall polarity. For smaller particles, surface sorption may become dominant over bulk partitioning, a phenomenon that probably is most relevant for polymer particles reaching the nano-scale (Velzeboer et al. 2014; Koelmans et al. 2015). Third, ingestion will bring plastic particles inside the gastrointestinal tract (GIT) of marine organisms where they will stay for a period of time depending on the biology of the species. Whether chemicals are being transferred inside the GIT primarily depends on the gradient between chemical fugacity in the plastic and in the relevant organisms' tissue, which for POPs and hydrophobic chemicals in general, especially is the lipids. If there is no gradient, plastic will pass the GIT and leave the organism without any chemical transfer. The possibility that there is a positive gradient between plastic and lipids receives a lot of speculation in the literature because it would imply an increase of exposure to plastic-associated chemicals compared to a scenario without ingestion of microplastic (Teuten et al. 2007, 2009; Hammer et al. 2012; Chua et al. 2014; Rochman 2015). However, the reverse, i.e. a gradient towards plastic, may be evenly likely (Gouin et al. 2011; Koelmans et al. 2013a; Chua et al. 2014). This potential uptake pathway assumes that microplastics are not decomposed in the relative short GIT residence time of hours to days for most species. Only for really large items that cause obstruction and blockage of the GIT as is observed for instance for birds, decomposition may become relevant. In such a scenario, however, physical harm would probably cause stress and mortality earlier than that related to chemical release. Consequently, an essential difference in chemical risk originating from contaminated plastic *versus* that of contaminated food as a diet component is that pre-equilibrated food is digested, which leads to an immediately increased concentration (fugacity) inside the GIT, whereas pre-equilibrated plastic may leave the

GIT unchanged in many cases. If pre-equilibrated food and plastic are ingested as a mixture, the pulse exposure due to decomposition of the food inside the GIT will cause a gradient from the gut and biota lipids towards the plastic. This means that plastic ingestion can suppress biomagnification and that plastic ingestion in fact may clean the organism (Gouin et al. 2011; Koelmans et al. 2013a). Because regular biomagnification increases with trophic level, the gradient between biota lipids and plastic would be larger for higher trophic levels, leading to more transfer from biota lipids to the plastic. Finally, regardless of whether microplastic increases or attenuates bioaccumulation, the actual importance of plastic ingestion also depends on whether the percentage of chemical transfer due to microplastic ingestion is substantial compared to that of the other uptake pathways, such as the transfer from digested food and uptake from water. This importance in turn depends on the residence times and ingestion rates of plastic and food items in the GIT and the exchange kinetics between these items and gut fluids. In summary, the effects of microplastic on bioaccumulation can be understood from (a) changes in external exposure driven by competitive partitioning processes and (b) by changes in 'internal' exposure due to microplastic acting as a source or a sink depending on initial concentrations in plastic and biota lipids that determine the direction of the gradient. Several authors have provided mathematical process descriptions and parameters to quantify the processes mentioned and to unify them in an integrated model framework. Below, the most important approaches are provided and reviewed.

11.2.1 Equilibrium Partitioning

Addition of clean microplastic to a closed contaminated system will cause a gradient towards the plastic and thus lower the concentrations in the compartments present, for instance water and biota, until a new equilibrium is established. Adding contaminated plastic will cause the opposite process and lead to higher concentrations in water and biota. The kinetics of such a systems response is well-understood and depends on the response times for the individual exchange processes, i.e. water-sediment, water-biota and water-plastic. In general, the slowest process will be rate-determining. For polymers in water, kinetics depends on the resistances to transfer, which are the resistance due to polymer diffusion and the resistance due to the undisturbed boundary layer (UBL) surrounding plastic particles. For hydrophobic chemicals and plastic particles >1 mm, it is generally assumed that the UBL resistance dominates and transfer can be described by (Schwarzenbach et al. 2003):

$$\frac{dC_{PL}}{dt} = k_1 C_W - k_2 C_{PL} \tag{11.1}$$

where t = time and k_1 (L kg^{-1} d^{-1}) and k_2 (d^{-1}) are first-order rate constants that can be related to the thickness of the UBL and aqueous diffusivity of the chemical, and C_{PL} (μg/kg) and C_W (μg/L) are concentrations in plastic and water, respectively. The presence of biofilms on the plastic may slow down the exchange kinetics. In systems with excess of water and sediment, C_W will not decrease due to sorption to plastic and can be assumed constant, i.e. $C_{W,0}$ such that:

$$C_{PL} = C_{W,0}\frac{k_1}{k_2}(1 - e^{-k_2 t})$$ (11.2)

where k_1/k_2 is the plastic to water partition coefficient $K_{P,PL}$ (L/kg), which may differ for different types of plastics. For chemicals that are less hydrophobic, exchange may be driven by polymer diffusion, a process that follows Fick's 2nd law of diffusion, which for spherical particles reads (Schwarzenbach et al. 2003; Teuten et al. 2009; Endo et al. 2013):

$$\frac{dC_{PL}}{dt} = \frac{D_{eff}}{r^2}\frac{\delta}{\delta r}\left(r^2\frac{\delta C_{PL}}{\delta t}\right)$$ (11.3)

Here, the key parameters are D_{eff} (m^2 d^{-1}), the effective polymer diffusion coefficient and r(m), the radius of the plastic particle. Although fundamentally different, the modelling of the two regimes can be unified using the approximation (Schwarzenbach et al. 2003; Koelmans et al. 2013a, b):

$$k_2 \cong 23\, D_{eff}/r^2$$ (11.4)

In practice, half-lives (i.e. $t_{1/2} = 0.693/k_2$) for desorption from microplastic particles in seawater have been reported as one day to years (Teuten et al. 2007; Endo et al. 2013; Bakir et al. 2014; Rochman et al. 2013a) depending on the chemical, the plastic, stirring conditions, presence of dissolved organic matter and measurement method. Because of the environmental persistence of microplastics, most particles will have resided in the water for years or decades and thus can generally be assumed to be close to sorption equilibrium.

The effect of addition of plastic on the aqueous chemical concentration in a simple closed sediment—water system can be calculated from a mass balance, thus: $C_W^1(V_W + M_{SED}K_{P,SED}) = C_W^2(V_W + M_{SED}K_{P,SED} + M_{PL}K_{P,PL})$, which translates into:

$$\frac{C_W^2}{C_W^1} = \frac{V_W + M_{SED}K_{P,SED}}{V_W + M_{SED}K_{P,SED} + M_{PL}K_{P,PL}}$$ (11.5)

in which C_W^1 and C_W^2 are the chemical concentrations in water before and after the addition of microplastic (μg/L), M_{SED} and M_{PL} are masses of sediment and plastic (kg), and $K_{P,SED}$ is the sediment-water equilibrium partition coefficient. For the sake of simplicity, only sediment is considered here, but for hydrophobic chemicals, similar terms $M_i K_{P,i}$ should be added for other important compartments 'i' such as phytoplankton and dissolved organic matter (DOC). Gouin et al. (2011) provided the most elaborate analysis in this respect by also including the air compartment using air-volume and air-water partition coefficients. It follows from

Eq. (11.5) that addition of plastic will only be important if the term $M_{PL}K_{P,PL}$ in the denominator of (11.5) adds substantially to the terms $V_W + M_{SED}K_{P,SED}$ and similar terms $M_i K_{P,i}$ for phytoplankton and DOC.

11.2.2 Decomposition and Disintegration

Decomposition, disintegration or (bio)degradation have been reported to occur at time scales of years to decades (Andrady 2011). Recent laboratory studies report degradation of 1–1.75 % of low density PE mass in 30 d, for micro-organisms isolated from marine waters and with high microbial densities (Harshvardhan and Jha 2013). If surface oxidation or surface degradation is the rate-limiting step, overall degradation can be assumed to depend on the amount of surface area that is available. With ongoing degradation, the surface area per unit of volume will increase due to increased surface roughness, as well as reduced particle size. The shrinking-particle theory (e.g. Di Toro et al. 1996) accounts for this change in size and for mono-disperse spherical particles would predict:

$$V_t = V_0 \left(1 - \frac{2k_s t}{d_0} \right)^\alpha \tag{11.6}$$

in which V_t (m³) is the particle volume at time t, V_0 (m³) is the initial particle volume, d_0 (m) is the initial particle diameter (spheres) or thickness (polymer films), α is a particle shape factor ($\alpha = 3$ for spheres and $\alpha = 1$ for thin films) and k_s is the apparent shrinking-rate constant (m³ m^{-2} d^{-1}). Calibration of the model on the ~1 % PE mass loss in 30 d observed for thin films deployed by Harshvardhan and Jha (2013) (with $\alpha = 1$ and assuming an initial thickness of 25.4 µm (1 mil) for their PE film), would yield a low value for k_s of 4.2×10^{-9} m³ m^{-2} d^{-1}. It can be assumed that loss of polymer equates to loss of chemical held by that volume of polymer. The time scales at which these decomposition processes occur, however, probably are orders of magnitude longer than the time scales of plastic-water partitioning or transfer inside the organisms' gut (see below). This implies that decomposition is not directly relevant for bioaccumulation assessment.

11.2.3 Bioaccumulation

Bioaccumulation can be modelled using traditional approaches that use a mass balance of uptake and loss processes (e.g. Thomann et al. 1992; Hendriks et al. 2001) (Fig. 11.1). Extensions of these models to account for uptake from contaminated particles as diet components were first provided by Sun et al. (2009) and Janssen et al. (2010). Koelmans et al. (2013a, b, 2014b) modelled bioaccumulation of hydrophobic chemicals ($dC_{B,t}/dt$; µg × kg^{-1} d^{-1}) from an environment containing plastic using:

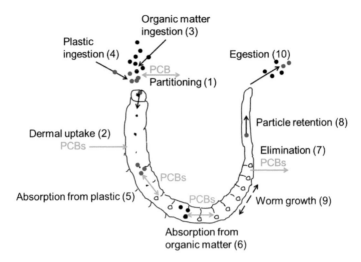

Fig. 11.1 Schematic representation of processes required for plastic-inclusive bioaccumulation modeling (example for PCBs accumulation in a lugworm *Arenicola marina*): *1* Partitioning between plastic, sediment and water, *2* dermal uptake, *3* organic matter (food, biofilm) ingestion, *4* microplastic ingestion, *5* absorption from plastic, *6* absorption from organic matter, *7* elimination, *8* particle retention, *9* worm growth, *10* particle egestion (sediment and plastic). Same or similar process descriptions can be used for other marine/aquatic organisms. Reprinted with permission from Koelmans et al. (2013a). Copyright 2013 American Chemical Society

$$\frac{dC_{B,t}}{dt} = k_{derm}C_W + IR\left(S_{FOOD}a_{FOOD}C_{FOOD} + S_{PL}C_{PLR,t}\right) - k_{loss}C_{B,t} \quad (11.7)$$

where the first term quantifies dermal (for fish; including gills) uptake from water. The second term quantifies uptake from the diet and exchange with plastic particles. The third term quantifies overall loss due to elimination and egestion. The first and third term can be parameterised following traditional approaches with k_{derm} (L × kg × d^{-1}) and k_{loss} (d^{-1}), first-order rate constants for dermal uptake and overall loss through elimination and egestion. In the second term, IR (g × g^{-1} × d^{-1}) represents the mass of food ingested per unit of time and organism dry weight, a_{FOOD} is the absorption efficiency from the diet, S_{FOOD} and S_{PL} are the mass fractions of food and plastic in ingested material, respectively ($S_{FOOD} + S_{PL} = 1$) and C_{FOOD} is the chemical concentration in the diet. The product $a_{FOOD} \times C_{FOOD}$ quantifies the contaminant concentration that is transferred from food, i.e. prey, to the organism during gut passage. The plastic particles may contain a biofilm (BF), which may also carry chemicals. The biofilm would contribute to the pool of digestible organic matter and may therefore be covered either by the sediment term or by an optional additional term in Eq. 11.7, similar to the sediment ingestion term (e.g. *IR* × S_{BF} × a_{BF} × C_{BF}). Where regular bioaccumulation models assume digestion of diet components and thus assume a certain fixed chemical absorption efficiency, Koelmans et al. (2013a, b, 2014b) assumed plastic not to degrade in the short time scale of gut passage. The transferred concentration from plastic during gut passage ($C_{PLR,t}$, µg/kg) thus was modelled to be dependent on the concentrations in plastic and biota lipids, the kinetics of transfer between plastic and lipids and the GIT residence time (GRT) (see Koelmans et al. 2013a, b for detailed derivation):

$$C_{PLR,t} = \frac{k_{1G}C_{PL} - k_{2G}C_{L,t}}{k_{1G} + \frac{M_{PL}}{M_L}k_{2G}}\left(1 - e^{-\left(k_{1G} + \frac{M_{PL}}{M_L}k_{2G}\right)GRT}\right) \tag{11.8}$$

in which k_{1G} and k_{2G} (d^{-1}) are forward and backward first-order rate constants describing the transport between plastic and biota lipids inside the GIT. If the numerator term $k_{1G}C_{PL} - k_{2G}C_{L,t}$ in Eq. 11.8 is positive, transfer from the plastic to biota lipids occurs, whereas opposite transfer ('cleaning by plastic') occurs when the term is negative. Various authors have provided these k values at simulated gut conditions, showing about an order of magnitude enhancement of transfer rates in artificial gut fluids up to $k_{1G} = 10$–12 d^{-1} (Teuten et al. 2007; Bakir et al. 2014). GRT is gut residence time (d), C_{PL} and $C_{L,t}$ (µg/kg) are the chemical concentrations in the ingested plastic particle and the biota lipids at the moment of ingestion and M_{PL} and M_L are the mass of plastic and lipids in the organism, respectively (kg). If C_W is constant in time (Eqs. 11.2 and 11.5) and C_{PL} is estimated by Eq. 11.2, an analytical solution to Eqs. 11.7 and 11.8 is available that calculates the body burden at steady state (C_B^{SS}) (Koelmans et al. 2014b):

$$C_B^{SS} = \frac{k_{derm}C_{W,0} + IR(S_{FOOD}a_{FOOD}C_{FOOD} + S_{PL}k_{1G}C_{PL}A_{PL})}{IRS_{PL}k_{2G}A_{PL}/f_{lip} + k_{loss}} \text{ and } A_{PL} = \frac{1 - e^{-\left(k_{1G} + \frac{M_{PL}}{M_L}k_{2G}\right)GRT_t}}{k_{1G} + \frac{M_{PL}}{M_L}k_{2G}} \tag{11.9}$$

Note, that Eq. 11.9 accounts for all uptake and loss pathways and can be used to assess the relative importance of plastic ingestion as an uptake pathway compared to other pathways such as food ingestion and dermal uptake, as well as the importance of chemical loss by plastic egestion compared to regular loss mechanisms.

11.3 Model-Based Assessment of Implications and Risks of Plastic-Associated Chemicals

Various authors used the aforementioned concepts to assess the effects and importance of plastic-associated chemicals on chemical partitioning and bioaccumulation. This section reviews these studies. Teuten et al. (2007) modelled the effect of adding 'clean' plastic to a sediment-water system (1.5 kg sediment, 0.4 L water, 1.5 g lugworm A. *marina*) contaminated with phenanthrene as a model compound using an equilibrium partitioning approach (Eq. 11.5). They concluded that plastic addition would reduce bioavailability due to scavenging of phenanthrene by the plastic. The effect was small (13 %) and depended on factors such as partition coefficients of the plastic and sediment, which also follows from Eq. 11.5. In another scenario, they assumed chemical concentrations in microplastic to be much higher in the sea-surface micro layer (SML) than in the water column. Subsequent settling and exposure of sediment biota to these enriched microplastic particles would lead to increased bioaccumulation. Both scenarios, however, used equilibrium partitioning concepts only and did not yet consider sediment and/or plastic ingestion as a possible uptake pathway. This implies that attenuation of bioaccumulation was not accounted for. There may also be some uncertainty related to the acclaimed enrichment in the SML. Analysis of the SML by Hardy et al. (1988, 1990) did not use passive samplers that would have detected the truly dissolved concentrations, but used analysis of total concentrations after filtration. Filtration is known to be insufficient in removing DOC and colloids present (e.g. Gschwend and Wu 1985). Because the SML is enriched with DOC, organic colloids, micro-organisms or oil films that act as 'extracting agents', this explains the enhanced apparent concentrations in the SML (e.g. Wurl et al. 2006). This is also consistent with SML concentration enrichment factors usually being higher for coastal areas and bays that have higher DOC levels, and for more hydrophobic chemicals. The *truly dissolved* chemical concentrations in the SML, however, would still be equal or close to those in the bulk of the water column, thus preventing enrichment of concentrations in microplastic. Furthermore, if an enhancement of concentrations in microplastics compared to the water column would still occur, desorption would probably attenuate the gradient upon settling in the water column and burial in the sediment.

Gouin et al. (2011) also used equilibrium partitioning concepts to define the chemical distribution of POPs among air, water, sediment and plastic, and used steady-state bioaccumulation modelling to assess their subsequent fate in the food web. Instead of considering one chemical their analysis spanned a wide range of chemical hydrophobicities and air-water partition coefficients. A model environment was defined representative of a coastal marine ecosystem with a realistic input of plastic debris. Mass-balance equations were used to construct chemical space diagrams. Data analysis showed that partitioning to PE was negligible (<0.1 % of chemical mass). Only if it was assumed that the present estimate of PE abundance was enhanced by three orders of magnitude and that the water contained no organic matter (i.e. DOC or phytoplankton) PE would became important (>1 % sorption to PE) for POPs with $LogK_{ow} > 5$. This implies that present plastic loadings were calculated to be insufficient to cause a meaningful redistribution of POPs from the oceanic environment to the plastic. Furthermore, DOC and phytoplankton that compete with plastic for POP distribution should be accounted for in order to assess whether future accumulation of plastic could lead to a substantial redistribution of POPs. Gouin et al. (2011) also discussed effects of PE presence on bioaccumulation by piscivorous fish, by including contaminated PE as a diet component in an elaborate food web bioaccumulation model. A steady-state approach was used that did not yet consider the kinetics of desorption from the plastic inside the gut, in relation to gut retention time. This means that the direction of an effect of PE ingestion would be calculated correctly but that its magnitude may have been overestimated because the model could not account for the extent of non-equilibrium in the gut. Interestingly, the authors found a counterintuitive *decrease* in predicted body burden upon an increase in PE in the diet. This was explained by the fact that without plastic, food organic matter is digested leading to high concentrations in the gut that subsequently are transferred to the organisms' lipids. In the presence of plastic, however, which is not degraded, a gradient from lipids towards plastic exists, leading to cleaning of the organism by the plastic.

Koelmans et al. (2013a, b) presented a general POP bioaccumulation model framework for marine aquatic organisms combining Eqs. 11.1–11.9, which was implemented for *A. marina* (Fig. 11.1). The model accounted for dilution of exposure concentration by sorption of POPs to plastic (POP 'dilution'), increased bioaccumulation by ingestion of plastic containing POPs ('carrier'), and decreased bioaccumulation by ingestion of clean plastic ('cleaning'). Kinetics in the gut were explicitly taken into account. The model was evaluated against bioaccumulation data from laboratory bioassays with PS microplastic. Further scenarios included PE microplastic, nano-sized plastic and open marine systems. Scenario studies assumed equilibrium of organisms and plastics prior to ingestion, as would occur for POPs in the environment. Model analysis showed that PS will have a decreasing effect on bioaccumulation, governed by dilution. For stronger sorbents such as polyethylene, the dilution, carrier and cleaning mechanism were more substantial. In closed laboratory bioassay systems, dilution and cleaning dominated, leading to decreased bioaccumulation. Also, in open marine systems a decrease was predicted due to a cleaning mechanism that counteracts biomagnification, similar to that recognized earlier by Gouin et al. (2011). However, the differences were considered too small to be relevant from a risk assessment perspective.

Pollution by POPs is diffuse, which implies that POPs will be always present at background concentrations, often at solid phase—water equilibrium (Van Noort and Koelmans 2012). In the early life stages of organisms, POP concentrations in the organism will be in equilibrium with the ambient water too, which implies that ingestion of polluted microplastic will coincide with the ingestion of polluted food, rendering the contribution of microplastic relatively unimportant. For additives, however, plastic ingestion by marine organisms may potentially be more relevant than for diffusely spread POPs because the plastic could still be a source of the additives (Teuten et al. 2009; Hammer et al. 2012; Koelmans et al. 2014b). Furthermore, compared to worms, leaching of additives or residual monomers may be more relevant for larger and longer-lived species, with longer gut retention times, such as fish. Two recent controlled laboratory studies confirmed that dietary exposure of organisms to microplastic pre-adsorbed with POPs or additives leads to chemical transfer from the microplastic to the organism (Browne et al. 2013; Rochman et al. 2013b). A remaining question, however, is what the relative importance of this microplastic uptake pathway is under natural conditions, where other pathways like dermal uptake, uptake via the gills or consumption of natural prey play a role. Furthermore, it is plausible that in actual marine systems, background chemical concentrations in biota may already exceed the concentrations that microplastic ingestion would be able to explain, in which case no gradient for transfer would exist.

In a follow up study, Koelmans et al. (2014b) used the same biodynamic model as was used for POPs to assess the potential of leaching of nonylphenol (NP) and bisphenol A (BPA) in the intestinal tracts of lugworm (*A. marina*) and cod (*Gadus morhua*). Parameters for the lugworm were based on Besseling et al. (2013). Parameters for cod were based on actual abundances of microplastic particles in the cod GIT as observed by Foekema et al. (2013). The resulting model was validated against the data provided by Browne et al. (2013) for leaching of NP from PVC to *A. marina*. Then, the model was used to calculate the body burdens that could be explained from plastic ingestion, which were compared to NP and BPA body burdens actually measured in the field. Uncertainty in the most crucial parameters was accounted for by probabilistic modelling. The conservative analysis showed that plastic ingestion by the lugworm indeed results in chemical transfer to the organism, but yields NP and BPA concentrations that stay below the lower ends of global NP and BPA concentration ranges in the lugworm, and therefore is unlikely to constitute a relevant exposure pathway. A similar comparison showed that plastic ingestion is also likely to constitute a negligible exposure pathway for cod.

Note that the key model concepts of chemical transfer in the intestinal tract or segments of the intestinal tract as condensed in Eqs. 11.7–11.9 are also applicable to higher marine organisms. They would only need different parameterizations and different initial boundary conditions.

11.4 Summarizing Discussion and Recommendations

This chapter discussed the present state of the art in modelling chemical transfer between microplastic and biota in relation to the experimental data available. Whereas the experimental data and field observations serve as best available proof of the actual occurrence of transfer processes that have been speculated on in the literature for a long time, model analysis has helped to understand why these effects occur, and to quantify their magnitude and direction. General prognostic risk assessments regarding plastic-associated chemicals will need simulation models for the same reasons why models are needed in general PBT assessment (Weisbrod et al. 2009).

Generally, the present experimental studies and model studies are consistent in that they can predict up to a factor of two to three increase in bioaccumulation if microplastic is the only source of the chemical and the only pathway of uptake. Conversely, they predict a decrease in bioaccumulation when chemical dilution outcompetes transfer in the gut. If more environmentally relevant scenarios are considered, i.e. with pre-equilibrated systems and all exposure pathways are accounted for, ingestion of microplastics seems to be much less important than the existing pathways. This does not mean that the hazards of plastic-associated chemicals are less than anticipated, but it may imply that the relevance of plastic ingestion as an additional exposure pathway may be less relevant than what has been assumed in the literature (e.g. Teuten et al. 2007, 2009; Hammer et al. 2012; Browne et al. 2013; Chua et al. 2014), at least for POPs.

Chemical transfer effects should not be studied or interpreted from chemical principles alone assuming biota to be a constant factor. Plastic ingestion may cause physical stress, for instance due to blockage of the GIT or decreased overall food quality (Lusher 2015), which in turn may affect ingestion rates, lipid contents, growth rates and in turn kinetic parameters for chemical transfer. Distinguishing between these chemical and biological effect mechanisms is an important challenge when interpreting bioaccumulation data from the laboratory or the field.

Although considerable progress has been made over the past years, there still is only a hand full of bioaccumulation studies addressing transfer from microplastic, typically of a 'proof of principle' nature. The processes at play seem to be well understood, their parameterisation, however, may need more work. While diffusion parameters and partition coefficients for pristine polymers are available, chemical exchange kinetics for microplastics under conditions of weathering, degradation and biofilm formation in the marine environment are poorly understood. Chemical exchange in the GIT has been investigated using artificial gut fluids, but dedicated dietary exposure experiments may provide better parameterisations for a wider range of chemicals. Hazard assessment of plastic-associated chemicals should ideally not only focus on particular biota and chemicals, but also use a systems approach accounting for all exposure pathways, including food web magnification and chemical mixtures. It is most plausible that marine organisms experience a trade-off between negative effects of chemical transfer from additives to the organism, and positive effects of attenuation of POP bioaccumulation, upon ingestion of microplastic (Koelmans et al. 2014b). In this respect, experimental model-validation studies using contaminated organisms and clean plastic may be as important to advance the science as most present studies that use an inverse gradient. Finally, a better quantitative understanding is needed with respect to the role of microplastic ingestion in the chemical transfer of degradable compounds. As recently pointed out by Rochman et al. (2013b), degradable compounds such as PAH and PBDEs are known to biomagnify less from prey due to degradation in the water column or metabolization by the organism or by prey species lower in the marine food web (e.g. Di Paolo et al. 2010). Because these chemicals would be preserved by sorption to microplastic, this could increase the relative role of microplastic ingestion as a relevant pathway for these chemicals. This means that the aforementioned effect of suppression of bioaccumulation of POPs would be less relevant for these degradable compounds.

References

Andrady, A. L. (2011). Microplastics in the marine environment. *Marine Pollution Bulletin, 62*, 1596–1605.

Andrady, A.L. (2015). Persistence of plastic litter in the oceans. In M. Bergmann, L. Gutow & M. Klages (Eds.), *Marine anthropogenic litter* (pp. 57–72), Berlin: Springer.

Bakir, A., Rowland, S. J., & Thompson, R. C. (2014). Enhanced desorption of persistent organic pollutants from microplastics under simulated physiological conditions. *Environmental Pollution, 185*, 16–23.

Besseling, E., Wegner, A., Foekema, E. M., van den Heuvel-Greve, M. J., & Koelmans, A. A. (2013). Effects of microplastic on performance and PCB bioaccumulation by the lugworm *Arenicola marina* (L.). *Environmental Science and Technology, 47*, 593–600.

Browne, M. A., Niven, S. J., Galloway, T. S., Rowland, S. J., & Thompson, R. C. (2013). Microplastic moves pollutants and additives to worms, reducing functions linked to health and biodiversity. *Current Biology, 23*, 2388–2392.

Chua, E. M., Shimeta, J., Nugegoda, D., Morrison, P. D., & Clarke, B. O. (2014). Assimilation of Polybrominated diphenyl ethers from microplastics by the marine amphipod, *Allorchestes compressa*. *Environmental Science and Technology, 48*, 8127–8134.

Derraik, J. G. B. (2002). The pollution of the marine environment by plastic debris: a review. *Marine Pollution Bulletin, 44*(9), 842–852.

Di Paolo, C., Gandhi, N., Bhavsar, S., Van den Heuvel-Greve, M., & Koelmans, A. A. (2010). Black carbon-inclusive multichemical modeling of PBDE and PCB biomagnification in estuarine food webs. *Environmental Science and Technology, 44*, 7548–7554.

Di Toro, D. M., Mahony, J. D., & Gonzalez, A. M. (1996). Particle oxidation model of synthetic FeS and sediment acid-volatile sulfide. *Environmental Toxicology and Chemistry, 15*, 2156–2167.

Endo, S., Yuyama, M., & Takada, H. (2013). Desorption kinetics of hydrophobic organic contaminants from marine plastic pellets. *Marine Pollution Bulletin, 74*, 125–131.

Foekema, E. M., De Gruijter, C., Mergia, M. T., Murk, A. J., van Franeker, J. A., & Koelmans, A. A. (2013). Plastic in North Sea Fish. *Environmental Science and Technology, 47*, 8818–8824.

Gouin, T., Roche, N., Lohmann, R., & Hodges, G. (2011). A thermodynamic approach for assessing the environmental exposure of chemicals absorbed to microplastic. *Environmental Science and Technology, 45*(4), 1466–1472.

Gschwend, P. M., & Wu, S. C. (1985). On the constancy of sediment-water partition coefficients of hydrophobic organic pollutants. *Environmental Science and Technology, 19*, 90–96.

Hale, S. E., Martin, T. J., Goss, K. U., Arp, H. P. H., & Werner, D. (2010). Partitioning of organochlorine pesticides from water to polyethylene passive samplers. *Environmental Pollution, 158*(7), 2511–2517.

Hammer, J., Kraak, M. H., & Parsons, J. R. (2012). Plastics in the marine environment: the dark side of a modern gift. *Reviews of Environmental Contamination and Toxicology, 2012*(220), 1–44.

Hardy, J. T., Coley, J. A., Antrim, L. D., & Kiesser, S. L. (1988). A hydrophobic large-volume sampler for collecting aquatic surface microlayers: Characterization and comparison with the glass plate method. *Canadian Journal of Fisheries and Aquatic Sciences, 45*, 822–826.

Hardy, J. T., Crecelius, E. A., Antrim, L. D., Kiesser, S. L., & Broadhurst, V. L. (1990). Aquatic surface microlayer contamination in Chesapeake Bay. *Marine Chemistry, 28*, 333–351.

Harshvardhan, K., & Jha, B. (2013). Biodegradation of low-density polyethylene by marine bacteria from pelagic waters, Arabian Sea, India. *Marine Pollution Bulletin, 77*, 100–106.

Hendriks, A. J., Van der Linde, A., Cornelissen, G., & Sijm, D. T. H. M. (2001). The power of size. 1. Rate constants and equilibrium ratios for accumulation of organic substances related to octanol-water partition ratio and species weight. *Environmental Toxicology and Chemistry, 20*, 1399–1420.

Hirai, H., Takada, H., Ogata, Y., Yamashita, R., Mizukawa, K., Saha, M., et al. (2011). Organic micropollutants in marine plastics debris from the open ocean and remote and urban beaches. *Marine Pollution Bulletin, 62*(8), 1683–1692.

Janssen, E. M. L., Croteau, M. N., Luoma, S. N., & Luthy, R. G. (2010). Measurement and modeling of polychlorinated biphenyl bioaccumulation from sediment for the marine polychaete *Neanthes arenaceodentata* and response to sorbent amendment. *Environmental Science & Technology, 44*, 2857–2863.

Koelmans, A. A., Besseling, E., & Shim, W. J. (2015). Nanoplastics in the aquatic environment. In M. Bergmann, L. Gutow & M. Klages (Eds.) *Marine anthropogenic litter* (pp. 329–344). Berlin: Springer.

Koelmans, A. A., Besseling, E., Wegner, A., & Foekema, E. M. (2013a). Plastic as a carrier of POPs to aquatic organisms: A model analysis. *Environmental Science & Technology, 47*, 7812–7820.

Koelmans, A. A., Besseling, E., Wegner, A., & Foekema, E. M. (2013b). Correction to plastic as a carrier of POPs to aquatic organisms: A model analysis. *Environmental Science & Technology, 47*, 8992–8993.

Koelmans, A. A., Gouin, T., Thompson, R., Wallace, N., & Arthur, C. (2014a). Plastics in the marine environment. *Environmental Toxicology and Chemistry, 33*, 5–10.

Koelmans, A. A., Besseling, E., & Foekema, E. M. (2014b). Leaching of plastic additives to marine organisms. *Environmental Pollution, 187*, 49–54.

Kühn, S., Bravo Rebolledo, E. L., & van Franeker, J. A. (2015). Deleterious effects of litter on marine life. In M. Bergmann, L. Gutow & M. Klages (Eds.), *Marine anthropogenic litter* (pp. 75–116), Berlin: Springer.

Lusher, A. (2015). Microplastics in the marine environment: Distribution, interactions and effects. In M. Bergmann, L. Gutow & M. Klages (Eds.), *Marine anthropogenic litter* (pp. 245–312), Berlin: Springer.

Ogata, Y., Takada, H., Mizukawa, K., Hirai, H., Iwasa, S., Endo, S., et al. (2009). International pellet watch: Global monitoring of persistent organic pollutants (POPs) in coastal waters. 1. Initial phase data on PCBs, DDTs, and HCHs. *Marine Pollution Bulletin, 58*(10), 1437–1446.

Rochman, C. M. (2015). The complex mixture, fate and toxicity of chemicals associated with plastic debris in the marine environment. In M. Bergmann, L. Gutow & M. Klages (Eds.), *Marine anthropogenic litter* (pp. 117–140), Berlin: Springer.

Rochman, C. M., Hoh, E., Hentschel, B. T., & Kaye, S. (2013a). Long-term field measurement of sorption of organic contaminants to five types of plastic pellets: Implications for plastic marine debris. *Environmental Science and Technology, 47*, 1646–1654.

Rochman, C. M., Hoh, E., Kurobe, T., & Teh, S. J. (2013b). Ingested plastic transfers hazardous chemicals to fish and induces hepatic stress. *Scientific Reports, 3*(3263), 1–7.

Schwarzenbach, R. P., Gschwend, P. M., & Imboden, D. M. (2003). *Environmental organic chemistry* (2nd ed.). Wiley-Interscience, London.

Teuten, E. L., Rowland, S. J., Galloway, T. S., & Thompson, R. C. (2007). Potential for plastics to transport hydrophobic contaminants. *Environmental Science and Technology, 41*, 7759–7764.

Teuten, E. L., Saquing, J. M., Knappe, D. R. U., Barlaz, M. A., Jonsson, S., Björn, A., et al. (2009). Transport and release of chemicals from plastics to the environment and to wildlife. *Philosophical Transactions of the Royal Society B, 364*(1526), 2027–2045.

Thomann, R. V., Connolly, J. P., & Parkerton, T. F. (1992). An equilibrium model of organic chemical accumulation in aquatic food webs with sediment interaction. *Environmental Toxicology and Chemistry, 11*, 615–629.

Van Noort, P. C. M., & Koelmans, A. A. (2012). Non-equilibrium of organic compounds in sediment-water systems. Consequences for risk assessment and remediation measures. *Environmental Science and Technology, 46*, 10900–10908.

Velzeboer, I., Kwadijk, C. J. A. F., & Koelmans, A. A. (2014). Strong sorption of PCBs to nanoplastics, microplastics, carbon nanotubes and fullerenes. *Environmental Science and Technology, 48*, 4869–4876.

Weisbrod, A., Woodburn, K., Koelmans, A. A., Parkerton, T., McElroy, A., & Borgå, K. (2009). Evaluation of bioaccumulation using *in-vivo* laboratory and field studies. *Integrated Environmental Assessment and Management, 5*, 598–623.

Wurl, O., Karuppiah, S., & Obbard, J. P. (2006). The role of the sea-surface microlayer in the air-sea gas exchange or organochlorine compounds. *Science of the Total Environment, 369*, 333–343.

Chapter 12
Nanoplastics in the Aquatic Environment. Critical Review

Albert A. Koelmans, Ellen Besseling and Won J. Shim

Abstract A growing body of literature reports on the abundance and effects of plastic debris, with an increasing focus on microplastic particles smaller than 5 mm. It has often been suggested that plastic particles in the <100 nm size range as defined earlier for nanomaterials (here referred to as 'nanoplastics'), may be emitted to or formed in the aquatic environment. Nanoplastics is probably the least known area of marine litter but potentially also the most hazardous. This paper provides the first review on sources, effects and hazards of nanoplastics. Detection methods are in an early stage of development and to date no nanoplastics have actually been detected in natural aquatic systems. Various sources of nanoplastics have been suggested such as release from products or nanofragmentation of larger particles. Nanoplastic fate studies for rivers show an important role for sedimentation of heteroaggregates, similar to that for non-polymer nanomaterials. Some prognostic effect studies have been performed but effect thresholds seem higher than nanoplastic concentrations expected in the environment. The high surface area of nanoplastics may imply that toxic chemicals are retained by nanoplastics, possibly increasing overall hazard. Release of non-polymer nanomaterial additives from small product fragments may add to the hazard of nanoplastics. Because

A.A. Koelmans (✉) · E. Besseling
Aquatic Ecology and Water Quality Management Group, Department of Environmental Sciences, Wageningen University, 6700 AA Wageningen, The Netherlands
e-mail: bart.koelmans@wur.nl

A.A. Koelmans · E. Besseling
IMARES—Institute for Marine Resources & Ecosystem Studies, Wageningen UR, 1970 AB IJmuiden, The Netherlands

W.J. Shim
Oil and POPs Research Group, Korea Institute of Ocean Science and Technology, Geoje 656-834, South Korea

W.J. Shim
Marine Environmental Chemistry and Biology, University of Science and Technology, Daejeon 305-320, South Korea

© The Author(s) 2015
M. Bergmann et al. (eds.), *Marine Anthropogenic Litter*,
DOI 10.1007/978-3-319-16510-3_12

of the presence of such co-contaminants, effect studies with nanoplastics pose some specific practical challenges. We conclude that hazards of nanoplastics are plausible yet unclear, which calls for a thorough evaluation of nanoplastic sources, fate and effects.

12.1 Introduction

Today, pollution with plastic debris and plastic fragments has been recognized as a major water quality problem in fresh and marine water systems. Various recent reviews address the sources, abundance and negative effects of plastic litter (e.g. Derraik 2002; Andrady 2011; Hammer et al. 2012; Koelmans et al. 2014a), including several other chapters in this volume (Browne 2015; Galgani et al. 2015; Thompson 2015). Science in this field is evolving rapidly, with initial studies mainly focusing on detection and abundance of >5 mm macroplastic in marine ecosystems and biota, followed by an increasing focus on <5 mm microplastics ranging down to the μm-scale. Implications of nanometre-sized plastic particles ('nanoplastics'), constitute a very recent area of the environmental sciences. Nanoplastics are of specific interest because of their nano-specific properties, which fundamentally differ from those of the same polymer type in bulk form (Klaine et al. 2012). A clear definition of what should be named a 'nanoplastic' has not yet been provided. For the sake of this review we suggest to follow the definition used for non-polymer nanomaterials, implying that a plastic particle is said to be nano-sized if it is <100 nm in at least one of its dimensions (Klaine et al. 2012). This links the name of the size class to the most convenient scale to actually express this size (i.e. nanometre), it assures a focus on the nano-specific properties and thus their associated hazards, it avoids confusion with the broad scientific field of nano-EHS, and it ensures that a discussion of regulatory implications of nanoplastics may benefit from the past and present developments in the regulation of other manufactured nanomaterials. It must be noted that the classification of plastic particles is not a trivial issue. Earlier, microplastic has been defined as all particles <5 mm, thus automatically including nanometre-sized plastic particles (Arthur et al. 2009). Another recent definition uses <20 μm as a criterion to classify nanoplastics (Wagner et al. 2014), similar to the cut off used by plankton ecologists for nanoplankton. This definition thus includes micrometre-sized particles. Furthermore, it must be stressed that in the fields of nanotechnology and material science the term 'nanoplastics' is already used for those plastics that have nanoscale additives to give the material specific properties (e.g. Bussière et al. 2013). In this chapter on environmental implications, we classify nanoplastic (NP) as particles <100 nm for the reasons stated.

NPs is probably the least known area of marine litter but potentially also the most hazardous. Various sources of NPs have been suggested such as release from products or formation from larger particles ('nanofragmentation') (Andrady 2011; Shim et al. 2014; Cózar et al. 2014). Detection methods are in an early stage of development but to date no NPs have been detected in natural aquatic systems. Some first prognostic bioaccumulation and effect studies have been performed (Brown et al. 2001; Ward and Kach 2009; Bhattacharya et al. 2010; Wegner et al. 2012; Lee et al. 2013; Casado et al. 2013; Besseling et al. 2014b) but there is no systematic effect assessment for relevant aquatic species let alone for the community or ecosystem level. Apart from physiological consequences, NPs might also have chemical effects. The high surface area of NPs may cause exceptionally strong sorption affinities for toxic compounds (Velzeboer et al. 2014a), potentially leading to cumulative particle and chemical toxicity effects once NPs have passed cell membranes. Furthermore, if nanofragmentation is a relevant process, release of non-polymer nanoscale additives from the product fragments may further add to the overall hazard (Nowack et al. 2012).

The aim of this chapter is to present and critically discuss the literature on detection, sources, fate and effects of NPs. Because the literature on NPs is still limited, our synthesis builds on knowledge about bulk polymers i.e. micro- and macroplastics as well as on knowledge about non-polymer nanomaterials. Challenges in performing ecotoxicity tests with NPs are discussed and an outlook to future work and recommendations are provided. The potential effects of NP on human health are covered by Galloway (2015).

12.2 Sources, Detection and Occurrence of Nanoplastic

12.2.1 Sources of Nanoplastic

Primary sources of NPs may relate to release from products and applications, in which nanoplastics are used or formed and that result in emissions to the environment during the product life cycle. Product categories may include waterborne paints, adhesives, coatings, redispersible lattices, biomedical products, drug delivery, medical diagnostics, electronics, magnetics and optoelectronics. Recently, thermal cutting of polystyrene foam has been shown to emit nanometre-sized polymer particles, in the range of ~22–220 nm (Zhang et al. 2012). Many polymers undergo similar thermal treatments during their life cycle. 3-D printing has been shown to emit nanometre-sized polymer particles, in the range of ~11–116 nm, at considerable rates (Stephens et al. 2013). Polystyrene and polyethylene nanoparticles are easy to synthesize (e.g. Lu et al. 2009; Rao and Geckeler 2011), are used for research and other applications and thus will find their way into the environment. Several medical applications include polymeric nanoparticles, nanospheres and nanocapsules, used for drug delivery (Guterres et al. 2007), which are, however, biodegradable solid lipids. Although formally within scope, we argue that such nanoplastics are not likely to be hazardous because of their low persistence in the environment. Cosmetic products are often mentioned in the context of nanoplastics. However, recent product inventories show lowest sizes of ~4 μm present in exfoliating scrubs or skin cleansers (Fendall and

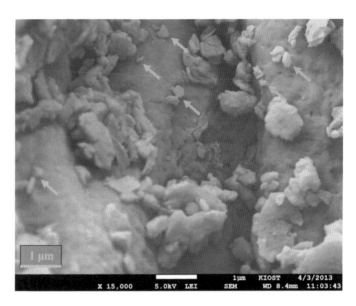

Fig. 12.1 Scanning electron microscopy image of micro- and nano-sized polystyrene particles attached on surface of polystyrene spherule, which were fragmented from the expanded polystyrene spherules by accelerated mechanical abrasion (tumbling at 113 rpm in a glass bottle) with glass beads (3 mm in diameter) for a month. Nanometre-sized particles are indicated by *yellow arrows*

Sewell 2009), rendering these products as an unlikely primary source of NPs. A second speculated source is fragmentation of microplastic to smaller-sized particles eventually reaching the nanoscale (Andrady 2011). Electrospinning of engineered plastics is used to produce mats with nanoscale fibres, which when applied in products might degrade further to the nanoscale (Lu et al. 2008). Polymers consist of a mixture of polymer chains of various lengths. The chains are chemically linked by weak secondary bonds (i.e. hydrogen or Van der Waals bonding) or by physical interaction through entanglement of chains, whereas there is void space in between the chains. The weak interactions are susceptible to breakage at a low energy level. This breakage brings embrittlement, which in combination with other external forces such as friction may cause formation of small particles in the nano-, micro- and millimetre size range, at the surface of the plastics. Shim et al. (2014) were the first to actually report fragmentation of expanded polystyrene (EPS) beads to micro- and nano-sized EPS in experiments involving a month of accelerated mechanical abrasion with glass beads and sand. Formation of nanometre-sized EPS was confirmed with scanning electron microscopy (SEM) and energy-dispersive X-ray spectroscopy (EDS) (Fig. 12.1). Without yet even taking UV exposure into account, these experimental conditions may already mimic conditions at beaches or river banks where prolonged abrasion of macro- and microplastics by sand particles possibly leads to the formation of NPs. The combination of photo-oxidation by UV exposure, high temperature and high humidity at beaches probably enhances fragmentation rates and reduces the size of the plastic particles. However, the occurrence and relative importance of this process still has to be validated in the field. Still, given the available information, we suspect that physical abrasion is a relevant source of NPs.

Although not proven, degradation of microplastics down to the <100 nanometre-scale may constitute a third source of NPs. Slow weathering by photodegradation is well known for all kinds of polymers (Sivan 2011), which is the reason that nano-composites use manufactured nano-particles (nanofillers) to increase the resistance to oxidation (e.g. Grigoriadou et al. 2011; Bussière et al. 2013). UV-B irradiation aided photo-oxidation of LDPE has been shown to lead to the formation of extractable oxygenated compounds as well as non-oxidised low-molecular weight hydrocarbons, which were utilized by bacteria leading to an LDPE mass loss of 8.4 % in 14 days (Roy et al. 2008). The LDPE films subsequently were too fragile to handle. In a recent environmental study, degradation of 1–1.75 % of PE mass was observed in the laboratory in 30 days, by micro-organisms isolated from marine waters present at high densities (Harshvardhan and Jha 2013). Koelmans (2015) suggested that a surface degradation based particle shrinking model may be applied to assess the time dependence of the loss of plastic volume. Using this model and laboratory volume loss rate data from Harshvardhan and Jha (2013), we calculated the time scales required to reach the 100 nm nanoscale as a function of initial plastic particle size. It appears that if oxidation/degradation of the plastic surface would be the rate-limiting process, the rather optimal conditions in the laboratory still would predict that ca. 320 years are needed to bring 1 mm (1000 μm) microplastics to the 100 nm nanoscale (Fig. 12.2). In the oceans, degradation can be assumed to proceed

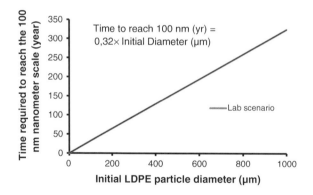

Fig. 12.2 Time required to reach the nanoscale (100 nm) by joint photo-oxidation and biodegradation at the polymer surface, as a function of initial microplastic particle size. The scenario calculation assumes particle shrinking due to photo-oxidation and biodegradation only, and neglects embrittlement and erosion. The reaction rate is proportional to the surface area with rate constant, k' as in $dV(t) = -k'A(t)dt$ with V (m^3), and A (m^2) are particle volume and surface area, respectively. The 'Lab scenario' is based on a mass loss of ~1 % low density polyethylene (LDPE) per month as observed under laboratory conditions by Harshvardhan and Jha (2013). It appears that a particle of 1000 μm (1 mm) diameter requires about 320 years to reach a diameter of 100 nm. In the oceans, degradation can be assumed to proceed much slower due to limited availability of light, oxygen and bacteria

much slower due to limited availability of light, oxygen and bacteria. Microbial or photodegradation at the water surface thus may contribute to the formation of smaller particles, yet reaching the nanoscale may take a long time. We are not aware of studies showing NP formation due to these processes. Nano-fragmentation thus may involve two mechanisms; (1) direct nano-fragmentation may take place at the surface of macro- and microplastics (major process) and further gradual size-reductions may take place due to degradation (minor process). The different time scales of the two processes imply that embrittlement followed by physical abrasion of microplastics probably is the most important process explaining the formation of NPs.

12.2.2 Detection and Occurrence of Nanoplastic

We are not aware of studies reporting established analytical methods to detect nanoplastics in marine or freshwater. Under controlled conditions in the laboratory, several methods that apply to nanomaterials in general are also useful for nanoplastic fate and effect research, such as UV-VIS spectrometry, electron microscopy, field flow fractionation (FFF) or dynamic light scattering (DLS) techniques, each having their advantages and flaws (Von der Kammer et al. 2012). Shim et al. (2014) used SEM-EDS to confirm the presence of nanoplastics in abrasion experiments. In their effect study with mussels (*Mytilus edulis*), Wegner et al. (2012) used multiple wavelength UV-VIS as a proxy to detect pink-dyed nanoparticles and used dynamic light scattering (DLS) to track the actual size of the bioavailable aggregates over time. Velzeboer et al. (2014a) used transmission EM and conventional light microscopy to characterise pristine nanopolystyrene particles and aggregates, respectively. A recent study applied FFF coupled to multi-angle light scattering with pyrolysis to discriminate between various plastic types in spiked natural surface water samples (Kools et al. 2014).

Since separation, concentration and identification of NPs in environmental samples is still difficult, the actual occurrence of NPs is still a matter of speculation even though recent literature takes it as a fact. In his review, Andrady (2011) stated that there is little doubt that nanoscale particles are produced during weathering of plastic debris, but acknowledges that they are not yet quantified. The evidence is circumstantial in that abrasion indeed seems to show formation of nanoplastics in the laboratory (Shim et al. 2014). A study by Cózar et al. (2014) identified a deficiency of plastic particles at the lower end of an expected size distribution in the oceans and argued that nanofragmentation might be a plausible explanation for this deficiency.

12.3 Fate of Nanoplastic

Because NPs have not yet been measured in aquatic systems, only prognostic assessments of NP fate are possible. Freshwater carries plastics from land-based sources to the sea, which renders fate modelling of microplastics and NPs an important area of research. In the literature, several processes have been identified as being important to address when modelling the fate of nanomaterials in freshwater, and a range of elaborate fate models are currently available (Gottschalk et al. 2013; Meesters et al. 2014; Quik et al. 2014). We argue that these models can be used for nanoplastics too, as long as some specific differences relating to densities, biofilm formation and attachment efficiencies are accounted for. For aquatic behaviour of nano-materials such as NPs, homo- and hetero-aggregation, advective flow, sedimentation, re-suspension, photo- and biodegradation, and sediment burial are important processes to consider (Quik et al. 2014; Besseling et al. 2014a). Velzeboer et al. (2014a) used pristine 60 nm polystyrene particles and observed a wide range of aggregate sizes, i.e. 199.3 ± 176.3 nm (range 100–500 nm) after 28 days, using TEM. Bhattacharya et al. (2010) measured substantial binding or heteroaggregation of 20 nm polystyrene particles with freshwater phytoplankton cells. Because of their low density, it is often assumed that substantial fractions of the total load of plastic particles from riverine sources reach the sea (Cózar et al. 2014; Wagner et al. 2014). However, it is plausible that organic matter fouling and subsequent hetero-aggregation with suspended solids, algae or detritus will cause settling and several recent reports indeed show presence of microplastics in the sediments (Zbyszewski and Corcoran 2011; Zbyszewski et al. 2014; Imhof et al. 2013; Wagner et al. 2014; Free et al. 2014). This process is relevant especially for NPs, because hetero-aggregation is particularly important at the nanometre-scale. In freshwater, burial is important to consider as it may be a loss process for nanomaterials from the biologically relevant sediment top layer (Koelmans et al. 2009). The loss processes including photo- or biodegradation have been discussed in the previous section. Besseling et al. (2014a) presented the first spatially explicit NP fate model that accounted for all the aforementioned processes. The model was implemented for a 40-km river stretch and showed the dependence of NP retention on nano- and microplastic particle size, density and attachment efficiencies. Simulations showed that settling of 100 nm NPs was stimulated by fast orthokinetic heteroaggregation, whereas for microplastics >0.1 mm Stokes settling dominated.

In marine systems, the same processes occur, although flow patterns, residence times and the nature of natural colloids and suspended solids (marine snow) are very different. Attachment efficiencies will be higher than in freshwaters due to the higher ionic strength. Collision frequencies however, will be lower due to much lower concentrations of natural colloids and solids in the water column. This trade-off has not yet been quantified for NPs. Wegner et al. (2012) were the first to measure and model the homoaggregation of 30 nm polystyrene particles in seawater and found rapid formation of 1000 nm aggregates within 16 minutes. Attachment efficiencies of 1 were required to explain the experimental observations (Wegner et al. 2012). Velzeboer et al. (2014a) used pristine 60 nm carboxylated polystyrene particles and observed a wide range of aggregate sizes, i.e. 361.1 ± 465.1 nm (TEM, range 100–500 nm) after 28 days (Fig. 12.3), which thus were larger than those observed in freshwater, as mentioned above. The 40 nm carboxylated polystyrene particles used by Della Torre et al. (2014) formed aggregates of 1764 ± 409 nm in natural seawater, whereas their 50 nm

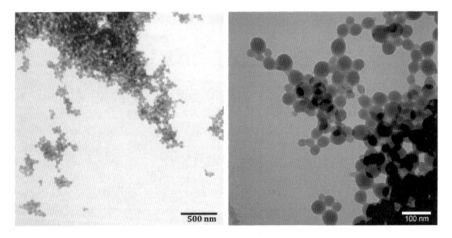

Fig. 12.3 Transmission electron microscopy images of 70 nm nano-sized polystyrene aggregates in freshwater (*left*) and seawater (*right*). Note that the TEM-based data may reflect exact in situ conditions to a lower extent because of the TEM preparation procedure. Reprinted with permission from Velzeboer et al. (2014a). © 2014 American Chemical Society

amino modified polystyrene remained dispersed at the nanoscale (89 ± 2 nm), although the authors report that these particles also partly aggregated with time. Another difference compared to freshwaters relates to the density of seawater, which is higher at lower temperature and higher salinity and thus increases with depth, an increase that additionally depends on season and location. The density of NP aggregates will also vary depending on polymer type, NP surface chemistry, extent of organic matter fouling and the thickness and nature of the biofilm once aggregates are formed. This means that settling of NP aggregates occurs until they reach seawater density and thereafter remain adrift in the water column (Cózar et al. 2014). Small changes in either aggregate or seawater density may cause slow upward or downward transport. Models that specifically simulate NP behaviour in the marine environment have not been published yet. However, because marine NP behaviour probably is behaviour of NP aggregates (Velzeboer et al. 2014a, b), Smoluchowski-Stokes based marine biogeochemical models can be applied such as those applied previously for settling of organic and mineral particles (e.g. Burd and Jackson 2009; Barkmann et al. 2010).

12.4 Bioaccumulation and Effects

12.4.1 Bioaccumulation and Effects of Nanoplastics

A handful of studies have investigated the accumulation or effects of NPs. As for membrane passage, Rossi et al. (2014) used molecular simulations to assess the effect of nano-sized polystyrene on the properties of model biological membranes and concluded that the NPs could permeate easily into lipid membranes, which may affect cellular functions. Experimental validation would still be required to assess the actual relevance of this pathway. In this respect, Salvati et al. (2011) showed that carboxylated nanopolystyrene with sizes ranging from 40 to 50 nm entered cells irreversibly, by different endocytosis pathways. Inflammation responses have been observed in rat lung tissue in response to 64 nm polystyrene particles, showing that a low-toxicity material, such as polystyrene, can have inflammatory potential when present in nano-size (Brown et al. 2001). This study used an air-inhalation exposure scenario and the question remains to what extent this can be translated to aquatic systems, where aggregation would limit the concentrations of free NPs and direct inhalation of air-dispersed NPs does not occur. Bhattacharya et al. (2010) showed that adsorption of 1.8–6.5 mg/L of 20 nm polystyrene particles (yet present as agglomerates) hindered algal photosynthesis, possibly through reduction of light intensity and of air flow by the nanoparticles, and stimulated Reactive Oxygen Species (ROS) production. Ward and Kach (2009) showed that mussels (Mytilus edulis) and oysters (Crassostrea virginica) take up 100 nm PS beads, especially when incorporated into aggregates. They concluded that the direct bioavailability of freely dispersed NPs was very low and that capture and ingestion were the dominant exposure pathways for these species. Wegner et al. (2012) showed that mussels reduced their filter-feeding activity in response to 100 mg/L 30 nm nanopolystyrene. In two-generation chronic toxicity tests, Lee et al. (2013) showed nanopolystyrene ingestion by copepods (Tigriopus japonicus) and detected mortality of nauplii and copepodites for 50 nm (yet partly aggregated) polystyrene particles at concentrations of 12.5 mg/L (F0 generation) and 1.25 mg/L (next generation). Della Torre et al. (2014) observed severe developmental effects of amino-modified polystyrene nanoparticles in the early development of sea urchin (Paracentrotus lividus) embryos, with EC_{50} values of 3.85 and 2.61 mg/L at 24 and 48 h post fertilization. Kashiwada (2006) reported sorption of 39.4 nm nanopolystyrene to the chorion of medaka (Oryzias latipes) eggs and uptake into the yolk and gallbladder during embryonic development, whereas adults accumulated the NPs mainly in the gills and intestine yet also in the brain, testis, liver and blood. It was thus suggested that the NPs were capable of passing the blood–brain barrier. The acute (24 h) toxicity to medaka eggs was zero and 35.6 % for 1 and 30 mg/L NPs, respectively, although toxicity increased with higher salinity.

We are aware of three studies that use freshwater species. Cedervall et al. (2012) showed that 25 nm nanopolystyrene particles were transported through an aquatic food chain from green algae (Scenedesmus sp.), through water fleas (Daphnia magna) to carp (Carassius carassius) and other fishes, and affected lipid metabolism and behaviour of the fish. The effects were mechanistically explained from the chemistry and dynamics of the protein corona surrounding the NPs. Because it was a feeding study, effects could not be linked to NP concentration in the water. Casado et al. (2013) investigated the effects of 55 and 110 nm polyethyleneimine polystyrene nanoparticles on algae (Pseudokirchneriella subcapitata), crustaceans (Thamnocephalus platyurus; Daphnia magna), bacteria (Vibrio fischeri) and rainbow trout (Oncorhynchus mykiss) cell lines (cytotoxicity). Effects were detected for the in vivo species with EC_{50} values between 0.54 and 5.2 mg/L, whereas EC_{50} values for cytotoxicity were between ~60 and 87 mg/L. Besseling et al. (2014b) reported that 70 nm polystyrene particles reduced the growth of algae (Scenedesmus obliquus) at high particle concentrations, and *malformed offspring* of *Daphnia* at a concentration of 32 mg/L. The effects on Daphnia were studied with and without fish (Perca fluviatilis) kairomones in the water and the effect of the kairomones appeared to be stronger in the presence of 1.8 mg/L nanoplastic. This suggests that nanoplastics might interfere with the *chemical communication* among species, which would cause subtle behavioural disturbances in finding a mate or food, or in the avoidance of predators such as fish. Such effects may be taking place at low concentrations that are not easy to detect using standard toxicological tests but that may result in changes in the food web in exposed ecosystems over time.

In summary, the limited literature provides some evidence of effects of NPs to marine and freshwater organisms, yet at relatively high concentrations, i.e. higher than ~0.5 mg/L NPs. There are currently no NP environmental concentrations to which this value can be compared, but the lowest NP effect concentration of 0.54 mg/L (Casado et al. 2013) is about four to six orders of magnitude higher than the 0.4–34 ng/L microplastic concentrations found in freshwaters in the USA (Eriksen et al. 2013) and Europe (Besseling et al. 2014c), but almost similar to the highest concentration estimated for marine water (i.e. 0.51 mg/L, see Besseling et al. 2014b; Lopez Lozano and Mouat 2009). However, because of the limited data, the uncertainties in these numbers and the absence of actual NP exposure data, these comparisons should be interpreted with caution.

12.4.2 Implications of Chemicals and Nanofillers Associated with Nanoplastics

Various kinds of additives are added during the manufacturing of plastics to increase its durability. Furthermore, residual monomers may remain in the plastic. For NPs in particular, the high surface area may cause exceptionally strong sorption affinities for 'external' toxic compounds (Velzeboer et al. 2014a), which implies that they will always be loaded with hydrophobic toxicants or trace metals (Rochman 2013a, 2014; Holmes et al. 2014). It can be hypothesized that the presence of such additives and absorbed chemicals might lead to increased exposure to these toxicants. In the laboratory, transfer and negative effects of such co-contaminants have indeed been shown upon ingestion of microplastic particles, but only in scenarios where clean organisms were exposed to plastics with rather high concentrations (Rochman et al. 2013b; Browne et al. 2013; Chua et al. 2014), thus forcing a maximum *fugacity gradient* upon the organism. Under more realistic natural exposure scenarios where organisms as well as the media water, sediment and plastic were brought at or close to equal chemical fugacity, no or limited (i.e. within a factor of two) increases or decreases in chemical transfer of toxicants were found (Besseling et al. 2013). Several studies even showed beneficial effects of microplastic ingestion by *reducing bioaccumulation* due to sorption of chemicals to the plastic (Teuten et al. 2007; Gouin et al. 2011; Koelmans et al. 2013a, b; Chua et al. 2014). These different outcomes illustrate how the 'carrier effects' of microplastic depend on the initial boundary conditions of the test, which determine the direction of mass transfer between ingested or bio-accumulated plastic and tissue. This is consistent with recent model analyses that systematically explored these exposure scenarios (Gouin et al. 2011; Koelmans et al. 2013a, b, 2014b; Koelmans 2015). While the actual risk caused by chemical transfer due to microplastic ingestion may thus be of limited importance, exposure to NPs may still constitute a real hazard. Because of the surface effect, it may be possible that NPs retain organic toxic chemicals or heavy metals at higher concentrations than microplastics, thus leading to a *fugacity gradient to organism tissue* once ingested. If NPs are capable of permeating membranes, passing cell walls, translocate and/or reside in epithelial tissues for prolonged times (Kashiwada 2006; Cedervall et al. 2012; Rossi et al. 2014), the combination of particle and chemical toxicity may yield *unforeseen risks*. These hypotheses need to be experimentally validated, while also accounting for the possibly *low bioavailability* of NPs due to aggregation. During nanofragmentation, release of non-polymer nanoscale additives from the polymer nanocomposite product fragments may further add to the overall hazard (Nowack et al. 2012; Schlagenhauf et al. 2014). The smaller the additives, the better the improvement of polymer durability, which explains the addition of engineered nanoparticles such as carbon nanotubes (Grigoriadou et al. 2011; Bussière et al. 2013; Schlagenhauf et al. 2014). Although beneficial for their application, these additives increase the persistence of plastics in the environment and once degraded, may increase the overall risk due to an additional emission of nanomaterials.

12.5 Specific Challenges in Nanoplastic Effect Research

Several specific problems may arise when using NPs in aquatic tests or whole sediment toxicity tests with or without co-contaminants present. At present, it is not possible to detect NPs in the environment or to isolate sufficient quantities from the environment for effects research. This implies that manufactured NPs need to be used. This promotes uniformity of tests, but only commercially available polymer types (i.e. polystyrene beads) with limited size and shape (i.e. sphere only) can be tested, whereas NPs in the environment will include many different polymers of varying size and shape. Manufactured NPs may behave differently from environmental NPs because of these different properties. Manufactured NPs come with additives, monomers or oligomers of the component molecules of the plastics, or come with dispersants that are either deliberately added or that are just by-products of the manufacture process. Polystyrene, for instance, release styrene monomers (Saido et al. 2014), which may add to the overall toxicity. If desired, such hydrophobic chemicals may be extracted from NP dispersions prior to testing, for instance using sequential Empore disk extractions (Koelmans et al. 2010). Commercial NPs are often delivered with a biocide to prevent bacterial growth during delivery and storage, which makes them useless for NP toxicity testing. Dispersants such as the surfactant sodium dodecyl sulphate (SDS) are often used. Although this helps to keep the NPs freely dispersed, dispersant concentrations should be kept far below toxicity thresholds and they should be included in the controls (Handy et al. 2012). Alternatively, the NPs can be dialysed towards clean water in order to reduce the concentrations of unwanted chemicals (e.g. Cedervall et al. 2012). NP surfaces are sometimes modified (functionalized) to maximize dispersion of otherwise hydrophobic NPs. This further raises the question what relevant exposure conditions are. On the one hand, a free dispersion may be preferred to achieve the level of control and constant nominal exposure concentration required from a regulatory perspective, and to obtain comparability of test results. On the other hand, a realistic test might aim at mimicking natural conditions as closely as possible, allowing for the formation of aggregates. All effect studies discussed in the previous section report the initial use of freely dispersed pristine NPs, yet acknowledge aggregate formation later on. This implies that aggregate formation and aggregate properties should be monitored during the tests. Several other challenges relating to the nanoscale of the particles are similar to those that were previously discussed for non-polymer manufactured nanomaterials (see Handy et al. 2012).

12.6 Implications and Recommendations

To date, the occurrence of NPs in the aquatic environment has not been proven and thus has to be considered a plausible hypothesis. Using manufactured NPs, some first effect tests have shown ingestion as well as negative effects of NPs on freshwater as well as marine species. Still, the toxicity thresholds seem higher than concentrations that are expected in the environment based on a worst-case assumption of conservative breakdown of microplastics present at currently known concentrations. However, we argue that potential impacts of NPs should not be considered in isolation. NPs might constitute an ecological stressor that adds to many other anthropogenic stressors such as trace metals, organic contaminants and non-polymer nanomaterials. Consequently, the question arises what contribution NPs make to the existing pool of other nano-sized materials. Natural nanoparticles have been shown to be ubiquitous in the environment, including hazardous ones (Wiesner et al. 2011). It has been suggested that engineered nanoparticles may account for only a negligible contribution to the concentrations of natural nanoparticles including soots, clays or other colloids that are already present (Koelmans et al. 2009). Future research may primarily focus on the sources, formation rates and exposure levels of NPs and on the fate of the particles in aquatic systems.

Methods to detect NPs in drinking and in natural waters are urgently needed. Prognostic screening-level effects tests may be performed in order to quantify the hazard once environmental concentrations are known. This research would benefit enormously from harmonisation and uniformity in classification of NPs and in methodologies used.

References

Andrady, A. L. (2011). Microplastics in the marine environment. *Marine Pollution Bulletin, 62*, 1596–1605.

Arthur C., Baker J. & Bamford H. (Eds.). (2009). *Proceedings of the international research workshop on the occurrence, effects, and fate of microplastic marine Debris*, Tacoma, Washington, USA, September 9–11, 2008. Technical Memorandum NOS-OR&R-30. National Oceanic and Atmospheric Administration, Silver Spring, MD, USA.

Barkmann, W., Schäfer-Neth, C., & Balzer, W. (2010). Modelling aggregate formation and sedimentation of organic and mineral particles. *Journal of Marine Systems, 82*, 81–95.

Besseling, E., Foekema, E. M. & Koelmans, A. A. (2014c). Preliminary investigation of microplastic in the management area of Water Board Rivierenland (In Dutch). Wageningen University, Wageningen, The Netherlands (pp. 1–18). http://edepot.wur.nl/299787.

Besseling, E., Quik, J. T. K. & Koelmans, A. A. (2014a). Modeling the fate of nano- and microplastics in freshwater systems. May 2014, SETAC Annual Meeting, Basel, Switzerland.

Besseling, E., Wang, B., Lurling, M., & Koelmans, A. A. (2014b). Nanoplastic affects growth of *S. obliquus* and reproduction of *D. magna. Environmental Science and Technology, 48*, 12336–12343.

Besseling, E., Wegner, A., Foekema, E. M., van den Heuvel-Greve, M. J., & Koelmans, A. A. (2013). Effects of microplastic on performance and PCB bioaccumulation by the lugworm *Arenicola marina* (L.). *Environmental Science and Technology, 47*(1), 593–600.

Bhattacharya, P., Lin, S., Turner, J. P., & Ke, P. C. (2010). Physical adsorption of charged plastic nanoparticles affects algal photosynthesis. *Journal of Physical Chemistry C, 114*, 16556–16561.

Brown, D. M., Wilson, M. R., MacNee, W., Stone, V., & Donaldson, K. (2001). Size-dependent proinflammatory effects of ultrafine polystyrene particles: a role for surface area and oxidative stress in the enhanced activity of ultrafines. *Toxicology and Applied Pharmacology, 175*, 191–199.

Browne, M. A. (2015). Sources and pathways of microplastic to habitats. In M. Bergmann, L. Gutow & M. Klages (Eds.), *Marine anthropogenic litter* (pp. 229–244). Springer: Berlin.

Browne, M. A., Niven, S. J., Galloway, T. S., Rowland, S. J., & Thompson, R. C. (2013). Microplastic moves pollutants and additives to worms, reducing functions linked to health and biodiversity. *Current Biology, 23*, 2388–2392.

Burd, A. B., & Jackson, G. A. (2009). Particle aggregation. *Annual Review Marine Science, 1*, 65–90.

Bussière, P. O., Peyroux, J., Chadeyron, G., & Therias, S. (2013). Influence of functional nanoparticles on the photostability of polymer materials: Recent progress and further applications. *Polymer Degradation and Stability, 98*, 2411–2418.

Casado, M., Macken, A., & Byrne, H. (2013). Ecotoxicological assessment of silica and polystyrene nanoparticles assessed by a multitrophic test battery. *Environment International, 51*, 97–105.

Cedervall, T., Hansson, L. A., Lard, M., Frohm, B., & Linse, S. (2012). Food chain transport of nanoparticles affects behaviour and fat metabolism in fish. *PLoS ONE, 7*(2), e32254.

Chua, E. M., Shimeta, J., Nugegoda, D., Morrison, P. D., & Clarke, B. O. (2014). Assimilation of polybrominated diphenyl ethers from microplastics by the marine amphipod, *Allorchestes compressa*. *Environmental Science and Technology, 48*, 8127–8134.

Cózar, A., Echevarría, F., Ignacio González-Gordillo, J., Irigoien, X., Úbeda, B., Hernández-León, S., et al. (2014) Plastic debris in the open ocean. *Proceedings of the National Academy of Sciences, 111*(28), 10239–10244.

Della Torre, C., Bergami, E., Salvati, A., Faleri, C., Cirino, P., Dawson, K. A., et al. (2014). Accumulation and embryotoxicity of polystyrene nanoparticles at early stage of development of sea urchin Embryos *Paracentrotus lividus*. *Environmental Science and Technology, 48*, 12302–12311.

Derraik, J. G. B. (2002). The pollution of the marine environment by plastic debris: A review. *Marine Pollution Bulletin, 44*(9), 842–852.

Eriksen, M., Mason, S., Wilson, S., Box, C., Zellers, A., Edwards, W., et al. (2013). Microplastic pollution in the surface waters of the Laurentian Great Lakes. *Marine Pollution Bulletin, 77*, 177–182.

Fendall, L. S., & Sewell, M. A. (2009). Contributing to marine pollution by washing your face: Microplastics in facial cleansers. *Marine Pollution Bulletin, 58*, 1225–1228.

Free, C. M., Jensen, O. P., Mason, S. A., Eriksen, M., Williamson, N. J., & Boldgiv, B. (2014). High-levels of microplastic pollution in a large, remote, mountain lake. *Marine Pollution Bulletin, 85*, 156–163.

Galgani, F., Hanke, G. & Maes, T. (2015). Global distribution, composition and abundance of marine litter. In M. Bergmann, L. Gutow & M. Klages (Eds.), *Marine anthropogenic litter* (pp. 29–56). Berlin: Springer.

Galloway, T. S. (2015). Micro- and nano-plastics and human health. In M. Bergmann, L. Gutow & M. Klages (Eds.), *Marine anthropogenic litter* (pp. 347–370). Berlin: Springer.

Gottschalk, F., Sun, T., & Nowack, B. (2013). Environmental concentrations of engineered nanomaterials: Review of modeling and analytical studies. *Environmental Pollution, 181*, 287–300.

Gouin, T., Roche, N., Lohmann, R., & Hodges, G. (2011). A thermodynamic approach for assessing the environmental exposure of chemicals absorbed to microplastic. *Environmental Science and Technology, 45*(4), 1466–1472.

Grigoriadou, I., Paraskevopoulos, K. M., Chrissafis, K., Pavlidou, E., Stamkopoulos, T. G., & Bikiaris, D. (2011). Effect of different nanoparticles on HDPE UV stability. *Polymer Degradation and Stability, 96*, 151–163.

Guterres, S. S., Marta, P. A., & Adriana, R. P. (2007). Polymeric nanoparticles, nanospheres and nanocapsules, for cutaneous applications. *Drug Target Insights, 2*, 147–157.

Hammer, J., Kraak, M. H., & Parsons, J. R. (2012). Plastics in the marine environment: the dark side of a modern gift. *Reviews of Environmental Contamination and Toxicology, 220*, 1–44.

Handy, R. D., Cornelis, G., Fernandes, T., Tsyusko, O., Decho, A., Sabo-Attwood, T., et al. (2012). Ecotoxicity test methods for engineered nanomaterials: Practical experiences and recommendations from the bench. *Environmental Toxicology and Chemistry, 31*, 15–31.

Harshvardhan, K., & Jha, B. (2013). Biodegradation of low-density polyethylene by marine bacteria from pelagic waters, Arabian Sea, India. *Marine Pollution Bulletin, 77*, 100–106.

Holmes, L. A., Turner, A., & Thompson, R. C. (2014). Interactions between trace metals and plastic production pellets under estuarine conditions. *Marine Chemistry, 167*, 25–32.

Imhof, H. K., Ivleva, N. P., Schmid, J., Niessner, R., & Laforsch, C. (2013). Contamination of beach sediments of a subalpine lake with microplastic particles. *Current Biology, 23*, 867–868.

Kashiwada, S. (2006). Distribution of nanoparticles in the see-through medaka (*Oryzias latipes*). *Environmental Health Perspectives, 114*, 1697–1702.

Klaine, S. J., Koelmans, A. A., Horne, N., Handy, R. D., Kapustka, L., Nowack, B., et al. (2012). Paradigms to assess the environmental impact of manufactured nanomaterials. *Environmental Toxicology and Chemistry, 31*, 3–14.

Koelmans, A. A. (2015). Modeling the role of microplastics in bioaccumulation of organic chemicals to marine aquatic organisms. A Critical Review. In M. Bergmann, L. Gutow, M. Klages (Eds.), *Marine anthropogenic litter* (pp. 313–328). Berlin: Springer.

Koelmans, A. A., Besseling, E., & Foekema, E. M. (2014b). Leaching of plastic additives to marine organisms. *Environmental Pollution, 187*, 49–54.

Koelmans, A. A., Besseling, E., Wegner, A., & Foekema, E. M. (2013a). Plastic as a carrier of POPs to aquatic organisms. A model analysis. *Environmental Science and Technology, 47*, 7812–7820.

Koelmans, A. A., Besseling, E., Wegner, A., & Foekema, E. M. (2013b). Correction to plastic as a carrier of POPs to aquatic organisms. A model analysis. *Environmental Science and Technology, 47*, 8992–8993.

Koelmans, A. A., Gouin, T., Thompson, R. C., Wallace, N., & Arthur, C. (2014a). Plastics in the marine environment. *Environmental Toxicology and Chemistry, 33*, 5–10.

Koelmans, A. A., Nowack, B., & Wiesner, M. (2009). Comparison of manufactured and black carbon nanoparticle concentrations in aquatic sediments. *Environmental Pollution, 157*, 1110–1116.

Koelmans, A. A., Poot, A., De Lange, H. J., Velzeboer, I., Harmsen, J., & Van Noort, P. C. M. (2010). Estimation of *in situ* sediment to water fluxes of polycyclic aromatic hydrocarbons, polychlorobiphenyls and polybrominated diphenylethers. *Environmental Science and Technology, 44*, 3014–3020.

Kools, S. A., Bauerlein, P., Siegers, W., Cornelissen, E. & De Voogt, P. (2014). *Detection and analysis of plastics in the watercycle.* In abstract book 24th annual meeting SETAC 2014.

Lee, K. W., Shim, W. J., Kwon, O. Y., & Kang, J.-H. (2013). Size-dependent effects of micro polystyrene particles in the marine copepod *Tigriopus japonicus. Environmental Science and Technology, 47*, 11278–11283.

Lopez Lozano, R., & Mouat, J. (2009). *Marine litter in the North-East Atlantic region* (pp. 1–120). London, United Kingdom: OSPAR Commission.

Lu, S., Qu, R., & Forcada, J. (2009). Preparation of magnetic polymeric composite nanoparticles by seeded emulsion polymerization. *Materials Letters, 63*, 770–772.

Lu, J. W., Zhang, Z. P., Ren, X. Z., Chen, Y. Z., Yu, J., & Guo, Z. X. (2008). High-Elongation fiber mats by electrospinning of polyoxymethylene. *Macromolecules, 41*, 3762–3764.

Meesters, J., Koelmans, A. A., Quik, J. T. K., Hendriks, A. J., & Van de Meent, D. (2014). Multimedia modeling of engineered nanoparticles with simplebox 4 nano: Model definition and evaluation. *Environmental Science and Technology, 48*, 5726–5736.

Nowack, B., Ranville, J., Diamond, S., Gallego-Urrea, J., Metcalfe, C., Rose, J., et al. (2012). Potential scenarios for nanomaterial release and subsequent alteration in the environment. *Environmental Toxicology and Chemistry, 31*, 50–59.

Quik, J. T. K., de Klein, J. J. M. & Koelmans, A. A. (2014). Spatially explicit fate modelling of nanomaterials in natural waters. May 2014, SETAC Annual Meeting, Basel, Switzerland.

Rao, J. P., & Geckeler, K. E. (2011). Polymer nanoparticles: Preparation techniques and size-control parameters. *Progress in Polymer Science, 2011*(36), 887–913.

Rochman, C. M., Hentschel, B. T., & Teh, S. J. (2014). Long-term sorption of metals is similar among plastic types: Implications for plastic debris in aquatic environments. *PLoS ONE, 9*(1), e85433.

Rochman, C. M., Hoh, E., Hentschel, B. T., & Kaye, S. (2013a). Long-term field measurement of sorption of organic contaminants to five types of plastic pellets: Implications for plastic marine debris. *Environmental Science and Technology, 47*, 1646–1654.

Rochman, C. M., Hoh, E., Kurobe, T., & Teh, S. J. (2013b). Ingested plastic transfers hazardous chemicals to fish and induces hepatic stress. *Scientific Reports, 3*(3263), 1–7.

Rossi, G., Barnoud, J., & Monticelli, L. (2014). Polystyrene nanoparticles perturb lipid membranes. *The Journal of Physical Chemistry Letters, 5*, 241–246.

Roy, P. K., Titus, S., Surekha, P., Tulsi, E., Deshmukh, C., & Rajagopal, C. (2008). Degradation of abiotically aged LDPE films containing pro-oxidant by bacterial Consortium. *Polymer Degradation and Stability, 93*(2008), 1917–1922.

Saido, K., Koizumi, K., Sato, H., Ogawa, N., Kwon, B. G., Chung, S.-Y., et al. (2014). New analytical method for the determination of styrene oligomers formed from polystyrene decomposition and its application at the coastlines of the North-West Pacific Ocean. *Science of the Total Environment, 473–474C*, 490–495.

Salvati, A., Aberg, C., dos Santos, T., Varela, J., Pinto, P., Lynch, I., et al. (2011). Experimental and theoretical comparison of intracellular import of polymeric nanoparticles and small molecules: Toward models of uptake kinetics. *Nanomedicine: Nanotechnology, Biology and Medicine, 7*, 818–826.

Schlagenhauf, L., Nüesch, F., & Wang, J. (2014). Release of carbon nanotubes from polymer nanocomposites. *Fibers, 2*, 108–127.

Shim, W.J., Song, Y.K., Hong, S.H., Jang, M. & Han, G.M. (2014). Producing fragmented micro- and nano-sized expanded polystyrene particles with an accelerated mechanical abrasion experiment. May 2014, SETAC Annual Meeting, Basel, Switzerland.

Sivan, A. (2011). New perspectives in plastic biodegradation. *Current Opinion in Biotechnology, 2011*(22), 422–426.

Stephens, B., Azimi, P., El Orch, Z., & Ramos, T. (2013). Ultrafine particle emissions from desktop 3D printers. *Atmospheric Environment, 79*, 334–339.

Teuten, E. L., Rowland, S. J., Galloway, T. S., & Thompson, R. C. (2007). Potential for plastics to transport hydrophobic contaminants. *Environmental Science and Technology, 41*, 7759–7764.

Thompson, R. C. (2015). Microplastics in the marine environment: Sources, consequences and solutions. In M. Bergmann, L. Gutow, M. Klages (Eds.) *Marine anthropogenic litter* (pp. 185–200). Berlin. Springer.

Velzeboer, I., Kwadijk, C. J. A. F., & Koelmans, A. A. (2014a). Strong sorption of PCBs to nanoplastics, microplastics, carbon nanotubes and fullerenes. *Environmental Science and Technology, 48*, 4869–4876.

Velzeboer, I., Quik, J. T. K., van de Meent, D., & Koelmans, A. A. (2014b). Rapid settling of nanomaterials due to hetero-aggregation with suspended sediment. *Environmental Toxicology and Chemistry, 33*, 1766–1773.

Von der Kammer, F., Ferguson, P. L., Holden, P., Masion, A., Rogers, K., Klaine, S. J., et al. (2012). Analysis of nanomaterials in complex matrices (environment and biota): General considerations and conceptual case studies. *Environmental Toxicology and Chemistry, 31*, 32–49.

Wagner, M., Scherer, C., Alvarez-Muñoz, D., Brennholt, N., Bourrain, X., Buchinger, S., et al. (2014). Microplastics in freshwater ecosystems: What we know and what we need to know. *Environmental Sciences Europe*, 26, 12 http://www.enveurope.com/content/26/1/12.

Ward, J. E., & Kach, D. J. (2009). Marine aggregates facilitate ingestion of nanoparticles by suspension-feeding bivalves. *Marine Environment Research, 68*, 137–142.

Wegner, A., Besseling, E., Foekema, E. M., Kamermans, P., & Koelmans, A. A. (2012). Effects of nanopolystyrene on the feeding behaviour of the blue mussel (*Mytilus edulis* L.). *Environmental Toxicology and Chemistry, 31*, 2490–2497.

Wiesner, M. R., Lowry, G. V., Casman, E., Bertsch, P. M., Matson, C. W., Di Giulio, R. T., et al. (2011). Meditations on the ubiquity and mutability of nano-sized materials in the environment. *ACS Nano, 5*, 8466–8470.

Zbyszewski, M., & Corcoran, P. L. (2011). Distribution and degradation of fresh water plastic particles along the beaches of Lake Huron, Canada. *Water, Air, and Soil Pollution, 220*, 365–372.

Zbyszewski, M., Corcoran, P. L., & Hockin, A. (2014). Comparison of the distribution and degradation of plastic debris along shorelines of the Great Lakes, Norht America. *Journal of Great Lakes Research, 2014*(40), 288–299.

Zhang, H., Kuo, Y.-Y., Gerecke, A. C., & Wang, J. (2012). Co-release of hexabromocyclododecane (HBCD) and nano- and microparticles from thermal cutting of polystyrene foams. *Environmental Science and Technology, 46*, 10990–10996.

Part IV
Socio-economic Implications of Marine Anthropogenic Litter

Chapter 13
Micro- and Nano-plastics and Human Health

Tamara S. Galloway

Abstract Plastics are highly versatile materials that have brought huge societal benefits. They can be manufactured at low cost and their lightweight and adaptable nature has a myriad of applications in all aspects of everyday life, including food packaging, consumer products, medical devices and construction. By 2050, however, it is anticipated that an extra 33 billion tonnes of plastic will be added to the planet. Given that most currently used plastic polymers are highly resistant to degradation, this influx of persistent, complex materials is a risk to human and environmental health. Continuous daily interaction with plastic items allows oral, dermal and inhalation exposure to chemical components, leading to the widespread presence in the human body of chemicals associated with plastics. Indiscriminate disposal places a huge burden on waste management systems, allowing plastic wastes to infiltrate ecosystems, with the potential to contaminate the food chain. Of particular concern has been the reported presence of microscopic plastic debris, or microplastics (debris ≤ 1 mm in size), in aquatic, terrestrial and marine habitats. Yet, the potential for microplastics and nanoplastics of environmental origin to cause harm to human health remains understudied. In this article, some of the most widely encountered plastics in everyday use are identified and their potential hazards listed. Different routes of exposure to human populations, both of plastic additives, microplastics and nanoplastics from food items and from discarded debris are discussed. Risks associated with plastics and additives considered to be of most concern for human health are identified. Finally, some recent developments in delivering a new generation of safer, more sustainable polymers are considered.

T.S. Galloway (✉)
College of Life and Environmental Sciences, University of Exeter,
Stocker Road, Exeter EX4 4QD, UK
e-mail: t.s.galloway@exeter.ac.uk

© The Author(s) 2015
M. Bergmann et al. (eds.), *Marine Anthropogenic Litter*,
DOI 10.1007/978-3-319-16510-3_13

13.1 Introduction

If a visitor from 50 years ago were to turn up today, one of the first things he would notice (other than how much heavier we all were), would be how much plastic there is everywhere. We use plastics to wrap our food, we drink from plastic containers, cook with plastic utensils, deliver drugs to patients through plastic tubing. We increasingly use plastics and polymer composites in construction. Worldwide annual production of plastics is estimated to be in the region of 300 million tonnes. Plastic demand in the European Union alone for 2010 was estimated at 46.4 million tonnes, consisting of two main types: plastics used for packaging of food and consumer items, with the second group constituting plastics used in the construction industry (PlasticsEurope 2013). With overall recycling rates at around 57.9 %, this corresponds to around 24.7 million tonnes of plastic debris entering the waste stream each year. Waste disposal includes littering, land fill and the sewerage system and ultimately, a significant proportion of plastic waste ends up in the sea. Jambeck et al. (2015) estimated that 4.8 to 12.7 million tonnes of plastic waste entered the ocean in 2010.

Whilst these figures are alarming in terms of volume, it is not yet clear how this large scale and ubiquitous use affects human health. As an example, some 14.5 million tonnes per annum of plastic is used in the food packaging industry alone (European Plastics Converters). On the positive side, improvements in food packaging can prevent bacterial infections, such as Salmonella and other food borne disease (Hanning et al. 2009; European Commission 2014), and can prevent wastage and aid distribution. Conversely, migration of contaminants from food packaging into food is considered the main route of exposure of human populations to contaminants associated with plastics (Grob et al. 2006), with only a small fraction of the thousands of substances that may be present having been subject to extensive testing (Claudio 2012). Whilst rigorous standards are in place to regulate food-contact substances in terms of migration into food (EFSA 2011), it is less clear how these guidelines offer protection once the plastics themselves have been discarded to the environment. With only limited information available about rates of degradation and fragmentation, leaching of chemicals into environmental matrices, and entry into the food chain, it is almost impossible to estimate the cumulative risks of chronic exposure to plastics and their additives.

One way around this problem is to determine what chemicals are actually present in the human body. Human biomonitoring involves measuring the concentrations of environmental contaminants and/or their metabolites in human tissues or body fluids, such as blood, breast milk, saliva or urine. Biomonitoring is considered a gold standard in assessing the health risks of environmental exposures because it can provide an integrated measure of an individual's exposure to contaminants from multiple sources (Sexton et al. 2004). This approach has shown that chemicals used in the manufacture of plastics are certainly present in the human population. For some chemicals, their widespread presence in the general population at concentrations capable of causing harm in animal models has raised public health concerns (Talsness et al. 2009; Melzer and Galloway 2010). The National Health and Nutrition Examination Survey (NHANES) is a program of studies designed to assess the health and nutritional status of adults and children in the United States and represents one of the most comprehensive human bio-monitoring programs yet undertaken (http://www.cdc.gov/nchs/nhanes.htm). Of interest for this article, NHANES reports on several chemicals associated with the use or production of plastics, including bisphenol A, phthalates, styrene, acrylamide, triclosan and brominated flame retardants, and their concentrations in the general population.

This review considers the kinds of plastics in widespread, everyday use and the potential hazards they may cause. It reviews the routes of uptake of micro and nanoplastics into humans through the food chain and the potential consequences for human health. Health risks associated with microplastics and plastic-associated chemicals are discussed. Lastly, some new developments in alternative low toxicity polymers and novel nanocomposite materials are described and their potential benefits to human health discussed.

13.2 What Kinds of Plastics Are in Use?

The term plastic is used to describe plastic polymers, to which various additives are added to give desirable properties to the final product (OECD 2004). The demand for plastics in Europe alone is estimated to be 45.9 million tonnes in 2012 (PlasticsEurope 2013), with plastics demand by industry segment shown in Table 13.1. As can be seen, packaging, which includes food and beverage packaging, is the single largest category by a considerable margin. Plastics are generally divided into two types: thermoplastic, which soften on heating and can be remoulded, and thermosetting, in which case cross-linking in the polymers means they cannot be re-softened and remoulded. With reference to these properties, plastics can be further classified into seven different groupings based on their ease of recycling. Table 13.2 lists some examples of products made from these seven different plastics groups and the demand for different resin and polymer types based on this classification system (for Europe). As can be seen, the

Table 13.1 Plastics demand by industry segment in Europe, 2012

Industry segment	Volume (millions of tonnes)	Percentage of total
Packaging	18.1	39.4
Building and construction	9.32	20.3
Automative	3.76	8.2
Electronics and electrical	3.03	6.6
Agriculture	1.93	4.2
Other (furniture, health and safety, sport, consumer and household appliances, etc.)	10.3	22.4
Total (demand for 2012)	45.9	100.0

Figures are derived from PlasticsEurope (2013)

Table 13.2 European plastic demand by resin type

Code	Resin type	Example product	Volume of demand (millions of tonnes)	% of total European demand	% recycled[a]
1	PET polyethylene terephthalate	Soft drink bottle, polyester fibre	2.98	6.5	20
2	PE-HD polyethylene high density	Plastic bottle, plastic bag, bottle cap	5.51	12.0	11
3	PVC polyvinyl chloride	Water proof boot, window frame, plumbing pipe	4.91	10.7	0
4	PE-LD polyethylene low density	Wire cable, plastic bag, bucket, soap dispenser bottle, plastic tube	8.03	17.5	6
5	PP polypropylene	Stationary folder, plant pot, bags, industrial fibre	8.63	18.8	1
6	PS. PSE polystyrene	Food container, plastic cup, glasses frame, car bumper	3.40	7.4	1
7	O other (PC Polycarbonate, PLA polyamide, styrene, SAN acrylonitrile, acrylic plastics, PAN/ polyacrylonitrile, bioplastics)	Drink bottle, consumer item, clothing, medical equipment	9.82	19.8	0
Total			45.9	100.0	39

Figures are for 2012 and are derived from PlasticsEurope (2013). [a]Recycling figures derived from Engler (2012)

main classification group (code 7) makes up 19.8 % of total European demand, yet has a 0 % recycling rate. The second most commonly used plastic, polypropylene (18.8 % of demand), has a 1 % recycling rate.

13.3 Plastics and Human Health

Plastic polymers are generally considered to be inert and of low concern to human health, and health risks relating to their use are attributed to the presence of the wide range of plastic additives they may contain, together with residual monomers that may be retained within the polymer structure (Araujo et al. 2002). Plastics are synthesised from monomers, which are polymerised to form macromolecular chains. A range of additional chemicals may be added during the manufacturing process, including initiators, catalysts and solvents. Additives that can alter the nature of the final plastic include stabilisers, plasticisers, flame retardants, pigments and fillers. Additives are not bound to the polymer matrix and because of their low molecular weight, these substances can leach out of the plastic polymer (Crompton 2007) into the surrounding environment, including into air, water, food or body tissues.

There are thousands of additives in routine use in the synthesis of plastic products. As comprehensively reviewed by Lithner et al. (2011), certain plastics types typically contain more additives than other types. Polyvinylchloride (PVC) is the polymer associated with the use of most additives, including heat stabilisers to keep the polymer stable during production, and plasticisers such as phthalates to allow flexibility (Lithner et al. 2011). Plasticisers may constitute a high percentage (up to 80 %) of the weight of the final product (Buchta et al. 2005). Polypropylene is highly sensitive to oxidation and typically contains significant amounts of anti-oxidants and UV stabilisers (Zweifel 2001). Other chemicals that may leach from plastics include nonylphenol from polyolefins, brominated flame retardants from acrylonitrile-butadienestyrene (ABS) or urethane foam and bisphenol A (BPA) from polycarbonate. The rate at which these substances are released from the product is governed by many factors, including the size and volatility of the additive, the per-meability of the polymer itself (migration is greater for highly permeable polymers), and the temperature and pH of the surrounding medium (air, water, soil, body tissues) (Zweifel 2001).

Plastics may also pose a hazard due to the release of the constituent monomers themselves (Lithner et al. 2011). Most of the plastics in everyday use are highly resistant to microbial degradation. Instead, degradation and release of polymers is ultimately caused by exposure to abiotic factors such as ultraviolet (UV) light, heat, mechanical and/or chemical abrasion (Andrady 2015). Breaking of the chem-ical bonds in the polymer backbone leads to chain scission and depolymerisation; chain stripping occurs when side chains are broken and released. All of these pro-cesses proceed at different rates under different environmental conditions, e.g. variations in temperature and oxygen, and proceed at different rates for differ-ent polymer types, with polyesters, polycarbonate and polyurethane more prone to depolymerisation for example than polyethylene or polypropylene (Nicholson 1996; La Mantia 2002). It is therefore extremely difficult to predict the risks asso-ciated with exposure to plastics and their additives, given the vast complexity and variability of the available product combinations, their varied uses and eventual environmental distribution once discarded.

Lithner et al. (2011) addressed this complex problem by conducting a comprehensive hazard ranking of plastic polymers based on their chemical com-position. They studied 55 of the most widely used polymer types with global production volumes of >10,000 tonnes per year. A model for ranking the hazard of each polymer was developed by ranking the constituent monomer chemicals according to internationally agreed criteria for identifying physical, environ-ment and health risks. The polymer types that received the highest and the low-est hazard rankings according to this criteria are shown in Table 13.3. Table 13.4 shows the ranking for polymer types commonly reported in plastic and micro-plastic litter.

Table 13.3 Ranking of some plastic polymer types based on hazard classification of constituent monomers, adapted from Lithner et al. (2011)

Polymer	Monomer(s)/additives	Relative hazard score[a]	Recycling code	Constituents measured in NHANES?
Polymers with the highest relative hazard scores				
Polyurethane PUR as a flexible foam	Propylene oxide	13,844	6	
	Ethylene oxide			
	Toluene-diisocyanate			
Polyacrylamide PAN with co-monomers	Acrylonitrile	12,379	7	Acrylamide
	Acrylamide			
	Vinyl acetate			
Polyvinylchloride PVC, plasticised	With plasticiser	10,551	3	Benzyl butyl phthalate (BBP)
	Benzyl butyl phthalate (BBP) at 50 wt%			
Polyvinylchloride, PVC, unplasticised		10,001	3	
Polyurethane, PUR as a rigid foam	Propylene oxide	7384	6	
	4,4′-methylenediphenyl diisocyanate (MDI)			
	Cyclopentane			
Epoxy resins DGEBPA	**Bisphenol A**	7139	7	Bisphenol A
	Epichlorohydrin			
	4,4′-methylenedianaline			
Modacrylic	Acrylonitrile	6957		
	Vinylidene chloride			
Acrylonitrile-butadiene-styrene ABS	**Styrene**	6552	7	Styrene
	Acrylonitrile			
	1,3 butadiene			
Styrene- acrylonitrile SAN	**Styrene**	2788	7	Styrene
	Acrylonitrile			
High impact polystyrene HIPS	**Styrene**	1628		Styrene
Polymers with the lowest relative hazard scores				
Low density polyethylene LDPE	Ethylene	11	4	
High density polyethylene HDPE	Ethylene	11	2	
Polyethylene terephthalate PET	Terephthalic acid	4	1	
Polyvinyl acetate PVA	Vinyl acetate	1		
Polypropylene PP	Propylene	1	5	

[a]Relative hazard score derived from different constituent monomers. Higher ranking = greater hazard

Table 13.4 Plastics identified in microplastic debris and their relative hazard ranking

Polymer type	Density g/cm^3	Relative hazard score[a]
Polyethylene (low, high density)	0.917–0.965	11
Polypropylene	0.9–0.91	1
Polystyrene	1.04–1.1	1628–30
Polyamide		63–50
polyethylene teraphthalate	1.37–1.45	4
Polyvinylchloride	1.16–1.58	10,551–5001

[a]Relative hazard score derived from different constituent monomers. Higher ranking = greater hazard

Adapted from Hidalgo-Ruz et al. (2012) and Lithner et al. (2011)

As noted by the authors, the hazard ranking does not imply that the polymers themselves are hazardous, but rather that release of hazardous substances or degradation products may occur during the product lifecycle, i.e. from production through use of the product and its eventual discard to waste or into the environment. From this point of view, Table 13.3 also identifies polymers that may contain compounds that are currently the subject of biomonitoring activities under the NHANES program. Note that NHANES also monitors compounds that may be present in multiple, diverse items including many different types of plastics and plastics products, such as the microbial agent triclosan and the UV screen and printing ink additive benzophenone.

The polymers ranked as most hazardous were those produced from monomers classified as carcinogenic, mutagenic or both, leading to high hazard rankings for polyurethanes, polyvinylchloride, epoxy resins and styrenic polymers. One limitation of this approach noted by the authors was the lack of available chemical safety data for many of the substances they were considering. In particular, there was no hazard class available for chemicals suspected of being endocrine disruptors, including bisphenol A, phthalates, and epichlorohydrin. This toxicity endpoint was therefore not included in the hazard assessment. This represents a major limitation in our current ability to predict the risks associated with plastics associated chemicals, since so many of these are recognised to have endocrine disrupting abilities (Koch and Calafat 2009). Despite these limitations, this study represents an extremely useful attempt to identify those polymer types that could be a cause for concern due to the environmental and health effects of their constituent monomers.

13.4 Micro- and Nanoplastics

13.4.1 Occurrence of Micro- and Nanoplastics in the Environment

In addition to larger items of plastic litter, concern has been raised that microscopic plastic debris (microplastic) (<1 mm) may also be detrimental to the environment and to human health (Thompson et al. 2004; Cole et al. 2011).

Microplastics have been studied mostly in the context of the marine environment, and have been found to be a major constituent of anthropogenic marine debris. They consist of small plastic items, such as exfoliates in cosmetics, or fragments from larger plastic debris, including polyester fibres from fabrics, polyethylene fragments from plastic bags and polystyrene particles from buoys and floats (reviewed by Cole et al. 2011).

There is sparse information available on the presence of microplastics in environments other than the oceans, for example in terrestrial soils or freshwater environments. The presence of microplastic particles (Dubaish and Liebezeit 2013) and synthetic polymer fibres (Zubris and Richards 2005) has been reported in sewage sludge and in the soils to which they had been applied (Zubris and Richards 2005), where they were still detectable five years after application. A study of surface waters in the southern North Sea found microplastics and microfibres in all of the samples that were tested, with an increasing gradient towards land sources (Dubaish and Leibezeit 2013). Browne et al. (2011) showed that the polyester and acrylic fibres used in clothing closely resembled those found in coastal sediments that receive sewage discharges, suggesting that sewage effluents represent a significant source of microfibres from the washing of clothes, and that these are not wholly retained during wastewater treatment.

A study of beach sediments around Lake Garda, a subalpine lake in Italy, found microplastics at abundances of up to 1108 ± 983 microplastic particles/m^2 (Imhof et al. 2013), which is similar to the contamination levels reported for the Great Lakes in the USA (Zbyszewski and Corcoran 2011). These levels of contamination most likely originate from landfill, litter and wastewater sources, and are within the range of values reported for the abundance of plastic particles found in marine coastal sediments (0.21–77,000 particles/m^2), albeit at the lower end of exposures (Hidalgo-Ruz et al. 2012). This does, however, indicate that microplastics are present in both agricultural soils and freshwater sites. Knowledge on the occurrence of nanoplastics in aquatic environments and biota is extremely limited because no methods exist for the reliable detection of nanoplastics in samples (Koelmans et al. 2015).

13.4.2 Micro- and Nanoplastics and Human Health

In terms of human health risks, microplastics as contaminants in the wider environment represent a concern because it has been shown that they can be ingested by a wide range of aquatic organisms, both marine and freshwater, and thus have the potential to accumulate through the food chain. Aquatic organisms for which ingestion of microplastics has been documented in the field include those from across the marine food web, including turtles, seabirds, fish, crustaceans and worms (reviewed by Wright et al. 2013). Laboratory studies have confirmed that many other organisms have the capacity to ingest microplastics including zooplankton (Cole et al. 2013; Setälä et al. 2012). The majority of studies have documented microplastics in the guts of organisms, an organ that is not generally consumed directly by humans.

Exceptions to this include shellfish such as mussels, clams and some shrimps that are eaten whole or with their gut. The risk of ingesting microplastics contained within other tissues depends on the degree to which uptake of microplastics and translocation and redistribution and retention within other body tissues occurs. This concept is discussed further below, in relation to human ingestion.

In addition to the potential for ingestion to cause adverse biological effects due to gut blockages and/or damage, or the reduction in energy assimilation (Wright et al. 2013), the large surface area of microplastics means that environmental pollutants may sorb to the surface of the particles, with the potential to be transferred into body tissues once ingested. For a more comprehensive coverage of the uptake of microplastics by wildlife organisms, and the transfer to tissues of hydrophobic pollutants adsorbed from the surrounding environment, the reader is referred to excellent recent reviews (e.g. Engler 2012) and to other chapters in this issue (Koelmans 2015; Lusher 2015). Despite this concern, there is currently no available information to evidence the uptake or biological effects of microplastics originating from marine or terrestrial debris and subsequently ingested by humans through the food chain.

13.4.3 Ingestion of Micro- and Nanoplastics and Uptake Across the Gut

Whilst the potential clearly exists for microplastics to be present in food items, there is currently no evidence for the unintentional ingestion or subsequent translocation and uptake of microplastics into the human body through the diet. There is, however, a huge interest worldwide in the use of micro- and nanospheres as pharmaceutical drug delivery systems through oral, intravenous and transcutaneous routes (Kim et al. 2010), and in the migration of nanopolymers from packaging materials into food (EFSA 2011; Lagaron and Lopez-Rubio 2011). Based on these growing and fast moving fields, an enhanced understanding of the mechanistic pathways by which micro- and nanoparticles could enter the human body is starting to emerge, although many aspects of this field remain to be elucidated.

Following oral ingestion, the gut mucosa represents an important barrier, which has evolved to allow efficient uptake of nutritious items, whilst excluding potentially harmful substances or organisms. Significant uptake of microplastics into the body through this route is in theory then limited to particles that can enter the body through exploitation of existing routes. Following oral ingestion, uptake of inert particles across the gut has been widely studied (O'Hagan 1996). The 'persorption' of starch particles as large as 150 μm through the tips of the villi was described in detail by Volkheimer (1977). According to his observations, persorption of particles can occur as a passive process in areas of the gut where the intestinal mucosa is covered by a single layer of epithelium. Persorbed particles were detectable in the lumen of blood and lymph vessels within minutes, and were eventually eliminated in the urine, confirming that the translocation of relatively large, inert particles from the gut to other body fluids is possible (Volkheimer 1977).

Aside from this observation, digestive absorption of smaller particles proceeds through pinocytosis and vesicular phagocytic processes for materials in the nano and micron range. Particle size is one of the most important factors in determining the extent and pathway of uptake. Smaller particles are generally favoured over larger ones. For example, polystyrene microspheres of 50–100 nm were more readily absorbed across the Peyer's patches and the villi of the gut than larger particles of 300–3000 nm (Jani et al. 1992; Florence and Hussain 2001). On the other hand, the uptake of ultrafine polylysine dendrimers of 2.5 nm was lower than that of larger polystyrene particles of 100 nm–3 μm, suggesting that size is not the only deciding factor (Florence et al. 2000). Indeed, a combination of size, surface charge and hydrophilicity all contribute to uptake affinity (as discussed by Awaad et al. 2012). The predominant site of uptake for micron-scale particles in the gut is reported to be through gut-associated lymphatic tissue (GALT), specifically by the Microfold (M) cells of the Peyer's patches. M cells are specialised epithelial cells that lack the microvilli found on other gut epithelial cells and instead have broader (micro)folds and a thinner luminal surface that allows them to actively take up particulate matter from the intestine. The reported efficiency of this uptake varies depending on the study method, species and particle type. Uptake of polystyrene microspheres through the gut by this route was higher in species such as rabbits, which have a high abundance of M cells (Pappo et al. 1989), and was enhanced when food was also present, probably due to the delayed transit time through the gut (Ebel 1990). As an alternative route, uptake by enterocytes appears to be limited to a size range of around 100 nm (Jani et al. 1992). Awaad et al. (2012) used fluorescent organosilica particles, histological examination and quantitative analysis to confirm an optimal size range of around 100 nm for uptake of particles through the M cells of the Peyer's patches, with smaller and larger particles less likely to be taken up. They also identified two alternative uptake pathways by which nanoparticles passed between (paracellular-E uptake) or through (transcellular-E uptake) enterocytes in the Peyer's patches. These two pathways have previously been described as major mechanisms for larger particles of >1 μm outside of the Peyer's patches (Kreuter 1991), but had not previously been described in relation to nanoparticle uptake by the Peyer's patches.

Garrett et al. (2012) used a novel bio-imaging technique, multimodal nonlinear optical microscopy, to document uptake of polymeric nanoparticles by enterocytes in the mouse gut in vivo. They studied a novel amphipathic polymer specifically designed for drug delivery, ammonium palmitoyl glycol chitosan (GCPQ) of 30–50 nm in diameter and showed that after uptake by enterocytes, particles accumulated at the base of the villi. From there, they passed into the blood stream and were transported to the liver, where they were detectable in the hepatocytes and intracellular spaces, before recirculating through the bile to the small intestine (Garrett et al. 2012) to be excreted with faecal matter. This is similar to previous results for larger micron-scale polystyrene and latex particles, suggesting that both micron and nano-scale polymers are treated in a similar manner (Jani et al. 1996), with uptake across the gut, recirculation and eventual elimination through faecal matter and urine (Fig. 13.1).

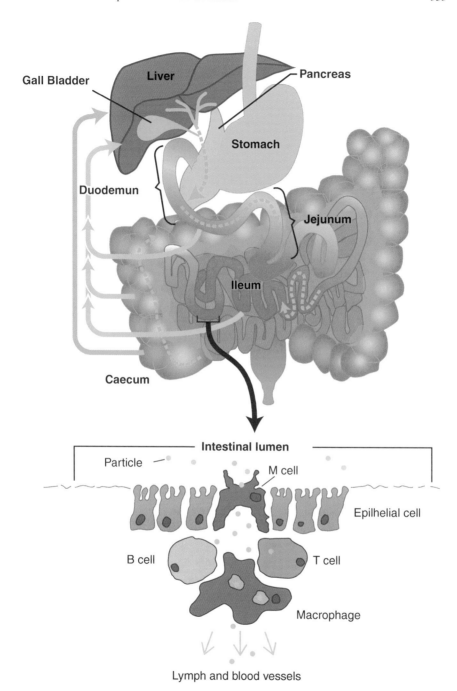

Fig. 13.1 A diagram illustrating a proposed recirculation pathway for polymer nanoparticles (ammonium palmitoyl glycol chitosan) after oral administration. The nanoparticles are taken up into the blood from the gut through M cells, and from there through the lymphatic system (shown in *yellow*) and into the liver and gall bladder. Particles are then re-released into the gut together with bile (shown in *green*) before excretion in faeces and urine. Adapted from Garrett et al. (2012)

This information is of high interest in terms of drug delivery, yet it also suggests that ample opportunity exists, following ingestion, for micro- and nanoplastics in food or water to enter, circulate and bioaccumulate within the body.

13.4.4 Interaction of Microspheres and Nanoparticles with Cells and Tissues

The behaviour of nano- and microplastics after they have entered the circulation from the gut is not fully understood, but has been the subject of study in relation to food packaging materials and nanomedicines. Certainly, in vivo behaviour will be dependent on numerous factors, such as the physico-chemical properties of the particles (size, surface charge, aspect ratio, porosity, surface corona) and the physiological state of the individual. Risk assessments of manufactured nanomaterials including titanium dioxide (Wang et al. 2007) and carbon (Poland et al. 2008) have shown comparable results to those shown above for nanopolymers, with uptake across the gut into the circulation and redistribution to the liver and spleen. Circulation time is highly dependent on the surface characteristics of the particle, with hydrophilic and positively charged particles showing enhanced circulation times (Silvestre et al. 2011).

13.4.5 Interactions with Biological Materials and Cells

Interaction of nanopolymers with cells and tissues has again been the subject of intensive study. Because of their surface properties, nanopolymers are predicted to adsorb macromolecules such as proteins and lipids from the surrounding body fluids onto their surface, in a process influenced by surface energy, charge and specific affinity for certain biomolecules. The resulting 'corona' will then influence the resulting behaviour and toxicity of the particle (Lundqvist et al. 2008; Tenzer et al. 2013). This process has been extensively studied for polymers intended for therapeutic use particularly using polystyrene as a model polymer, but little or nothing is known of how protein coronas may form on the types of polymers most commonly found in environmental debris.

The results from mechanistic studies of different types of particle show that the potential for cytotoxicity of circulating particles in vivo to cells and tissues is related to many factors, including size, shape, solubility, surface charge, surface reactivity and energy band structure (Nel et al. 2006; Burello and Worth 2011). For example, it would be reasonable to hypothesize that particles with a high abundance of reactive surface groups would be capable of denaturing surrounding lipids and proteins. As an illustration of this, the toxicity of silica nanoparticles in vivo was attributed to proton donating silanol groups on the surface of the particles, leading to denaturation of membrane proteins and subsequent membrane damage. In this case, the reactivity of the surface hydrogen of silica bonds with membrane proteins led to their abstraction from the membrane, with subsequent membrane damage and distortion leading to haemolytic symptoms following exposure (Pandurangi et al. 1990).

Surface charge is also a strong attributing factor for toxicity (Geys et al. 2008). In inhalation studies in rats, the toxicity of acrylic ester nanopolymers in the size range 50–1500 nm was found to be low, and this was attributed to their anionic surface charge (Ma-Hock et al. 2012). Studies in which the surface charge of stearylamine-polylactic acid (PLA) polymers was modified from positive to negative showed that cationic particles showed higher pulmonary toxicity (Harush-Frenkel et al. 2010). This was attributed both to a higher localisation of cationic particles in the lung and to enhanced cellular uptake. Overall, the interaction of cationic polymers with the negatively charged cell surface has been proposed as a cause of their higher cytotoxicity (Fischer et al. 2003).

Translocation of nanopolymers into diverse tissues and cell types presents another point at which toxicity may occur. Translocation is dependent on interactions with the cell membrane and is most likely to proceed, as for uptake by enterocytes in the gut, through pinocytic, phagocytic and receptor-mediated endocytosis (Fruijter-Polloth 2012). A study, which measured the uptake rates of individual polystyrene microspheres into human astrocytes and lung carcinoma cells in culture found that the uptake rate differed for particles of different sizes, implying that there are differences in the mechanisms involved. Particles with a diameter of 40 nm showed higher uptake rates than either 20 or 100 nm particles. Since the van der Waals force between a sphere and a surface is proportional to the diameter of the sphere (Israelachvili 1992), it could be predicted that larger particles would be taken up faster. The conclusion was that the endocytic mechanism for internalisation of 40 nm particles exhibited faster kinetics, providing a privileged size gap for 40 nm particles (Varela et al. 2012).

Phagosomes containing particles may fuse with endosomes following internalisation, leading to accumulation of particles in lysosomes. Depending on the dose and type of particle, this has the potential to overwhelm lysosomal capacity and interfere with programmed cell death and pathways of cellular breakdown of pathogens (Fruijter-Polloth 2012). The numerous additional modes of toxicity that may result are again dependent on particle and cell type, and include the potential for oxidative damage, inflammation and accumulation in diverse tissue types (Silvestre et al. 2011; Nel et al. 2006, 2009). In theory, all organs may be at risk following chronic exposure to nanopolymers, including the brain, testis and reproductive organs, prior to their eventual excretion in urine and faeces (Jani et al. 1996; Garrett et al. 2012). Distribution to the foetus in utero is also a possibility that cannot be excluded. Given the long-term persistence of many polymer types, more research is required to adequately assess the risks that accumulation of micro- and nanoplastics in the body may pose.

13.5 Assessing the Risks that Micro- and Nanoplastics Pose to Human Health

13.5.1 Leaching of Toxic Chemicals from Plastics

As discussed previously, plastics can contain complex mixtures of additives to enhance their physical properties, which can leach from the polymer into the surrounding milieu. Leaching will occur primarily at the surface of the plastic particle, with the possibility of constant diffusion of chemicals from the core of the particle to the surface. Thus, leaching from plastic particles could present a long-term source of chemicals into tissues and body fluids, despite the fact that many of these chemicals are not persistent and have short half lives in the body (Engler 2012). Plastics additives of concern to human health include phthalates, bisphenol A, brominated flame retardants, triclosan, bisphenone and organotins.

The potential migration of polymer constituents and additives into food and drinks is considered to be a major route of exposure of the human population and as might be expected is subject to extensive legislation. The measurement of migration levels is typically estimated from measurements using different solvents to simulate the receiving environment (e.g. foodstuffs), or can be estimated using partitioning models that consider aspects including the desportion rates from the polymers, dimensions of the polymer framework and dimensions of the diffusing molecules (Helmroth et al. 2002). The European Food Standards Agency has a total migration limit of 10 mg/dm^2 for additives within plastics intended for packaging use, with a more stringent migration limit of 0.01 mg/kg for certain chemicals of concern (Commission Directive 2007/19/CE that modifies Directive 2002/72/CE). This means that for an average 60 kg adult who consumes 3 kg of foods and liquids per day, exposures to individual substances from food packaging could be up to 250 μg/kg body weight per day (Muncke 2011).

13.5.2 Bisphenol a and Human Health

There is very little information on the leaching of additives into biological tissues directly, but one chemical monomer that has received considerable attention in relation to its human health effects is bisphenol A (Fig. 13.2). Bisphenol A (BPA) was first synthesised in the 1930s as a synthetic estrogen (Dodds and Lawson 1936) and is now a high-production volume chemical used as a monomer in the production of polycarbonate plastic and in the epoxy resins lining food and beverage cans. There are numerous studies showing that BPA can migrate out of polycarbonate (reviewed

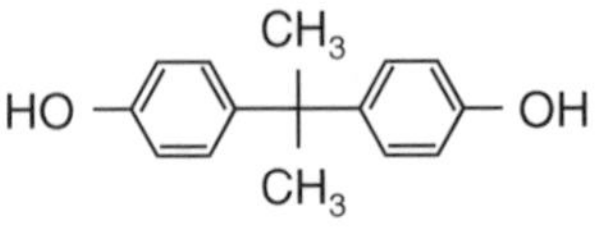

Fig. 13.2 Bisphenol A

in Guart et al. 2013) and contaminate foodstuffs and drinks, and oral ingestion is considered the major route of exposure of the human population (Calafat et al 2008). Additional routes of exposure are predicted from the inhalation of household dusts and dermal uptake from printed materials (Ehrlich et al. 2014). BPA undoubtedly enters the human body, with studies showing exposure of >95 % of populations in USA, Europe and Asia (Galloway et al. 2010; Vandenberg et al. 2010).

Bisphenol A exerts its biological activity predominantly through interaction with steroid hormone receptors, showing both estrogenic and antiandrogenic activity and suppressing aromatase activity (Bonefeld-Jørgensen et al. 2007, Lee et al 2003). Additional receptor-mediated effects reported in various model systems include binding to the orphan estrogen-related receptor ERRα (Okada et al. 2008), thyroid hormone disruption (Moriyama et al. 2002), altered pancreatic beta cell function (Ropero et al. 2008) and obesity promoting effects (Newbold et al. 2008). There is growing evidence from epidemiological and laboratory studies that exposure to BPA at levels found in the general population, around 0.2–20 ng/ml (values given for urinary BPA), is associated with adverse human health effects, including the onset of obesity and cardiovascular disease (Lang et al. 2008; Melzer et al. 2010, 2012; Cipelli et al. 2013) and with numerous reproductive and developmental outcomes. These include increases in abnormal penile/urethra development in males, an increase in hormonally-mediated cancers including breast and prostate cancers, neurobehavioural disorders including autism and early sexual maturation in females (reviewed by vom Saal et al. 2007; Hengstler et al. 2011; Rochester 2013).

Whether the release of BPA from ingested micro- or nanoplastics directly into the body contributes to human exposure remains unknown. The current tolerable daily intake is 0.05 mg/kg/day (EFSA 2006) and compared with this, the median exposure of the general adult population globally has been estimated from human biomonitoring or urinary BPA to be 0.01–0.12 μg/kg/day (EFSA 2015). The concentrations of BPA in plasma are higher than would be predicted only from this level of exposure to BPA through food and drink (Mielke and Gundert-Remy 2009), and it is therefore plausible that other routes of exposure could occur, e.g. from ingestion of plastic particles containing BPA, which subsequently leaches into tissues. BPA can certainly be absorbed across body surfaces other than the gut. Gayrard et al. (2013) showed that BPA can be absorbed with relatively high efficiency sublingually, an effect likely enhanced by its low molecular weight and moderate water solubility, allowing it to penetrate the sublingual membrane.

There are no studies in humans of the transfer of BPA from plastic directly into tissues, but the potential for BPA to leach from ingested polycarbonate into aquatic species was explored by Koelmans et al. (2014) who used biodynamic modeling to calculate the relative contribution of plastic ingestion to total exposure to chemicals residing in the ingested plastic. They estimated plastic:lipid exchange coefficients for a range of plastic particle sizes for two species, fish and sediment-dwelling worms. They proposed that a continuous ingestion of plastic containing 100 mg/kg BPA would lead to a very low steady-state concentration of 0.044 ng/kg BPA in fish and 60 μg/kg (normalized to lipid) in worms. Whilst this represents a substantial exposure pathway, the risk of exposure through this route was considered low in comparison with other pathways of exposure, based on the reported abundance of microplastics.

13.5.3 Safer Alternatives to BPA

Concern over exposure to BPA and its potential to cause harmful effects has led to worldwide efforts to formulate alternative polymer materials. This is a technically challenging area, largely because polycarbonate is such a useful material. It is an optically clear, strong and heat resistant plastic and hence has a wide range of uses. One promising formulation is a copolyester called Tritan^{TM}, which contains three different monomers, dimethyl terephthalate, cyclohexane dimethanol and tetramethyl cyclobutanediol (Eastman 2010). Studies have shown that it has a low migration potential, both for its constituent monomers and for the additives that are present in the polymer matrix. More importantly, the constituents and the leacheates from the polymer showed neither hormonal nor toxic activity. In a study by Guart et al. (2013), the leacheate from Tritan^{TM} and from polycarbonate bottles into water was collected and tested in a number of in vitro bioassays, including for estrogenic, (anti) androgenic activity and for retinoic acid and vitamin D type activities. The Tritan^{TM} leacheates showed no activity at any concentration, whereas the leacheate from polycarbonate showed estrogenic and antiandrogenic activity at higher concentrations (Guart et al. 2013). These findings are interesting, as they show the potential for newer, safer polymer alternatives to reduce unintended exposure of the human population.

13.5.4 Novel Polymer Formulations

In assessing the physical risks posed by ingestion of nano- or microplastics that unintentionally enter the food chain, much information and guidance can be gained from existing risk assessments performed for food packaging. For example, the European Food Safety Authority (EFSA) has produced detailed guidance for assessing the risks of exposure to nanomaterials, including nanocomposites, biopolymers and other complex materials, from their applications in the food chain (EFSA 2011). As detailed by EFSA, there are huge uncertainties that are associated with detecting, identifying and characterising different micro-, and nanoparticles and polymers in complex matrices such as food, even when the likely constituent substances are known, and these problems are multiplied where rates and sources of contamination remain unknown. In general, however, the considerations suggested by EFSA provide a useful framework applicable to the risks posed by microplastics and nanoplastics as contaminants in food.

Based on the guidance provided by EFSA, it can be predicted that the risks posed by micro- or nanopolymers to human health will be determined by the chemical composition and physico-chemical properties of the particles themselves, their potential for uptake and interactions with tissues, and the likely potential exposure levels. Actual information on migration rates of nanoparticles into food or food stimulants is sparse. Simon et al. (2008) derived a theoretical model to estimate migration rates of nanoparticles from a polymer matrix. The model predicted that migration from polymers of low dynamic viscosity would be limited to particles of <1 nm in diameter, even where the interaction between particle and polymer was negligible. These estimates are in accord with results from Schmidt et al. (2009) who used a combination of field flow fractionation and analytical chemistry techniques to study the migration of nanoparticles out of polylactic acid (PLA). Whilst migration out of the matrix did definitely occur, the resulting nanoparticle concentrations were well below the recommended migration limits. Migration may also be higher into acidic matrices (Mauricio-Inglesias et al. 2010). Table 13.5 compiles some of the indicators identified by EFSA to have the potential to lead to toxicity following uptake of nanoparticles in the diet. These include high levels of reactivity, complex morphologies, the ability to interact with biomolecules, stability and presence of toxic additives. Accordingly, one might expect the greatest hazard to human health to come from ingestion of complex, high aspect ratio nano-scale fibres, synthesised from mixed substances of variable persistence.

13.5.5 Nanopolymers and Nanofillers

There are many technological advances in the development of complex biocomposites and nanopolymers that are relevant for consideration here. Nanocomposites are complex macromolecular materials containing small quantities of nanoscale additives, or nanofillers. The most commonly employed nanofillers for food packaging (the most common type of plastic litter) are nanoclays. Other common nanofillers include nanocellulose fibres, carbon nanotubes, metals and oxides. Nanofillers are intended to enhance or improve the inherent properties of the polymer, including factors such as mechanical strength, thermal and ultraviolet stability, and gas and vapour barrier properties (Lagaron and Lopez-Rubio 2011). The high surface-to-volume ratio of nanofillers enhances their inherent chemical and mechanical properties compared with larger-scale versions of the same material, whilst allowing them to disperse within polymers without introducing structural defects. For example, addition of nanoclays can enhance the oxygen barrier properties of plastics, which makes them particularly attractive for keeping food from spoiling (Lagaron and Lopez-Rubio 2011).

The addition of nanomaterials into polymers can also cut down on the need for large amounts of additives, for example by acting as antioxidants or antimicrobial agents themselves (De Azedero 2013). In relation to plastic discarded to the environment, a major added benefit of nanofillers is that they may also be able to reduce the unintended migration of additives out of polymers (de Abreu et al. 2010).

Table 13.5 Indicators of potential toxicity for nanoparticles contained within food packaging

Characteristic of particle	Details	Example
High levels of reactivity	Catalytic, chemical or biological reactivity	
Complex morphology	Rigid, long tubes or fibres, high aspect ratios, hard fissures or edges, high porosity, mixed composites containing substances of diverse persistence and character	
Ability to interact with biomolecules	Binding or interaction with enzymes, DNA, steroid receptors, signal transduction pathways	
Stability: ability to undergo complex transformations	Polymer ageing, changes to surface properties, porosity, metabolites, changes or loss of coating (e.g. protein surface corona)	
Presence of antimicrobials	Release of biocides into surrounding tissues, unintended consequences for gut *flora*	

Adapted from EFSA (2011)

The migration of various polymer additives, including triclosan and diphenyl butadiene from polyamide into food stimulants was found to be up to six times lower when nanoclays were added to the polyamide. The nanoclay particles were thought to slow down the rate of migration of the additives due to their layering within the polymer matrix, creating a tortuosity effect (Fig. 13.3) (de Abreu et al. 2010). Thus, new advances in nanotechnology may bring unintended benefits in terms of the reduced leaching of their additives and hence the environmental safety of the polymers that contain them.

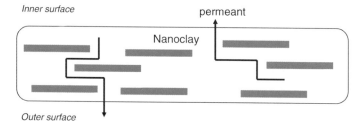

Fig. 13.3 Tortuosity effect of nanoclay in limiting the diffusion of permeants through polymers (adapted from Ray and Okamato 2003)

13.6 Conclusions and Future Work

This short account has identified some of the most widely encountered plastics in everyday use and illustrated some of the attempts that have been made to assess their potential hazards to human health. Different routes of exposure to human populations, both of plastic additives, micro- and nanoplastics from food items and from discarded debris are discussed in relation to the existing literature for nanomedicines and nanocomposite packaging materials, for which an increasing body of knowledge exists. It is clear that our understanding of the potential contamination of the human population by micro- or nanoplastics sourced from the environment is in its infancy, leaving many questions unanswered:

- Does significant bioaccumulation and trophic transfer for micro- and nanoplastics occur in the environment? If so, what species are most at risk?
- How does ageing of plastics affect their physico-chemical properties and subsequent toxicity?
- Following ingestion, does uptake of micro- and nanoplastics occur? Do proteins bind to the surface of the particles to form a protein corona? How does this vary for different plastic litter types and what cell types are most vulnerable to toxicity?
- What methods should we be using for locating, identifying and quantifying micro- and nanoplastics in complex matrices including biological tissues? Techniques mentioned in this chapter include field flow fractionation, multi-angled light scattering (MALS), inductively coupled plasma mass spectrometry (ICP-MS) and non-linear optical bioimaging. Further development of suitable methods for extracting micro- and nanoplastics from biological materials and for studying them in situ remains a compelling research gap for the future.

Acknowledgements TG gratefully acknowledges financial support from grants EU FP7 Cleansea Grant Agreement 308370 and NERC NE/L007010/1 during the preparation of this chapter.

References

Andrady, A. L. (2015). Persistence of plastic litter in the oceans. In M. Bergmann, L. Gutow, M.
  Klages (Eds.), *Marine anthropogenic litter* (pp. 57–72). Berlin: Springer.
Araujo, P., Sayer, C., Poco, J., & Giudici, R. (2002). Techniques for reducing residual monomer
  content in polymers: A review. *Polymer Engineering and Science, 42*, 1442–1468.
Awaad, A., Nakamura, M., & Ishimura, K. (2012). Imaging of size-dependent uptake and identi-
  fication of novel pathways in mouse Peyer's patches using fluorescent organosilica particles.
  Nanomedicine: Nanotechnology Biology and Medicine, 8, 627–636.
Bonefeld-Jørgensen, E. C., Long, M., Hofmeister, M. V., & Vinggaard, A. M. (2007). Endocrine-
  disrupting potential of bisphenol A, bisphenol A dimethacrylate, 4-n-nonylphenol, and 4-n-octyl-
  phenol *in vitro*: new data and a brief review. *Environmental Health Perspectives, 115*, 69.
Browne, M. A., Crump, P., Niven, S., Teuten, E., Tonkin, A., Galloway, T. S., et al. (2011).
  Accumulation of microplastic on shorelines worldwide: sources and sinks. *Environmental
  Science and Technology, 45*, 9175–9179.
Buchta, C., Bittner, C., Heinzl, H., Höcker, P., Mocher, M., Mayerhofer, M., et al. (2005).
  Transfusion-related exposure to the plasticizer di(2-ethylhexyl)phthalate in patients receiving
  plateletpheresis concentrates. *Transfusion, 45*, 798–802.
Burello, E., & Worth, A. P. (2011). A theoretical framework for predicting the oxidative stress
  potential of oxide nanoparticles. *Nanotoxicology, 5*, 228–235.
Calafat, A. M., Ye, X., Wong, L.-Y., Reidy, J. A., & Needham, L. L. (2008). Exposure of the
  US population to bisphenol A and 4-tertiary-octylphenol: 2003–2004. *Environmental Health
  Perspectives, 116*, 39–45.
Cipelli, R., Harries, L., Okuda, K., Yoshihara, S., Melzer, D., & Galloway, T. S. (2013).Bisphenol
  A modulates the expression of Estrogen-Related Receptor-α in T-Cells.*Reproduction, 147*,
  419–426.
Claudio, L. (2012). Food packaging and public health. *Environmental Health Perspectives,
  120*(6), A233–A237.
Cole, M., Lindeque, P., Goodhead, R., Moger, J., Halsband-Lenk, C., & Galloway, T. S.
  (2013). Microplastic ingestion by zooplankton. *Environmental Science and Technology, 47*,
  6646–6655.
Cole, M., Lindeque, P., Halsband-Lenk, C., & Galloway, T. S. (2011). Microplastic as a contaminant in the marine environment: a review. *Marine Pollution Bulletin, 62*, 2588–2597.
Crompton, T. (2007). *Additive migration from plastics into foods. A guide for analytical chemistry*. Shrewsbury: iSmithers Rapra Publishing.
de Abreu, D. A. P., Cruz, J. M., Angulo, L., & Losada, P. P. (2010). Mass transport studies of different additives in polyamide and exfoliated nanocomposite polyamide films for food indus-
  try. *Packaging Technology and Science, 23*, 59–68.
De Azeredo, H. M.C. (2013). Antimicrobial nanostructures in food packaging. *Trends in Food
  Science and Technology, 30*, 56–69.
Dodds, E. C., & Lawson, W. (1936). Synthetic estrogenic agents without the phenanthrene
  nucleus. *Nature, 137*, 996.
Dubaish, F., & Liebezeit, G. (2013). Suspended microplastics and black carbon particles in the
  Jade system, southern North Sea. *Water, Air Soil Pollution, 224*, 1–8.
Eastman. (2010). Eastman material data sheet. http://www.eastman.com, http://www.
  eastman.com/lierature.center/T/TRS252.pdf
Ebel, J. P. (1990). A method for quantifying particle absorption from the small intenstine.
  Pharmaceutical Research, 7, 848–851.

EFSA. (2006). Opinion of the scientific panel on food additives, flavourings, processing aids and materials in contact with food on a request from the commission related to 2,2-bis(4-hydroxyphenyl)propane (bisphenol A). *EFSA Journal, 428*, 13–75.

EFSA. (2011). Scientific opinion: Guidance on the risk assessment of the application of nanoscience and nanotechnologies in the food and feed chain. *EFSA Journal, 9*, 2140.

EFSA (2015) Scientific Opinion on the risks to public health related to the presence of bisphenol A (BPA) in foodstuffs. *EFSA Journal, 13(1)*, 3978.

Ehrlich, S., Calafat, A. M., Hunblet, O., Smith, T., & Hauser, R. (2014). Handling of thermal receipts as a source of exposure to bisphenol A. *JAMA, 311*, 859–860.

Engler, R. E. (2012). The complex interaction between marine debris and toxic chemicals in the ocean. *Environmental Science and Technology, 46*, 12302–12315.

European Commision. (2014). EC guidance document. Guidance to hazard analysis and critical control point (HACCP) application and to regulation 852/2004 and 853/2004. http://europa.eu/comm/food/food/biosafety/hygienelegislation/index_en.htm. Accessed Feb 2014.

Fischer, D., Li, Y., Ahlemeyer, B., Krieglstein, J., & Kissel, T. (2003). *In vitro* cytotoxicity testing of polycations: Influence of polymer structure on cell viability and hemolysis. *Biomaterials, 24*, 1121–1131.

Florence, A. T., Sakthivel, T., Toth, I. (2000). Oral uptake and translocation of a polylysine dendrimerwith a lipid surface. *Journal Control Release, 65*, 253-259.

Florence, A. T., & Hussain, N. (2001). Transcytosis of nanoparticle and dendrimer delivery systems, evolving vistas. *Advanced Drug Delivery Reviews, 50*, S69–S89.

Fruijtier-Pölloth, C. (2012) The toxicological mode of action and the safety of synthetic amorphoussilica- a nanostructured material. *Toxicology, 294*, 61–79.

Galloway, T. S., Cipelli, R., Guralnik, J., Ferrucci, L., Bandinelli, S., Corsi, A. M., et al. (2010). Daily bisphenol A excretion and associations with sex hormone concentrations: Results from the InCHIANTI adult population study. *Environmental Health Perspectives, 118*, 1603–1608.

Garrett, N. L., Lalatsa, A., Uchegbu, I., Schätzlein, A., & Moger, J. (2012). Exploring uptake mechanisms of oral nanomedicines using multimodal nonlinear optical microscopy. *Journal of Biophotonics, 5*, 458–468.

Gayrard, V., Lacroix, M. Z., Collet, S. H., Viguié, C., Bousquet-Melou, A., Toutain, P. L., et al. (2013). High bioavailability of bisphenol A from sublingual exposure. *Environ Health Perspect, 121*, 951–956.

Geys, J., Nemmar, A., Verbeken, E., Smolders, E., Ratoi, M., Hoylaerts, M. F., et al. (2008). Acute toxicity and prothrombotic effects of quantum dots: Impact of surface charge. *Environmental Health Perspectives, 116*, 1607–1613.

Grob, K., Biedermann, M., Scherbaum, E., Roth, M., & Rieger, K. (2006). Food contamination with organic materials in perspective: Packaging materials as the largest and least controlled source? A view focusing on the European situation. *Critical Reviews in Food Science and Nutrition, 46*, 529–536.

Guart, A., Wagner, M., Mezquida, A., Lacorte, S., Oehlmann, J., & Borrell, A. (2013). Migration of plasticisers from Tritan™ and polycarbonate bottles and toxicological evaluation. *Food Chemistry, 141*, 373–380.

Hanning, I. B., Nutt, J. D., & Ricke, S. C. (2009). Salmonellosis outbreaks in the United States due to fresh produce: Sources and potential intervention measures. *Foodborne Pathogens and Disease, 6*, 635–648.

Harush-Frenkel, O., Bivas-Benita, M., Nassar, T., Springer, C., Sherman, Y., Avital, A., et al. (2010). A safety and tolerability study of differently-charged nanoparticles for local pulmonary drug delivery. *Toxicology and Applied Pharmacology, 246*, 83–90.

Helmroth, E., Rijk, R., Dekker, M., & Jongen, W. (2002). Predictive modelling of migration from packaging materials into food products for regulatory purposes. *Trends in Food Science and Technology, 13*, 102–109.

Hengstler, J. G., Foth, H., Gebel, T., Kramer, P. J., Lilienblum, W., Schweinfurth, H., et al. (2011). Critical evaluation of key evidence on the human health hazards of exposure to bisphenol A. *Critical Reviews in Toxicology, 41*, 263–291.

Hidalgo-Ruz, V., Gutow, L., Thompson, R. C., & Thiel, M. (2012). Microplastics in the marine environment: A review of the methods used for identification and quantification. *Environmental Science and Technology, 46*, 3060–3075.

Imhof, H. K., Ivleva, N. P., Schmid, J., Niessner, R., & Laforsch, C. (2013). Contamination of beach sediments of a subalpine lake with microplastic particles. *Current Biology, 23*, R867–R868.

Israelachvili, J. (1992). *Intermolecular and surface forces* (2nd ed.). London: Academic Press.

Jambeck, J. R., Geyer, R., Wilcox, C., Siegler, T. R., Perryman, M., Andrady, A., et al. (2015). Plastic waste inputs from land into the ocean. *Science, 347*, 768-771.

Jani, P. U., Florence, A. T., & McCarthy, D E. (1992). Further histological evidence of gastrointestinal absorption of polystyrene nanospheres in the rat. *International Journal of Pharmaceutics, 84*, 245–252.

Jani, P. U., Nomura, T., Yamashita, F., Takakura, A., Florence, A. T., & Hashida, M. (1996). Biliary excretion of polystyrene microspheres with covalently linked FITC fluorescence after oral and parenteral administration to male wistar rats. *Journal of Drug Targeting, 4*, 87–93.

Kim, B., Rutka, J., & Chan, W. (2010). Nanomedicine. *New England Journal of Medicine, 363*, 2434–2443.

Koch, H. M., & Calafat, A. M. (2009). Human body burdens of chemicals used in plastics manufacture. *Philosophical Transactions of the Royal Society B, 364*, 2063–2078.

Koelmans, A. A. (2015). Modeling the role of microplastics in bioaccumulation of organic chemicals to marine aquatic organisms. Critical review. In M. Bergmann, L. Gutow, & M. Klages (Eds.), *Marine anthropogenic litter* (pp. 313–328). Berlin: Springer.

Koelmans, A. A., Besseling, E., & Foekema, E. M. (2014). Leaching of plastic additives to marine organisms. *Environmental Pollution, 187*, 49–54.

Koelmans, A. A., Besseling, E., & Shim, W. J. (2015). Nanoplastics in the aquatic environment. In M. Bergmann, L. Gutow, & M. Klages (Eds.), *Marine anthropogenic litter* (pp. 329–344). Berlin: Springer.

Kreuter, J. (1991). Peroral administration of nanoparticles. *Advanced Drug Delivery Reviews, 7*, 71-86.

La Mantia, F. (2002). *Handbook of plastics recycling*. Shrewsbury: iSmithers Rapra Publishing.

Lagaron, J. M., & Lopez-Rubio, A. (2011). Nanotechnology for bioplastics: Opportunities, challenges and strategies. *Trends in Food Science and Technology, 22*, 611–617.

Lang, I. A., Galloway, T. S., Scarlett, A., Henley, W. E., Depledge, M., Wallace, R. B., et al. (2008). Association of urinary bisphenol A concentration with medical disorders and laboratory abnormalities in adults. *JAMA, 300*, 1353–1355.

Lee, H. J., Chattopadhyay, S., Gong, E. Y., Ahn, R. S., & Lee, K. (2003). Antiandrogenic effects of bisphenol A and nonylphenol on the function of androgen receptor. *Toxicological Sciences, 75*, 40–47.

Lithner, D., Larsson, Å., & Dave, G. (2011). Environmental and health hazard ranking and assessment of plastic polymers based on chemical composition. *Science of the Total Environment, 409*, 3309–3324.

Lundqvist, M., Stigler, J., Elia, G., Lynch, I., Cedervall, T., & Dawson, K. A. (2008). Nanoparticle size and surface properties determine the protein corona with possible implications for biological impacts. *PNAS, 105*, 14265–14270.

Lusher, A. (2015). Microplastics in the marine environment: distribution, interactions and effects. In M. Bergmann, L. Gutow, & M. Klages (Eds.), *Marine anthropogenic litter* (pp. 245–308). Berlin: Springer.

Ma-Hock,L., Landsiedel, R., Wiench, K., Geiger, D., Strauss, V. Gröters, S., et al. (2012). Short term rat inhalation study with aerosols of acrylic ester-based polymer dispersions containing a fraction of nanoparticles. *International Journal of Toxicology, 31*, 46–57.

Mauricio-Inglesias, M., Peyron, S., Guillard, V., & Gontard, N. (2010). Wheat gluten nanocomposite films as food contact materials: Migration tests and impact of a novel food stabilising technology (high pressure). *Journal of Applied Polymer Science, 116*, 2526–2535.

Melzer, D., & Galloway, T. S. (2010). Burden of proof. *New Scientist*, October 2010, (pp. 26–27).

Melzer, D., Rice, N. E., Lewis, C., Henley, W. E., Galloway, T. S. (2010). Association of urinary bisphenol. a concentration with heart disease: evidence from NHANES 2003/06. *PLoS ONE, 5*, e8673

Melzer, D., Osborne, N. J., Henley, W. E., Cipelli, R., Young, A., Money, C., et al. (2012). Urinary bisphenol A: Concentration and risk of future coronary artery disease in apparently healthy men and women. *Circulation, 125*, 1482–1490.

Mielke, H., & Gundert-Remy, U. (2009). Bisphenol A levels in blood depend on age and exposure. *Toxicology Letters, 190*, 32–40.

Moriyama, K., Tagami, T., Akamizu, T., Usui, T., Saijo, M., Kanamoto, N., et al. (2002). Thyroid hormone action is disrupted by bisphenol A as an antagonist. *The Journal of Clinical Endocrinology and Metabolism, 87*, 5185–5190.

Muncke, J. (2011). Endocrine disrupting chemicals and other substances of concern in food contact materials: An updated review of exposure, effect and risk assessment. *The Journal of Steroid Biochemistry and Molecular Biology, 127*, 118–127.

Nel, A. E., Xia, T., Mädler, L., & Li, N. (2006). Toxic potential of materials at the nanolevel. *Science, 311*, 622–627.

Nel, A. E., Madler, L., Velegol, D., Xia, T., Hoek, E. M. V., Somasundaran, P., et al. (2009). Understanding bio physicochemical interactions at the nano- bio- interface. *Nature Materials, 8*, 543-557.

Newbold, R. R., Padilla-Banks, E., Jefferson, W. N. &, Heindel, J. J. (2008). Effects of endocrine disruptors on obesity. *International Journal of Andrology, 31*, 201–208.

Nicholson, J. W. (1996). *The chemistry of polymers* (p. 2006). Cambridge: The Royal Society of Chemistry.

O'Hagan, D. T. (1996). The intestinal uptake of particles and the implications for drug and antigen delivery. *Journal of Anatomy, 189*, 477–482.

OCED. (2004). Emission scenario document on plastic additives. Series on emission scenario documents No. 3. Paris. Environmental Directorate. OECD Environmental Health and Safety Publications.

Okada, H., Tokunaga, T., Liu, X., Takayanagi, S., Matsushima, A., & Shimohigashi, Y. (2008). Direct evidence revealing structural elements essential for the high binding ability of bisphenol A to human estrogen-related receptor. *Environmental Health Perspectives, 116*, 32–38.

PlasticsEurope. (2013). Plastics, the facts 2013. www.plasticseurope.org. Accessed 21 Jan 2014.

Pandurangi, R. S., Seehra, M. S., Razzaboni, B. L., & Bolsaitis, P. (1990). Surface and bulk infrared modes of crystalline and amorphorous silica particles: a study of the relation of surface structure to cytotoxicity of respirable silica. *Environmental Health Perspectives, 86*, 327–336.

Pappo, J., Ermak, T. H., & Steger, H. J. (1989). Monoclonal antibody directed targeting of fluorescent microspheres to Peyer's patches M cells. *Immunology, 73*, 277–280.

Poland, C. A., Duffin, R., Kinloch, I., Maynard, A., Wallace, W. A. H., Seaton, A., et al. (2008). Carbon nanotubes introduced into the abdominal cavity of mice show asbestos-like pathogenicity in a pilot study. *Nature Nanotechnology, 23*, 423–428.

Ray, S. S., & Okamoto, M. (2003). Polymeric/layered silicate nanocomposites: A review from preparation to processing. *Progress in Polymer Science, 28*, 1539–1641.

Rochester, J. R. (2013). Bisphenol A and human health: A review of the literature. *Reproductive Toxicology, 42C*, 132–155.

Ropero, A. B., Alonso Magdalena, P., García García, E., Ripoll, C., Fuentes, E., & Nadal, A. (2008). Bisphenol A disruption of the endocrine pancreas and blood glucose homeostasis. *International Journal of Andrology, 31*, 194–200.

Schmidt, B., Petersen, J. H., Bender Koch, C., Plackett, D., Johansen, N. R., Katiyar, V., et al. (2009). Combining asymmetrical flow field-flow fractionation with light-scattering and inductively coupled plasma mass spectrometric detection for characterization of nanoclay used in biopolymer nanocomposites. *Food Additives and Contaminants A, 26*, 1619–1627.

Setälä, O., Fleming-Lehtinen, V., & Lehtiniemi, M. (2012). Ingestion and transfer of microplastics in the planktonic food web. *Environmental Pollution, 185*, 77–83.

Sexton, K., Needham, L., & Pickles, J. (2004). Human biomonitoring of environmental chemicals. *American Science, 92*, 38–45.

Silvestre, C., Duraccio, D., & Cimmino, S. (2011). Food packaging based on polymer nanomaterials. *Progress in Polymer Science, 36*, 1766–1782.

Šimon, P., Chaudrey, Q., & Bakoš, D. (2008). Migration of engineered nanoparticles from polymer packaging to food—a physicochemical view. *Journal of Food and Nutrition Research, 47*, 105–113.

Talsness, C. E., Andrade, A. J. M., Kuriyama, S. N., Taylor, J. A., & vom Saal, F. S. (2009). Components of plastic: Experimental studies in animals and relevance for human health. *Philosophical Transactions of the Royal Society B, 364*, 2079–2096.

Tenzer, S., Docter, D., Kuharev, J., Musyanovych, A., Fetz, V., Hecht, R., et al. (2013). Rapid formation of plasma protein corona critically affects nanoparticle pathophysiology. *Nature Nanotechnology, 8*, 772–781.

Thompson, R. C., Olsen, Y., Mitchell, R. P., Davis, A., Rowland, S. J., John, A. W. G., et al. (2004). Lost at sea, where is all the plastic? *Science, 304*, 838.

Vandenberg, L. N., Chauhoud, I., Heindel, J. J., Padmanabhan, V., Paumgartten, F. J., & Schoenfelder, G. (2010). Urinary, circulating and tissue biomonitoring studies indicate widespread exposure to bisphenol A. *Environmental Health Perspectives, 118*, 1055–1070.

Varela, J. A., Bexiga, M. G., Åberg, C., Simpson, J. C., & Dawson, K. A. (2012). Quantifying size-dependent interactions between fluorescently labeled polystyrene nanoparticles and mammalian cells. *Journal of Nanobiotechnology, 10*, 39.

Volkheimer, G. (1977). Passage of particles through the wall of the gastrointenstinal tract. *Environmental Health Perspectives, 9*, 215–225.

vom Saal, F. S., Akingbemi, B. T., Belcher, S. M., Birnbaum, L. S., Crain, D. A., Eriksen, M., et al. (2007). Chapel Hill bisphenol A expert panel consensus statement: Integration of mechanisms, effects in animals and potential to impact human health at current levels of exposure. *Reproductive Toxicology, 24*, 131–138.

Wang, J., Zhou, G., Chen, C., Yu, H., Wang, T., Ma, Y., et al. (2007). Acute toxicity and biodistribution of different sized titanium dioxide particles in mice after oral administration. *Toxicology Letters, 168*, 176–185.

Wright, S. L., Thompson, R. C., & Galloway, T. S. (2013). The physical impacts of microplastics on marine organisms: A review. *Environmental Pollution, 178*, 483–492.

Zbyszewski, M., & Corcoran, P. L. (2011). Distribution and degradation of fresh water plastic particles along the beaches of Lake Huron, Canada. *Water, Air, and Soil Pollution, 220*, 1–8.

Zubris, K. A. V., & Richards, B. K. (2005). Synthetic fibers as indicator of land application of sludge. *Environmental Pollution, 138*, 201–211.

Zweifel, H. (2001). *Plastics additives handbook* (5th ed.). Munich: Carl Hanser Verlag.

Chapter 14
The Economics of Marine Litter

Stephanie Newman, Emma Watkins, Andrew Farmer,
Patrick ten Brink and Jean-Pierre Schweitzer

Abstract This chapter aims to provide an overview of research into quantifying the economic impacts of marine litter. From an environmental economics perspective it introduces the difficulties in measuring the economic costs of marine litter; reviews those sectors where these costs are notable; and considers policy instruments, which can reduce these costs. Marine litter is underpinned by dynamic and complex processes, the drivers and impacts of which are multi-scalar, transboundary, and play out in both marine and terrestrial environments. These impacts include economic costs to expenditure, welfare and lost revenue. In most cases, these are not borne by the producers or the polluters. In industries such as fisheries and tourism the costs of marine litter are beginning to be quantified and are considerable. In other areas such as impacts on human health, or more intangible costs related to reduced ecosystem services, more research is evidently needed. As the costs of marine litter are most often used to cover removing debris or recovering from the damage which they have caused, this expenditure represents treatment rather than cure, and although probably cheaper than inaction do not present a strategy for cost reduction. Economic instruments, such as taxes and charges addressing the drivers of waste, for instance those being developed for plastic bags, could be used to reduce the production of marine litter and minimise its impacts. In any case, there remain big gaps in our understanding of the harm caused by marine litter, which presents difficulties when attempting to both quantify its economic costs, and develop effective and efficient instruments to reduce them.

Keywords Marine litter · Environmental economics · Economic instruments · Plastic bags · Waste · Polluter pays

S. Newman (✉) · E. Watkins · A. Farmer · P.t. Brink · J.-P. Schweitzer
Institute for European Environmental Policy, 11 Belgrave Road, IEEP Offices, Floor 3,
London SW1V 1RB, UK
e-mail: SNewman@ieep.eu

M. Bergmann et al. (eds.), *Marine Anthropogenic Litter*,
DOI 10.1007/978-3-319-16510-3_14

14.1 Introduction

In addition to the environmental and health issues discussed in previous book sections (Galloway 2015; Kühn et al. 2015), marine litter can cause a range of economic impacts that both increase the costs associated with marine and coastal activities, and reduce the economic benefits derived from them. This chapter aims to provide an overview of the results of research performed to date, which has attempted to quantify the economic impact of marine litter. It provides a brief analysis of the marine litter problem from an environmental economics perspective, and discusses the use and design of economic-based policy instruments to tackle the problem.

14.2 Estimating the Economic Impacts of Marine Litter

Measuring the full economic cost of marine litter is complex due to the wide range of economic, social and environmental impacts, the range of sectors impacted by marine litter and the geographic spread of those affected. Some of the impacts are easier to evaluate in economic terms because they are more direct, such as increased marine litter cleaning costs. Others are more complex, for example, the less direct and/or more intangible values such as the impacts of ecosystem deterioration or reductions in quality of life. Furthermore, the spatial and temporal complexity of the impacts related to marine litter result in costs, which may not always be immediate or conspicuous but are nevertheless significant for sustainability (National Research Council 2008). As regards ecosystem degradation, it is useful to differentiate between impacts on biodiversity (species and habitats) and the impact on the ecosystem services flowing from the ecosystem (e.g. provisioning services such as food provision, regulating services such as water and waste purification; and cultural services such as tourism and recreation). As regards economic costs it is important to differentiate between actual economic costs linked to expenditure (e.g. costs of cleanup of beaches; costs associated with damage to or loss of fishing gear or obstruction of motors; eventual cost of hospitalisation from marine debris related health impacts), economic costs of loss of output or revenue (e.g. loss of revenue from fish or loss of income from tourism) and assessment of welfare costs in economic terms (e.g. health impacts from marine debris; assessing the economic value of loss of cultural values such as recreation or landscape aesthetics).

While marine litter has become an increasingly important issue in policy discussions, there is only a very sketchy (albeit growing) body of knowledge on the costs of the impacts. Because of a lack of recording even the direct economic costs of marine litter tend not to be measured (Mouat et al. 2010). Furthermore, even though there is a growing interest in ecosystem services (Costanza et al. 1997; MA 2005; TEEB 2010, 2011) little research has been done to date on the economic cost of marine litter on ecosystem service provision. Having said this, evaluations of marine ecosystem services, which are estimated at €16.5 trillion in one study (Costanza et al. 1997), suggest that even fractional deterioration in provision would represent a significant cost (Beaumont et al. 2007; Galparsoro et al. 2014).

Thus far, studies undertaken to estimate the economic impacts of marine litter have generally focused on the direct losses borne by economic activities adversely affected by the presence of marine litter in the environment, within which they operate and rely upon (see Hall 2000; Mouat et al. 2010; MacFayden 2009; McIlgorm et al. 2011). Largely, such studies have not taken into account the often intangible costs of any social and ecological impacts. Some early studies allude to the need for research to explore these costs. For instance, Kirkley and McConnell (1997, p. 185) call for strategies, which account for the economics related to lost ecological functions driven by marine litter. The intricacy of developing such strategies can be illustrated with the example of alien invasive species. Marine litter provides additional opportunities for marine organisms to travel (including alien invasive species) up to threefold (Barnes 2002). Given that the introduction of alien invasive species can have a detrimental impact on marine ecosystems and biodiversity (Kiessling et al. 2015) and can result in serious economic losses to many marine industries, any estimates, which exclude such ecological impacts, will inevitably fall seriously short of the true cost of the marine litter problem. For example, the introduction of the carpet sea squirt (*Didemnum vexillum*) in Holyhead Harbour (Wales, U.K.) resulted in an eradication and monitoring program over a decade starting in 2009, which was expected to cost €670,000. This expenditure was economically justified as allowing the species to spread unpredated and smother organisms and marine habitats would have cost the local mussel fisheries up to €8.6 million alone over 10 years (Holt 2009). Goldstein et al. (2014) recorded the ciliate pathogen *Halofolliculina* (known to cause skeletal eroding band disease in corals) on floating plastic debris in the western Pacific and suggested that the spread of the disease to Caribbean and Hawaiian corals may be due to rafting on the enormous quantities of litter reported from the area. Increased coral mortality or the introduction of other pathogens via floating marine debris may lead to economic costs, for example through decreased revenues due to falling numbers of visiting tourists.

Despite their partial coverage, the studies that are available provide sufficient information to draw a number of important conclusions. The main economic sectors, which have been identified from the literature as being affected by marine litter are agriculture, aquaculture, fisheries, commercial shipping and recreational boating, coastal municipalities, coastal tourism sector and the emergency rescue services (Hall 2000; Mouat et al. 2010). The economic impacts affecting these sectors are described and quantified where possible (Hall 2000; Mouat et al. 2010; McIlgorm et al. 2011; Jang et al., 2014; Antonelis 2011). They also make attempts at aggregating economic impacts across sectors to provide regional cost estimates. Mouat et al. (2010) provide an estimate of marine litter costs for the Shetland (U.K.) economy, of €1–1.1 million on average per year, an estimate, which consists of actual expenditures and, in some cases, estimated lost income. This is only a single case study, and the sectors affected on Shetland would be affected to varying degrees in other coastal areas. However, these findings clearly demonstrate that the economic impact of marine litter on coastal communities can be extremely high. McIlgorm et al. (2011) calculated the costs of marine litter for 21 economies in the Asia-Pacific region. Similarly to Hall (2000) and Mouat et al. (2010), they consisted of such losses as those from entangled ship propellers, lost fishing time, and tourism losses from deterring visitors, but not the cost of any harm to ecosystem services or other non-market values (McIlgorm et al. 2011). In total, McIlgorm et al. (2011) estimated the cost of marine litter in the Asia-Pacific region to be in the region of €1 billion per year to marine industries, equivalent to 0.3 % of the gross domestic product for the marine sector of the region.

14.2.1 Beach Cleaning, Tourism and Recreation

Coastal municipalities are impacted economically by marine litter primarily through the direct cost of keeping beaches clear of litter and its wider implications for tourism and recreation. Direct costs include the collection, transportation and disposal of litter, and administrative costs such as contract management. Ensuring that beaches are clean, attractive and safe for visitors is prioritised by municipalities when the economic case for protecting the local economy and tourism industry justifies the costs of removing the litter. In areas where coastlines make a significant contribution to the economy, the costs incurred through marine litter can be substantial.

In the U.K., the cost of removing beach litter to all coastal municipalities is estimated to be in the region of €18–19 million (Mouat et al. 2010). This equates to an average cost per municipality of €146,000 (Mouat et al. 2010). The majority of this cost was accounted for by labor costs. Mouat et al. (2010) also calculated the cost annually per km of coastline. Although the average cost of litter removal was between €7,000 and €7,300 per km per year, there was a lot of variation, with costs ranging from €171 to €82,000 per km per year (Mouat et al. 2010). Higher costs correlated with more intense cleaning operations on small areas of coastline, particularly in tourist areas. In Belgium and The Netherlands, the total cost of beach litter removal was estimated to be €10.4 million per year, at an average of €200,000 per municipality per year (Mouat et al. 2010). Per km, the cleaning costs came to €34,000 per year on average, again with great variation (e.g. from €600 to €97,300 in Den Haag) (Mouat et al. 2010). This average is much greater than that in the U.K. as municipalities in Belgium and The Netherlands removed litter from a much higher proportion of their coastline (because it is more densely populated). The great variation in amounts spent by municipalities on different beaches reflects the variation in importance of different stretches of coastline to the tourism industry. Of course many areas of coastline worldwide do not have anything spent on them to provide a litter cleanup service.

It is important to recognize, however, that beach cleaning is not necessarily performed by municipalities alone, and that voluntary organisations tend to play a large role in removing litter (see Hidalgo-Ruz and Thiel 2015). This is an economic impact on society that comprises operational expenditure, financial assistance or some sort of 'in kind' assistance such as materials or insurance, and the value of volunteers' time. There may also be an opportunity cost where volunteer time could be spent servicing the community in other ways. Mouat et al. (2010) estimated the value of volunteers' time in two annual beach clean operations in the U.K. at which a substantial quantity of litter from the U.K. coastline was collected, to be around €131,000. As this estimate includes neither financial assistance nor operational management costs, it is likely to be a substantial underestimate.

In coastal municipalities, particularly those where beaches contribute significantly to the local economy, the indirect economic impacts of marine litter are more important. A few studies have attempted to calculate the costs incurred to coastal areas as a result of marine litter. Jang et al. (2014) considered the economic impact of a single marine litter event in South Korea, in which heavy rainfall resulted in an unusually high level of marine litter to be washed on the beaches of Goeje Island, a popular tourist destination. Based on government figures and a number of surveys they assessed multiple economic effects of marine litter such as lost expenditure on hotels and lodging, which could be influenced by marine litter. Lost expenditure was expressed as the product of decreased visitors and average visitor expenditure, in this case between €23 and €29 million of lost revenue in 2011 compared to 2010, as a result of over 500,000 fewer visitors to the island. In 2013, a study of 31 beaches in Orange County (California, USA) considered how marine debris influences their decision to go to the beach, and at what expense (Leggett et al. 2014). It applied the *travel cost model*, which estimates the value people derive from recreation at a particular site based on the utility they expect to experience in relation to alternative sites. They showed that marine debris had a significant impact on residents' beach choices, and that a 75 % reduction in marine litter at six popular beaches generated over €40 million in additional benefits to Orange County residents over just 3 months. These two studies clearly demonstrate the value people place in the clean marine and coastal environments and the potential for costs to communities, which derive utility for the services which these environments provide. This is an interesting finding given that many of these visitors may also be responsible for some of the pollution, by littering during their beach visits, for example.

14.2.2 Shipping and Yachting

The shipping and yachting industries also experience economic impacts as a result of marine litter pollution, with harbors and marinas incurring the cost of removing marine litter from their facilities in order to keep them safe and attractive to users, and vessels experiencing interference with propellers, anchors, rudders and blocked intake pipes and valves (Mouat et al. 2010). On occasion, some of these vessel encounters pose navigational hazards that require the rescue services to become involved, thereby increasing costs dramatically.

Mouat et al. (2010) estimated that removing marine litter costs U.K. ports and harbors on average €2.4 million per year. But this can range from €0 (as not all harbors surveyed in this study took action to remove marine litter, and thereby incurred no direct costs) to almost €73,000 per year for individual harbors (Mouat et al. 2010). Higher costs tended to correlate with larger and busier harbors. Disposal and manual removal of floating debris were observed to make up the bulk of these costs, as dredging to remove items off the seabed, although expensive, is not performed very commonly. There is no estimate for the cost of removing marine litter to the U.K. marina industry as a whole, but data from a small sample indicate that it could be costly, with one marina reporting an annual bill of €39,000 (Mouat et al. 2010). When factoring in the cost of undertaking rescue operations the cost of marine litter to shipping and yachting rises further. An estimate for the U.K. Royal National Lifeboat Institution in 2008 calculated that 286 rescue operations to vessels with tangled propellers cost between €830,000 and €2,189,000 (Mouat et al. 2010).

14.2.3 Fisheries

The fishing sector is more commonly viewed as a source of marine litter, but it is also subject to economic costs itself. Direct economic impacts faced by the sector arise from the need to repair or replace gear that has been damaged or lost due to encounters with marine litter; repairing vessels with tangled propellers (Fig. 14.1), anchors, rudders, blocked intake pipes, etc.; loss of earnings due to time diverted to deal with marine litter encounters; and loss of earnings from reduced or contaminated catches resulting from marine litter encounters including ghost fishing

Fig. 14.1 Diver removes derelict rope wrapped around the propeller of a ship (*left* photo: NOAA Marine Debris program). Derelict rope crab pot "ghost fishing" at 1,091 m in Astoria Canyon, off Oregon (*right* © 2006 MBARI)

(Fig. 14.1). Wallace (1990) reported that in the Eastern US, over 45 % of the commercial fishers had their propellers caught, over 30 % had their gear fouled, and almost 40 % had their engine cooling system clogged by plastic debris at some point in time. The sector also experiences indirect losses of earnings due to the impact of loss and abandoned fishing gear on fish stocks (MacFayden et al. 2009; Sheavly and Register 2007).

There are potentially also costs associated with loss of value of fisheries resources (provisioning services under the ecosystem service nomenclature), whether through reductions in fish and shellfish numbers or reduced value due to impacts on quality of fish and shellfish (e.g. through ingested plastics or contamination with persistent organic pollutants, POPs). The body of literature describing the contamination of commercially exploited fish and shellfish by microplastic ingestion is growing rapidly, as is the literature analysing the consequences of this contamination on the health of individuals and populations (Galloway 2015; Lusher 2015; Rochman 2015). However, as yet there have been no economic assessments to estimate the costs of these impacts.

Derelict fishing gear (DFG) constitutes a considerable portion of marine litter and can result in economic losses for fisheries. DFG includes any equipment, which can catch (shell-)fish, which is lost by fisheries, including trawl nets, gill nets, traps, cages and pots (National Research Council 2008). As a result of their functional design, DFG can continue to trap marine life after they have been lost (a phenomenon known as ghost fishing). Increasingly durable materials used in fishing equipment means that it can continue to ghost fish for some time; in this way it presents particular challenges as marine waste. Fisheries incur costs, firstly in having to replace the fishing gear they have lost at sea, and secondly in a reduction in their potential harvestable catch, and indeed the sustainability of that catch (Butler et al. 2013; Arthur et al. 2014; Bilkovic 2014). One study in Puget Sound, Washington, estimated that over 175,000 dungeness crab (*Metacarcinus magister*) were killed each year by derelict fishing traps, equivalent to around €586,000 or 4.5 % of the average annual harvest (Antonelis 2011).

Mouat et al. (2010) focused on estimating the direct economic impact of marine litter on Scottish fishing vessels (i.e. costs of repairs and direct losses in earnings, not indirect losses due to ghost fishing) and estimated that on average marine litter costs each fishing vessel between €17,000 and €19,000 per year (Mouat et al. 2010). Two-thirds of this cost (€12,000) was incurred through time lost clearing litter from nets (calculated using the average value of 1 h's fishing time as estimated by vessels surveyed during this project). Aggregated, this costs the Scottish fishing industry as a whole between €11.7 and 13 million every year (Mouat et al. 2010). To put this in perspective, marine litter knocks 5 % off the fleets' total annual revenue. This is clearly a substantial cost to an industry that is already under high pressure and important in coastal communities.

Similar to voluntary beach cleanup operations, there are a number of 'Fishing-for-Litter' schemes in operation whereby fishermen voluntarily agree to collect the litter, which they catch in their nets during their normal fishing activity, and dispose of this safely on the quayside at designated waste disposal sites (Fig. 14.2).

Fig. 14.2 OSPAR Fishing for Litter program. From *left to right* catch with a tyre from a *Nephrops* trawler in the Clyde (U.K.) (Photo: M. Bergmann, AWI); fisher from *FV Andrea* sorting litter into bags provided by Fishing for Litter scheme (Photo: G. Lengler, NABU, DSD); portside container for litter collected by fishers (Photo: K. Detloff, NABU)

These schemes are currently in operation in the U.K., Sweden, Denmark, The Netherlands, Belgium and Germany, and potentially other EU countries as there is EU support to fund such operations (KIMO International 2013; European Commission 2011).

The fishers benefit from being involved as they reduce the volume of litter accumulating in the oceans and on beaches, and they thereby reduce the amount of time they spend untangling litter from nets and reduce the risks of other marine litter related costs described above. Best-practice guidelines indicate that Fishing-for-Litter schemes also have the benefit of changing the culture within the industry to adopt good waste-management practices (OSPAR 2007). These schemes are not without costs to operate, however, as coordinators are needed to promote and run the programs, and there are of course costs associated with waste disposal (OSPAR 2007). These costs will vary depending on the country and the amounts of litter collected, but the schemes in Europe to date demonstrate that Fishing-for-Litter is a significant and cost-effective measure to reduce litter (OSPAR 2007).

14.2.4 Aquaculture

Marine litter can result in costs to the aquaculture industry, through entangling propellers and blocking intake pipes, and time spent removing debris from and around fish farm operations. Mouat et al. (2010) surveyed finfish and shellfish aquaculture producers in Scotland and estimated that marine litter costs the sector on average €156,000 per year, which amounted to approximately €580 per year per producer. Ninety percent of this cost was due to time spent untangling fouled propellers on workboats and repairs. Removing marine litter from aquaculture sites was less of an issue overall, but this was highly variable, and in some areas it was a regular problem. These figures demonstrate that in comparison to other sectors such as fisheries, and even agriculture, the direct cost imposed by marine litter on aquaculture is relatively low.

14.2.5 Agriculture

As a terrestrial economic activity, agriculture is not the most obvious sector to suffer economic losses because of marine litter. Indeed, similarly to the fishing sector it is more frequently seen as a source of marine litter (de Stephanis et al. 2013). However, in some locations, debris can blow, drift or get washed up on coastal farmland, causing damage to property, equipment and presenting a risk to livestock through ingestion and entanglement. These impacts may all lead to economic losses in addition to the cost of preventative litter removal. The Shetland Isles are one such location where marine debris litters agricultural lands, due to the prevalence of strong winds. Through interviewing farmers from Shetland with land by the coast, Hall (2000) estimated annual losses to be €500 per farmer, and a total of €770,000 for islands as a whole. This comprised time spent cleaning land, clearing ditches, freeing entangled animals, additional vet bills, and repairs to fences damaged by litter. Other losses to farmers not factored into this estimate were the loss of seaweed as a fertilizer due to plastic entanglement, and limits on the practice of grazing livestock on seaweed on beaches. Mouat et al. (2010) estimated that marine litter cost each smallholding an average of €841 per year and the agricultural industry of the Shetland Islands as a whole a total of approximately €252,000 (based on the assumption that 25 % of the 1,200 active crofters were operating in areas subject to marine litter damage). The cost to small-scale agricultural producers is of particular concern given that they have small profit margins. Although the scope of these estimates is restricted to one Scottish archipelago and the costs of marine litter to coastal agricultural businesses elsewhere is unknown, anecdotal evidence from the south of England and Sweden suggests it is a big problem in other coastal regions, too (Mouat et al. 2010). Clearly, more research would be required to determine the impacts and costs of marine litter in other coastal agricultural enterprises.

14.2.6 Human Health

Whilst the impact of marine litter on human health is a relatively new area of research this does not negate its potential for generating economic and welfare costs. At a local level ocean collisions with marine litter can seriously injure or kill mariners (Gold et al. 2013). This is particularly the case with impacts between smaller vessels and larger objects, such as semi-submerged lost shipping containers, which is a known danger to recreational sailors and fishers.

Plastic pollution poses a number of more nuanced risks, which could directly and indirectly impact on human health (Teuten et al. 2009; Thompson et al. 2009; Gold et al. 2013; UNEP 2014: p. 50; Galloway 2015). Firstly, the physical and chemical properties of polymers lends to their ability to facilitate the accumulation of contaminants already present in sea water. Industrial and agricultural chemicals, including polychlorinated biphenyls (PCBs), dichlorodiphenyltrichloroethane (DDT), and aqueous metals, have been linked to health impacts such as disease and reproductive abnormalities (Teuten et al. 2009). Marine litter acts as nuclei of accumulation for such toxins, which become several orders of magnitude more concentrated on the surface of plastics (EPA 2011). For instance, plastics can contain up to 1 million times the concentration of PCBs in contrast to sea water (Gold et al. 2013: p. 5; EPA 2013). Secondly, chemicals used in the production of polymers can increase local concentrations of harmful toxins which are also known to impact on health. Additives such as bisphenol A (BPA) and flame retardants, such as polybrominated diphenyl ethers (PBDEs), commonly found in plastic waste, can dissociate in the environment and are linked to endocrine disruption in both wildlife and humans (Gold et al. 2013). Both sources of chemicals increase the potential for bioaccumulation of toxins within food chains when marine litter is ingested by smaller organisms.

Thirdly, plastics could facilitate the transmission of, and act as a vector for, viral and bacterial diseases in areas where they would not naturally occur. In some locations plastic marine litter has developed its own habitat, the "plastisphere", supporting organisms which differ from those in the surrounding water (Gold et al. 2013). Lippsett (2013) found a plastic sample dominated by bacteria, which cause cholera and gastrointestinal disease. Consequently invasive species and foreign substrates linked to marine litter could pose significant health threats. The economics of such health risks are difficult to formulate but figures are urgently needed to assess these additional cost associated with marine litter.

14.2.7 Summary

So far we have presented the results of recent research estimating the direct costs of marine litter to the key coastal and maritime economic sectors affected. Estimates of economic impacts on a national or regional scale are hard to come by, however. It is clear that we require more monitoring of the costs associated with marine litter, both in terms of direct costs incurred on losses to outputs and income, and in terms of assessing health, ecosystem services, wellbeing, and welfare impacts.

What is also very clear from this review is that marine litter exerts substantial economic impacts on coastal sectors, and that the polluters or producers of plastics do not pay for these costs. Furthermore, those who do pay are often operating with tight budgets, such as municipalities, small-scale agricultural businesses and fisheries. In addition, the costs described and quantified in the literature consist primarily of the costs of cleaning up marine debris or recovering from marine litter damage. As explained above, the cost of cleaning up is justified by the even higher costs of inaction. However, this expenditure does not address the underlying issue, and does not act to prevent litter from entering the marine environment in the first place. When prioritizing action and the allocation of funds, the costs of damage and clean up needs to be weighed against the cost of prevention.

14.3 Marine Litter and Economic Incentives

From an environmental economics perspective, marine litter arises, like other waste or pollution problems, through market failure. The marginal price of goods on the market, and that of disposable plastics in particular, does not reflect the full marginal cost to society of producing that good. In other words, there is an external cost to society not borne by the producer (or consumer) as demonstrated in the previous section. Furthermore, clean seas and beaches are public goods, which are vulnerable to free-riding, whereby those disposing of waste inappropriately benefit from the good without paying the full cost, thereby causing contamination and degradation of the marine environment.

Like other environmental problems, marine litter can be prevented and controlled using measures that limit and control this sort of behavior (command and control measures), by awareness raising and other information tools, and by using market-based measures that aim to encourage a change in behavior by altering the economic incentives in place and/or to raise revenue to bring the market price in line with the social cost.

There are a range of market-based instruments that can be used to address marine litter. Landfill taxes, if set at adequately high levels, can disincentivize the final disposal of waste and help to incentivize recycling and recovery, reducing the risk of waste reaching the marine environment (although care should be taken to set taxes at a level that does not give significant encouragement to illegal dumping of waste). Product taxes and charges can be used to discourage the consumption of certain products that frequently end up as marine litter, such as plastic bags, packaging and fishing tackle. Infrastructure charges, for example, for the use of port-waste facilities, help to ensure that waste management infrastructures and facilities are developed and maintained. Deposit-refund schemes, which are most often applied to packaging items such as bottles, can encourage return and reuse by consumers, and therefore reduce the number of such items ending up as litter. Hardesty et al. (2014) evaluated the effectiveness of South Australia's container deposit scheme in reducing waste lost to beaches and reported a threefold reduction in this dominant plastic item in the environment. Direct investment in infrastructure, such as rubbish bins and secure waste collections on beaches and in coastal areas, can help to keep coastal areas free of litter and reduce the risk of items reaching the seas. Such investment can be financed for example by tourist taxes or car parking fees. High fees and fines for littering, illegal waste disposal and fly-tipping help to dissuade behaviors that result in waste escaping from formal waste management processes, reducing the risk of waste reaching the marine environment as litter (ten Brink et al. 2009).

The following sections of this chapter focus on a small selection of market-based instruments, which can potentially have an impact on the amount of litter that reaches the marine environment: landfill taxes, instruments addressing plastic bags (including charges/taxes and bans), producer responsibility schemes, and fees for the use of port waste reception facilities.

14.3.1 Landfill Taxes/Levies

A significant proportion of marine litter originates from land-based sources; a global figure of 80 % is frequently cited, although the origins of this are unclear (Arthur et al. 2014) and figures may vary considerably regionally. The National Marine Debris Monitoring Program, which analyzed marine litter on US beaches,[1] determined that 49 % was from land-based sources and 18 % from ocean-based sources, with a further 33 % for which the source could not be identified) (Ocean Conservancy 2007). Up to 95 % of the litter found on Australian beaches comes from suburban streets through the stormwater system (Clean Up Australia 2009). For this reason, measures to promote improved waste management on land have an important role to play in preventing land-based waste from reaching the seas.

One of the most common economic instruments used in the waste sector is the application of a tax or levy on waste sent to landfill. Landfill taxes/levies can help to tackle marine litter by increasing the price of landfill to encourage the diversion of waste to other forms of treatment that are higher up in the waste hierarchy, including closed-loop waste-management processes such as recovery, recycling or reuse. If lightweight items in particular, such as many small packaging items, can be kept out of landfills, this eliminates the risk of them being blown by the wind from the surface of landfills, preventing them from reaching water courses and eventually entering the sea. It should be noted that landfill taxes can incentivise illegal landfilling and fly-tipping as a means of tax avoidance. Estimates of the amount of marine litter that comes from landfill and fly-tipping are limited; one estimate from the Scottish Government is that around 1.6 % of marine litter comes from fly-tipping incidents (Scottish Government 2013). To stop these unchecked methods of waste disposal from resulting in more waste being blown or washed into rivers and seas, landfill taxes should be accompanied by measures such as the closure of illegal landfills and enforcing fines on those who fly-tip or dump illegally. Producer-responsibility schemes can also help promote recycling (see below).

---

[1]The NMDMP ran from September 2001–September 2006. The US was divided into nine coastal regions; within each region a random selection of between 12 and 23 beach sites was chosen for surveying (175 sites in total). Over 600 volunteers conducted surveys at 28-day intervals, covering a 500-m stretch of beach at each study site, and collected and recorded the various marine debris items found.

Landfill taxes are typically charged per tonne of waste landfilled, and there are often different rates for active (e.g. biodegradable) and inert (e.g. mineral/construction) waste, to reflect the varying environmental impacts of different wastes. A brief overview of the use of landfill taxes is provided here. It should be noted that this is not intended to be a fully comprehensive review of the global situation.

In **Europe**, many countries have introduced landfill taxes since the EU Landfill Directive (1999/31/EC) entered into force. The directive aims to encourage the prevention, recycling and recovery of waste by limiting its final disposal through landfills, including setting targets to reduce the amount of biodegradable waste sent to landfills and associated methane emissions. In addition, Annex I sets out requirements for the location (e.g. with regard to the proximity of water bodies and coastal waters) and technical specifications (e.g. design features to avoid pollution of soils and waters from landfills, including from wind-blown waste) of landfill sites. There are currently 20 countries[2] in **Europe** that tax waste sent to landfills. From 1995 to 2012, the number of EU countries implementing a landfill tax rose from 7 to 20 (Watkins et al. 2012). Over the same period, the amount of municipal waste sent to landfill decreased from around 63–33 % (Eurostat 2014a). In the majority of cases, the tax is collected by state tax authorities or regional institutions. However, only in some countries (Bulgaria, Finland, Belgium (Wallonia), France, Poland, Portugal, Spain and the United Kingdom) does part of the revenue go towards waste management and environmental initiatives (Fischer et al. 2012). The level of the tax varies considerably between countries: Watkins et al. (2012) identified a wide range of tax rates for municipal solid waste, from €3 t^{-1} in Bulgaria to €107 t^{-1} in The Netherlands. Higher landfill taxes tend to result in lower proportions of municipal waste being sent to landfills and higher rates of recycling and composting. The majority of countries with total landfill charges[3] of less than €40 t^{-1} send over 60 % of their municipal waste to landfill. Countries are much more likely to meet a 50 % recycling target once landfill charges (or the cost of the cheapest disposal option) approach €100 t^{-1} (Watkins et al. 2012). It should be noted that the best performing countries in terms of diverting waste from landfill usually also have other measures in place, such as bans on the landfilling of certain types of waste.

The **Australian** state of Victoria has a levy of €17 t^{-1} for rural municipal waste, €29 t^{-1} for rural industrial waste, €34 t^{-1} for metropolitan municipal and industrial waste. All levies are paid into the Environment Protection Fund, with revenues from the levy contributing to improved waste management including upgrading of kerbside recycling systems, developing markets for recycled materials, and

---

[2]Austria, Belgium (Flanders and Wallonia), Bulgaria, the Czech Republic, Denmark, Estonia, Finland, France, Ireland, Italy, Latvia, the Netherlands, Norway, Poland, Portugal, Slovenia, Spain (Andalusia, Catalonia, Madrid and Murcia), Sweden, Switzerland and the United Kingdom.

[3]Total landfill charge defined as tax plus 'gate fee' charged by landfill operator for receiving the waste.

studies into waste minimization, handling and disposal, organics recycling and lit-ter control (EPA Victoria 2013). The levy in the Sydney metropolitan area has risen sharply since 2006/07 and is planned to reach €84 t^{-1} by 2015/16 (New Zealand Ministry for the Environment 2014). The state of Western Australia applies a levy of €20 t^{-1} (increasing to €39 t^{-1} from 1 January 2015) for putrescible waste and €6 t^{-1} (€28 t^{-1} from 1 January 2015) for inert waste deposited in metropolitan landfills. The levy is paid by the owner of the landfill receiving the waste, but they may pass the cost on to customers. Not less than 25 % of revenues will be spent on initiatives to manage, reduce, re-use or recycle waste and to monitor/measure waste, and 7 % provided to the Office of the Environmental Protection Authority to assist in service delivery (Wastenet 2014). Evidence from Australia on the impact of landfill levies is somewhat mixed. The Western Australian Local Government Association claims there is limited evidence that a levy directly disincentives land-fill, and that a lack of accompanying investment in waste management can actually be detrimental to waste diversion activities due to reduced expenditure on recy-cling infrastructure (WALGA 2012). The waste levy in Sydney helped to increase recycling by making waste recovery more financially attractive than landfill; the total quantity of waste to landfill was lower in 2010/11 than 2002/03, and waste recycled more than doubled over the same time period (New Zealand Ministry for the Environment 2014). Conversely, following the removal of a €25 t^{-1} landfill levy in Queensland, there was a 20 % reduction in recycling (Ritchie 2014), which indicates that landfill taxes may indeed encourage recycling.

New Zealand applies a tax of €6 t^{-1} to any waste deposited at a waste disposal facility. This is paid by disposal facility operators, but they may pass the cost on to the households/businesses that generate the waste. The levy's primary aim has been to raise revenue for waste minimization and recycling projects, but the levy was set at a relatively low level to avoid illegal dumping and to reduce the impact on businesses and households. Half of the revenues go to territorial authorities to assist with waste minimization. The rest (minus administration costs) is paid into a national waste minimization fund (New Zealand Ministry for the Environment 2013). A recent review of the effectiveness of the waste disposal levy found that the levy is estimated to be applied to only 30 % of total waste disposed of to land, and that the amount of waste landfilled has increased by around 6 % between 2010 and 2013. Revenues have supported a broad range of waste minimization initia-tives, although the funding outcomes should be more effectively measured and monitored (New Zealand Ministry for the Environment 2014).

14.3.2 Plastic Bag Initiatives

Plastic is the most common litter type found in the marine environment. In European regional seas, for example, plastics comprise more than half of the marine litter. More than half of the plastic fraction of marine litter consists of plastic pack-aging waste, with bottles and bags being the most frequently found items (European

Commission 2013a). Plastics account for some 78 % by number of pieces, and 67 % by weight, of washed-up materials on the coasts of the Northwest Pacific Region (Kanehiro 2012). Lightweight plastic bags are particularly prone to becoming marine litter since they are seen by consumers as single-use, are often disposed of carelessly, are frequently not accepted in household recycling collections, and are easily blown by the wind into drains, water courses and the marine environment. During the Ocean Conservancy's 2013 International Coastal Cleanup,[4] 6 % of the total litter items found were plastic bags (grocery or other plastic bags) (Ocean Conservancy 2014). In China in 2012, plastic bags comprised 23 % of drifting marine litter found in the sea, and 59 % of that found on beaches (Meng and Chen 2013). In Shanghai, they accounted for between 15 and 29 % of all marine litter items recorded during coastal surveys between 2008 and 2012 (ICC 2013).

European Commission guidance for EU Member States on developing waste prevention programs, published in October 2012, suggests that **plastic bags** can be effectively targeted by waste prevention activities (European Commission 2012). Many countries have already taken specific action to tackle plastic bags: over 30 countries have introduced taxes/fees, and over 30 have introduced bans for single-use carrier bags, or bans on bags with certain characteristics, such as those made from plastic of less than a certain thickness (Earth Policy Institute 2013). In some countries, there is a mix of bans and charges, since this is an area of policy that is often dealt with at a local or city level. The following paragraphs summarise several of these initiatives, with a focus on those where information is available on their impacts.

In **Europe**, several countries have introduced taxes or charges on single-use (disposable) plastic carrier bags. Denmark has applied a charge for plastic and paper carrier bags since 1993 (the charge depends on the weight and material). The year after the tax was introduced saw an initial reduction in bag use of 60 % (Earth Policy Institute 2013).

Ireland introduced a €0.15 levy per general purpose plastic bag in 2002, and increased the levy to €0.22 in 2007. The levy led to an immediate decrease in plastic bag use from an estimated 328 bags per capita per year to 21 bags per capita. Although per capita consumption increased again to 31 bags during 2006, an increase in the levy in 2007 led to a further reduction to 18 bags in 2010 (Department of Environment Community and Local Government 2013). Plastic bags constituted 0.3 % of litter pollution nationally in 2012 compared to an estimated 5 % in 2001 prior to the introduction of the levy (National Litter Pollution Monitoring System 2013). In 2001 (pre-levy) around 17 plastic bags were found per 500 m of coastline. This figure fell to around 10 bags in 2002 (the year the levy was introduced), 5 bags in 2003, and 2 in 2012 (Doyle and O'Hagan 2013).

In the U.K., a €0.06 levy on the use of single-use carrier bags (plastic and paper) was introduced in Wales in 2011, and in Northern Ireland in 2013; retailers in

---

[4]Ocean Conservancy's (2013) International Coastal Cleanup involved nearly 650,000 volunteers at over 5,500 beach/coastal sites covering a total length of 12,914 miles in 92 countries and locations.

Scotland introduced the same charge in 2014, and supermarkets and larger stores in England will adopt the same charge in 2015. One year after the introduction of the **Welsh** charge, a 70–96 % decrease was observed in bag use at food retailers and a 68–75 % decrease at fashion retailers (Welsh Government 2012). While in 2010 (pre-charge), 0.35 million thin-gauge carrier bags were distributed in Wales; in 2011 (the year the charge was introduced), 0.27 billion bags were distributed and one year after the introduction of the charge the figure dropped significantly to 0.07 billion bags (WRAP 2013). This represents an 81 % reduction in only three years whereas no reduction was observed in any other nation of the U.K. over the same time period, which suggests that the charge has significantly contributed to the reduction in Wales. During International Coastal Cleanup days, 435 plastic bags were found in Wales in 2011, and 292 in 2012 (Ocean Conservancy 2012, 2013). These data only relate to a single day each year and therefore present a limited picture, but the reduction in plastic bag use due to the charge may have had at least some impact on the number of bags found.

In November 2013, the European Commission put forward a proposal for **European Union** legislation to reduce the use of lightweight plastic bags. The proposal would amend the existing Packaging and Packaging Waste Directive (1994/62/EC) (see Chen 2015), requiring all EU Member States to take action to reduce the consumption of lightweight plastic bags, but allowing them to choose the most appropriate measures to do this (European Commission 2013b). The proposal will be discussed in the European Parliament and Council in 2015.

Several nations in **Africa** have either totally banned the use of plastic bags (Eritrea in 2005; Somalia/Somaliland in 2005; Tanzania in 2006/Zanzibar in 2008; Democratic Republic of Congo in 2011) or banned the use of bags below a minimum thickness (Kenya and Uganda, both 2007) (see also Chen 2015). Botswana (2007) established a minimum thickness for bags and required retailers to apply a minimum levy to thicker bags (many retailers charged more than the minimum), to fund government environmental projects. A study of four retail chains 18 months after implementation of the charge showed that bag use had fallen by 50 % (Earth Policy Institute 2013). Morocco has implemented a minimum thickness standard. In 2004, South Africa implemented a tax of €0.002/bag for thicker bags, to run concurrently with a minimum thickness ban (Miller 2012). A portion of the tax funds environmental projects. Whilst bag use decreased by 90 % when the measures were first introduced, consumption has slowly risen again since (Earth Policy Institute 2013).

One-hundred and thirty-two cities and counties in the **US**, with a combined population of over 20 million people, now have plastic bag bans or fees. In California, the state with the largest number of anti-bag measures in place, plastic bag purchases by retailers fell from around 48,000 tonnes in 2008 to around 28,000 tonnes in 2012, a decrease of around 42 %. The Department of Public Works reported that the January 2012 ban on plastic bags in retail stores and the €0.08 charge for paper bags, implemented in unincorporated areas of Los Angeles County, had led to a sustained 90 % reduction in single-use bag use at large stores by December 2013. A €0.04 charge for plastic and paper carryout bags at all food/

alcohol retailers in Washington DC, introduced in January 2010, led to four out of five households using fewer bags, with two thirds of residents reporting seeing less plastic bag litter after the tax came into effect (Larsen and Venkova 2014). It is estimated that the bag tax in Washington DC reduced grocery bag sales by somewhere between 67 and 80 % after two years (Beacon Hill Institute 2012).

In **Asia**, several countries have taken action against plastic bags, including complete bans (Bangladesh: 2002; Papua New Guinea: 2009), minimum thickness bans (Taiwan: 2001) or taxes (Taiwan, tax of €0.02–0.08 since 2003; Hong Kong, tax of €0.05 since 2009). The Taiwanese charge increased the number of people who regularly took used plastic bags to reuse when they went shopping (rising from 18 % in 2001 to 72 % in 2006), and the Hong Kong charge successfully reduced plastic bag use by 75 % in affected stores (Earth Policy Institute 2013). In the first year following the 2008 ban on the provision of free plastic bags in all shops in **China**, the National Development and Reform Commission (NDRC) estimated that supermarkets had reduced plastic bag usage by 66 % (amounting to 40 billion fewer bags used) (Block, n.d.). Raw data, however, do not indicate that this reduction in usage translated into a reduction in bags found as marine litter. International Coastal Cleanup (ICC) surveys in the Nanhui District of Shanghai found 129 plastic bags in September 2007 (pre-ban), 286 in September 2008 (year of ban introduction), then (post-ban) an average of 400 per site per cleanup in 2009, an average of 245 per site per cleanup in 2010, and an average of 1,294 per site per cleanup in 2012 (NOWPAP DINRAC 2014). It is not clear whether these fluctuations (and significant increase in 2012) are due to variations in the number of volunteers participating in the cleanups, illegal selling of bags, ineffective implementation of the ban, or bags from non-Chinese sources being washed up on the coast. Indeed, bans in Bhutan (1999), India (attempted first in 1999 and several times subsequently) and Bangladesh (2002) have largely failed to bring about a decrease in use due to poor implementation and enforcement (Earth Policy Institute 2013).

In **Australia,** South Australia banned plastic bags in 2009, the Northern Territory and Australian Capital Territory followed suit in 2011, and Tasmania banned very thin plastic bags in 2013. Several cities and towns have also introduced voluntary bans. The South Australian ban purportedly encouraged customers to bring their own bags more often (Earth Policy Institute 2013). New Zealand has had a voluntary retailer levy of €0.03–0.06 since 2009 (Miller 2012).

There are also myriad voluntary initiatives worldwide undertaken by individual retailers to attempt to limit the number of disposable bags they hand out to customers, ranging from eliminating disposable bags altogether to charging for disposable bags or selling reusable bags. In Spain, a voluntary agreement between Catalonia's Waste Agency, regional and national business groups, plastic bag manufacturers, food distributors and supermarkets led to a 40 % decrease in the consumption of single-use plastic bags between 2007 and 2011, and 87 % reduction (equal to 1 billion individual bags) in annual supermarket plastic bag use (Earth Policy Institute 2013).

From the examples outlined here, it is clear that bans and charges have had varying degrees of success, ranging from no discernible impact (failed bans in

Bhutan, India and Bangladesh) to reductions in bag use of over 90 % (charges/taxes in Ireland, Wales and parts of Los Angeles County). Sometimes impressive initial results are not sustained over time (Ireland, South Africa). This picture indicates that there is no one-size-fits all solution to the issue of plastic bags and that measures must be tailored to address different consumer/business behavior in different countries. It perhaps also suggests a need for measures that can be adapted to respond to failures following initial successes (e.g. increase in charges), and highlights the need for full implementation and proper enforcement of measures such as charges and bans to ensure their success.

14.3.3 Packaging Producer Responsibility in the EU

Extended producer responsibility (EPR) is an environmental policy approach in which a producer's responsibility for a product is extended to the post-consumer stage of a product's life cycle, meaning that they are responsible (financially and/or logistically) for dealing with the product when it becomes waste. This concept has been widely implemented in the EU over the past 20 years, with the introduction of a great variety of EPR schemes and the creation of producer-responsibility organisations (PROs), collective entities set up by producers or through legislation to meet the recovery and recycling obligations of individual producers.

Waste packaging forms a significant proportion of marine litter. Food wrappers, plastic and glass drinks bottles, bottle caps and drinks cans all regularly feature in the top ten most frequently found items during marine litter surveys; together these items comprised 31 % of all items found during the Ocean Conservancy's 2013 International Coastal Cleanup. When plastic and paper shopping bags, which may also be classed as packaging, are added, this figure increases to 37 % (Ocean Conservancy 2014). More than half of the plastic fraction of marine litter consists of plastic packaging waste (European Commission 2013a). This section therefore focuses on EPR schemes that deal with waste packaging.

All EU Member States have implemented EPR for packaging waste, since it is targeted by the Packaging and Packaging Waste Directive, 94/62/EC. In the 27 EU Member States in 2011 (prior to the accession of Croatia), 64 % of waste packaging was recycled (this figure includes composting for biodegradable packaging), and 77 % was recovered (this figure includes incineration with energy recovery). Within these figures there are of course variations between countries, ranging from only 41 % recycling in Poland to 80 % in Belgium, and only 45 % recovery in Malta to 97 % in Germany (Eurostat 2014b). Successful EPR schemes and the associated recycling infrastructures, including doorstep recycling collections, play a significant role in achieving high recycling and recovery rates and diverting packaging waste away from final disposal. Capturing packaging waste in closed-loop collection and recycling systems reduces the risk of items reaching the seas and becoming marine litter.

Typically in packaging EPR schemes, producers pay a fee to a PRO based on the amount of packaging they place on the market (for example, a fee per tonne of paper/card, glass, aluminium packaging, with the fees for each material typically being different). These fees are then be used to cover, or contribute to, the cost of collection and treatment of waste packaging. Basing producers' contributions on the actual amount of packaging they place on the market ensures they pay their 'fair share' of the cost of waste management; this is the application of the producer-pays principle in practice, internalising the end-of-life costs into the cost of the product. This can incentivize producers to reduce the amount of packaging they place on the market, since this decreases the fees they pay.

A recent study (BIO by Deloitte et al. 2014) that looked at packaging EPR schemes in seven EU countries found that fees paid by producers ranged from just over €1 per capita per year in the U.K. to almost €20 per capita per year in Austria. The wide variation was primarily due to the different levels of cost coverage: fees from the purchase of Packaging Recovery Notes (PRN) in the U.K. were estimated to cover only 10 % of the total cost of the system, whereas in most of the other schemes reviewed, 100 % of the net costs of collection and treatment of separately collected waste were covered. Discussions with stakeholders during the study did not provide a consensus on whether producers should also finance the costs of dealing with packaging that is littered by consumers; measures that more directly target consumer behavior, such as deposit-refund schemes for packaging, have the potential to reduce littering. The more expensive schemes were not necessarily found to be the best in terms of recycling and recovery levels achieved. The highest recycling rates were achieved in Belgium: 85 % for household packaging and 82 % for commercial and industrial packaging, with costs per capita per year of around €8 and just over €1 respectively.

Many factors have an impact on the costs and performance of EPR schemes. Collection costs are typically higher in areas with lower population density. The historical development and quality of waste collection and treatment infrastructure is important, since economies of scale can be achieved through greater sorting and treatment capacities; EPR schemes can help to trigger infrastructure development and to finance improvements and maintenance. The value of secondary materials on national markets can be important, and can be influenced both by the demand for secondary raw materials and by the provision of high-quality materials once a recycling industry is in place. Citizens' awareness of separate collection schemes, and their willingness to participate, is also crucial, and investment in public communication can help EPR schemes to succeed. Other waste policy instruments, including those discussed in this chapter (disposal taxes) and others (pay-as-you-throw schemes, deposit-refund schemes, etc.) can complement EPR schemes and increase the efficiency of the general waste management system (BIO by Deloitte et al. 2014).

14.3.4 Charges for Port Reception Facilities

With the abundance of ships using the oceans for shipping, transportation, tourism, military purposes and other maritime industries, there is a tremendous amount of waste being generated at sea. It is therefore important for countries and their ports to provide adequate reception facilities for all of the types of ships that frequent those ports and all the types of waste they produce. It is also important to create proper incentives to encourage ships to use the correct facilities at ports, rather than dump waste into the sea. International regulations ratified by the signing members of MARPOL (73/78), the International Convention for the Prevention of Pollution from Ships, govern what kinds of waste can be discharged overboard and where they can be discharged. MARPOL prohibits the disposal of plastics at sea, as well as other garbage (though there are exceptions for food waste), and requires signatories to the convention to ensure that adequate reception facilities for ship-generated waste and cargo residues are established and that they are able to receive such waste from ships calling at the port without causing any undue delay. Although the convention provides a number of recommendations on how reception facilities can be established it does not specify how waste should be handled, and leaves all organisational issues to the responsible port authority. Compliance with the convention requires the efficient collection of waste from ships, and what happens thereafter is regulated by national legislation (for more details see Chen 2015).

One of the most important factors in incentivizing ship waste delivery is the waste fee system. Handling and disposing of waste is costly to ports (with the exception of oily waste, which, due to rising oil prices, has recently become economical to collect and recycle) (Øhlenschlæger et al. 2013). Like the other services provided by ports, and in coherence with the polluter-pays principle, the cost of waste collection should rightly be recovered through the collection of port fees. However, high fees for waste collection can act as a disincentive to ships to discharge their waste at port, when they can throw their waste unseen overboard for free (if they can get away with it). It is therefore necessary to strike a balance between cost-recovery of waste handling and not discouraging disposal at port.

Because of the increasing number of illegal discharges in the Baltic Sea, in the late 1990s the HELCOM Convention provided a number of recommendations regarding the introduction of a 'No Special Fee' or 'indirect fee' in Baltic ports, leading to the introduction of a 100 % indirect fee system for solid garbage waste ('household' waste, oily waste from machinery space and sewage) by several Baltic Sea ports (Gothenburg, Copenhagen, Klaipeda, Helsinki and Stockholm) (Øhlenschlæger et al. 2013). An indirect fee means that the cost of delivering solid garbage waste to port reception facilities is included in the fee paid by all ships visiting the port, irrespective of the quantities discharged, and is not specified on the invoice (Ikonen 2013).

The no-special-fee system effectively prevents cost from becoming a disincentive for using port reception facilities. Given that all ships will pay the fee regardless of use, they therefore all contribute to the financing of waste collection facilities, and this approach also has the benefit of reducing the fee. The administrative burden on port operators also appears lower when a 100 % indirect-fee system is adopted, because operators simply collect waste without the ports' financial administrations needing to calculate fees based on the actual amounts of waste delivered (Øhlenschlæger et al. 2013). One negative effect that has been observed, however, is that ships are incentivized to deliver oily waste at each port, rather than accumulating the waste on board and delivering only when slop tanks are full (Ikonen 2012; Øhlenschlæger et al. 2013). This results in a costly and inefficient situation whereby smaller amounts of oily waste are being delivered at each port. For solid garbage waste this is not really a problem as more frequent deliveries of smaller amounts of waste are almost as convenient and cheap to dispose of as less frequent larger quantities (Øhlenschlæger et al. 2013). In addition, as the fee is not proportional to the amount of waste produced, it does not encourage waste reduction on board vessels (Ikonen 2012).

There is no evidence for the effectiveness of the no-special-fee system implemented in the Baltic on trends in waste delivered in ports, as quantities of waste delivered to ports is influenced by many factors, and it is almost impossible to detect illegal dumping of solid waste because essentially you would have to catch a perpetrator red-handed. However, there are data on detected incidences of illegal oil spills in the Baltic (which can be observed using aerial surveys), which indicate a decline in illegal spills following the introduction of the no-special-fee (HELCOM 2012; Ikonen 2013; Øhlenschlæger et al. 2013). Given that solid waste is easier to deliver than oily waste and is often delivered at the same time, it is reasonable to assume that the number of illegal waste discharges at sea also dropped over this period (Øhlenschlæger et al. 2013).

14.4 Choosing Economic Instruments

Economic instruments to tackle marine litter should be designed so as to deliver three objectives:

1. Minimize production of marine litter.
2. To minimize the harm caused by marine litter.
3. To avoid unintended consequences from the application of the instrument.

Achieving all of these objectives is a challenge. This chapter has provided examples of the use of economic instruments to reduce different types of waste that contribute to marine litter or to target specific sources of such waste. However, there is a difference between reducing waste arisings and managing disposal, which is the focus of economic instruments currently applied to waste, and addressing the harm caused by marine litter.

Reducing the quantity of marine litter may depend on targeting key sources. For example, where waste enters the environment affects its ability to contribute to marine litter. A plastic bag dropped from a ship is more likely to become marine litter than one dropped on coastal land which is, in turn, more likely to become marine litter than one dropped 100 km inland.

Targeting the economic instruments to address marine litter that causes the most harm is particularly problematic. Marine litter causes different types of impacts and the harm arising from these varies—ghost fishing, suffocation by plastic bags, introduction of toxic substances—such impacts may be unique to some types of waste or focused around particular types of waste. In contrast, the impact of marine litter on tourism due to the presence of litter on beaches is largely a factor of its total quantity (although some types of waste are particularly unpleasant or unsanitary).

In addressing marine litter, economic instruments can be used to reduce the impacts of such litter in a variety of ways. Such instruments may:

- Incentivize industries to use less plastic (packaging) either through economic disincentives/subsidies (internalizing external cost);
- Target waste arisings generally—such as with a landfill tax;
- Target specific types of waste—such as plastic bags;
- Target sources of waste most problematic for marine litter—such as shipping;
- Target individual types of marine litter—such as to reduce ghost fishing;
- Pay for the collection of litter;
- Target the toxicity of litter;
- Discourage polluting behavior.

Economic instruments have been adopted for some of these types of waste/litter. However, the toxicity of waste/litter is usually addressed through regulation controlling the quality of products or materials. The use of a regulatory approach on this issue in Europe, for example, is strongly linked to the single EU market. However, differential taxes or charges for products with materials that would have different toxicities in water are theoretically possible.

All instruments can have unintended consequences, that is impacts other than those for which the instrument is designed. The most obvious are costs to businesses, administrations or individuals. Economic instruments may have such costs—charges are an obvious cost, but administrations may incur costs to administer an instrument. However, where charges or taxes are levied these can be used to pay for their administration or contribute in other ways (e.g. funding awareness raising to ensure compliance, monitoring of instrument efficiency).

The choice of economic instrument also needs to consider the acceptance of the instrument by those affected. An instrument that results in additional costs (a tax, charge, etc.) may be resisted by some stakeholders. For example, 'pay-as-you-throw' schemes are strongly opposed by some communities, but not others. However, those same communities may welcome a reward scheme to encourage 'good' behavior funded by local taxes (yet this still results in costs to people). Acceptance may change over time: for example, where plastic bag taxes were introduced early resistance has largely disappeared as communities have seen the benefits of the schemes.

Finally, it is worth noting that there is discussion on the use of economic instruments to manage litter on beaches, i.e. the financing of its removal. For example, Birdir et al. (2013) undertook a willingness-to-pay study of beach litter in Turkey. Their conclusions suggested local taxes and collection boxes as means to fund beach cleanups. However, while it is appropriate to consider how such services are funded, these are not economic instruments to tackle the problem at source.

14.5 Conclusions

Marine litter is a complex problem to address, which exerts significant economic costs, often borne not by the polluter but by coastal and marine industries such as fisheries, aquaculture, tourism, etc. (some of which also contribute significantly to marine litter). Economic instruments have a potentially important role to play in addressing marine litter, with initiatives in place in several countries proving that they can lead to significant reductions in waste entering the environment (ten Brink et al. 2009).

The development of effective and efficient instruments requires a strong link between the behavior change driven by the instrument and the harm caused by marine litter. However, there are several areas where there is a lack of sufficient information to make this link. At the heart of this is the problem of understanding the harm caused by marine litter. The harm caused by some forms of litter is known, however, there are large gaps in this understanding.

While the presence of litter in the marine environment and even its ingestion, etc., in species is documented, it is not clear what impact it is having on critical populations of marine organisms or indeed species higher up the food chain (including humans). Further, while some specific types of litter are identified as having some impacts (discarded nets, plastic bags, etc.), the impacts of other types of litter are currently poorly understood, which is most notable with the debate on micro- and nanoplastics. In relation to socio-economic impacts, impacts on tourism from beach litter are documented, but a quantitative link between the impact and levels of litter is poorly understood (Ballance et al. 2000).

These links between types, quantities and sources of marine litter and their varied impacts are important to understand if targeted economic instruments are to be developed. Otherwise an instrument may lead to a reduction in litter, but with a limited reduction in impact.

References

Antonelis, K., Huppert, D., Velasquez, D., & June, J. (2011). Dungeness crab mortality due to lost traps and a cost-benefit analysis of trap removal in Washington State Waters of the Salish Sea. *North American Journal of Fisheries Management, 31*(5), 880–893.

Arthur, C., Sutton-Grier, A. E., Murphy, P., & Bamford, H. (2014). Out of sight but not out of mind: Harmful effects of derelict traps in selected U.S. coastal waters. *Marine Pollution Bulletin, 86*, 19–28.

Ballance, A., Ryan, P. G., & Turpie, J. K. (2000). How much is a clean beach worth? The impact of litter on beach users in the Cape Peninsula, South Africa. *South Africa Journal of Science, 96*, 5210–5213.

Barnes, D. K. A. (2002). Invasions by marine life on plastic debris. *Nature, 416*, 808–809.

Beacon Hill Institute at Suffolk University (2012). Two years of the Washington, D.C. Bag tax: An analysis. Retrieved December 9, 2014 from http://s3.amazonaws.com/atrfiles/files/files/BHI_Report.pdf.

Beaumont, N. J., Austen, M., Atkins, J. P., Burdon, D., Degraer, S., & Dentinho, T. P. (2007). Identification, definition and quantification of goods and services provided by marine biodiversity: Implications for the ecosystem approach. *Marine Pollution Bulletin, 54*(3), 253–265.

Bilkovic, D. M., Havens, K., Stanhope, D., & Angstadt, K. (2014). Derelict fishing gear in Chesapeake Bay, Virginia: Spatial patterns and implications for marine fauna. *Marine Pollution Bulletin, 80*, 114–123.

BIO by Deloitte et al. (2014). Development of guidance on extended producer responsibility (EPR): Final report. Retrieved September 5, 2014 from http://ec.europa.eu/environment/waste/pdf/target_review/Guidance%20on%20EPR%20-%20Final%20Report.pdf.

Birdir, S., Unal, O., Birdir, K., & Williams, A. T. (2013). Willingness to pay as an economic instrument for coastal management: Cases from Mersin, Turkey. *Tourism Management, 36*, 279–283.

Block, B. (n.d). *China reports 66-percent drop in plastic bag use*. Eye on Earth, Worldwatch Institute. Retrieved November 29, 2013 from http://www.worldwatch.org/node/6167.

Butler, J. R. A, Gunn, R., Berry, H. L., Wagey, G.A., Hardesty, B.D., Wilcox, C. (2013). A value chain analysis of ghost nets in the Arafura Sea: Identifying trans-boundary stakeholders, intervention points and livelihood trade-offs. *Journal of Environmental Management, 123*, 14–25.

Chen, C.-L. (2015). Regulation and management of marine litter. In M. Bergmann, L. Gutow, M. Klages (Eds.), *Marine anthropogenic litter*. Springer, Berlin.

Clean Up Australia. (2009). *Cigarette butts factsheet*. Retrieved September 2, 2014 from http://www.cleanup.org.au/PDF/au/cleanupaustralia_cigarette_buts_factsheet.pdf.

Costanza, R., d'Arge, R., de Groot, R., Farber, S., Grasso, M., & Hannon, B. (1997). The value of the world's ecosystem services and natural capital. *Nature, 387*, 253–260.

de Stephanis, R., Giménez, J., Carpinelli, E., Gutierrez-Exposito, C., & Cañadas, A. (2013). As main meal for sperm whales: Plastics debris. *Marine Pollution Bulletin, 69*(1–2), 206–214.

Department of Environment Community and Local Government, Ireland. (2013). *Plastic bags*. Retrieved November 29, 2013 from http://www.environ.ie/en/Environment/Waste/PlasticBags/.

Doyle, T. K. & O'Hagan, A. (2013). *The Irish 'plastic bag levy': A mechanism to reduce marine litter?* Paper Presented at International Conference on Prevention and Management of Marine Litter in European Seas, Berlin. Retrieved November 29, 2013 from http://www.marine-litter-conference-berlin.info/userfiles/file/online/Plastic%20Bag%20Levy_Doyle.pdf.

Earth Policy Institute. (2013). *Data for plan b update 123*. Retrieved September 5, 2014 from http://www.earth-policy.org/datacenter/xls/update123_all.xlsx.

EPA [US Environmental Protection Agency]. (2011). *Marine debris in the North Pacific: A summary of existing information and identification of data gaps*. Retrieved October 27, 2014 from U.S. Environmental Protection Agency Web site: http://www.epa.gov/region9/marine-debris/pdf/MarineDebris-NPacFinalAprvd.pdf.

EPA. (2013). *Marine debris impacts*. U.S. Environmental Protection Agency. Retrieved October 27, 2014 from http://water.epa.gov/type/oceb/marinedebris/md_impacts.cfm.

EPA Victoria. (2013). *Landfill and prescribed waste levies*. Retrieved November 29, 2013 from http://www.epa.vic.gov.au/your-environment/waste/landfills/landfill-and-prescribed-waste-levies.

European Commission. (2011). *Reducing plastic marine litter in Mediterranean: A "Fishing for Litter" campaign in France*. Press release—20/5/2011. Retrieved December 8, 2014 from http://ec.europa.eu/fisheries/news_and_events/press_releases/2011/20110520/index_en.htm.

European Commission. (2012). *Preparing a waste prevention programme: Guidance document*. Retrieved December 9, 2014 from http://ec.europa.eu/environment/waste/prevention/pdf/Waste%20prevention%20guidelines.pdf.

European Commission. (2013a). *Integration of results from three marine litter studies*. Retrieved December 9, 2014 from http://ec.europa.eu/environment/marine/pdf/Integration%20of%20results%20from%20three%20Marine%20Litter%20Studies.pdf.

European Commission. (2013b). *Proposal for a directive of the european parliament and of the council amending directive 94/62/EC on packaging and packaging waste to reduce the consumption of lightweight plastic carrier bags, COM(2013)761*. Retrieved November 29, 2013 from http://ec.europa.eu/environment/waste/packaging/pdf/proposal_plastic_bag.pdf.

Eurostat. (2014a). *Municipal waste generation and treatment, by type of treatment method, kg per capita (Code: tsdpc240)*. Retrieved September 4, 2013 from http://epp.eurostat.ec.europa.eu/portal/page/portal/waste/data/main_tables.

Eurostat. (2014b). *Packaging waste* (Code: env_waspac). Retrieved September 5, 2014 from http://epp.eurostat.ec.europa.eu/portal/page/portal/waste/key_waste_streams/packaging_waste.

Fischer, C., Lehner, M., & McKinnon, D. L. (2012). *Overview of the use of landfill taxes in Europe*. ETC/SCP working paper 1/2012. Retrieved December 9, 2014 from http://scp.eionet.europa.eu/publications/WP2012_1/wp/WP2012_1.

Galloway, T. S. (2015). Micro- and nano-plastics and human health. In M. Bergmann, L. Gutow, & M. Klages (Eds.), *Marine anthropogenic litter*. Berlin: Springer.

Galparsoro, I., Borja, A., & Uyarra, C. (2014). Mapping ecosystem services provided by benthic habitats in the European North Atlantic Ocean. *Frontiers in Marine Science, 1*(23), 1–14.

Gold, M., Mika, K., Horowitz, C., Herzog, M., & Leitner, L. (2013). Stemming the tide of plastic marine litter: A global action agenda. *Pritzker Environmental Law and Policy Briefs, 5*, UCLA. pp. 24.

Goldstein, M. C., Carson, H. S., & Eriksen, M. (2014). Relationship of diversity and habitat area in North Pacific plastic-associated rafting communities. *Marine Biology, 161*, 1441–1453.

Hall, K. (2000). *Impacts of marine debris and oil: Economic and social costs to coastal communities*. Kommunenes Internasjonale Miljøorganisasjon (KIMO), Lerwick, U.K. pp. 86 website: http://www.kimointernational.org/WebData/Files/Karensreport.pdf.

Hardesty, B., Wilcox, C., Lawson, T., Lansdell, M., & van der Velde, T. (2014). Understanding the effects of marine debris on wildlife. A final report to Earthwatch Australia. CSIRO, Australia. Available from: https://publications.csiro.au/rpr/download?pid=csiro:EP147352&dsid=DS1.

HELCOM. (2012). Annual 2011 HELCOM report on illegal discharges observed during aerial surveillance. Retrieved December 2, 2014, from HELCOM (Baltic Marine Environment Protection Commission—Helsinki Commission) website: http://helcom.fi/Lists/Publications/HELCOM%20Report%20on%20illegal%20discharges%20observed%20during%20aerial%20surveillance%20in%202011.pdf.

Hidalgo-Ruz, V., Thiel, M. (2015). The contribution of citizen scientists to the monitoring of marine litter. In M. Bergmann, L. Gutow, & M. Klages (Eds.), *Marine anthropogenic litter* (pp. 433–451), Berlin: Springer.

Holt, R. (2009). *The carpet sea squirtDidemnum vexillum: eradication from Holyhead Marina*. Presentation to the Scottish Natural Heritage Conference 'Marine Non-native Species: Responding to the threat', 27 Oct 2009. Battleby, U.K.

Ikonen, M. (2012). *No-special-fee system for ships in the Baltic Sea ports*. Paper Presented at Joint Workshop on No-special-fee System to Ship-Generated Wastes in the Baltic Sea Area, Copenhagen/Malmö. Retrieved December 9, 2014 from http://www.baltic.org/files/2338/Mirja_Ikonen_5_Nov_2012.pdf.

Ikonen, M. (2013). *No-special-fee system for ships in the Baltic Sea ports*. Paper Presented at International Conference on Prevention and Management of Marine Litter in European Seas, Berlin. Retrieved December 9, 2014 from http://www.marine-litter-conference-berlin.info/userfiles/file/online/No-special-fee%20system%20for%20ships%20in%20the%20Baltic%20Sea%20ports_Ikonen.pdf.

International Coastal Cleanup. (ICC). (2013). *Summary report, summary card and items collected*. Retrieved September 5, 2014 from http://dinrac.nowpap.org/ICC_Results.htm.

Jang, Y. C., Hong, S., Lee, J., Lee, M. J., & Shim, W. J. (2014). Estimation of lost tourism revenue in Geoje Island from the 2011 marine debris pollution event in South Korea. *Marine Pollution Bulletin, 81*, 49–54.

Kanehiro, H. (2012). *Global problem of marine pollution by plastic litter*. Paper presented at Expert Consultation Meeting on Environmental Challenges Related to Transboundary Marine

Pollution, Seoul. Retrieved February 10, 2014 from http://www.neaspec.org/sites/default/files/Session1_Marine_litter_KANEHIRO_0.pdf.

Kiessling, T., Gutow, L., & Thiel, M. (2015). Marine litter as a habitat and dispersal vector. In M. Bergmann, L. Gutow, & M. Klages (Eds.), *Marine anthropogenic litter* (pp. 141–181). Berlin: Springer.

KIMO International. (2013). *Fishing for litter*. Retrieved December 8, 2014 from http://www.kimointernational.org/FishingforLitter.aspx.

Kirkley, J. & McConnell, K. E. (1997). Marine debris: Benefits, costs and choices. In J. M. Coe, & D. B. Rogers (Eds.), Marine debris: sources, impacts, and solutions, New York: Springer, pp. 171–185.

Kühn, S., Bravo Rebolledo, E. L., & van Franeker, J. A. (2015). Deleterious effects of litter on marine life. In M. Bergmann, L. Gutow, & M. Klages (Eds.), *Marine anthropogenic litter* (pp. 75–116). Berlin: Springer.

Larsen, J. & Venkova, S. (2014). Plastic bag bans spreading in the United States. Retrieved September 5, 2014 from Earth Policy Institute Web site: http://www.earth-policy.org/plan_b_updates/2014/update122.

Leggett, C., Scherer, N., Curry, M. & Bailey, R. (2014). Assessing the economic benefits of reductions in marine debris: A pilot study of beach recreation in Orange County, California. Final report: June 15, 2014, from National Oceanic and Atmospheric Administration. Cambridge, USA. pp. 44. http://marinedebris.noaa.gov/sites/default/files/MarineDebrisEconomicStudy.pdf.

Lippsett, L. (2013). Behold the 'plastisphere'. *Oceanus Magazine*, 50(2) Woods Hole Oceanographic Institution. Retrieved October 27, 2014 from http://www.whoi.edu/oceanus/feature/behold-the-plastisphere.

Lusher, A. (2015). Microplastics in the marine environment: distribution, interactions and effects. In M. Bergmann, L. Gutow, & M. Klages (Eds.), *Marine anthropogenic litter* (pp. 245–308). Berlin: Springer.

MA. (2005). *General synthesis report*, World Resources Institute, Washington, DC. pp. 137, www.millenniumassessment.org.

Macfadyen, G., Huntington, T. & Cappell, R. (2009). *Abandoned, lost or otherwise discarded fishing gear*. UNEP Regional Seas Reports and Studies No. 185; FAO Fisheries And Aquaculture Technical Paper No. 523. UNEP/FAO, Rome. pp. 88.

McIlgorm, A., Campbell, H. F., & Rule, M. J. (2011). The economic cost and control of marine debris damage in the Asia-Pacific region. *Ocean and Coastal Management, 54*, 643–651.

Meng, Q. & Chen, H. (2013). *Marine litter management in China*. Paper Presented at NOWPAP ICC, Okinawa. Retrieved February 10, 2014 from http://dinrac.nowpap.org/document-ICC-2013.php.

Miller, R. M. (2012). Plastic shopping bags: An analysis of policy instruments for plastic bag reduction. Thesis, University of Utrecht, pp. 66. Retrieved December 9, 2014 from http://igitur-archive.library.uu.nl/student-theses/2012-0828-200707/Thesis-%20writing.pdf.

Mouat, J., Lozano, R. L. & Bateson, H. (2010). *Economic Impacts of marine litter*. KIMO International, pp. 105. Retrieved November 29, 2013 from http://www.seas-at-risk.org/1mages/Economic%20impacts%20of%20marine%20litter%20KIMO.pdf.

National Litter Pollution Monitoring System, Ireland. (2013). *Litter monitoring body system results2012*. Dublin, p. 27. Retrieved November 29, 2013 from http://www.litter.ie/Reports/Systems%20Survey%20Report%202012.pdf.

National Research Council. (2008). *Tackling marine debris in the 21st century*. Washington, DC: National Academies Press, pp. 224.

New Zealand Ministry for the Environment. (2013). *Waste disposal levy FAQs*. Retrieved November 29, 2013 from http://www.mfe.govt.nz/issues/waste/waste-disposal-levy/faq.html.

New Zealand Ministry for the Environment. (2014). *Review of the effectiveness of the waste disposal levy, 2014 in accordance with section 39 of the Waste Minimisation Act 2008*. Retrieved September 5, 2014 from http://www.mfe.govt.nz/publications/waste/waste-disposal-levy-review/waste-disposal-levy-review-2014-pdf.pdf.

North West Pacific Action Plan Data and Information Network Regional Activity Center (NOWPAP DINRAC). (2014). International Coastal Cleanup (ICC) Summary Report,

Summary Card and Items Collected. Retrieved February 11, 2014 from http://dinrac.nowpap. org/MarineLitter.php?page=ICC_results.

Ocean Conservancy. (2014). *Turning the tide on trash: 2014 Report*. Retrieved September 5, 2014 from http://www.oceanconservancy.org/our-work/marine-debris/icc-data-2014.pdf.

Ocean Conservancy. (2013). *International coastal cleanup 2012: Ocean trash index*. Retrieved February 11, 2014 from http://www.oceanconservancy.org/our-work/international-coastal-cleanup/2012-ocean-trash-index.html.

Ocean Conservancy. (2012). *International coastal cleanup: 2012 data release*. Retrieved February 11, 2014 from http://www.oceanconservancy.org/our-work/marine-debris/check-out-our-latest-trash.html.

Ocean Conservancy. (2007). *Nationalmarine debris monitoring program: Final program report, data analysis and summary*. Retrieved August 27, 2014 from http://www.unep.org/regionalseas/marinelitter/publications/docs/NMDMP_REPORT_Ocean_Conservancy__2_.pdf.

Øhlenschlæger, J. P., Newman, S. & Farmer, A. (2013). *Reducing ship generated marine litter— Recommendations to improve the EU port reception facilities directive*. Report produced for Seas At Risk. Institute for European Environmental Policy, London.

OSPAR Commission. (2007). Guidelines on how to develop a fishing-for-litter project. Reference number: 2007–10.

Ritchie, M. (2014). *The state of waste 2014*. ResourceRecovery.biz, April 16, 2014. Retrieved September 5, 2014 from http://www.resourcerecovery.biz/features/state-waste-2014.

Rochman, C. M. (2015). The complex mixture, fate and toxicity of chemicals associated with plastic debris in the marine environment. In M. Bergmann, L. Gutow, & M. Klages (Eds.), *Marine anthropogenic litter* (pp. 117–140). Berlin: Springer.

Scottish Government. (2013). *Marine litter strategy, national litter strategy: Strategic environmental assessment environmental report*. Retrieved September 5, 2014 from http://www.scotland.gov.uk/Publications/2013/07/9297/5.

Sheavly, S. B., & Register, K. M. (2007). Marine debris and plastics: Environmental concerns, sources, impacts and solutions. *Journal of Polymers and the Environment, 15*, 301–305.

TEEB. (2010). The economics of ecosystems and biodiversity: ecological and economic foundations. In P. Kumar (Ed.), *Earthscan*, London.

TEEB (2011) The economics of ecosystems and biodiversity (TEEB) in national and international policy making an output of TEEB. In P. ten Brink (Ed.), *Earthscan*, IEEP: London.

ten Brink, P., Lutchman, I., Bassi, S., Speck, S., Sheavly, S., Register, K., et al. (2009). Guidelines on the use of market-based instruments to address the problem of marine litter. Institute for European Environmental Policy (IEEP), Brussels, Belgium, and Sheavly Consultants, Virginia Beach, Virginia, USA.

Teuten, E. L., Saquing, J. M., Knappe, D. R. U., Barlaz, M. A., Jonsson, S., & Björn, A. (2009). Transport and release of chemicals from plastics to the environment and to wildlife. *Philosophical Transactions of the Royal Society B, 364*(1527), 2027–2045.

Thompson, R. C., Moore, C. J., vom Saal, F. S., & Swan, S. H. (2009). Plastics, the environment and human health: current consensus and future trends. *Philosophical Transactions of the Royal Society B, 364*, 2153–2166.

UNEP (2014). UNEP year book: Emerging issues in our global environment. Chapter 8: Plastic debris in the ocean. United Nations Environment Programme, Nairobi, Kenya. pp. 49–53.

WALGA (2012). *Background paper: Landfill levy*. Retrieved September 5, 2014 from http://www.wastenet.net.au/Assets/Documents/Content/Information/Background_Paper_Levy_Final_amended_March_2012.pdf.

Wallace, B. (1990). How much do commercial and recreational fishermen know about marine debris and entanglement? Part 1. In R. S. Shomura, M. L. Godfrey, (Eds.), *Proceedings of the Second International Conference on Marine Debris* April 2–7, 1989, Honolulu, Hawaii. NOAA-TM-NMFS-SWFSC-154. Washington, DC: Dept of Commerce, National Oceanic and Atmospheric Administration, National Marine Fisheries Service; pp. 1140–1148.

Wastenet. (2014). *Landfill levy*. Retrieved September 5, 2014 from http://www.wastenet.net.au/landfill-levy.aspx.

Watkins, E., Dominic Hogg, D., Mitsios, A., Mudgal, S., Neubauer, A., Reisinger, H., et al. (2012). Use of economic instruments and waste management performances: Final report. http://ec.europa.eu/environment/waste/pdf/final_report_10042012.pdf.

Welsh Government. (2012). Reduction in single-use carrier bags. Retrieved November 29, 2013 from http://wales.gov.uk/topics/environmentcountryside/epq/waste_recycling/substance/carrierbags/reduction/?lang=en.

WRAP. (2013). *UK voluntary carrier bag monitoring—2013*. Retrieved February 11, 2014 from http://www.wrap.org.uk/sites/files/wrap/Carrier%20bags%20results%20%282012%20data%29.pdf.

Chapter 15
Regulation and Management of Marine Litter

Chung-Ling Chen

Abstract This chapter aims to provide an overview of the regulation and management instruments developed at international, regional and national levels to address marine litter problems, put forward the potential gaps in the existing management body and suggest solutions. While not covering the gamut of all relevant instruments, a number of existing instruments, including specific management measures contained therein, were profiled as illustration. The management measures illustrated are either on a mandatory or voluntary basis and provide a general, snapshot picture of the management framework of marine litter. They can be broadly divided into four categories: preventive, mitigating, removing and behavior-changing. The preventive and behavior-changing measures are particularly important in addressing marine litter at its root. The former schemes include source reduction, waste reuse and recycling, containing debris at points of entry into receiving waters and land-based management initiatives (e.g. restriction of the use of plastic bags, establishment of extended producer responsibility). The latter schemes aid people's engagement in the other three types of measures, including education campaigns and activities raising awareness (e.g. Fishing for Litter). The potential gaps include limits of existing instruments in addressing plastic marine litter, deficiencies in the legislation and a lack of enforcement of regulations, poor cooperation among countries on marine litter issues and insufficient data on marine litter. To fill these gaps, recommendations are proposed, including establishment of a new international instrument targeted to the plastic marine litter problem, amending existing instruments to narrow exceptions and clarify enforcement standards, establishing national marine litter programe, enhancing participation and cooperation of states with regard to international/regional initiative, and devising measures to prevent marine litter from fishing vessels.

C.-L. Chen (✉)
Institute of Ocean Technology and Marine Affairs,
Department of Hydraulic and Ocean Engineering,
National Cheng Kung University, Tainan City, Taiwan
e-mail: chungling@mail.ncku.edu.tw

© The Author(s) 2015
M. Bergmann et al. (eds.), *Marine Anthropogenic Litter*,
DOI 10.1007/978-3-319-16510-3_15

395

Keywords Marine litter · Management · Regulation · Plastic · Source reduction · Behavior-changing

15.1 Introduction

Marine litter (also called marine debris) has long been on the political and public agenda. It is recognized as a worldwide rising pollution problem affecting all the oceans and coastal areas of the world (Galgani et al. 2015; Ryan 2015; Thompson 2015). The increasing production and use of durable synthetic materials such as plastics[1] has led to a gradual, but significant accumulation of litter in the marine environment, making it ever more difficult to tackle (Barnes et al. 2009; Kühn et al. 2015). Moreover, the high-profile reports of garbage patches found in the North Pacific and North Atlantic regions (Pichel et al. 2007; Law et al. 2010; Howell et al. 2012) further propel an intensified international drive to address the ongoing problem of marine litter. Indeed, the model simulations suggest that debris accumulates in a number of convergence zones or gyres where they remain for many years (UNEP 2013).

Marine litter is defined as "any persistent, manufactured or processed solid material discarded, disposed of or abandoned in the marine and coastal environment" (UNEP 2005, 2009). It is largely associated with diverse human activities occurring both on land and at sea, and is concomitant with the increasing use of synthetic materials, industrialization and urbanization of coastal areas, and inadequate disposal practices. Generally it can be said that the problem of marine litter is rooted in the prevailing production and consumption pattern and the way we dispose of and manage waste. Marine litter originates from three main sources: land-based, riverine and ocean-based sources (Galgani et al. 2015; Browne 2015; Jambeck et al. 2015). The former include public littering, poor waste management practices, industrial activities, sewage related debris and storm water discharge, all of which can be transported via rivers (Morritt et al. 2014; Free et al. 2014; Hoellein et al. 2014). The latter include fishing activities, shipping, marine leisure industry, and offshore oil and hydrocarbon industries (Mouat et al. 2010). In particular, derelict fishing gear[2] has become a serious concern with the intensified fishing effort in the world's oceans and the increasing durability of fishing gear (Macfadyen et al. 2009; Bilkovic et al. 2014).

It is widely documented that marine litter has a wide range of adverse environmental, economic, social and public health and safety impacts (Newman et al. 2015). They are illustrated by marine litter injuring or killing wildlife by ingestion and/or entanglement (Jones 1995; Bugoni et al. 2001; Donohue et al. 2007; Allen et al. 2012; Bond et al. 2013; Baulch and Perry 2014; Kühn et al. 2015), altering ecosystems by introducing non-native species (Barnes 2002; CBD 2012; Kiessling et al. 2015),

---

[1]Since 1950, global plastics production has continued the growth pattern by 9 % per annum. From 1.7 million t in 1950, total global production reached 288 million t in 2012 (PlasticsEurope 2013).

[2]Derelict fishing gear is often referred to ALDFG, which is a collective term for fishing gear that has been abandoned, lost or otherwise discarded (Macfadyen et al. 2009).

threatening sensitive habitats (e.g. corals, salt marsh) by moving along the seabed (derelict fishing gear) (Donohue et al. 2001; Arthur et al. 2014), posing risks to human health and safety (e.g. hazards to navigation) (Taylor et al. 2014), entailing economic costs to coastal towns/communities, fisheries, tourism, and other maritime industries (Ballance et al. 2000; Mouat et al. 2010; Jang et al. 2014; Newman et al. 2015). For instance, the total number of turtles entangled by the 8,690 derelict fishing nets sampled in northern Australia was estimated to be between 4,886 and 14,600 (Wilcox et al. 2014). The estimate of damage cost from marine litter across the 21 Pacific Rim economics is €949 million annually in total, €273 million for the fishing industry, €209 million for the shipping industry and €467 million for marine tourism (Mcllgorm et al. 2011). In addition to these negative impacts, there is a growing concern about microplastics as they increase the risk of plastics entering food webs (Lusher 2015). If ingested microplastics have the potential to transfer toxic substances to the food chain, posing a threat to the health of humans and ecosystems (Teuten et al. 2009; Thompson et al. 2009; Rochman 2015).

To minimize the negative impacts, a plethora of instruments has been developed at international, regional and national levels to prevent, reduce and manage marine litter. They represent a wide range of international, regional and national efforts devoted to combat marine litter. The goal of this article is to provide an overview of these instruments, to identify the potential gaps in the existing management body and suggest solutions.

As it is impossible and impractical to cover the gamut of all relevant instruments in detail within the scope of this chapter, I will first consider the general mechanisms of the instruments and refer to specific ones as illustration when appropriate. This approach has the advantage of providing a general, snapshot picture of the management framework of marine litter, while also laying out the specifics of certain instruments, including the management measures contained therein. It should also be noted that marine litter is an issue of, or related to, broader topics, such as marine environmental protection, changes in biodiversity, rafting of invasive species, water quality and hazardous waste, waste and sewage water management as well as eco design and producer responsibility. The instruments addressing these broader issues would also be applicable to marine litter, although not specifically mentioned. However, as such instruments are large in scope and may not encompass the specifics of marine litter management, I will focus on those that specifically address marine litter.

15.2 Instruments of Marine Litter at International, Regional and National Levels

15.2.1 General Mechanisms of Instruments

As previously mentioned, a large number of instruments at international, regional and national levels have been adopted to tackle marine litter problems. These instruments comprise conventions, agreements, regulations, strategies, action plans, programs and guidelines. They contain specific management measures that are either compulsory or voluntary.

There are two basic types of instruments at the international level, in terms of their connection with regional or national instruments. The first comprises those, which are explicitly transposed into regional or national ones, usually in the form of regional agreements or national legislations. Similar texts can also be found in the instruments at the regional or national level. Examples include international instruments such as Annex V[3] of MARPOL 73/78,[4] the London Protocol and the Action Plan on tackling the inadequacy of port reception facilities (PRFs). The corresponding regional or national instruments transposed from international ones include: the European Union (EU) PRF Directive, the Annex IV of the Helsinki Convention, the United States (US) Marine Plastic Pollution Research and Control Act, the United Kingdom (UK) Merchant Shipping (Prevention of Pollution by Sewage and Garbage from Ships) Regulations 2008, and various other national legislations. The second type comprises instruments, which are not explicitly transposed into regional or national schemes. These instruments mostly serve as global guiding instruments encouraging regional bodies or countries to follow the actions proposed therein, or as a platform for the states concerned to engage in coordination and cooperation in marine litter issues. The most prominent examples are perhaps a series of initiatives developed by the United Nations Environment Programme (UNEP), including the Regional Sea Programme (RSP), Guidelines on survey and monitoring of marine litter, Guidelines on the use of market-based and economic instruments and the Honolulu Strategy.

As for the instruments at the regional or national level that lack a clear link traced back to international instruments, they are devised by their own respective regional bodies or nations to deal with marine litter problems. These instruments usually consist of regional agreements, regional or national programs, legislations, or activities dealing with specific aspects of marine litter problems. Examples include the Barcelona Convention, the Guideline for monitoring marine litter on the beaches in the OSPAR[5] Maritime Area, the EU Marine Strategy Framework Directive, the CCAMLR[6] Marine Debris Program, the US National Marine Debris Program, numerous coastal cleanup activities, and various national legislations relevant to marine litter.

---

[3]Regulations for the Prevention of Pollution by Garbage from Ships.

[4]International Convention for the Prevention of Marine Pollution from Ships, 1973 as modified by the Protocol of 1978, known as MARPOL 73/78.

[5]Commission for the Protection of the Marine Environment of the Northeast Atlantic.

[6]Commission for the Conservation of Antarctic Marine Living Resources.

15.2.2 Examples of Instruments on Marine Litter

This section presents examples of instruments at international, regional and national levels to illustrate the current regulation and management of marine litter.

15.2.2.1 International Instruments

United Nations Convention on the Law of the Sea (UNCLOS)

The UNCLOS is one of the most important agreements related to the use of the oceans. The convention entered into force in 1994 and comprises 320 articles and nine annexes. It established a comprehensive regime for the law of the sea by governing all aspects of the oceans from geopolitical delimitations to environmental control, scientific research, economic and commercial activities, technology and the settlement of disputes relating to ocean matters (Roberts 2010). In particular, articles 192–237 of Part XII are dedicated to the protection and preservation of the marine environment. While the provisions do not explicitly refer to marine litter, they place a general obligation on states to protect and preserve the marine environment, which can be used in the context of marine litter regulation.

Annex V of MARPOL 73/78

Annex V of MARPOL 73/78 is the major international instrument addressing ocean-based litter pollution from ships and was developed under the auspices of the international Maritime Organization (IMO). Annex V was recently revised in 2011 and came into force in 2013. The revised Annex V provides an updated framework for the control of garbage generated by ships. It imposes a general ban on discharges of all garbage from ships at sea, except for a few clearly defined circumstances.[7] These circumstances are associated with the types of garbage that can be disposed of, specifications of the distances from the coast, discharge of garbage within or outside special areas,[8] the manner in which they may be disposed of, and en route requirements for allowable discharge.[9] The updated disposal regulations

---

[7]Revised Annex V, reg. 3.

[8]Revised Annex V, reg 1: Special areas refer to a sea area where for recognized technical reasons in relation to its oceanographic and ecological condition and to the particular character of its traffic the adoption of special mandatory methods for the prevention of sea pollution by garbage is required. The special areas of Annex V are the Mediterranean Sea, the Baltic Sea, the Red Sea, the Gulfs area, the North Sea, Antarctica and the Wider Caribbean.

[9]Revised Annex V, reg 1.: En route means that the ship is underway at sea on a course or courses, including deviation from the shortest direct route, which as far as practicable for navigational purposes, will cause any discharge to be spread over as great an area of the sea as is reasonable and practicable.

are summarized in Table 15.1. Other major changes include expanding the requirements for placards and garbage management plans to fixed and floating platforms,[10] and reduction of the minimum tonnage limit for garbage management plans from 400 gross tonnage (GT) to 100 GT.[11]

Major provisions remaining unchanged include: the obligation to provide a Garbage Record Book (GRB) for ships ≥400 GT or ships certified to carry ≥15 persons,[12] and the provision of adequate reception facilities at ports without causing undue delay to ships.[13] A GRB is to record each discharge made at sea or a reception facility, or a completed incineration, including date, time, ship position, category of the garbage and the estimated amount discharged or incinerated.[14] The GRB is subject to inspection by the competent authority of a party to MARPOL 73/78 when the ship is in port.[15]

London Protocol

The London Protocol (LP) is a major instrument dealing with dumping of wastes and other matter at sea. The discharge of garbage during normal operations as regulated in the Annex V of MARPOL 73/78 is not considered as dumping.[16] In 1996, the protocol was adopted to further modernize the 1972 London Convention[17] and eventually replace it. The protocol entered into force in 2006. While the goal of the 1972 convention is to regulate pollution by dumping, the goal of the Protocol is to stop waste dumping at sea (Louka 2006). Namely, the protocol is more restrictive in regulating wastes dumping than the 1972 convention by introducing a reverse listing approach. This approach is, in essence, to prohibit the dumping of any wastes or other matter except for the materials listed in Annex I.[18] Dumping of these materials (such as dredged material, sewage sludge, fish wastes, vessels and platforms, inert, inorganic geological material) requires a permit and parties shall adopt measures to ensure that the issuance of permits and permit conditions comply with Annex II.[19] In addition, the protocol prohibits incineration of wastes at sea and the export of wastes to countries for dumping or

---

[10]Revised Annex V, reg. 10.1.

[11]Revised Annex V, reg. 10.2.

[12]Revised Annex V, reg. 10.3.

[13]Revised Annex V, reg. 8.1 The relevant regulations on port reception at ports are also seen in Annex I, II, IV, and VI.

[14]Revised Annex V, reg. 10.3.1 and 10.3.2.

[15]Revised Annex V, reg. 10.5.

[16]LP, reg. art. 1.4.2.

[17]Convention on the Prevention of Marine Pollution by Dumping of Wastes and Other Matter.

[18]London Protocol, art. 4.1.1.

[19]London Protocol, art. 4.1.2.

Table 15.1 Summary of discharge provisions of the revised MARPOL Annex V

Type of garbage	Ships outside special areas[a]	Ships within special areas[a]	Offshore platforms and all ships within 500 m of such platforms
Food wastes comminuted or ground[b]	Discharge permitted ≥3 nm from the nearest land and en route	Discharge permitted ≥12 nm from the nearest land and en route	Discharge permitted ≥12 nm from the nearest land
Food wastes not comminuted or ground	Discharge permitted ≥12 nm from the nearest land and en route	Discharge prohibited	Discharged prohibited
Cargo residues[c] not contained in wash water	Discharge permitted ≥12 nm from the nearest land and en route	Discharge prohibited	Discharge prohibited
Cargo residues[c] contained in wash water	Discharge permitted ≥12 nm from the nearest land and en route	Discharge only permitted in specific circumstances[d] and ≥12 nm from the nearest land and en route	Discharge prohibited
Cleaning agents and additives[c] contained in cargo hold wash water	Discharge permitted	Discharge only permitted in specific circumstances[d] and ≥12 nm from the nearest land and en route	Discharge prohibited
Cleaning agents and additives[c] contained in deck and external surfaces wash water	Discharge permitted	Discharge permitted	Discharge prohibited
Animal carcasses	Discharge permitted as far from the nearest land as possible and en route	Discharge prohibited	Discharge prohibited
All other garbage including plastics, domestic wastes, cooking oil, incinerator ashes, operational wastes, and fishing gear	Discharge prohibited	Discharge prohibited	Discharge prohibited
Mixed garbage	When garbage is mixed with or contaminated by other substances prohibited from discharge or having different discharge requirements, the more stringent requirements shall apply		

Source Resolution MEPC.201(62) Amendments to the Annex of MARPOL 73/78 (entered into force on 1 January 2013)

Note

[a] According to reg. 1.14, special areas are the Mediterranean Sea area, the Baltic Sea area, the Black Sea area, the Red Sea area, the Gulfs area, the North Sea area, the Antarctica area and the Wider Caribbean Region

[b] According to reg. 4.1.1, 5.2, 6.1.1, comminuted or ground food wastes shall be capable of passing through a screen with openings no greater than 25 mm

[c] These substances must not be harmful to the marine environment

[d] According to reg. 6.1.2, the discharge shall only be allowed if: (a) both the port of departure and the next port of destination are within the special area and the ship will not transit outside the special area between these ports; and (b) if no adequate reception facilities are available at those ports

incineration at sea.[20] The protocol is to supersede the convention for the state parties that ratified it and will eventually replace the convention as more and more parties ratify.

Action Plan on Tackling the Inadequacy of PRFs

In 2006, the Marine Environment Protection Committee of the IMO approved the Action Plan on tackling the inadequacy of PRFs. The plan was developed to contribute to the effective implementation of MARPOL 73/78 and to promote quality and environmental consciousness among administrations and the shipping industry. It covers standardized reporting, information on PRFs, equipment technology, types and amount of wastes, regulatory matters, technical cooperation and assistance.[21]

UNEP Regional Sea Programme

The UNEP Regional Sea Programme and Global Programme of Action (GPA[22]) embarked in 2003 on the development of a Global Initiative on Marine Litter. This initiative has succeeded in organizing and implementing regional activities on marine litter around the world. Activities focusing on managing marine litter were arranged through individual agreements in 12 Regional Seas.[23] The main activities include: a review and assessment of the status of marine litter in the region, organization of a regional meeting of national authorities and experts on marine litter, preparation of a regional action plan for the management of marine litter, and participation in a regional cleanup day within the framework of the International Coastal Cleanup Campaign.[24] This regional initiative also provides a platform for the establishment of partnerships, cooperation and coordination of activities for the

---

[20]London Protocol, art. 5 and 6.

[21]Further information on this Plan is available at www.imo.org/ourwork/environment/pollutionpr evention/portreceptionfacilities.

[22]Global Programme of Action for the Protection of the Marine Environment from Land-Based Activities. The GPA, adopted in 1995, is a programme that addresses the impacts of land-based sources and activities on coastal and marine environment and human well-being. Litter is one of nine source categories of the GPA and as such is important for its implementation (UNEP 2009).

[23]Baltic Sea, Black Sea, Caspian, East Asian Seas, Eastern Africa, Mediterranean, Northwest Pacific, Northeast Atlantic, Red Sea, Gulf of Aden, South Asian Seas, Southeast Pacific and Wider Caribbean.

[24]International Coastal Cleanup (ICC) is the world's largest volunteer effort to clean up beaches and waterways, with its many global public and private partners. The ICC is organized by Ocean Conservancy (a US-based NGO) and has been operating since 1986. It annually hosts cleanup activities around the world. In 2012, the ICC mobilized >560,000 volunteers to clean coastal beaches and inland waterways in 97 countries and locations, and a total of 4.5 million kg of trash were collected on the shoreline of 28,485 km (Ocean Conservancy 2013).

control and sustainable management of marine litter. The main partners include Regional Sea Conventions and Action Plans, government representatives, UN agencies, relevant bodies, donor agencies, the private sector and NGOs (UNEP 2009).

UNEP/IOC Guidelines on Surveying and Monitoring of Marine Litter

The UNEP developed, in cooperation with the intergovernmental Oceanographic Commission (IOC), guidelines on surveying and monitoring of marine litter in order to provide a long-term platform for scientific monitoring. Four sets of operational guidelines were developed: comprehensive assessments of beach, benthic and floating litter, and rapid assessments of beach litter. The first three sets target the collection of highly resolved data to support the development and/or evaluation of mitigation strategies, while the last aims to raise public awareness of and educate about marine litter issues (Cheshire et al. 2009).

UNEP Guidelines on the Use of Market-Based and Economic Instruments

The UNEP developed guidelines on the use of market-based and economic instruments. This report serves as a practical reference to decision makers on how to select, apply and implement related economic tools. Tools include deposit-refund programs on plastic and glass bottles, plastic bag tax, incentives to fishers for reporting and removing debris, subsidies, tourist taxes, car park fees, and waterfront business charges (Ten Brink et al. 2009; Newman et al. 2015).

UNEP/FAO Abandoned, Lost or Otherwise Discarded Fishing Gear

A report commissioned by the UNEP and Food Agriculture Organization (FAO) identified reasons for fishing gear being abandoned, lost or otherwise discarded, reviewed existing measures to reduce derelict fishing gear, and proposed recommendations for future action (Macfadyen et al. 2009). A variety of existing measures have been presented, including gear marking, port-state measures,[25] onshore collection, payment for retrieved gear, better locating and reporting lost gear, disposal and recycling, and awareness raising schemes.

Honolulu Strategy

The UNEP and the US National Oceanic and Atmospheric Administration (NOAA) co-organized the Fifth International Marine Debris Conference in 2011, where the Honolulu Strategy was formulated. This strategy can be regarded as a global

---

[25]Port state measures help to address illegal, unregulated and unreported (IUU) fishing, which is a significant contributor to derelict fishing gear problems.

framework on possible actions to combat marine litter. It contains three goals, 19 strategies and numerous specific actions, serving as a useful and practical reference for concerned parties to take actions at national levels (UNEP/NOAA 2011).

UNEP Global Partnership of Marine Litter

The most recent initiative was to establish a Global Partnership of Marine Litter (GPML) in June 2012 by the UNEP. The GPML builds on the Honolulu Strategy. It is a global partnership, acting as a "coordinating forum" for all stakeholders (international, regional, national and local organizations) working in the area of marine litter prevention and management. The forum assists stakeholders to complement each other's efforts, to avoid duplication and to optimize the efficiency and efficacy of their resources.[26]

15.2.2.2 Regional Instruments

EU PRF Directive

In response to MARPOL 73/78, which requires party states to ensure the provision of adequate PRFs, the EU adopted the Port Reception Facility (PRF) Directive aimed at reducing the input of ship-generated waste to the sea. The directive came into force in 2002 and key requirements include: member states are obliged to ensure the availability of PRFs to meet the needs of ships, ports to develop and implement a waste reception and handling plan, a reporting requirement for the master of a ship regarding the delivery of waste, implementation of a cost-recovery system, and establishment of an enforcement scheme (EU 2000). A study by the European Maritime Safety Agency (EMSA) shows that there was an increase in the-total delivery from 2004 to 2008 for oily waste and from 2004 to 2009 for garbage for European ports and the decrease, experienced in 2009 and 2010, for oily waste and garbage, respectively, is thought to be a result of the financial crisis and thus a decrease in the number of calls to the ports (ship/cargo traffic) (EMSA 2012).

EU Marine Strategy Framework Directive

Multiple initiatives exist to tackle marine debris in the EU. Among them, perhaps the most relevant is the Marine Strategy Framework Directive (MSFD) (EU 2008), the environmental pillar of the EU Integrated Maritime Policy. This directive is an integral policy instrument for the protection of the marine environment for the European Community, following an ecosystem-based, adaptive and integrated

---

[26]Further information on the GPML is available at www.gpa.unep.org/index.php/global-partnership-on-marine-litter.

approach to the management of human activities, which have an impact on the marine environment. The directive establishes a framework, within which member states shall take necessary measures to achieve or maintain good environmental status (GES) in the marine environment by 2020.[27] Marine litter is listed as the tenth of 11 qualitative descriptors for determining GES, which states that the properties and quantities of marine litter do not cause harm to the coastal and marine environment.[28]

To achieve GES, each member state should define GES as well as environmental targets and put in place its own marine strategy to protect its waters. In relation to this, two criteria and associated indicators for marine debris that define GES have been identified, serving as a reference for member states to follow. One criterion is characteristics of litter in the marine and coastal environment, and the associated indicators are trends in the amount of litter on beaches, in the water column and on the seafloor as well as trends in the amount, distribution and where possible, composition of microparticles (particularly microplastics). The other criterion deals with the impacts of litter on marine life, and the associated indicator is marine litter taken up by marine organisms (EU 2010). Furthermore, the Technical Group on Marine Litter was established to support member states by providing technical and scientific recommendations for the implementation of MSFD requirements with regard to marine litter. The group continues to work on, among other concerns, harmonizing monitoring tools (protocols) and strategies, defining and quantifying harm to the marine environment, assessing land- and sea-based sources from which marine litter enters the sea including riverine inputs, and developing a common understanding of appropriate operational/environmental targets (Galgani et al. 2013).

EU Initiatives on Land-Based Waste Management

The EU has a wide range of initiatives on land-based waste management, which could have a significant impact on the amount of waste in the marine environment. For example, the Packaging and Packaging Waste Directive outlines a range of requirements to reduce the impact of packaging waste on the environment. It contains provisions on the prevention of packaging waste, on the re-use of packaging and on the recovery and recycling of packaging waste (Interwies et al. 2013). Other initiatives include the Waste Framework Directive, the Landfill Directive and the Urban Waste Water Directive.

---

[27]The MSFD, art. 1.

[28]Annex I of the MSFD. The remaining descriptors include, to name a few, biological diversity, non-indigenous species, populations of commercially exploited fish and shellfish, eutrophication, introduction of energy.

Helsinki Convention and Its Associated Initiatives

The 1992 Helsinki Convention[29] is a regional instrument aimed at protecting the marine environment of the whole Baltic Sea area, including inland waters as well as the seawater itself and the seabed. Its Annex IV (Prevention of Pollution from Ships) contains Regulation 4 (Application of the Annexes of MARPOL 73/78) and Regulation 6 (Mandatory discharge of all wastes to a port reception facility), which can be used in the context of marine litter. According to Regulation 4 contracting parties shall apply the provisions of Annexes I–V of MARPOL 73/78. According to Regulation 6 ships shall discharge all ship-generated wastes before leaving port, which are not allowed to be discharged into the sea in the Baltic Sea in accordance with MARPOL 73/78 and the convention. In relation to this, the Commission (HELCOM[30]) has approved the strategy for PRFs for ship-generated wastes (also known as the Baltic Strategy). This strategy comprises a set of measures and regulations with the main goals to ensure ships' compliance with global and regional discharge regulations and to eliminate illegal discharges of all wastes from all ships. Over 210 PRFs are provided in ports located around the Baltic Sea. To encourage their use, a "no-special-fee" system has been designed, by which disposal fees are included in port charges (HELCOM 2012).

In addition, the Baltic Sea Action Plan adopted by the HELCOM includes an agreement to raise awareness of the negative environmental and economic effects of marine litter in the marine environment, including effects of "ghost fishing" of derelict fishing gear (BSAP 2007).

Recently, Ministerial Declaration 2013 was adopted at the HELCOM Copenhagen Ministerial Meeting. It was agreed to prevent and reduce marine litter from land- and sea-based sources, causing harmful impacts on coastal and marine habitats and species, and negative impacts on various economic sectors, such as fisheries, shipping or tourism, and decided to develop a regional action plan by 2015 at the latest with the aim of achieving a significant quantitative reduction of marine litter by 2025, compared to 2015, and to prevent harm to the coastal and marine environment (HELCOM 2013). It was specifically agreed that the regional action plan on marine litter should allow to, among others, carry out concrete measures for prevention and reduction of marine litter from its main sources, develop and test technology for removal of microplastics and nano-particles in municipal waste water treatment plants by 2020, develop common indicators and associated targets related to quantities, composition, sources and pathway of marine litter.

OSPAR Initiatives on Monitoring Marine Litter

Since 1998, OSPAR has monitored levels of beach litter (OSPAR 2010a). A pilot project (2000–2006) on monitoring marine beach litter in the OSPAR region

---

[29]Convention on the Protection of the Marine Environment of the Baltic Sea Area, known as Helsinki Convention.

[30]HELCOM is the governing body of the Helsinki Convention.

using the standardized method was conducted (OSPAR 2007). The guideline for monitoring marine litter on the beaches in the OSPAR Maritime Area was further adopted in 2010, providing practical advice, especially with standardized methodology and a photographic guide, for determining the nature and amount of litter (OSPAR 2010b).

In addition, monitoring of plastic ingestion by northern fulmar (*Fulmarus glacialis*) has been implemented by OSPAR (van Franeker et al. 2011). An Ecological Quality Objective (EcoQO) has been established that <10 % of northern fulmars should have >0.1 g plastics particles in the stomach samples of 50–100 beach fulmars from each of the 4–5 areas of the North Sea over a period ≥5 years (OSPAR 2010a). Meeting this objective would indicate a reduction of litter at sea. Between 2002 and 2006, the stomachs of 1090 beached fulmars from the North Sea were analyzed. The proportion of fulmars with >0.1 g plastic in the stomach ranged from 45 to >60 %. To meet the EcoQO, refinements may be needed on the implementation of the EU Directive on Port Reception Facilities and MARPOL Annex V, as well as specific measures on lost fisheries materials (OSPAR 2010a).

OSPAR Fishing for Litter

In 2007, OSPAR published Guidelines for the implementation of Fishing for Litter (FFL) projects in the OSPAR area. FFL has two main aims: first the physical removal of marine litter from the seabed (Fig. 15.1) and, second, to raise awareness within the fishing industry that it is not acceptable to throw litter overboard. Participating vessels are given large bags to store marine litter that collects in their nets during normal fishing activities. The concept of FFL has received support within the fishing industry with increasing numbers of vessels participating in this activity over the past seven years (OSPAR 2010a). Indeed, the 210 vessels registered for the FFL initiative in Scotland landed >700 t of marine litter at the participating harbors between 2011 and 2014 (KIMO 2014).

Fig. 15.1 OSPAR Fishing for Litter program. From *left* to *right:* catch with litter from a *Nephrops* trawler in the Clyde Sea(U.K.) (Photo: M. Bergmann); fisher from *FV Andrea* sorting litter from catch (Photo: G. Lengler, NABU, DSD); disposal of litter collected by fisher into portside Fishing for Litter container (Photo: K. Detloff, NABU)

In the light of recent weight estimates of 268,940 t of litter adrift in the oceans (Eriksen et al. 2014) this initiative could significantly help to reduce marine litter (although this figure did not include litter on the seabed). FFL initiatives are currently also realized in The Netherlands, Belgium, Germany, England, Ireland, Italy and Sweden.

Barcelona Convention

The Barcelona Convention[31] is a regional instrument aimed at protecting and promoting sustainable development of the Mediterranean marine and coastal environment. It was adopted in 1976 and amended in 1995 by the parties to the Mediterranean Action Plan (MAP).[32] Seven protocols to the convention establish the MAP legal framework and address specific aspects of conservation. The one most relevant to marine litter is the Land-based Sources and Activities Protocol (LBS Protocol). It states that parties undertake to eliminate pollution deriving from land-based sources and activities, in particular to phase out inputs of the substances that are toxic, persistent and liable to bioaccumulate listed in its Annex I,[33] including litter. In addition, the Dumping Protocol has relevance to marine litter. It states that dumping of wastes and other matter is prohibited, except for dredged material, food waste, platforms and other man-made structures, and inert geological materials.[34]

CCAMLR Marine Debris Program

The Commission for the Conservation of Antarctic Marine Resources (CCAMLR) has initiated the Marine Debris Program in its convention area. Specific measures were employed to reduce the amount of debris entering the marine system and to mitigate its impacts. The measures include monitoring marine debris, addressing the risk associated with entanglement of marine mammals in plastic packaging bands and the injury to seabirds caused by the discharge of hooks in offal, and educating fishers and fishing vessel operators about the potential impact of marine debris on marine wildlife. Members annually submit information on marine debris beach surveys, debris associated with seabird colonies, entanglements of marine mammals, and seabirds and marine mammals soiled with oil.[35]

---

[31]Convention for the Protection of the Marine Environment and the Coastal Region of the Mediterranean.

[32]This MAP is the first UNEP RSP.

[33]The LBS Protocol, art. 5.1.

[34]The Dumping Protocol, art. 4.

[35]Further information on CCAMLR marine debris initiatives is available at www.ccamlr.org.

15.2.2.3 National Instruments

US Marine Plastic Pollution Research and Control Act (MPPRCA)

The MPPRCA of 1987 is the national legislation of MARPOL Annex V (UNEP 2005). The Interagency Marine Debris Coordinating Committee (IMDCC) established by this Act engages in a holistic approach to marine litter. The committee develops and recommends comprehensive and multi-disciplinary approaches to reduce the sources and adverse impacts of marine debris on the nation's marine and coastal environment, natural resources, human health, public safety and the economy. The committee consists of several stakeholder agencies,[36] ensuring that these agencies increase their coordination to address marine debris (NOAA 2012).

US Marine Debris Program

The Marine Debris Program (MDP) is a national program to investigate and solve the problems that stem from marine debris, in order to protect and conserve the nation's marine environment, natural resources, industries, economy and people. It offers a holistic approach to marine litter and was established by the Marine Debris Research, Prevention, and Reduction Act of 2006 (MDRPRA), which was amended by the Marine Debris Act Amendments in 2012. The MDP serves as a centralized capability within NOAA, supporting national and international programs to research, prevent, and reduce the impacts of marine debris, coordinating activities within NOAA and with other federal agencies, as well as using partnerships to support projects carried out by state and local agencies, tribes, NGOs, academia and industry. The MDP has sponsored numerous programs, including Fishing for Energy, international coastal cleanups, monitoring and assessment projects, and collaboration with UNEP to provide technical assistance to countries in the wider Caribbean region. Among them, the project of Fishing for Energy was launched in 2008 and provided fishers no-cost disposal service for derelict fishing gear and recycled and converted it into renewable energy (Barry 2010).[37] Until May 2014, >1.1 million kg of fishing gear were collected at rubbish bins placed in 41 communities across the country. This generated enough electricity to power 183 homes for one year (NFWF 2014).

US National Marine Debris Monitoring Program

The National Marine Debris Monitoring Program (NMDMP) was developed to standardize marine debris data collection in the US by using a scientifically valid protocol to determine marine debris status and trends. This program was conducted

---

[36]NOAA serves as the Chair.

[37]Other initiatives under the MDP are available at www.marinedebris.noaa.gov.

over a five-year period between 2001 and 2006. The results indicate that land-based sources of marine debris account for 49 % of the debris surveyed nationally, in comparison to 18 % from ocean-based and 33 % from general sources (Sheavly 2010).

US Legislations Relevant to Marine Litter

Other legislations of relevance to marine litter could have a significant impact on the amount of waste in the ocean. For example, the Shore Protection Act aims to minimize trash, medical debris, and other harmful material from being deposited into coastal waters as a result of inadequate waste handling procedures by vessels transporting waste. The Beaches Environmental Assessment and Coastal Health Act aims to reduce the risk of diseases to users of the coastal recreation waters.[38]

UK Legislations on Garbage from Ships and PRFs

In the UK, the national legislation of Annex V is the Merchant Shipping (Prevention of Pollution by Sewage and Garbage from Ships) Regulation 2008 and the Merchant Shipping and Fishing Vessels (Port Waste Reception Facilities) Regulations 2003 and amendments. The former contains provisions on garbage disposal restriction, garbage management plans and record books, inspection, detention and offences. The latter requires all ports, terminals, harbors and marinas to provide adequate reception facilities for waste and prepare a waste management plan.[39]

UK Beach Cleanup and Awareness Campaigns

Numerous cleanup and awareness campaigns have been carried out in the UK, including the Marine Conservation Society's 'Beachwatch' and 'Adopt a Beach' campaigns (MCS 2013; UNEP 2005), and the Forth Estuary Forum's Coastal Litter campaign (Storrier and McGlashan 2006).

Scotland Marine Litter Strategy and National Litter Strategy

The Scottish Government and Marine Scotland recently initiated a process to advance the Marine Litter Strategy and the National Litter Strategy to jointly manage litter in Scotland's terrestrial (including inland waters), coastal and marine environments. Both strategies were initiated in response to the MSFD, cover the

---

[38]Further information on US legislations relevant to marine litter is available at www.water.epa.gov/type/oceb/marinedebris/lawsregs.cfm.

[39]Full text of regulations is at www.legislation.gov.uk.

period 2012–2020 and seek to prevent and/or reduce the incidence of litter through a combination of approaches: education and awareness, infrastructure and tools, and enforcement and deterrence (The Scottish Government 2013).

South Korea Initiatives on Marine Litter

Since 1999, South Korea has begun to develop comprehensive and field-oriented strategies to address marine litter at the national level. Diverse initiatives were put forward, including: cleanup operations, recycling or environmentally friendly disposal of material collected, underwater marine debris removal programs, development of a practical integrated system of marine debris, river basin marine debris management systems, a fishing gear buyback program, a national coastal monitoring and education system on marine debris, and relevant legal and institutional restructuring (Jung et al. 2010). In addition, South Korea introduced a gear-marking initiative in 2006, which helps to identify owners or users of the marked fishing gear and thus contributes to preventing fisheries-related marine litter being abandoned (Macfadyen et al. 2009).

The practical integrated system started in 1999 and aimed to reduce marine litter through technological innovations in prevention, deep-water survey, removal, treatment and recycling. For example, a floating debris containment boom was developed to prevent floating debris from entering the coastal waters through rivers or channels. Deep-water survey equipment (termed "Tow-Sled") was designed to examine benthic deep-sea derelict fishing gear at depths of 500–1000 m, which was adequate for the East Sea of Korea where the steep slope of the seabed provides a suitable habitat for snow crabs (Jung et al. 2010).

The fishing gear buyback program encouraged fishers to collect fishing gear or other marine debris (excluding that generated by the fishers' own ships) during fishing by offering monetary rewards based on the amount of debris collected (Cho 2009). The program has generated desirable results: between 2004 and 2008 almost 30,000 t of litter were collected and there was an annual increase in the amount of litter collected from 2,819 t in 2004 to 8,797 t in 2008 (Noh et al. 2010). In addition, the cost of this program (€1.5 million) was less than half of the cost incurred if the same volume of litter had been collected directly by the government (€3.1 million). The coastal cleanup programe was carried out at ports and harbors, seabed areas and coastline. It has provided supplementary job opportunities for local residents (mainly senior citizens): >46,000 residents were hired as workers (Han et al. 2010).

As for legal and institutional restructuring, the "National Basic Plan for the Marine Debris Management" was institutionalized in 2008 by most of the concerned central government agencies (Jung et al. 2010). The First Basic Plan to Manage Marine Debris was established for the period from 2009 to 2013 with a budget of ca €45 billion (Jang and Song 2013). This plan is referring the Marine Environment Management Law as its legal base[40] and sets two quantitative goals: reduce the amount of marine debris annu-

---

[40]Sentence 1, Article 24 of the Law states that Minister of Maritime Affairs and Fisheries should establish and implement the plan to treat the garbage at sea, which was flown to or generated at sea.

ally entering the ocean from 159,800 t (2007) to 127,840 t (2013) and increase the collection rate from 34 % (2007) to 45 % (2013). However, a study showed that this marine debris policy is not successful in dealing with the marine debris issue since the policy focuses on collecting debris already at sea rather than preventing it from entering the ocean initially and it is almost impossible to measure the debris flow, given countless non-point sources (Jang and Song 2013).

Taiwan Legislations Relevant to Marine Litter

A comprehensive national program to assess or remediate marine litter is currently not available in Taiwan, although marine litter is pervasive along its coastline. No clear integral mechanism exists for solving marine litter problems. Regulations governing the marine litter disposal fall under the management bodies. Specifically, the Fishing Harbor Act prohibits the discharge of litter to harbor areas. The Commercial Port Act regulates waste discharges at PRFs. The Marine Pollution Control Act is the national legislation of MARPOL 73/78 and London Protocol. The act regulates that waste shall remain on board or be discharged into reception facilities, unless specific conditions apply for legal discharge. However, thus far, specific conditions have yet to be promulgated. In addition, while the authority has already transposed the revised MARPOL Annex V into national law in 15 April 2013, no penalties in breach of this rule exist. Therefore, the relevant regulations have no deterrent effects and are difficult to enforce.

Taiwan Initiatives on Land-Based Waste Management

The plastic restriction policy and the compulsory garbage sorting policy are two major initiatives on land-based waste management. These two initiatives were intended to reduce the amount of waste and have a significant impact on the reduction of the volume of plastic waste. Since 1997, Taiwan has engaged in a waste-recycling campaign by collaborating with communities, recycling enterprises, municipal trash collection teams and the recycling fund. In 2006, a compulsory nation-wide garbage sorting program was initiated to further enhance the household recycling rate.[41] The recycling rate of 38 % in 2010 was high, a 100 % increase compared to 2002 (TEPA 2010). In 2002, the government started to implement the plastic restriction policy. Measures include restrictions of the use of plastic shopping bags and disposable plastic tableware in all government agencies and public facilities (e.g. department stores, shopping centers, supermarkets, convenience stores). Within three years of this policy's implementation, the number

---

[41]The recyclable materials include iron/ aluminum/plastic containers, paper tableware, batteries, tires, lubricants, IT objects, house appliances (televisions, washing machines etc.) and light bulbs.

and the weight of plastic carrier bags were reduced by 58 and 68 %, respectively. In addition, >80 % of shoppers carried shopping bags compared to <20 % prior to the policy, indicating that this policy has initiated a behavioral change toward the use of fewer plastic bags (TEPA 2011).

Taiwan Coastal Cleanup Activities

The project of cleaning the coastal environment has been in place since 1997 with an aim to keep the coastal environment tidy, particularly the relatively populated areas, by conducting regular cleanup activities and setting up adequate reception facilities. However, this project did not involve monitoring marine debris. In general, beach litter surveys around Taiwan have been conducted by civil groups (e.g. Taiwan Ocean Cleanup Alliance) without formal long-term commitments by the government. However, the surveyed areas were limited to a few coastal locations and the survey results were not considered by relevant authorities.

15.3 Types of Management Measures to Combat Marine Litter

It should be noted that the preceding description of international, regional and national instruments tackling marine litter presents a representative snapshot of a wide range of relevant instruments, rather than an exhaustive list. While such representative information is not complete, it shows that a basic framework for addressing marine litter is in place (Fig. 15.2) and provides an overall picture of the current management measures. Based on their principle purposes, the measures can be divided into four categories: preventive, mitigating, removing and behavior-changing (Table 15.2).

15.3.1 Preventive Measures

Preventive measures focus on avoiding the generation of debris, or preventing debris from entering the sea. Measures of this type include source reduction, waste reuse and recycling, waste conversion to energy,[42] portreception facilities, gear marking, debris contained at points of entry into receiving waters and various waste management initiatives on land. Product modification and improvement (e.g. through eco design) is an important method for source reduction. A variety of

---

[42]But during this process toxins are produced and even if they are filtered the toxic filters have to be disposed of.

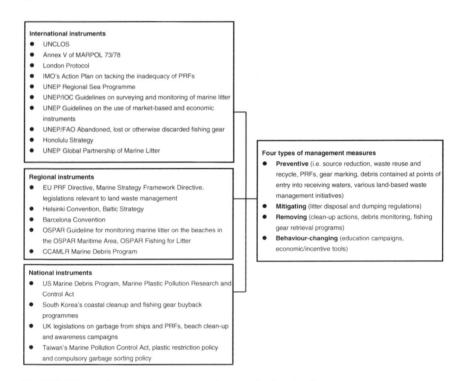

Fig. 15.2 The regulatory and management framework of marine litter

Table 15.2 Management schemes addressing marine litter

Types	Examples of measures
Preventive	Source reduction (e.g. eco design), waste reuse and recycling, waste converted to energy, port reception facilities, gear marking, debris contained at points of entry into receiving waters, various land-based waste management initiatives
Mitigating	Various debris disposal and dumping regulations, i.e. waste discharged outside certain distances from land, wastes not containing harmful substances to the marine environment allowed for discharge, prohibition of waste discharge into ecologically sensitive areas, prohibition of the disposal of certain types of garbage into seas
Removing	Beach and seafloor cleanup activities, derelict fishing gear retrieval programs, marine debris monitoring
Behavior-changing	Educational campaigns, economic/incentive tools

source reduction schemes are available, such as designing packaging such that the product can be refilled (e.g. shampoo bottles), maintaining and repairing durable products (e.g. bicycles), developing more concentrated products (e.g. laundry detergent) and electric messaging (Vaughn 2009). Other methods include the development of packaging material that is made from sustainable resources, the design of push-tap opening of metal beverage cans[43] and the design of lids of beverage bottles or containers attached to bottles with a leash (Gold et al. 2013). Restriction of the use of plastic bags is one of such measures, which is significant in the reduction of plastic waste. Bangladesh was the first nation to outlaw polythene bags in 2002 followed by Myanmar, China and a number of African countries including Eritrea, Mali, Mauritania, South Africa, Tanzania, Uganda and Kenya. What is more, the production of plastic bags has become a criminal offence in Mauritania, Mali, Somalia and Rwanda, which even searches the luggage of visitors upon arrival at its airports.

Based on the hierarchy of waste management, the strategies of preventing wastes from being formed in the first place is of paramount importance as are recycling, resource recovery and waste-to-energy approaches as less waste is generated and relatively low risks and costs are associated with waste management, compared to other strategies such as treatment and disposal (Cheremisinoff 2003). In this regard, extended producer responsibility (EPR) should be well established since it is a strategy to prevent wastes at source, promote product design for the environment and support the achievement of public recycling and materials management goals (OECD 2001) (see also Newman et al. 2015). Currently, consumers often do not have a chance to select a more environmentally friendly packaged/produced good as they are all packaged/manufactured with plastics. With EPR established, producers accept significant responsibility for the treatment or disposal of post-consumer products. It may take the form of a reuse, buy-back, or recycling program. The EU Waste Framework Directive establishes EPR and describes drivers for sustainable production taking into account the full life cycle of products (EU 2013). This directive encourages member states to take legislative or non-legislative measures in order to strengthen re-use and the prevention, recycling and other recovery operations of waste.

15.3.2 Mitigating Measures

Mitigating measures concern the ways that litter is disposed of. Methods of debris disposal are employed to minimize its adverse impact on the marine environment. These measures are largely command and control regulations, and overlap with preventive ones if they also involve preventing certain types of debris from entering the sea. Examples of such measures include prohibition of certain types of litter (e.g. plastics) discharged into seas or to coastal landfills, dumping regulations if dumping is allowed, prohibition of certain types of wastes discharged into ecologically sensitive areas, specifications of the distances from the land and of waste

---

[43]As opposed to the design of pull-tap opening, this design prevents taps from separating from beverage containers and thus the two could be retrieved together for recycling. It is noted that taps separating from cans could conveniently be thrown away and easily become marine litter items.

status for disposal (e.g. waste discharged ≥ 12 miles from the land and wastes not containing substances harmful to the marine environment), and prohibition of certain activities at sea (e.g. incineration of wastes at sea).

15.3.3 Removing Measures

Removing measures aim to remove debris already present in the marine environment. Beach cleanups are commonly employed for this but are time-consuming, costly (see Newman et al. 2015) and only capture a fraction of the overall debris. UK municipalities, for example, spend approximately €18 million each year removing beach litter, representing a 37 % increase in cost over the past decade (Mouat et al. 2010). In addition to beach cleanups, a few initiatives have employed divers to collect and monitor benthic marine debris, for example, in Hawaii (Donohue et al. 2001) and the Florida Keys (Watson 2012). In Fishing for Litter initiatives fishers remove all litter items collected during normal fishing operations and deposit them safely on the quayside to then be collected for disposal. Gear retrieval programs encourage fishers to retrieve derelict fishing gear at sea during fishing operations (e.g. Noh et al. 2010; Watson 2012). While monitoring marine debris is concerned with recording information on debris types, amounts and sources, it can be classified as removing measure since it often concomitantly involves the removal of debris. Monitoring is instrumental in devising effective management strategies to prevent specific types of litter from entering the sea. Importantly, long-term monitoring programmes enable us to assess the effectiveness of legislation and coastal management polices (Rees and Pond 1995) and have the potential to help management at individual sites and to generate large-scale pollution maps (from regional to global) to inform decision makers (Ribic et al. 2010).

15.3.4 Behavior-Changing Measures

Behavior-changing measures seek to influence behavior such that people engage in activities that help to reduce marine debris. Behavior-changing schemes are cross-cutting and aid the development and implementation of the above-mentioned three types of measures. Such schemes aim to encourage people to embrace the notion of waste as a resource and choose the products that generate lower quantities of litter (preventive), dispose of waste in a more environmentally sound way (mitigating) and participate in beach cleanups (removal). Education campaigns (Hartley et al. 2015), activities raising awareness such as Fishing for Litter initiatives and provision of incentives are examples of such measures. Behavior-changing schemes are fundamental in addressing marine debris at its root.

15.4 Potential Gaps in Marine Litter Management

As previously described, a basic regulatory and management framework addressing marine litter is in place and a number of regions and countries have taken management measures to tackle the issues. A few cases indicate that some of the management measures have generated desirable results, such as South Korea's fishing gear buyback programme, Taiwan's plastic restriction policy and compulsory garbage sorting policy, US Fish for Energy, OSPAR Fishing for Litter, EU PRF Directive, HELCOM Baltic Strategy (see previous sections). Despite this, marine litter continues to increase worldwide: on shorelines, in estuaries and mangroves, in oceanic gyres, and on seafloors, signalling that marine litter remains an abiding problem, particularly with respect to microplastics (Barnes et al. 2009; UNEP 2011; Lima et al. 2014; Mohamed Nor and Obbard 2014; Pham et al. 2014; Lusher 2015). There are complex reasons for this and, it is possible to identify a number of gaps in the current framework that prevent the effective control of marine litter.

- Limits of existing instruments in addressing plastic marine litter
  Gold et al. (2013) identified a number of limitations in existing international instruments in addressing marine litter, including their insufficient scope with respect to the main sources of plastic pollution, exemptions and lack of enforcement standards. For instance, UNCLOS acknowledges the existence of land-based sources but simply requests that countries address the problem through domestic means.[44] MARPOL Annex V exempts accidental loss of disposal of plastic resulting from damage to the ship or its equipment,[45] as well as ships <400 GT, a category to which most of the fishing vessels belong, from recoding garbage discharge operations in Garbage Record Books (GRBs).[46] However, GRBs are of utmost importance to ensure compliance with discharge regulations (HELCOM 2012).
  The lack of enforcement standards can be found in the terms used in the legal instruments. UNCLOS, for instance, requires only that nations "shall endeavor" to use the "best practical means" to reduce marine pollution "in accordance" with their capabilities. Similarly, the Helsinki Convention requires contracting parties to take "all appropriate" measures to prevent and eliminate pollution. This leaves room for interpretation for countries with differing legal systems, environmental circumstances and capacities (Gold et al. 2013).
- Deficiencies in the legislation and a lack of implementation and enforcement of regulations and management measures
  The implementation and enforcement of regulations and management measures at national levels is a key component to combat marine litter. However, a number

---

[44]UNCLOS, art. 207 (concerning pollution from land-based sources).

[45]Revised Annex V, reg. 7.

[46]Revised Annex V, reg. 10.

of cases below show that international initiatives have not yet been transposed into national management schemes; or where they have there is a lack of enforcement, insufficient implementation, insufficient penalties to deter violators, or a lack of clarity in legislation leaving room for interpretation. These all represent major obstacles to the effective control of marine litter. For instance, the UNEP (2009) pointed out that at the national level, only the Wider Caribbean and Northwest Pacific regions have countries with specific national legislation addressing marine litter. The revised MARPOL Annex V has not yet been transposed into national law in countries such as Germany (UBA 2013) and thus there is no legal footing to implement this revised Annex V at the national level. The IMO Global Integrated Shipping Information System (GISIS) shows that there are numerous reported cases of alleged inadequacy of reception facilities.[47] In the US, as of 1995, <10 % of cases put to trial under MARPOL Annex V have resulted in penalties[48] and each of the penalized cases was fined an average of €4,560, an amount far too low to serve as a deterrent (Gold et al. 2013). In Taiwan, no penalties exist for the violation of the Annex V. The EU PRF Directive is vague at defining the fee/cost recovery system. The transposition of the directive into national legislation leaves room for different solutions on how to introduce incentives for waste delivery at ports. The use of different waste-fee systems by EU ports creates confusion among ship owners and operators (EMSA 2012; Øhlenschlæger et al. 2013).

- Poor cooperation and insufficient participation of states in international/regional initiatives

  Despite the fact that numerous international and regional initiatives already exist and provide a platform for cooperation and coordination of marine debris issues, a few cases indicate that cooperative action on marine litter has lagged behind, or the participation of states in these initiatives was insufficient. This would leave a loophole in the global/regional efforts, given the fact that marine debris is a transboundary issue. For example, there are no legal instruments in place dedicated to the management of marine litter as yet in the Black Sea, even though the *Bucharest Convention*[49] contains several articles pertaining to marine debris (Interwies et al. 2013). Some regional seas do not even participate in the UNEP Global Initiative, such as west central and southern Africa, northeast Pacific, Pacific and the ROPME[50] sea area (UNEP 2009). Countries bordering these regional seas might lack appropriate waste-management schemes because of economic constraints, although a number of African countries have recently banned the use of plastic bags.

---

[47]Detailed information is available at www.gisis.imo.org.

[48]Most often, the US Coast Guard chose to settle violations with a warning, dismissal, or referral of the case to the ship's flag state (Gold et al. 2013).

[49]Convention for the Protection of the Black Sea Against Pollution.

[50]Regional Organization for the Protection of the Marine Environment. The ROPME sea area is surrounded by Bahrain, I.R. Iran, Iraq, Kuwait, Oman, Qatar, Saudi Arabia and the UAE.

- Insufficient data on marine litter

  Despite the existing schemes against marine litter, our current knowledge of the quantities and the degradation of litter in the marine environment and its potential physical and chemical impacts on marine life are scarce (Galgani et al. 2013). Our knowledge gaps in terms of the biological consequences of micro-plastics exposure, economic and social impacts of marine debris have been mentioned (see other chapters). These gaps hinder the ability to prioritize mitigation efforts and to assess the effectiveness of implementation measures (The Scottish Government 2012). Specific data gaps were identified in a number of studies. For instance, very little data exist on quantities, trends, sources and sinks of marine litter in the west Indian Ocean region and very little is known about the extent and nature of the problem in the east Asian Seas region (GESAMP 2010). In European seas, data gaps were identified, including amounts and composition, transport, origin and impacts of marine litter on the seafloor, in the water column and rivers (Interwies et al. 2013). In addition, illegal, unreported and unregulated fishing activities and their contribution to litter generation, quantities and impacts of derelict fishing gear and micro-particles were referred to. Further data are needed in relation to large-scale and long-term monitoring across countries and environments, smaller-scale dynamics that affect plastic movement and accumulation, and trophic transfer dynamics of persistent organic pollutants via plastics through the marine food web (USEPA 2011).

15.5 Recommendations

In view of the above and taking into account the relevant information that has been put forth in the literature, recommendations for improvement are made as follows:

- Development of a new international instrument to tackle the marine litter problem

  Given that the scope of existing international law fails to match the scale and severity of the marine litter problem, Gold et al. (2013) urged the global community to develop a new multilateral agreement similar to the Montreal Protocol on Substances that Deplete the Ozone Layer. A set of elements were proposed to be included in such an agreement, including regulation of disposal of plastic litter from both ocean- and land-based sources, incorporating tracking, monitoring, reporting and enforcement standards and mechanisms, banning the most common or deleterious types of plastic litter, calling for a phase-out of all plastics that are not recycled at a rate of 75 % or higher by a certain date.

- Amending existing instruments to narrow exceptions and clarify enforcement standards

  Given the long time required to reach and implement a new agreement, Gold et al. (2013) recommended modifications to existing policy to eliminate some of the gaps. For example, amendment of the current vessel size and tonnage limitations in Annex V for requirements respecting placards, garbage management plans, and garbage record books is recommended so that fewer vessels are exempted.[51] Macfadyen et al. (2009) suggested amending Annex V so as to provide specific guidance on reasonable accidental losses of fishing gear. Regarding the vague definition of the fee systems in the EU PRF Directive, Øhlenschlæger et al. (2013) recommended implementation of a 100 % indirect fee system[52] for all European ports.

- Establishment of comprehensive national marine litter programmes
  Marine litter is a transboundary governance problem as it crosses scale, sectors and social divisions (Hastings and Potts 2013). To solve this problem, each state should develop a national marine litter programme (or a similar management scheme). This would constitute a high-level political commitment that could be a driver for relevant actions to be undertaken and ensure that marine litter issues are reflected in all policymaking. Such programmes have the potential to tackle the previously mentioned deficiencies. They should not only aim to reduce litter, but also quantify the sources of litter from land and ocean and promote a culture change with a view to consider "waste as a resource". To ensure its effective implementation, such programmes should have clear objectives, develop an efficient and integrated regulatory and management system, implement a suit of actions related to monitoring and research, infrastructure, education, incentive schemes, and enforcement and compliance, and establish public-private partnership/community involvement. In particular, such programmes should focus on long-range land-based waste management plans that lead to full collection and disposal services since the management of solid wastes on land directly affects quantifies of marine litter (Liffmann et al. 1997).
- Enhancing participation and cooperation of states in international/regional initiatives
  The transboundary nature of marine litter underlines that the problem is global in scale and international in impact. In this regard, national measures alone are insufficient to control marine debris, and international/regional cooperation is required. An empirical long-term litter monitoring study in the Southern Ocean showed that ocean-based litter monitoring needs to be integrated at an international or regional level (Edyvane et al. 2004). A wide range of international/regional initiatives on marine litter (such as UNEP RSP, GPA and GPML and various regional sea instruments) have established a platform for concerned states to engage in cooperation; participation and cooperation should be enhanced and strengthened both in terms of the number of participating states and the substantiality of cooperation.

---

[51]According to Revised Annex V, reg. 10, ships ≥12 min length are required to display placards, ships ≥100 GT to follow garbage management plans, and ships ≥400 GT to use garbage record books.

[52]This fee is paid by all ships calling at a port irrespective of the amount of waste disposed of at PRFs. It can effectively prevent cost from becoming a disincentive for using PRFs and has been implemented in ports such as Copenhagen (Denmark) and Stockholm (Sweden) (see Newman et al. 2015).

This would promote a dialogue among states on good practices in marine litter management and allow for substantial coordination and cooperation in research and developing and implementing more effective and practical management measures, such as the standardization of litter monitoring methods, the technologies for solid waste management, the waste notification system and the fee system for ship-generated waste. Moreover, this would help less wealthy countries to advance solid waste and sewage management through technical and financial assistance and training provided by more experienced countries and international organizations (Liffmann et al. 1997).

- Strengthening management measures on fishing vessels
Although many studies suggest that fisheries are an important source of marine litter, most fishing vessels are exempt from the discharge regulations of Annex V of MARPOL 73/78 because of their low tonnage. In addition to the previous recommendations to amend Annex V to narrow exceptions, I propose two approaches based on the area where fishing vessels operate. For vessels, which work solely in national waters, management measures at national levels should be specifically devised and strengthened. For example, Murray and Cowie (2011) demonstrated the presence of plastic microfibres shed from fishing net protectors in the intestines of >80 % of the commercially harvested prawns, an issue that could be well addressed by gear regulations. Arthur et al. (2014) found that the number of crabs caught per derelict fishing trap per year ranged from 4 to 76 in selected US coastal waters. This issue could be addressed by designing traps (e.g. escape panels) that allow species to escape when traps become derelict, thus rendering derelict traps "non-fishing". Kim et al. (2014) estimated that 11,500 t of traps and 38,500 t of gill-nets are abandoned annually in Korean waters and suggested incentive programmes for fishermen to use eco-friendly gear designs.

In addition, several measures could be adopted, including developing waste recycling practice among fishers, installing adequate PRFs, encouraging environmental education, promoting lost gear recovery, encouraging the use of environmentally friendly gear, promoting spatial management to reduce gear conflict and improving gear marking (Cho 2009; Macfadyen et al. 2009; Chen and Liu 2013; Gold et al. 2013; Arthur et al. 2014). Some of these measures may also apply to other types of small vessels (e.g. pleasure crafts), which are also exempt from Annex V.

For vessels operating on the high seas, numerous regional fishery bodies (RFBs)[53] have been established to manage and conserve fisheries resources based on geographical areas or fish species. They are generally established by coastal states and fishing nations with a common interest in overcoming collective-action problems related to the management of transboundary stocks (Sydnes 2001). Many have management mechanisms in place to regulate fishing activities, such as CCLAMR, International Commission for the Conservation of Atlantic Tunas,

---

[53]A full list of RFBs is available at www.fao.org/fishery/rfb/search/en.

Indian Ocean Tuna Commission, Western and Central Pacific Fisheries Commission, Northwest Atlantic Fisheries Organization, to name but a few. Taking advantage of the fully fledged management mechanisms, RFBs could take further actions to integrate fishery-related debris reduction into their wider management regime. To enable RFBs to adopt appropriate actions, it is advisable that the FAO, the lead organization of fisheries management and conservation, takes a lead in this initiative by providing guidance on effective and practical measures. In relation to this, some progress is being made to deal with derelict fishing gear by proposing a list of recommendations in a UNEP/FAO Technical Paper. The recommendations include both international and national actions, including developing an action plan on adequacy of PRFs for fisheries waste, amending Annex V, and formulating a global action plan to address the waste of fishing gear (Macfadyen et al. 2009). In addition, Gold et al. (2013) suggested that RFBs should adopt management standards to minimize the impacts of gear loss and move toward the replacement of plastic and synthetic gear with biodegradable nets and traps to minimize ghost fishing and entanglement.

15.6 Conclusion

The problem of marine litter is complex, as it is rooted in our prevailing production and consumption patterns and the way we dispose of and manage waste. Tackling this problem necessitates the inclusion of a vast amount of activities, sectors and sources that cannot be addressed by a single measure. A global reduction of the production of plastic waste/products through extended producer responsibility should be at the heart of all management solutions as this would ultimately be reflected in decreased inputs into our oceans. A variety of instruments at international, regional and national levels has been developed. In this chapter, the general mechanisms of instruments were analyzed and a number of them, including specific management measures contained therein, were profiled as illustration. The measures on marine litter are either on a mandatory or voluntary basis. In addition, based on the principle purposes, management measures were broadly divided into four categories: preventive, mitigating, removing and behavior-changing. This chapter further identified the potential gaps in existing frameworks and offered recommendations for improvement. The recommendations include establishment of a new international instrument targeted to the plastic marine litter problem, amending existing instruments to narrow exceptions and clarify enforcement standards, establishing comprehensive national marine litter programmes, enhancing participation and cooperation of states with regard to international/regional initiatives, and devising measures to prevent marine litter from fishing vessels.

As with other environmental problems, marine litter could be prevented and controlled through an effective collaboration of education and outreach programmes, strong regulations and policies, effective enforcement, and adequate support infrastructure. Based on this perspective, I hope that the current regulatory and management framework, potential gaps identified and recommendations made, will contribute to better management of marine debris. Last but not least, it is envisaged that through the ongoing efforts to combat marine litter, a shared vision for "litter-free marine environments" would be realized among all of the various actors and stakeholders concerned.

References

Allen, R., Jarvis, D., Sayer, S., & Mills, C. (2012). Entanglement of grey seals *Halichoerus grypus* at a haul out site in Cornwall, UK. *Marine Pollution Bulletin, 64,* 2815–2819.

Arthur, C., Sutton-Grier, A. E., Murphy, P., & Bamford, H. (2014). Out of sight but not out of mind: Harmful effects of derelict traps in selected US coastal waters. *Marine Pollution Bulletin, 86,* 19–28.

Ballance, A., Ryan, P. G., & Turpie, J. K. (2000). How much is a clean beach worth? The impact of litter on beach users in the Cape Peninsula, South Africa. *South African Journal of Science, 96,* 5210–5213.

Barnes, D. K. A. (2002). Biodiversity: Invasions by marine life on plastics debris. *Nature, 416,* 808–809.

Barnes, D. K. A., Galgani, F., Thompson, R. C., & Barlaz, M. (2009). Accumulation and fragmentation of plastic debris in global environments. *Philosophical Transactions of the Royal Society B, 364,* 1985–1998.

Barry, T. (2010). Fishing for energy: A public-private partnership approach to preventing and reducing derelict fishing gear. In C. Morishige (Ed.), Marine debris prevention projects and activities in the Republic of Korea and United States: A compilation of project summary reports (pp. 41–50). NOAA Technical Memorandum NOS-OR&R-36.

Baulch, S., & Perry, C. (2014). Evaluating the impacts of marine debris on cetaceans. *Marine Pollution Bulletin, 80,* 210–221.

Bilkovic, D., Havens, K., Stanhope, D., & Angstadt, K. (2014). Derelict fishing gear in Chesapeake Bay, Virginia: Spatial patterns and implications for marine fauna. *Marine Pollution Bulletin, 80,* 114–123.

Bond, A. L., Provencher, J. F., Elliot, R. D., Ryan, P. C., Rowe, S., Jones, I. L., et al. (2013). Ingestion of plastic marine debris by Common and Thick-billed Murres in the northwestern Atlantic from 1985 to 2012. *Marine Pollution Bulletin, 77,* 192–295.

Browne, M. A. (2015). Sources and pathways of microplastic to habitats. In M. Bergmann, L. Gutow & M. Klages (Eds.), *Marine anthropogenic litter* (pp. 229–244). Berlin: Springer.

BSAP. (2007). *Baltic sea action plan.* Retrieved August 15, 2014, from http://helcom.fi/baltic-sea-action-plan/contents.

Bugoni, L., Krause, L., & Petry, M. V. (2001). Marine debris and human impacts on sea turtles in Southern Brazil. *Marine Pollution Bulletin, 42,* 1330–1334.

CBD [Convention of Biological Diversity]. (2012). Impacts of marine debris on biodiversity: Current status and potential solutions. CBD Technical Series No. 67. Montreal: Secretariat of the CBD/the Scientific and Technical Advisory Panel—GEF.

Chen, C. L., & Liu, T. K. (2013). Fill the gap: Developing management strategies to control garbage pollution from fishing vessels. *Marine Policy, 40,* 34–40.

Cheremisinoff, N. P. (2003). *Handbook of solid waste management and waste minimization technologies*. Amsterdam: Butterworth-Heinemann.

Cheshire, A. C., Adler, E., Barbière, J., Cohen, Y., Evans, S., Jarayabhand, S., et al. (2009). UNEP/IOC guidelines on survey and monitoring of marine litter. UNEP Regional Seas Reports and Studies, No. 186; IOC Technical Series No. 83.

Cho, D. O. (2009). The incentive program for fishermen to collect marine debris in Korea. *Marine Pollution Bulletin, 58*, 415–417.

Donohue, M. J., & Foley, D. G. (2007). Remote sensing reveals links among the endangered Hawaiian monk seal, marine debris, and El Niño. *Marine Mammal Science, 23*, 468–473.

Donohue, M. J., Boland, R. C., Sramek, C. M., & Antonelis, G. A. (2001). Derelict fishing gear in the Northwestern Hawaiian Islands: Diving surveys and debris removal in 1999 confirm threat to coral reef ecosystems. *Marine Pollution Bulletin, 42*, 1301–1312.

Edyvane, K. S., Dalgetty, A., Hone, P. W., Higham, J. S., & Wace, N. M. (2004). Long-term marine litter monitoring in the remote Great Australian Bight, South Australia. *Marine Pollution Bulletin, 48*, 1060–1075.

EMSA [European Maritime and Safety Agecny]. (2012). Study on the delivery of ship-generated waste and cargo residues to port reception facilities in EU ports. Reference No. EMSA/OP/06/2011.

EPA [US Environmental Protection Agency]. (2011). *Marine debris in the North Pacific: A summary of existing information and identification of data gaps*. Retrieved August 15, 2014, from http://www.epa.gov/region9/marine-debris/pdf/MarineDebris-NPacFinalAprvd.pdf.

Eriksen, M., Lebreton, L. C. M., Carson, H. S., Thiel, M., Moore, C. J., Borerro, J. C., et al. (2014). Plastic pollution in the world's oceans: More than 5 trillion plastic pieces weighing over 250,000 Tons Afloat at Sea. *PLoS ONE, 9*, e111913.

EU [European Union]. (2000). *Directive 2000/59/EC of the European Parliament and of the Council of 27 November 2000 on PRFs for ship-generated waste and cargo residues*. Retrieved August 15, 2014, from http://eur-lex.europa.eu/LexUriServ/LexUriServ.do?uri=CONSLEG:2000L0059:20081211:EN:PDF.

EU. (2008). *Directive 2008/56/EC of the European Parliament and of the Council of 17 June 2008 establishing a framework for community action in the field of marine environmental policy* (Marine Strategy Framework irective). Retrieved August 15, 2014, from http://eur-lex.europa.eu/LexUriServ/LexUriServ.do?uri=OJ:L:2008:164:0019:0040:EN:PDF.

EU. (2010). *Commission decision of 1 September 2010 on criteria and methodological standards on good environmental status of marine waters (2010/477/EU)*. Retrieved August 15, 2014, from http://eur-lex.europa.eu/LexUriServ/LexUriServ.do?uri=OJ:L:2010:232:0014:0024:EN:PDF.

EU. (2013). *Green paper on a European strategy on plastic waste in the environment*. COM (2013) 123 final. Retrieved August 15, 2014, from http://ec.europa.eu/environment/waste/pdf/green_paper/green_paper_en.pdf.

Free, C. M., Jensen, O. P., Mason, S. A., Eriksen, M., Williamson, N. J., & Boldgiv, B. (2014). High-levels of microplastic pollution in a large, remote, mountain lake. *Marine Pollution Bulletin, 85*, 156–163.

Galgani, F., Hanke, G., Werner, S., & De Vrees, L. (2013). Marine litter within the European marine strategy framework directive. *ICES Journal of Marine Science, 70*, 1055–1064.

Galgani, F., Hanke, G., & Maes, T. (2015). Global distribution, composition and abundance of marine litter. In M. Bergmann, L. Gutow, & M. Klages (Eds.), *Marine anthropogenic litter* (pp. 29–56). Berlin: Springer.

GESAMP [Joint Group of Experts on the Scientific Aspects of Marine Environmental Protection]. (2010). In T. Bowmer & P. J. Kershaw (Eds.), *Proceedings of the GESAMP International Workshop on Plastic Particles as a Vector in Transporting Persistent, Bioaccumulating and Toxic Substances in the Oceans*. GESAMP Rep. Stud. No. 82.

Gold, M., Mika, K., Horowitz, C., Herzog, M., & Leitner, L. (2013). Stemming the tide of plastic marine litter: A global action agenda. Pritzker Policy Brief 5.

Han, S. G., Kim H. K., Kim S. D., & Noh, H. J. (2010). South Korea coastal cleanup program for marine litter. In C. Morishige (Ed.), *Marine debris prevention projects and activities in the Republic of Korea and United States: A compilation of project summary reports* (pp. 9–15). NOAA Technical Memorandum NOS-OR&R-36.

Hartley, B. L., Thompson, R. C., & Pahl, S. (2015). Marine litter education boosts children's understanding and self-reported actions. *Marine Pollution Bulletin, 90*, 209–217.

Hastings, E., & Potts, T. (2013). Marine litter: Progress in developing an integrated policy approach in Scotland. *Marine Policy, 42*, 49–55.

HELCOM. (2012). *Clean seas guide: Information for mariners*. Retrieved August 15, 2014, from http://helcom.fi/Lists/Publications/Clean%20Seas%20Guide%20-%20Information%20for%20Mariners.pdf.

HELCOM. (2013). *HELCOM Copenhagen ministerial declaration: Taking further actions to implement the Baltic Sea Action Plan—Reaching good environmental status for a healthy Baltic Sea*. Retrieved August 15, 2014, from http://helcom.fi/helcom-at-work/ministerial-declarations.

Hoellein, T., Rogas, M., Pink, K., Gasior, J., & Kelly, J. (2014). Anthropogenic litter in urban freshwater ecosystems: Distribution and microbial interactions. *PLoS ONE, 9*, e98485.

Howell, E. A., Bograd, S. J., Morishige, C., Seki, M. P., & Polovina, J. J. (2012). On North Pacific circulation and associated marine debris concentration. *Marine Pollution Bulletin, 65*, 16–22.

Interwies, E., Görlitz, S., Stöfen A., Cools, J., van Breusegem, W., Werner, S., et al. (2013). *Issue Paper to the International Conference on Prevention and Management of Marine Litter in European Seas* (final version). Retrieved August 15, 2014, from http://www.marine-litter-conference-berlin.info/userfiles/file/Issue%20Paper_Final%20Version.pdf.

Jambeck, J. R., Geyer, R., Wilcox, C., Siegler, T. R., Perryman, M., Andrady, A., et al. (2015). Plastic waste inputs from land into the ocean. *Science, 347*, 768–771.

Jang, Y. C., & Song, B. J. (2013). A critical analysis of the rationality of South Korea's marine debris policy. *International Journal of Policy Studies, 4*, 83–105.

Jang, Y. C., Hong, S., Lee, J., Lee, M. J., & Shim, W. J. (2014). Estimation of lost tourism revenue in Geoje Island from the 2011 marine debris pollution event in South Korea. *Marine Pollution Bulletin, 81*, 49–54.

Jones, M. M. (1995). Fishing debris in the Australian marine environment. *Marine Pollution Bulletin, 30*, 25–33.

Jung, R. T., Sung, H. G., Chun, T. B., & Keel, S. I. (2010). Practical engineering approaches and infrastructure to address the problem of marine debris in Korea. *Marine Pollution Bulletin, 60*, 1523–1532.

Kiessling, T., Gutow, L., & Thiel, M. (2015). Marine litter as a habitat and dispersal vector. In M. Bergmann, L. Gutow, & M. Klages (Eds.), *Marine anthropogenic litter* (pp. 141–181). Berlin: Springer.

Kim, S.-G., Lee, W.-I., & Yuseok, M. (2014). The estimation of derelict fishing gear in the coastal waters of South Korea: Trap and gill-net fisheries. *Marine Policy, 46*, 119–122.

KIMO. (2014). *Fishing for Litter 2014 spring newsletter*. Retrieved August 15, 2014, from http://www.kimointernational.org/WebData/Files/FFL%20Scotland/Spring%20Newsletter%2014.pdf.

Kühn, S., Bravo Rebolledo, E. L., & van Franeker, J. A. (2015). Deleterious effects of litter on marine life. In M. Bergmann, L. Gutow, & M. Klages (Eds.), *Marine anthropogenic litter* (pp. 75–116). Berlin: Springer.

Law, K. L., Morét-Ferguson, S., Maximenko, N. A., Proskurowski, G., Peacock, E. E., Hafner, J., et al. (2010). Plastic accumulation in the North Atlantic Subtropical Gyre. *Science, 329*, 1185–1188.

Liffmann, M., Howard, B., O'Hara, K., & Coe, J. M. (1997). Strategies to reduce, control, and minimize land-source marine debris. In J. M. Coe & D. B. Rogers (Eds.), *Marine debris: Sources, impacts, and solutions* (pp. 381–390). Berlin: Springer.

Lima, A. R. A., Costa, M. F., & Barletta, M. (2014). Distribution patterns of microplastics within the plankton of a tropical estuary. *Environmental Research, 132*, 146–155.

Louka, E. (2006). *International environmental law: Fairness, effectiveness, and world order*. New York: Cambridge University Press.

Lusher, A. (2015). Microplastics in the marine environment: Distribution, interactions and effects. In M. Bergmann, L. Gutow, & M. Klages (Eds.), *Marine anthropogenic litter* (pp. 245–308). Berlin: Springer.

Macfadyen, G., Huntington, T., & Cappell, R. (2009). Abandoned, lost or otherwise discarded fishing gear. UNEP Regional Seas Reports and Studies, No. 185; FAO Fisheries and Aquaculture Technical Paper, No. 523.

McIlgorm, A., Campbell, H. F., & Rule, M. J. (2011). The economic cost and control of marine debris damage in the Asia-Pacific region. *Ocean & Coastal Management, 54*, 643–651.

MCS [Marine Conservation Society]. (2013). *Beachwatch big weekend 2013*. Retrieved August 15, 2014, http://www.mcsuk.org/downloads/pollution/beachwatch/latest2014/Beachwatch_Summary_Report_2013.pdf.

Mohamed Nor, N. H., & Obbard, J. (2014). Microplastics in Singapore's coastal mangrove ecosystems. *Marine Pollution Bulletin, 79*, 278–283.

Morritt, D., Stefanoudis, P. V., Pearce, D., Crimmen, O. A., & Clark, P. F. (2014). Plastic in the Thames: A river runs through it. *Marine Pollution Bulletin, 78*, 196–200.

Mouat, J., Lozano, R., & Bateson, H. (2010). *Economic impacts of marine litter*. Shetland, Scotland, UK: KIMO.

Murray, F., & Cowie, P. R. (2011). Plastic contamination in the decapod crustacean *Nephrops norvegicus* (Linnaeus, 1758). *Marine Pollution Bulletin, 62*, 1207–1217.

Newman, S., Watkins, E., Farmer, A., ten Brink, P., & Schweitzer, J.-P. (2015). The economics of marine litter. In M. Bergmann, L. Gutow, & M. Klages (Eds.), *Marine anthropogenic litter* (pp. 371–398). Berlin: Springer.

NFWF [National Fish and Wildlife Foundation]. (2014). *Fish for energy factsheet*. Retrieved August 15, 2014, from http://www.nfwf.org/fishingforenergy/Documents/FFE__Fact%20Sheet_2014_revised.pdf.

NOAA [National Oceanic and Atmospheric Administration]. (2012). *2010–2011 Progress report on the implementation of the MDRPRA*. Silver Spring: NOAA.

Noh, H.-J., Kim, H.-K., Kim, S.-D., & Han, S.-G. (2010). Buyback program for fishing gear and marine litter from fishery activities. In C. Morishige (Ed.), *Marine debris prevention projects and activities in the Republic of Korea and United States: A compilation of project summary reports* (pp. 3–8). NOAA Technical Memorandum NOS-OR&R-36.

Ocean Conservancy. (2013). *Working for clean beaches and clear water 2013 report*. Retrieved August 15, 2014, from http://www.oceanconservancy.org/our-work/international-coastal-cleanup/2012-ocean-trash-index.html.

OECD [Organisation for Economic Co-operation and Development]. (2001). *A guidance manual for governments*. Paris: OECD.

Øhlenschlæger, J. P., Newman, S., & Farmer, A. (2013). Reducing ship generated marine litter—Recommendations to improve the EU port reception facilities directive. Report produced for Seas At Risk. Institute for European Environmental Policy, London. Retrieved August 15, 2014, from http://www.ieep.eu/assets/1257/IEEP_2013_Reducing_ship_generated_marine_litter_-_recommendations_to_improve_the_PRF_Directive.pdf.

OSPAR. (2007). *OSPAR pilot project on monitoring marine beach litter: Monitoring of marine debris on beaches in the OSPAR region*. London: OSPAR Commission.

OSPAR. (2010a). *Quality status report 2010*. London: OSPAR Commission.

OSPAR. (2010b). *Guideline for monitoring marine litter on the beaches in the OSPAR maritime area*. London: OSPAR Commission.

Pham, C. K., Ramirez-Llodra, E., Alt, C. H. S., Amaro, T., Bergmann, M., Canals, M., et al. (2014). Marine litter distribution and density in European seas, from the shelves to deep basins. *PLoS ONE, 9*, e95839.

Pichel, W. G., Churnside, J. H., Veenstra, T. S., Foley, D. G., Friedman, K. S., Brainard, R. E., et al. (2007). Marine debris collects within the North Pacific subtropical convergence zone. *Marine Pollution Bulletin, 54*, 1207–1211.

PlasticsEurope. (2013). *Plastics-the facts 2013: An analysis of European plastics production, demand and waste data.* Brussels: PlasticsEurope.

Rees, G., & Pond, K. (1995). Marine litter monitoring programs—A review of methods with special reference to national surveys. *Marine Pollution Bulletin, 30,* 103–108.

Ribic, C. A., Sheavly, S. B., Rugg, D. J., & Erdmann, E. S. (2010). Trends and drivers of marine debris on the Atlantic coast of the United States 1997–2007. *Marine Pollution Bulletin, 60,* 1231–1242.

Roberts, J. (2010). *Marine environment protection and biodiversity conservation: The application and future development of the IMO's particularly sensitive sea area concept.* Berlin: Springer.

Rochman, C. M. (2015). The complex mixture, fate and toxicity of chemicals associated with plastic debris in the marine environment. In M. Bergmann, L. Gutow, & M. Klages (Eds.), *Marine anthropogenic litter* (pp. 117–140). Berlin: Springer.

Ryan, P. G. (2015). A brief history of marine litter research. In M. Bergmann, L. Gutow, & M. Klages (Eds.), *Marine anthropogenic litter* (pp. 1–25). Berlin: Springer.

Sheavly, S. B. (2010). *Nationalmarine debris monitoring program lessons learned.* (EPA 842-R-10-001). Retrieved August 15, 2014, from http://water.epa.gov/type/oceb/marinedebris/upload/lessons_learned.pdf.

Storrier, K. L., & McGlashan, D. J. (2006). Development and management of a coastal litter campaign: The voluntary coastal partnership approach. *Marine Policy, 30,* 189–196.

Sydnes, K. A. (2001). Establishing a regional fisheries management organization for the Western and Central Pacific tuna fisheries. *Ocean and Coastal Management, 44,* 787–811.

Taylor, J. R., DeVogelaere, A. P., Burton, E. J., Frey, O., Lundsten, L., Kuhnz, L. A., et al. (2014). Deep-sea faunal communities associated with a lost intermodal shipping container in the Monterey Bay National Marine Sanctuary, CA. *Marine Pollution Bulletin, 83,* 92–106.

Ten Brink, P., Lutchman, I., Bassi, S., Speck, S., Sheavly, S., Register, K., et al. (2009). *Guidelines on the use of market-based instruments to address the problem of marine litter.* Brussels, Belgium/Virginia Beach, Virginia, USA: Institute for European Environmental Policy (IEEP)/Sheavly Consultants.

TEPA [Taiwan Environmental Protection Administration]. (2011). *Three Rs: Resource recycling and sustainable reuse.* Retrieved August 15, 2014, from http://www.ttv.com.tw/event/2011//3R/reduce/p2.htm.

TEPA (2010). *Yearbook of waste recycling and reuse.* Taipei: EPA (in Chinese).

Teuten, E. L., Saqing, J. M., Knappe, D. R. U., Barlaz, M. A., Jonsson, S., Björn, A., et al. (2009). Transport and release of chemicals from plastics to the environment and to wildlife. *Philosophical Transactions of the Royal Society B, 364,* 2027–2045.

The Scottish Government. (2012). *Marine litter issues, impacts and actions.* Retrieved August 15, 2014, from http://www.scotland.gov.uk/Publications/2012/09/6461.

The Scottish Government. (2013). *Marine litter strategy/national litter strategy: Strategic environmental assessment environmental report.* Retrieved August 15, 2014, from http://www.scotland.gov.uk/Publications/2013/07/9297.

Thompson, R. C. (2015). Microplastics in the marine environment: Sources, consequences and solutions. In M. Bergmann, L. Gutow, & M. Klages (Eds.), *Marine anthropogenic litter* (pp. 185–200). Berlin: Springer.

Thompson, R. C., Moore, C. J., vom Saal, F. S., & Swan, S. H. (2009). Plastics, the environment and human health: Current consensus and future trends. *Philosophical Transactions of the Royal Society B, 364,* 2153–2166.

UBA [Umwelt Bundes Amt]. (2013). *Factsheet 3: Measures for the prevention of marine litter.* Retrieved August 15, 2014, from http://www.marine-litter-conference-berlin.info/userfiles/file/Factsheet%203%20Measures.pdf.

UNEP. (2005). *Marine litter: An analytical overview.* Nairobi: UNEP.

UNEP. (2009). *Marine litter: A global challenge.* Nairobi: UNEP.

UNEP. (2011). *UNEP year book 2011.* Nairobi: UNEP.

UNEP [United Nations Environment Programme]. (2013). *UNEP year book 2011: Emerging issues in our global environment*. Nairobi: UNEP.

UNEP/NOAA. (2011). *The Honolulu strategy: A global framework for prevention and management of marine debris*. Nairobi/Silver Spring, MD: UNEP/NOAA.

Van Cauwenberghe, L., Vanreusel, A., Mees, J., & Janssen, C. R. (2013). Microplastic pollution in deep-sea sediments. *Environmental Pollution, 182*, 495–499.

Van Franeker, J. A., Blaize, C., Danielsen, J., Fairclough, K., Gollan, J., Guse, N., et al. (2011). Monitoring plastic ingestion by the northern fulmar *Fulmarus glacialis* in the North Sea. *Environmental Pollution, 159*, 2609–2615.

Vaughn, J. (2009). *Waste management: A reference handbook*. Santa Barbara, CA: ABC-CLIO.

Watson, M. (2012). Marine debris along the florida keys reef tract-mapping, analysis and perception study. Open access theses. Paper 330. (http://scholarlyrepository.miami.edu/cgi/viewcontent.cgi?article=1330&context=oa_theses).

Wilcox, C., Heathcote, G., Goldberg, J., Gunn, R., Peel, D., & Hardesty, B. D. (2014). Understanding the sources and effects of abandoned, lost and discarded fishing gear on marine turtles in Northern Australia. *Conservation Biology*.

Chapter 16
The Contribution of Citizen Scientists to the Monitoring of Marine Litter

Valeria Hidalgo-Ruz and Martin Thiel

Abstract Citizen science projects are based on volunteer participation of untrained citizens who contribute information, data and samples to scientific studies. Herein we provide an overview of marine litter studies that have been supported by citizen scientists (n = 40) and compare these studies with selected studies conducted by professional scientists (n = 40). Citizen science studies have mainly focused on the distribution and composition of marine litter in the intertidal zone. Studies extended over regional, national and international scales, with time periods generally extending from less than one year to two years. Professional studies have also examined the distribution and composition of marine litter, but from intertidal, subtidal and pelagic zones, with some focusing exclusively on microplastics. These studies have been conducted over local, regional and international scales, usually for less than one year each. Both citizen science and professional studies on marine litter have been conducted mainly in the northern hemisphere, revealing a lack of information available on coastal regions of the southern hemisphere. A main concern of citizen science studies is the reliability of the collected information, which is why many studies include steps to ensure data quality, such as preparation of clear protocols, training of volunteers, *in situ* supervision by professional scientists, and revision of samples and data. The results of this comparative review confirm that citizen science can be a useful approach to increase the available information on marine litter sources, distribution and ecological impacts. Future studies should strive to incorporate

V. Hidalgo-Ruz · M. Thiel (✉)
Facultad Ciencias del Mar, Universidad Católica del Norte,
Larrondo 1281, Coquimbo, Chile
e-mail: thiel@ucn.cl

M. Thiel
Millennium Nucleus Ecology and Sustainable Management of Oceanic Island (ESMOI),
Coquimbo, Chile

M. Thiel
Centro de Estudios Avanzados en Zonas Áridas (CEAZA), Coquimbo, Chile

M. Bergmann et al. (eds.), *Marine Anthropogenic Litter*,
DOI 10.1007/978-3-319-16510-3_16

additional citizen scientists who frequent marine environments, for instance, divers and sailors, to improve our understanding of marine litter dynamics.

Keywords Citizen science · Marine litter · Professional studies · Volunteers · Data quality

16.1 Introduction

Large quantities of anthropogenic litter reach the marine environment, where they spread throughout all oceans and persist for many years (Derraik 2002; Barnes et al. 2009; Ryan et al. 2009; Eriksen et al. 2014). The accumulation of litter causes diverse impacts on marine biota, such as entanglement (Laist 1997; Moore et al. 2009), ingestion (e.g. van Franeker et al. 2011; Carson 2013; Cole et al. 2013), and dispersal of alien species (Barnes 2002; Masó et al. 2003; Kiessling et al. 2015). The extensive spreading of marine litter, even to the most remote regions of the world's oceans, makes litter distribution and abundance surveys difficult and time consuming (Ryan et al. 2009; Eriksen et al. 2014). Coastal regions, where a large fraction of marine litter is deposited, receive visits from a wide range of people, including, but not limited to, tourists, fishermen, and schoolchildren. Such coastal users have been recruited to support scientific beach surveys to quantify marine litter worldwide (e.g. Ogata et al. 2009; Ribic et al. 2010). These volunteers (here termed "citizen scientists") (Bonney et al. 2009) have participated in the collection, analysis and interpretation of data in a wide range of studies, determining litter distribution and abundance in the marine environment (Lindborg et al. 2012; Smith and Edgar 2014; Thiel et al. 2014).

Herein we provide a review of marine litter studies that have been supported by citizen scientists in order to evaluate their contribution to current knowledge on marine litter distribution, abundance, and interaction with marine biota. In particular, we compare the type and quality of data collected in these citizen science studies with those collected by professional scientists (scientists that have received a formal scientific education). Based on the results of this comparison, we offer recommendations for future marine litter surveys that are supported by citizen scientists.

16.2 Marine Litter Studies Supported by Citizen Scientists

People from a wide range of educational backgrounds have supported scientific studies on marine litter. Their interest to participate in this kind of investigation may vary depending on their own personal motivation, which may include being part of an environmental organization (e.g. marine conservation NGO, girl & boy scouts) or an educational project within a school (Fig. 16.1). For example, beach cleanup

Fig. 16.1 Examples of
citizen scientists participating
in studies on marine litter

campaigns are conducted in many countries, and can be a source of marine litter information supported by citizen scientists (e.g. Ribic 1998; Martin 2013).

For this review, marine litter studies were identified by searching the ISI Web of Knowledge and Google Scholar databases for papers using the keywords "citizen science" or "volunteer" with "marine litter", "marine debris" or "plastic debris". We thoroughly scanned the literature, identifying all studies in which volunteers had participated in sampling and/or sample processing. We only selected studies with a main focus on marine litter; studies which coincidentally also report interactions of litter with marine biota were not considered, unless these explicitly focused on litter aspects, such as plastic ingestion and entanglement by seabirds and marine mammals (Moore et al. 2009; van Franeker et al. 2011). At the time of writing, 40 marine litter studies were identified, which were based entirely or partly on data or samples contributed by citizen scientists (Appendix 1).

We also selected 40 peer-reviewed articles that were exclusively conducted by professional scientists, in order to compare those with and evaluate the importance, scope and quality of citizen-science studies. For the purpose of this review, we only included the most cited papers (according to ISI Web of Knowledge and Google Scholar databases) that have been published during the past 10 years (2004–2014). Keywords used to identify these studies were "marine litter", "marine debris", "plastic debris" and "beach survey". Review articles were not considered for this comparison between citizen science and professional studies (see Appendix 1 for the complete list of selected studies).

16.3 Comparison of Citizen Science and Professional Science Studies on Marine Litter

16.3.1 Research Topic

Research on marine litter has focused on six major topics: (1) Distribution and composition of marine litter, (2) interaction with marine biota, (3) toxic effects, (4) horizontal and vertical transport, (5) social aspects and (6) degradation of marine plastic litter. The majority of citizen science studies (68 %) examined the spatial distribution and composition of marine litter (Table 16.1). In these cases, citizens participated in beach cleanup activities or beach surveys of marine litter (e.g. Gregory 1991; Storrier et al. 2007), plastic beverage containers (Józwiak 2005) and small plastic debris (Hidalgo-Ruz and Thiel 2013). Apart from one study conducted by divers in subtidal environments (Smith and Edgar 2014), the intertidal zone was the only sampling zone.

One citizen-science study also covered the open ocean via a drifter experiment to investigate the pathways of litter from island areas in Hawaii; citizen scientists reported drifters that had stranded on local shores (Carson et al. 2013).

In most cases, data were registered on datasheets provided by an organization, but one study also created a smartphone application, which was used by personal phones and iPods (Martin 2013). Interaction of marine litter with biota was the second most common topic addressed by citizen science studies, but given the overwhelming proportion of studies on the distribution and composition of litter (68 %), this topic represented only 18 % of all studies (Table 16.1). These studies

Table 16.1 Comparison of research topics on marine litter, conducted by citizen scientists (N = 40) and professional scientists (N = 40)

Topic	Citizen science		Professional	
	No.	%	No.	%
Distribution and composition	27	67.5	18	45
Interaction with biota	7	17.5	14	35
Toxic effects	4	10.0	3	7.5
Transport	1	2.5	3	7.5
Social aspects	1	2.5	1	2.5
Degradation	0	0	1	2.5

focused on specific animal groups, mostly vertebrates: seabirds (van Franeker et al. 2011; Lindborg et al. 2012), marine mammals (Moore et al. 2009), fish (Carson 2013) and sea turtles (Bjorndal et al. 1994). For example, citizen scientists supported cleanups of derelict crab traps and quantified the species entangled by ghost fishing activity (Anderson and Alford 2014), or information on species affected by marine litter was documented on an interactive website (Hong et al. 2013). Other aspects of marine litter were examined in 15 % of all citizen science studies (Table 16.1). Persistent organic pollutants were determined in beached plastic pellets, which were collected by citizens (Ogata et al. 2009; Hirai et al. 2011; Heskett et al. 2012). Transport of marine litter was studied to determine the factors driving marine debris deposition on Hawaiian beaches (Morishige et al. 2007; Carson et al. 2013). A social study examined the behavior, education and preference of the general public to reduce littering on beaches (Eastman et al. 2013). Degradation of marine litter was not addressed by citizen scientists.

A large portion of professional studies was also based on the distribution and composition of marine litter (45 % of all studies) (Table 16.1). These studies examined beach litter from the intertidal zone (n = 11) (e.g. McDermid and McMullen 2004; Claessens et al. 2011), seafloor debris from the subtidal zone (n = 4) (e.g. Katsanevakis and Katsarou 2004), and pelagic plastic litter from the open ocean (n = 6) (Lattin et al. 2004; Pichel et al. 2007). In contrast to citizen science studies, a considerable number of the professional studies focused exclusively on microplastics (n = 11) (e.g. Thompson et al. 2004; Ng and Obbard 2006; Browne et al. 2010). This difference is likely due to the advanced techniques required for proper identification of microplastics (especially the smaller fraction of microplastics, 1 μm–1 mm; Löder and Gerdts 2015), which is unfeasible in citizen science studies (Hidalgo-Ruz and Thiel 2013). The interaction of marine litter with biota was addressed by a high proportion of studies (35 % of all studies) (Table 16.1). These studies focused mainly on plastic ingestion by both invertebrates and vertebrates (e.g. Graham and Thompson 2009; Boerger et al. 2010), but also on entanglement of pinnipeds in marine litter (Page et al. 2004; Boren et al. 2006), and on the impact of lost fishing gear on coral reefs (Chiappone et al. 2005). Toxic effects, transport, social aspects and degradation of marine litter were examined by 20 % of the professional studies (Table 16.1). These focused on the quantification of persistent organic pollutants (POP's) in plastics (Rios et al. 2007, 2010; Teuten et al. 2007), the temporal variability and dynamics of marine debris at the sea surface (e.g. Martinez et al. 2009; Law et al. 2010), the socio-economic characteristics of beach users and littering (Santos et al. 2005), and the relationship between composition, surface texture, and degradation of plastics (Corcoran et al. 2009).

16.3.2 Spatial Scale

Considering that marine litter is a global issue, the collection of data over extensive spatial scales is particularly important (Galgani et al. 2015). Professional research can address this issue, but requires a work-intensive sampling effort or the use of expensive equipment, such as buoys, aircraft, submersible vehicles and satellites (e.g. Pichel et al. 2007; Maximenko et al. 2012). These types of sophisticated surveys might be too expensive for citizen science projects, but with a reasonable budget, citizen science offers the opportunity to establish extensive networks of sampling stations on the ground. Citizen science studies have been conducted on the local (one sampling site), regional (several sampling sites), national and even international scale (Table 16.2). The more extensive citizen science studies were conducted by "The International Pellet Watch" project (see Ogata et al. 2009; Hirai et al. 2011; Heskett et al. 2012), in which volunteers from 17 countries have collected pellets from local beaches and sent them to Tokyo for laboratory analyses. This project has monitored the pollution status of persistent organic pollutants in the oceans since 2005, extending their sampling locations to several new places (Fig. 16.2).

In contrast, professional studies have been conducted relatively homogeneously over all spatial scales, with the exception of national surveys, which only represented two (5 %) of all professional studies (Table 16.2). Examples of local studies include Corcoran et al. (2009) on plastic degradation on Kauai, Hawaii, Graham and Thompson (2009) on the ingestion of plastics by sea cucumbers, and

Table 16.2 Comparison of the spatial scale of citizen science (N = 40), and professional studies (N = 40)

Spatial scale	Citizen science		Professional	
	No.	%	No.	%
Local	12	30	16	40
Regional	14	35	9	22.5
National	8	20	2	5.0
International	6	15	13	32.5

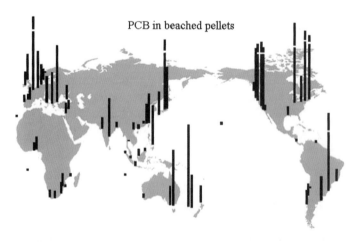

Fig. 16.2 PCB concentration on beached pellets from the volunteer-based global monitoring program "International Pellet Watch". Figure modified from: http://www.pelletwatch.org/ (access: July 2014)

Santos et al. (2005) on the relationship between beach users and litter generation at Cassino Beach, Rio Grande City, Brazil. Regional research examples are from Costa et al. (2010) on the distribution and composition of debris on beaches from Northeast Brazil, and Chiappone et al. (2005) on the impact of lost fishing gear on coral reefs in Florida, USA. A worldwide coverage was achieved by Browne et al. (2011) who determined the microplastics abundance (mainly from cloth fibres) at shorelines of 18 countries.

16.3.3 Temporal Scale

Citizen-science studies require a lot of organization. Accordingly, short-term studies are expected to be the most common among all citizen-science studies. Nevertheless, the time range of citizen-science studies vary from single events, up to a study of 27 years by van Franeker et al. (2011), who determined the abundance of ingested plastics by northern fulmars *Fulmarus glacialis* from the North Sea, as an indication of litter contamination. The majority of citizen-science studies (63 %) cover time periods ranging from less than 1–2 years, followed by studies between 5 and 10 years (20 %) (Table 16.3). Professional studies varied between single events up to a study on microplastics that compared recent samples with samples taken 40 years ago (Thompson et al. 2004). Interestingly, many professional studies were conducted only once, i.e. they spanned less than one year (53 %), whereas others ranged from 1 to 2 years (10 %) and 2 to 5 years (10 %), respectively. Three professional observational studies did not report the temporal scale of the investigation (Corcoran et al. 2009; Costa et al. 2010; Claessens et al. 2011) (Table 16.3).

16.3.4 Regions Where Studies Have Been Done

The problem of marine litter is widespread and has caused concern worldwide. However, global knowledge about marine litter is limited, because the majority of both citizen-science and professional studies on marine litter have been conducted in the northern hemisphere. Most citizen-science studies have been reported from Asia and South America (Fig. 16.3a). Professional studies have been conducted

Table 16.3 Comparison of the temporal scale of citizen-science (N = 40) and professional studies (N = 40)	No. of years	Citizen science		Professional	
		No.	%	No.	%
	<1	12	30	21	52.5
	1–2	13	32.5	4	10
	>2–5	4	10	4	10
	>5–10	8	20	2	5
	>10	3	7.5	6	15
	No information	–	0	3	7.5

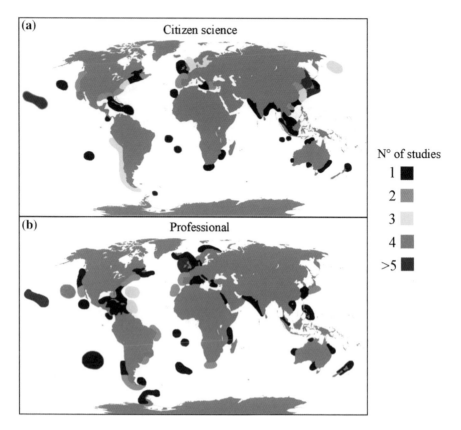

Fig. 16.3 World map with representation of the number of studies per ecoregion (limits of ecoregions after Spalding et al. 2007), for (**a**) citizen-science and (**b**) professional studies

mainly in Europe and the North Pacific Ocean (Hawaii and the North Pacific gyre) (Fig. 16.3b). This reveals a lack of information on coastal regions of the southern hemisphere, such as Africa and South America, except Chile. However, in the near future, the combination of citizen-science and professional studies can be the key to achieving global knowledge about litter sources and quantities, especially for regions of the world where this information is still needed. Therefore, citizen science studies could be a good approach to help filling the last missing gaps on the world map.

16.4 Data Collection and Quality Control of Citizen-Science Studies

A main concern of citizen-science studies is whether the collected data are reliable and comparable to professional studies. Four main aspects need to be considered in order to ensure or improve data quality: (1) preparation of easy and straightforward protocols,

(2) training of volunteers, (3) *in situ* supervision by professional participation, and (4) validation of data and samples (modified from Bonney et al. 2009). Of the citizen-science studies on marine litter examined, 55 % included at least one of these steps (e.g. Rosevelt et al. 2013; Anderson and Alford 2014; Gago et al. 2014).

16.4.1 Preparation of Easy and Straight-Forward Protocols

The studies that took measures to guarantee data quality, provided standardized protocols, guidelines and datasheets (e.g. Ribic 1998; Gago et al. 2014). In order to create clear protocols, some studies needed to adjust the sampling target to be easily identified by citizen scientists. For instance, Ribic et al. (2010) found that citizen scientists occasionally missed small pieces of debris (no specific size range was mentioned) in a monitoring program for beach litter. As a consequence of that observation, Hidalgo-Ruz and Thiel (2013) in a study focusing exclusively on small-plastic debris, decided to sample only items larger than 1 mm, which can be identified by the naked eye after sieving of sand. Once the sampling target is determined, the marine litter items likely to be found by citizen scientists can be photographed and included in preparatory materials. Photographs of marine litter items were used in 15 % of the studies (e.g. Moore et al. 2009; Anderson and Alford 2014).

16.4.2 Training of Volunteers

Data quality can also be improved by volunteer training (e.g. Storrier and McGlashan 2006; Smith et al. 2014). Indeed, 38 % of the citizen science studies examined here included a degree of training or preparation of the volunteers. Training could consist of a one-hour classroom preparation (e.g. Smith et al. 2014) or a brief introduction in the field just before the sampling activity (e.g. Moore et al. 2009). For instance, a study on ghost fishing by derelict crab traps (Anderson and Alford 2014) was preceded by a training period. Furthermore, during one study year, participants were asked to take photos of every trap and to identify the organisms in the traps. These photos were later examined by professional scientists who confirmed that the data recorded for each trap were accurate.

16.4.3 In Situ Supervision by Professionals

Scientists and survey monitors participated in the sampling activity in 43 % of the examined citizen-science studies. These professionals were in charge of assuring accuracy of debris classification, data recording and identification of missed/overlooked debris items (e.g. Ribic et al. 2011, 2012a). For example, in a study from South Korea on the impacts of marine debris on wildlife, experts from wildlife, nature and marine research institutes provided data quality assurance on a voluntary basis contributing pictures of dissections or autoradiography in order to demonstrate how animals were affected by the debris (Hong et al. 2013).

16.4.4 Validation of Data and Samples

Citizen-science studies can also incorporate a validation process in which the data gathered by volunteers are compared to data obtained by professional scientists. This comparative approach was applied by 18 % of the studies, which evaluated the quality of the citizen-science data by re-counting the litter items, also using a microscope to differentiate between biological and synthetic litter (Rosevelt et al. 2013). For instance, Hidalgo-Ruz and Thiel (2013) recounted small plastic particles in samples that had been counted by citizen scientists. In one case, it was found that glass shards had been misidentified as small plastic debris. Elimination of samples with this kind of obvious error from the analysis can substantially improve data quality. According to Lindborg et al. (2012), citizen scientists can dissect and analyze seabird boluses with high accuracy resulting in measurements of contamination rates similar to those obtained by professional scientists. Validation can also be done by scientists analyzing photographs of samples taken by volunteers (Moore et al. 2009). Technological equipment can be used to generate complementary data. For instance, Seino et al. (2009) used high-frequency ocean radar, airplanes and balloons to take photographs of marine litter, which were used to complement data collected by volunteers. Data quality control can also entail the elimination of erroneous data. For instance, in a user survey on beach littering, Eastman et al. (2013) explicitly reported the data that were dismissed for further analyses. These data were related to mistaken, non-sensical and incomplete surveys, such as when children were too young to accurately complete the survey, or data were from locations with characteristics that differed from the main surveyed area (Eastman et al. 2013).

A remaining 45 % of citizen-science studies had no data quality control. In certain studies, no specific validation step might be necessary because volunteers only gathered qualitative data during beach cleanup activities (n = 11) or citizens only participated in opportunistic sighting and sample collection of dead animals, bird boluses, pellets and drifter buoys found on beaches (n = 5). No data quality control was explicitly mentioned in the professional studies examined herein.

16.5 Recommendations for Citizen-Science Projects on Marine Litter

In order to carefully plan a citizen-science study, certain models for developing studies should be followed (Bonney et al. 2009). The research question should be easy to understand by participants and should incorporate strategies to motivate volunteers (Eastman et al. 2014). High levels of personal motivation, training procedures, and encouraging volunteers to describe any uncertainties to researchers resulted in improved accuracy achieved by citizen scientists (Lindborg et al. 2012). The time commitment of the participation of individuals and organizations should be respected. Accordingly, project leaders should concede ample time for the recruitment and training of volunteers. Sampling methods and data collection should be easy to manage with simple tools (e.g. transects, quadrats). Technology, such as smartphone applications and geo-referenced photos can be a novel tool to explore (e.g. Martin 2013). It is strongly recommended that a professional scientist demonstrates the tasks that citizen scientists will be performing in the field beforehand. Whenever possible, scientific surveys themselves should be supervised by scientists in order to ensure proper sampling and data collection. Participants should also be involved in the data evaluation and communication of results as a concluding activity, because this will enhance their commitment to the activity. Considering these recommendations, citizen scientists are capable to collect relevant data, even showing no significant difference with results gathered by experienced scientists (Thiel et al. 2014).

16.6 Outlook and Conclusions

The vast distribution of marine litter throughout the world requires extensive sampling efforts of research teams, and the available information is still limited to certain topics of research and regions of the world. In this respect, citizen-science projects have made important contributions to marine litter science. Collaborations with citizen scientists can be a useful approach to expand the understanding of marine litter in the world. Most studies have focused on the distribution and composition of marine litter, and beach cleanups are activities with the most active participation from citizen scientists.

Citizen science studies can cover a wide range of scales, from local to international range, single events to long-term multi-year projects. Through the use of citizen scientists, new research areas can be addressed in the future. Coastal marine litter may be monitored by citizen-science studies, which can also include other citizens related to the sea, such as local people, fishers, sport clubs and tourists. For instance, diver associations around the world can be trained to sample subtidal plastic debris, and new projects can be initiated with the help of sailing clubs, where long-distance travelers can survey floating marine debris by direct observation at sea, to study the distribution, composition and degradation of marine litter in the open ocean. Citizen scientists can help to determine local litter sources, thereby contributing to keeping coastal regions clean. Citizen-science projects can focus on interviewing mariners, coastal people and local governments, for the purpose of identifying ways to reduce marine litter deposition.

With proper coordination, citizen science can include several other topics, such as interaction with biota and toxic effects. Nevertheless, a main concern of marine citizen science is to assure the quality of the collected data. In general, studies should include several steps to ensure data quality, including clear protocols, training of volunteers, participation of professional scientists, and revision of samples and data. If these considerations are taken into account, citizen scientists not only can help with investigating the problem of marine litter, but they can become key allies in solving the problem of marine litter.

Acknowledgments   It seems ironic that most professional scientists studying marine litter have initially been trained in marine biology, microbiology, oceanography or related disciplines. By having redirected their research attention to the problem of marine litter, they have demonstrated that they are true citizens of the oceans! With their fascination for the sea, they join the thousands of enthusiastic citizens all over the world who have contributed to some of the research discussed herein. We thank all the schoolchildren and teachers who have participated in the citizen science program "Científicos de la Basura" and who continue to motivate us every day in this quest for a clean ocean. We are also grateful to two anonymous reviewers and to Annie Mejaes who did the final language check of the manuscript.

References

Anderson, J. A., & Alford, A. B. (2014). Ghost fishing activity in derelict blue crab traps in Louisiana. *Marine Pollution Bulletin, 79*, 261–267.

Barnes, D. K. A. (2002). Biodiversity: Invasions by marine life on plastic debris. *Nature, 416*, 808–809.

Barnes, D. K. A., & Milner, P. (2005). Drifting plastic and its consequences for sessile organism dispersal in the Atlantic Ocean. *Marine Biology, 146*, 815–825.

Barnes, D. K. A., Galgani, F., Thompson, R. C., & Barlaz, M. (2009). Accumulation and fragmentation of plastic debris in global environments. *Philosophical Transactions of the Royal Society B, 364*, 985–1998.

Bjorndal, K. A., Bolten, A. B., & Lagueux, C. J. (1994). Ingestion of marine debris by juvenile sea turtles in coastal Florida habitats. *Marine Pollution Bulletin, 28*, 154–158.

Boerger, C. M., Lattin, G. L., Moore, S. L., & Moore, C. J. (2010). Plastic ingestion by planktivorous fishes in the North Pacific Central Gyre. *Marine Pollution Bulletin, 60*, 2275–2278.

Bonney, R., Cooper, C. B., Dickinson, J., Kelling, S., Phillips, T., Rosenberg, K. V., et al. (2009). Citizen science: A developing tool for expanding science knowledge and scientific literacy. *BioScience, 59*, 977–984.

Boren, L. J., Morrissey, M., Muller, C. G., & Gemmell, N. J. (2006). Entanglement of New Zealand fur seals in man-made debris at Kaikoura, New Zealand. *Marine Pollution Bulletin, 52*, 442–446.

Bravo, M., Gallardo, M., Luna-Jorquera, G., Núñez, P., Vásquez, N., & Thiel, M. (2009). Anthropogenic debris on beaches in the SE Pacific (Chile): Results from a national survey supported by volunteers. *Marine Pollution Bulletin, 58*, 1718–1726.

Browne, M. A., Dissanayake, A., Galloway, T. S., Lowe, D. M., & Thompson, R. C. (2008). Ingested microscopic plastic translocates to the circulatory system of the mussel, *Mytilus edulis* (L.). *Environmental Science and Technology, 42*, 5026–5031.

Browne, M. A., Galloway, T. S., & Thompson, R. C. (2010). Spatial patterns of plastic debris along estuarine shorelines. *Environmental Science and Technology, 44*, 3404–3409.

Browne, M. A., Crump, P., Niven, S. J., Teuten, E., Tonkin, A., Galloway, T., et al. (2011). Accumulation of microplastic on shorelines woldwide: Sources and sinks. *Environmental Science and Technology, 45*, 9175–9179.

Carson, H. S. (2013). The incidence of plastic ingestion by fishes: From the prey's perspective. *Marine Pollution Bulletin, 74*, 170–174.

Carson, H. S., Lamson, M. R., Nakashima, D., Toloumu, D., Hafner, J., Maximenko, N., et al. (2013). Tracking the sources and sinks of local marine debris in Hawaii. *Marine Environmental Research, 84*, 76–83.

Chiappone, M., Dienes, H., Swanson, D. W., & Miller, S. L. (2005). Impacts of lost fishing gear on coral reef sessile invertebrates in the Florida Keys National Marine Sanctuary. *Biological Conservation, 121*, 221–230.

Claereboudt, M. R. (2004). Shore litter along sandy beaches of the Gulf of Oman. *Marine Pollution Bulletin, 49*, 770–777.

Claessens, M., Meester, S. D., Landuyt, L. V., Clerck, K. D., & Janssen, C. R. (2011). Occurrence and distribution of microplastics in marine sediments along the Belgian coast. *Marine Pollution Bulletin, 62*, 2199–2204.

Cole, M., Lindeque, P., Fileman, E., Halsband, C., Goodhead, R., Moger, J., et al. (2013). Microplastic ingestion by zooplankton. *Environmental Science and Technology, 47*, 6646–6655.

Corcoran, P. L., Biesinger, M. C., & Grifi, M. (2009). Plastics and beaches: A degrading relationship. *Marine Pollution Bulletin, 58*, 80–84.

Costa, M. F., do Sul, J. A. I., Silva-Cavalcanti, J. S., Araújo, M. C. B., Spengler, Â., & Tourinho, P. S. (2010). On the importance of size of plastic fragments and pellets on the strandline: a snapshot of a Brazilian beach. *Environmental Monitoring and Assessment, 168*, 299–304.

Dameron, O. J., Parke, M., Albins, M. A., & Brainard, R. (2007). Marine debris accumulation in the Northwestern Hawaiian Islands: An examination of rates and processes. *Marine Pollution Bulletin, 54*, 423–433.

Davison, P., & Asch, R. G. (2011). Plastic ingestion by mesopelagic fishes in the North Pacific Subtropical Gyre. *Marine Ecology Progress Series, 432*, 173–180.

Derraik, J. G. B. (2002). The pollution of the marine environment by plastic debris: A review. *Marine Pollution Bulletin, 44*, 842–852.

Eastman, L. B., Núñez, P., Crettier, B., & Thiel, M. (2013). Identification of self-reported user behavior, education level, and preferences to reduce littering on beaches—A survey from the SE Pacific. *Ocean and Coastal Management, 78*, 18–24.

Eastman, L., Hidalgo-Ruz, V., Macaya-Caquilpán, V., Nuñez, P., & Thiel, M. (2014). The potential for young citizen scientist projects: A case study of Chilean schoolchildren collecting data on marine litter. *Coastal and Marine Management in Latin America14*, 569–579.

Edyvane, K. S., Dalgetty, A., Hone, P. W., Higham, J. S., & Wace, N. M. (2004). Long-term marine litter monitoring in the remote Great Australian Bight, South Australia. *Marine Pollution Bulletin, 48*, 1060–1075.

Endo, S., Takizawa, R., Okuda, K., Takada, H., Chiba, K., Kanehiro, H., et al. (2005). Concentration of polychlorinated biphenyls (PCBs) in beached resin pellets: Variability among individual particles and regional differences. *Marine Pollution Bulletin, 50*, 1103–1114.

Eriksen, M., Lebreton, L. C. M., Carson, H. S., Thiel, M., Moore, C. J., Borerro, J. C., et al. (2014). Plastic pollution in the world's oceans: More than 5 trillion plastic pieces weighing over 250,000 Tons afloat at sea. *PLoS ONE, 9*, e111913.

Fendall, L. S., & Sewell, M. A. (2009). Contributing to marine pollution by washing your face: Microplastics in facial cleansers. *Marine Pollution Bulletin, 58*, 1225–1228.

Gago, J., Lahuerta, F., & Antelo, P. (2014). Characteristics (abundance, type and origin) of beach litter on the Galician coast (NW Spain) from 2001 to 2010. *Scientia Marina, 78*, 125–134.

Galgani, F., Hanke, G., & Maes, T. (2015). Global distribution, composition and abundance of marine litter. In M. Bergmann, L. Gutow & M. Klages (Eds.), *Marine anthropogenic litter* (pp. 29–56). Springer: Berlin.

Graham, E. R., & Thompson, J. T. (2009). Deposit- and suspension-feeding sea cucumbers (Echinodermata) ingest plastic fragments. *Journal of Experimental Marine Biology and Ecology, 368,* 22–29.

Gregory, M. R. (1991). The hazards of persistent marine pollution: Drift plastics and conservation islands. *Journal of the Royal Society of New Zealand, 21,* 83–100.

Heskett, M., Takada, H., Yamashita, R., Yuyama, M., Ito, M., Geok, Y. B., et al. (2012). Measurement of persistent organic pollutants (POPs) in plastic resin pellets from remote islands: Toward establishment of background concentrations for International Pellet Watch. *Marine Pollution Bulletin, 64,* 445–448.

Hidalgo-Ruz, V., & Thiel, M. (2013). Distribution and abundance of small plastic debris on beaches in the SE Pacific (Chile): A study supported by a citizen science project. *Marine Environmental Research, 87,* 12–18.

Hinojosa, I. A., & Thiel, M. (2009). Floating marine debris in fjords, gulfs and channels of southern Chile. *Marine Pollution Bulletin, 58,* 341–350.

Hirai, H., Takada, H., Ogata, Y., Yamashita, R., Mizukawa, K., Saha, M., et al. (2011). Organic micropollutants in marine plastics debris from the open ocean and remote and urban beaches. *Marine Pollution Bulletin, 62,* 1683–1692.

Hong, S., Lee, J., Jang, Y. C., Kim, Y. J., Kim, H. J., Han, D., et al. (2013). Impacts of marine debris on wild animals in the coastal area of Korea. *Marine Pollution Bulletin, 66,* 117–124.

Hong, S., Lee, J., Kang, D., Choi, H. W., & Ko, S. H. (2014). Quantities, composition, and sources of beach debris in Korea from the results of nationwide monitoring. *Marine Pollution Bulletin, 84,* 27–34.

Jackson, N. L., Cerrato, M. L., & Elliot, N. (1997). Geography and fieldwork at the secondary school level: An investigation of anthropogenic litter on an estuarine shoreline. *Journal of Geography, 96,* 301–306.

Jóźwiak, T. (2005). Tendencies in the numbers of beverage containers on the Polish coast in the decade from 1992 to 2001. *Marine Pollution Bulletin, 50,* 87–90.

Katsanevakis, S., & Katsarou, A. (2004). Influences on the distribution of marine debris on the seafloor of shallow coastal areas in Greece (Eastern Mediterranean). *Water, Air, and Soil Pollution, 159,* 325–337.

Kiessling, T., Gutow, L., & Thiel, M. (2015). Marine litter as a habitat and dispersal vector. In M. Bergmann, L. Gutow, & M. Klages (Eds.), *Marine anthropogenic litter* (pp. 141–181). Springer: Berlin.

Kordella, S., Geraga, M., Papatheodorou, G., Fakiris, E., & Mitropoulou, I. M. (2013). Litter composition and source contribution for 80 beaches in Greece, Eastern Mediterranean: A nationwide voluntary clean-up campaign. *Aquatic Ecosystem Health and Management, 16,* 111–118.

Kusui, T., & Noda, M. (2003). International survey on the distribution of stranded and buried litter on beaches along the Sea of Japan. *Marine Pollution Bulletin, 47,* 175–179.

Laist, D. W. (1997). Impacts of marine debris: Entanglement of marine life in marine debris including a comprehensive list of species with entanglement and ingestion records. In J. M. Coe, & D. B. Rogers (Eds.), *Marine debris: sources, impacts, and solutions* (pp. 99–139). New York: Springer.

Lattin, G. L., Moore, C. J., Zellers, A. F., Moore, S. L., & Weisberg, S. B. (2004). A comparison of neustonic plastic and zooplankton at different depths near the southern California shore. *Marine Pollution Bulletin, 49,* 291–294.

Law, K. L., Morét-Ferguson, S., Maximenko, N. A., Proskurowski, G., Peacock, E. E., Hafner, J., et al. (2010). Plastic accumulation in the North Atlantic subtropical gyre. *Science, 329,* 1185–1188.

Lazar, B., & Gračan, R. (2011). Ingestion of marine debris by loggerhead sea turtles, *Caretta caretta,* in the Adriatic Sea. *Marine Pollution Bulletin, 62,* 43–47.

Lindborg, V. A., Ledbetter, J. F., Walat, J. M., & Moffett, C. (2012). Plastic consumption and diet of Glaucous-winged gulls (*Larus glaucescens*). *Marine Pollution Bulletin, 64,* 2351–2356.

Löder, M. G. J., & Gerdts, G. (2015). Methodology used for the detection and identification of microplastics—A critical appraisal. In M. Bergmann, L. Gutow, & M. Klages (Eds.), *Marine anthropogenic litter* (pp. 201–227). Springer: Berlin.

Martin, J. M. (2013). Marine debris removal: One year of effort by the Georgia Sea turtle-center-marine debris initiative. *Marine Pollution Bulletin, 74,* 165–169.

Martinez, E., Maamaatuaiahutapu, K., & Taillandier, V. (2009). Floating marine debris surface drift: Convergence and accumulation toward the South Pacific subtropical gyre. *Marine Pollution Bulletin, 58,* 1347–1355.

Mascarenhas, R., Santos, R., & Zeppelini, D. (2004). Debris ingestion by sea turtle in Paraíba, Brazil. *Marine Pollution Bulletin, 49,* 354–355.

Masó, M., Garcés, E., Pagès, F., & Camp, J. (2003). Drifting plastic debris as a potential vector for dispersing Harmful Algal Bloom (HAB) species. *Scientia Marina, 67,* 107–111.

Maximenko, N., Hafner, J., & Niiler, P. (2012). Pathways of marine debris derived from trajectories of Lagrangian drifters. *Marine Pollution Bulletin, 65,* 51–62.

McDermid, K. J., & McMullen, T. L. (2004). Quantitative analysis of small-plastic debris on beaches in the Hawaiian archipelago. *Marine Pollution Bulletin, 48,* 790–794.

Moore, S. L., Gregorio, D., Carreon, M., Weisberg, S. B., & Leecaster, M. K. (2001). Composition and distribution of beach debris in Orange County, California. *Marine Pollution Bulletin, 42,* 241–245.

Moore, E., Lyday, S., Roletto, J., Litle, K., Parrish, J. K., Nevins, H., et al. (2009). Entanglements of marine mammals and seabirds in central California and the north-west coast of the United States 2001–2005. *Marine Pollution Bulletin, 58,* 1045–1051.

Morét-Ferguson, S., Law, K. L., Proskurowski, G., Murphy, E. K., Peacock, E. E., & Reddy, C. M. (2010). The size, mass, and composition of plastic debris in the western North Atlantic Ocean. *Marine Pollution Bulletin, 60,* 1873–1878.

Morishige, C., Donohue, M. J., Flint, E., Swenson, C., & Woolaway, C. (2007). Factors affecting marine debris deposition at French Frigate Shoals, northwestern Hawaiian islands marine national monument, 1990–2006. *Marine Pollution Bulletin, 54,* 1162–1169.

Murray, F., & Cowie, P. R. (2011). Plastic contamination in the decapod crustacean *Nephrops norvegicus* (Linnaeus, 1758). *Marine Pollution Bulletin, 62,* 1207–1217.

Ng, K. L., & Obbard, J. P. (2006). Prevalence of microplastics in Singapore's coastal marine environment. *Marine Pollution Bulletin, 52,* 761–767.

Ogata, Y., Takada, H., Mizukawa, K., Hirai, H., Iwasa, S., Endo, S., et al. (2009). International pellet watch: Global monitoring of persistent organic pollutants (POPs) in coastal waters. 1. Initial phase data on PCBs, DDTs, and HCHs. *Marine Pollution Bulletin, 58,* 1437–1446.

Oigman-Pszczol, S. S., & Creed, J. C. (2007). Quantification and classification of marine litter on beaches along Armação dos Búzios, Rio de Janeiro, Brazil. *Journal of Coastal Research, 23,* 421–428.

Page, B., McKenzie, J., McIntosh, R., Baylis, A., Morrissey, A., Calvert, N., et al. (2004). Entanglement of Australian sea lions and New Zealand fur seals in lost fishing gear and other marine debris before and after Government and industry attempts to reduce the problem. *Marine Pollution Bulletin, 49,* 33–42.

Pichel, W. G., Churnside, J. H., Veenstra, T. S., Foley, D. G., Friedman, K. S., Brainard, R. E., et al. (2007). Marine debris collects within the North Pacific subtropical convergence zone. *Marine Pollution Bulletin, 54,* 1207–1211.

Reddy, M. S., Basha, S., Adimurthy, S., & Ramachandraiah, G. (2006). Description of the small plastics fragments in marine sediments along the Alang-Sosiya ship-breaking yard, India. *Estuarine, Coastal and Shelf Science, 68,* 656–660.

Ribic, C. A. (1998). Use of indicator items to monitor marine debris on a New Jersey beach from 1991 to 1996. *Marine Pollution Bulletin, 36,* 887–891.

Ribic, C. A., Sheavly, S. B., Rugg, D. J., & Erdmann, E. S. (2010). Trends and drivers of marine debris on the Atlantic coast of the United States 1997–2007. *Marine Pollution Bulletin, 60,* 1231–1242.

Ribic, C. A., Sheavly, S. B., & Rugg, D. J. (2011). Trends in marine debris in the US Caribbean and the Gulf of Mexico 1996–2003. *Journal of Integrated Coastal Zone Management, 11,* 7–19.

Ribic, C. A., Sheavly, S. B., Rugg, D. J., & Erdmann, E. S. (2012a). Trends in marine debris along the US Pacific Coast and Hawai'i 1998–2007. *Marine Pollution Bulletin, 64,* 994–1004.

Ribic, C. A., Sheavly, S. B., & Klavitter, J. (2012b). Baseline for beached marine debris on Sand Island, Midway Atoll. *Marine Pollution Bulletin, 64*, 1726–1729.

Rios, L. M., Moore, C., & Jones, P. R. (2007). Persistent organic pollutants carried by synthetic polymers in the ocean environment. *Marine Pollution Bulletin, 54*, 1230–1237.

Rios, L. M., Jones, P. R., Moore, C., & Narayan, U. V. (2010). Quantitation of persistent organic pollutants adsorbed on plastic debris from the Northern Pacific Gyre's "eastern garbage patch". *Journal of Environmental Monitoring, 12*, 2226–2236.

Rosevelt, C., Los Huertos, M., Garza, C., & Nevins, H. M. (2013). Marine debris in central California: Quantifying type and abundance of beach litter in Monterey Bay, CA. *Marine Pollution Bulletin, 71*, 299–306.

Ross, J. B., Parker, R., & Strickland, M. (1991). A survey of shoreline litter in Halifax Harbour 1989. *Marine Pollution Bulletin, 22*, 245–248.

Ryan, P. G. (2008). Seabirds indicate changes in the composition of plastic litter in the Atlantic and south-western Indian Oceans. *Marine Pollution Bulletin, 56*, 1406–1409.

Ryan, P. G., Moore, C. J., van Franeker, J. A., & Moloney, C. L. (2009). Monitoring the abundance of plastic debris in the marine environment. *Philosophical Transactions of the Royal Society of London B, 364*, 1999–2012.

Santos, I. R., Friedrich, A. C., Wallner-Kersanach, M., & Fillmann, G. (2005). Influence of socio-economic characteristics of beach users on litter generation. *Ocean and Coastal Management, 48*, 742–752.

Santos, I. R., Friedrich, A. C., & Ivar do Sul, J. A. (2009). Marine debris contamination along undeveloped tropical beaches from northeast Brazil. *Environmental Monitoring and Assessment, 148*, 455–462.

Seino, S., Kojima, A., Hinata, H., Magome, S. N., & Isobe, A. (2009). Multi-sectoral research on East China Sea beach litter based on oceanographic methodology and local knowledge. *Journal of Coastal Research, 56*, 1289–1292.

Shimizu, T., Nakai, J., Nakajima, K., Kozai, N., Takahashi, G., Matsumoto, M., et al. (2008). Seasonal variations in coastal debris on Awaji Island, Japan. *Marine Pollution Bulletin, 57*, 182–186.

Smith, S. D., & Edgar, R. J. (2014). Documenting the density of subtidal marine debris across multiple marine and coastal habitats. *PLoS ONE, 9*, e94593.

Smith, S. D., Gillies, C. L., & Shortland-Jones, H. (2014). Patterns of marine debris distribution on the beaches of Rottnest Island, Western Australia. *Marine Pollution Bulletin, 88*, 188–193.

Spalding, M. D., Fox, H. E., Allen, G. R., Davidson, N., Ferdaña, Z. A., Finlayson, M., et al. (2007). Marine ecoregions of the world: A bioregionalization of coastal and shelf areas. *BioScience, 57*, 573–583.

Storrier, K. L., & McGlashan, D. J. (2006). Development and management of a coastal litter campaign: The voluntary coastal partnership approach. *Marine Policy, 30*, 189–196.

Storrier, K. L., McGlashan, D. J., Bonellie, S., & Velander, K. (2007). Beach litter deposition at a selection of beaches in the Firth of Forth, Scotland. *Journal of Coastal Research, 23*, 813–822.

Teuten, E. L., Rowland, S. J., Galloway, T. S., & Thompson, R. C. (2007). Potential for plastics to transport hydrophobic contaminants. *Environmental Science and Technology, 41*, 7759–7764.

Thiel, M., Penna-Díaz, M. A., Luna-Jorquera, G., Sala, S., Sellanes, J., & Stotz, W. (2014). Citizen scientists and marine research: Volunteer participants, their contributions and projection for the future. *Oceanography and Marine Biology: An Annual Review, 52*, 257–314.

Thompson, R. C., Olsen, Y., Mitchell, R. P., Davis, A., Rowland, S. J., John, A. W., et al. (2004). Lost at sea: where is all the plastic? *Science, 304*, 838.

Tourinho, P. S., Ivar do Sul, J. A., & Fillmann, G. (2010). Is marine debris ingestion still a problem for the coastal marine biota of southern Brazil? *Marine Pollution Bulletin, 60*, 396–401.

van Franeker, J. A., Blaize, C., Danielsen, J., Fairclough, K., Gollan, J., Guse, N., et al. (2011). Monitoring plastic ingestion by the northern fulmar *Fulmarus glacialis* in the North Sea. *Environmental Pollution, 159*, 2609–2615.

Whiting, S. D. (1998). Types and sources of marine debris in Fog Bay, Northern Australia. *Marine Pollution Bulletin, 36*, 904–910.

Young, L. C., Vanderlip, C., Duffy, D. C., Afanasyev, V., & Shaffer, S. A. (2009). Bringing home the trash: Do colony-based differences in foraging distribution lead to increased plastic ingestion in Laysan albatrosses? *PLoS ONE, 4*, e7623.

Appendices

Appendix 1: Citizen-Science and Professional Studies on Marine Litter

Reference	Topic	Locations
Citizen science studies		
Anderson and Alford (2014)	Interaction with biota	Lousiana, United States
Bjorndal et al. (1994)	Interaction with biota	Florida, United States
Bravo et al. (2009)	Distribution and composition	Chilean coast
Carson (2013)	Interaction with biota	Hawaii an islands
Carson et al. (2013)	Distribution and composition	Hawaii an islands
Eastman et al. (2013)	Social aspects	Chilean coast
Edyvane et al. (2004)	Distribution and composition	Anxious Bay, South Australia
Endo et al. (2005)	Toxic effects	Tokyo, Japan
Gago et al. (2014)	Distribution and composition	Galicia, Spain
Gregory (1991)	Distribution and composition	Hauraki Bay, New Zealand
Heskett et al. (2012)	Toxic effects	Canary, Oahu, Hawaii, Barbados, Cocos and St. Helena Islands
Hidalgo-Ruz and Thiel (2013)	Distribution and composition	Continental Chile, Easter Island
Hirai et al. (2011)	Toxic effects	North Pacific
Hong et al. (2013)	Interaction with biota	Korea
Hong et al. (2014)	Distribution and composition	Korea
Jackson et al. (1997)	Distribution and composition	New Jersey, United States
Jóźwiak (2005)	Distribution and composition	Poland
Kordella et al. (2013)	Distribution and composition	Eastern Mediterranean
Kusui and Noda (2003)	Distribution and composition	Japan, Russia
Lindborg et al. (2012)	Interaction with biota	Washington, United States
Martin (2013)	Distribution and composition	Jekyll Island
Moore et al. (2001)	Distribution and composition	California, United States
Moore et al. (2009)	Interaction with biota	West coast, United States
Morishige et al. (2007)	Transport	Hawaii an islands
Ogata et al. (2009)	Toxic effects	Global
Ribic (1998)	Distribution and composition	New Jersey, United States
Ribic et al. (2010)	Distribution and composition	East coast, United States
Ribic et al. (2011)	Distribution and composition	Gulf of Mexico

(continued)

(continued)

Reference	Topic	Locations
Ribic et al. (2012a)	Distribution and composition	West coast, United States, Hawaii
Ribic et al. (201b)	Distribution and composition	Midway Atoll
Rosevelt et al. (2013)	Distribution and composition	California, United States
Ross et al. (1991)	Distribution and composition	Nova Scotia, Canada
Seino et al. (2009)	Distribution and composition	East China Sea
Shimizu et al. (2008)	Distribution and composition	Awaji Island, Japan
Smith and Edgar (2014)	Distribution and composition	Australia
Smith et al. (2014)	Distribution and composition	Western Australia
Storrier and McGlashan (2006)	Distribution and composition	Forth Estuary, United Kingdom
Storrier et al. (2007)	Distribution and composition	Forth Estuary, United Kingdom
Van Franeker et al. (2011)	Interaction with biota	North Sea
Whiting (1998)	Distribution and composition	Australia
Professional studies		
Barnes and Milner (2005)	Interaction with biota	Atlantic Ocean
Boerger et al. (2010)	Interaction with biota	North Pacific oceanic gyre
Boren et al. (2006)	Interaction with biota	Kaikoura, New Zealand
Browne et al. (2008)	Interaction with biota	Cornwall, United Kingdom
Browne et al. (2010)	Distribution and composition	Tamar Estuary, United Kingdom
Browne et al. (2011)	Distribution and composition	Global
Chiappone et al. (2005)	Interaction with biota	Florida, United States
Claereboudt (2004)	Distribution and composition	Gulf of Oman
Claessens et al. (2011)	Distribution and composition	Belgium
Corcoran et al. (2009)	Degradation	Kauai Island, Hawaii
Costa et al. (2010)	Distribution and composition	Northeast Brazil
Dameron et al. (2007)	Distribution and composition	Hawaiian Islands
Davison and Asch (2011)	Interaction with biota	North Pacific gyre
Fendall and Sewell (2009)	Distribution and composition	Auckland, New Zealand
Graham and Thompson (2009)	Interaction with biota	Florida, Maine, United States
Hinojosa and Thiel (2009)	Distribution and composition	Fjords, Southern Chile
Katsanevakis and Katsarou (2004)	Distribution and composition	Greece
Lattin et al. 2004	Distribution and composition	California, United States
Law et al. (2010)	Transport	Atlantic Ocean, Caribbean sea
Lazar and Gracan (2011)	Interaction with biota	Adriatic Sea
Martinez et al. (2009)	Transport	South Pacific gyre
Mascarenhas et al. (2004)	Interaction with biota	Paraiba, Brazil
Maximenko et al. (2012)	Transport	Global
McDermid and McMullen (2004)	Distribution and composition	Hawaii an islands

(continued)

(continued)

Reference	Topic	Locations
Morét-Ferguson et al. (2010)	Distribution and composition	North Atlantic Ocean
Murray and Cowie (2011)	Interaction with biota	Clyde Sea, United Kingdom
Ng and Obbard (2006)	Distribution and composition	Singapore
Oigman-Pszczol and Creed (2007)	Distribution and composition	Rio de Janeiro, Brazil
Page et al. (2004)	Interaction with biota	Australia
Pichel et al. (2007)	Distribution and composition	North Pacific gyre
Reddy et al. (2006)	Distribution and composition	India
Rios et al. (2007)	Toxic effects	North Pacific gyre
Rios et al. (2010)	Toxic effects	North Pacific gyre
Ryan (2008)	Interaction with biota	Atlantic, southwestern Indian Oceans
Santos et al. (2005)	Social aspects	Rio Grande, Brazil
Santos et al. (2009)	Distribution and composition	Brazil
Teuten et al. (2007)	Toxic effects	United Kingdom
Thompson et al. (2004)	Distribution and composition	Plymouth, United Kingdom
Tourinho et al. (2010)	Interaction with biota	Brazil
Young et al. (2009)	Interaction with biota	North Pacific Ocean

Printed by Printforce, the Netherlands